W9-CNA-444

Data Communications

and Computer Networks

A Business User's Approach

Sixth Edition

Data Communications and Computer Networks

A Business User's Approach

Sixth Edition

▶ **Curt M. White**
DePaul University

COURSE TECHNOLOGY
CENGAGE Learning

Australia • Brazil • Japan • Korea • Mexico • Singapore • Spain • United Kingdom • United States

**Data Communications and Computer Networks,
A Business User's Approach, Sixth Edition**
Curt M. White

VP/Editorial Director: Jack Calhoun

Publisher: Joe Sabatino

Senior Acquisitions Editor: Charles McCormick, Jr.

Senior Product Manager: Kate Mason

Editorial Assistant: Nora Heink

Content Product Manager: Karunakaran Gunasekaran

Manufacturing Coordinator: Julio Esperas

Marketing Coordinator: Suellen Ruttkay

Senior Art Director: Stacy Jenkins Shirley

Cover Designer: Craig Ramsdell

Cover Image: iStock Photo

Compositor: Pre-Press PMG

For product information and technology assistance, contact us at
Cengage Learning Customer & Sales Support, 1-800-354-9706
For permission to use material from this text or product, submit all requests online at **cengage.com/permissions**
Further permissions questions can be emailed to
permissionrequest@cengage.com

Library of Congress Control Number: 2010920250

ISBN-13: 978-0-538-45261-8

ISBN-10: 0-538-45261-7

Course Technology
20 Channel Center Street
Boston, Massachusetts 02210
USA

Some of the product names and company names used in this book have been used for identification purposes only and may be trademarks or registered trademarks of their respective manufacturers and sellers.

Any fictional data related to persons or companies or URLs used throughout this book is intended for instructional purposes only. At the time this book was printed, any such data was fictional and not belonging to any real persons or companies.

Course Technology, a part of Cengage Learning, reserves the right to revise this publication and make changes from time to time in its content without notice.

Cengage Learning is a leading provider of customized learning solutions with office locations around the globe, including Singapore, the United Kingdom, Australia, Mexico, Brazil, and Japan. Locate your local office at:
international.cengage.com/region

Cengage Learning products are represented in Canada by Nelson -Education, Ltd.

Visit our corporate website at **cengage.com**

To learn more about Course Technology, visit **www.cengage.com/coursetechnology**

Purchase any of our products at your local college store or at our preferred online store **www.CengageBrain.com**

Printed in the United States of America
1 2 3 4 5 6 7 16 15 14 13 12 11 10

To Kathleen, Hannah Colleen, and
Samuel Memphis—it's never boring.

Brief Contents

Contents

Preface

Today's business world could not function without data communications and computer networks. Most people cannot make it through an average day without coming in contact with or using some form of computer network. In the past, this field of study used to occupy the time of only engineers and technicians, but it now involves business managers, end users, programmers, and just about anyone who might use a telephone or computer! Because of this, *Data Communications and Computer Networks: A Business User's Approach, Sixth Edition* maintains its business user's perspective on this vast and increasingly significant subject.

In a generic sense, this book serves as an owner's manual for the individual computer user. In a world in which computer networks are involved in nearly every facet of business and personal life, it is paramount that each of us understands the basic features, operations, and limitations of different types of computer networks. This understanding will make us better managers, better employees, and simply better computer users. As a computer network *user*, you will probably not be the one who designs, installs, and maintains the network. Instead, you will have interactions, either direct or indirect, with the individuals who do. Reading this book should give you a strong foundation in computer networks, which will enable you to work effectively with network administrators, network installers, and network designers.

Here are some of the many scenarios in which the knowledge contained in this book would be particularly useful:

> ▶ You work for a company and must deal directly with a network specialist. To better understand the specialist and be able to conduct a meaningful dialog with him or her, you need a basic understanding of the many aspects of computer networks.

> ▶ You are a manager within a company and depend on a number of network specialists to provide you with recommendations for the company's network. You do not want to find yourself in a situation in which you must blindly accept the recommendations of network professionals. To ensure that you can make intelligent decisions regarding network resources, you need to know the basic concepts of data communications and computer networks.

> ▶ You work in a small company, in which each employee wears many hats. Thus, you may need to perform some level of network assessment, administration, or support.

> ▶ You have your own business and need to fully understand the advantages of using computer networks to support your operations. To optimize those advantages, you should have a good grasp of the basic characteristics of a computer network.

> ▶ You have a computer at home or at work, and you simply wish to learn more about computer networks.

> ▶ You have realized that to keep your job skills current and remain a key player in the information technology arena, you must understand how different computer networks work and become familiar with their advantages and shortcomings.

Audience

Data Communications and Computer Networks: A Business User's Approach, Sixth Edition is intended for a one-semester course in business data communications for students majoring in business, information systems, management information systems, and other applied fields of computer science. Even computer science departments will find the book valuable, particularly if the students read the Details sections accompanying most chapters. It is a readable resource for computer network users that draws on examples from business environments.

In a university setting, this book can be used at practically any level above the first year. Instructors who wish to use this book at the graduate level can draw on the many advanced projects provided at the end of each chapter to create a more challenging environment for the advanced student.

Defining Characteristics of This Book

The major goal of this sixth edition is the same as that of the first edition: to go beyond simply providing readers with a handful of new definitions, and instead introduce them to the next level of details found within the fields of computer networks and data communications. This higher level of detail includes the network technologies and standards necessary to support computer network systems and their applications. This book is more than just an introduction to advanced terminology. It involves introducing concepts that will help the reader achieve a more in-depth understanding of the often complex topic of data communications. It is hoped that once readers attain this in-depth understanding, the topic of networks and data communications will be less intimidating to them. To facilitate this understanding, the book strives to maintain high standards in three major areas: readability, a balance between the technical and the practical, and currency.

Readability

Great care has been taken to provide the technical material in as readable a fashion as possible. Each new edition has received a complete rewrite, in which every sentence has been re-examined in an attempt to convey the concepts as clearly as possible. Given the nature of this book's subject matter, the use of terminology is unavoidable. However, every effort has been made to present terms in a clear fashion, with minimal use of acronyms and even less use of computer jargon.

Balance Between the Technical and the Practical

As in the very successful first edition, a major objective in writing *Data Communications and Computer Networks, Sixth Edition* was to achieve a good balance between the more technical aspects of data communications and its everyday practical aspects. Throughout each chapter, there are sections entitled "Details," which delve into the more specialized aspects of the topic at hand. Should readers not have time to explore this technical information, they can skip these Details sections without missing out on the basic concepts of the topic.

Current Technology

Because of the fast pace of change in virtually all computer-related fields, every attempt has been made to present the most current trends in data communications and computer networks. Some of these topics include:

- Latest wireless technologies
- Modern multiplexing techniques, such as discrete multitone and wavelength division multiplexing
- Switching in local area networks
- Advanced encryption standards and digital signatures
- Compression techniques
- Cable modems and DSL
- Current LAN network operating systems (Windows 2008 and Linux)
- Introduction to cloud computing
- Wi-Max wireless Internet service

It is also important to remember the many older technologies still in prevalent use today. Discussions of these older technologies can be found, when appropriate, in each chapter of this book.

Organization

The organization of *Data Communications and Computer Networks, Sixth Edition* roughly follows that of the TCP/IP protocol suite, from the physical layer to the upper layers. In addition, the book has been carefully designed to consist of 13 chapters in order to fit well into a typical 15- or 16-week semester (along with any required exams). While some chapters may not require an entire week of study, other chapters may require more than one week. The intent was to design a balanced introduction to the study of computer networks by creating a set of chapters that is cohesive but at the same time allows for flexibility in the week-to-week curriculum.

Thus, instructors may choose to emphasize or de-emphasize certain topics, depending on the focus of their curriculums. If all 13 chapters cannot be covered during one term, it is possible for the instructor to concentrate on certain chapters. For example, if the curriculum's focus is information systems, the instructor might concentrate on Chapters 1, 3, 4, 6–8, 10, 12, and 13. If the focus is on the more technical aspects of computer networks, the instructor might concentrate on Chapters 1–11. It is the author's recommendation, however, that all chapters be covered in some level of detail.

Features

To assist readers in better understanding the technical nature of data communications and computer networks, each chapter contains a number of significant features. These features are based on older, well-tested pedagogical techniques as well as some newer techniques.

Opening Case

Each chapter begins with a short case or vignette that emphasizes the main concept of the chapter and sets the stage for exploration. These cases are designed to spark readers' interest and create a desire to learn more about the chapter's concepts.

Learning Objectives

Following the opening case is a list of learning objectives that should be accomplished by the end of the chapter. Each objective is tied to the main sections of the chapter. Readers can use the objectives to grasp the scope and intent of the chapter. The objectives also work in conjunction with the end-of-chapter summary and review questions, so that readers can assess whether they have adequately mastered the material.

Details

Many chapters contain one or more Details sections, which dig deeper into a particular topic. Readers who are interested in more technical details will find these sections valuable. Since the Details sections are physically separate from the main text, they can be skipped if the reader does not have time to explore this level of technical detail. Skipping these sections will not affect the reader's overall understanding of a chapter's material.

In Action

At the end of each chapter's main content presentation is an In Action example that demonstrates an application of the chapter's key topic in a realistic environment. Although a number of In Action examples include imaginary people and organizations, every attempt was made to make the hypothetical scenarios as representative as possible of situations and issues found in real-world business and home environments. Thus, the In Action examples help the reader visualize the concepts presented in the chapter.

End-of-Chapter Material

The end-of-chapter material is designed to help readers review the content of the chapter and assess whether they have adequately mastered the concepts. It includes:

- A bulleted summary that readers can use as a review of the key topics of the chapter and as a study guide.
- A list of the key terms used within the chapter.
- A list of review questions that readers can use to quickly check whether or not they understand the chapter's key concepts.
- A set of exercises that draw on the material presented in the chapter.

> ▶ A set of Thinking Outside the Box exercises, which are more in-depth in nature and require readers to consider various possible alternative solutions by comparing their advantages and disadvantages.

> ▶ A set of Hands-On Projects that require readers to reach beyond the material found within the text and use outside resources to compose a response. Many of these projects lend themselves nicely to writing assignments. Thus, they can serve as valuable tools for instructors, especially at a time when more and more colleges and universities are seeking to implement "writing across the curriculum" strategies.

Glossary

At the end of the book, you will find a glossary that includes the key terms from each chapter.

Student Online Companion

The student online companion for this book can be found at *www.cengage.com/mis/white*. It contains a number of features, including:

> ▶ Hands-on labs that allow students to practice one or more of the chapter concepts on their schools' local area networks

> ▶ A complete set of PowerPoint lecture note slides

> ▶ A set of more in-depth discussions on topics such as $\times.21$, dial-up modems, ISDN, Dijkstra's Algorithm, SDLC, and BISYNC

> ▶ Suggestions for further readings on numerous topics within the book

This Web site also presents visual demonstrations of many of the key data communications and networking concepts introduced in this text. A visual demonstration accompanies the following concepts:

> ▶ Chapter One: Introduction to Computer Networks and Data Communications—Layer encapsulation example

> ▶ Chapter Four: Making Connections—RS-232 example of two modems establishing a connection

> ▶ Chapter Five: Making Connections Efficient: Multiplexing and Compression—Example of packets from multiple sources coming together for synchronous TDM and a second example demonstrating statistical TDM

> ▶ Chapter Six: Errors, Error Detection, and Error Control—Sliding window example using ARQ error control

> ▶ Chapter Seven: Local Area Networks—CSMA/CD example with workstations sending packets and collisions happening

> ▶ Chapter Seven: Local Area Networks—Two LANs with a bridge showing how bridge tables are created and packets routed; a second example shows one LAN with a switch in place of a hub

> ▶ Chapter Nine: Introduction to Metropolitan Area Networks and Wide Area Networks—Datagram network sending individual packets; and virtual circuit network first creating a connection then sending packets down a prescribed path

> ▶ Chapter Ten: The Internet—Domain Name System as it tries to find the dotted decimal notation for a given URL

Changes to the Sixth Edition

In order to keep abreast of the changes in computer networks and data communications, this sixth edition has incorporated many updates and additions in every chapter, as well as some reorganization of sections within chapters. Here's a summary of the major changes that can be found in each of the following chapters:

Chapter One, Introduction to Computer Networks and Data Communications, introduces an update on the many types of computer network connections, along with many of the major concepts that will be discussed in the following chapters, with an emphasis on the TCP/IP protocol suite followed by the OSI models. The topic of convergence has been introduced in this first chapter and will be revisited as needed in subsequent chapters.

Chapter Two, Fundamentals of Data and Signals, covers basic concepts that are critical to the proper understanding of all computer networks and data communications.

Chapter Three, Conducted and Wireless Media, introduces the different types of media for transmitting data. Newer sections on satellite bands and Category 7 wire are included, and the section on cellular telephones was updated to include the latest cell phone technologies.

Chapter Four, Making Connections, discusses how a connection or interface is created between a computer and a peripheral device, with an emphasis on the USB interface.

Chapter Five, Making Connections Efficient: Multiplexing and Compression, introduces the topic of compression. Lossless compression techniques such as run-length encoding are discussed, as well as lossy compression techniques such as MP3 and JPEG.

Chapter Six, Errors, Error Detection, and Error Control, explains the actions that can take place when a data transmission produces an error. The concept of arithmetic checksum, as it used on the Internet, is included.

Chapter Seven, Local Area Networks: The Basics, is devoted to the basic concepts of local area networks, including the most popular topologies and systems. The local area network switch has been given more prominence, as it is in the current industry.

Chapter Eight, Local Area Networks: Software and Support Systems, discusses the various network operating systems and other network software, with updated material on Microsoft, Linux, the MAC OS X Server, and Novell's version of NetWare.

Chapter Nine, Introduction to Metropolitan Area Networks and Wide Area Networks, introduces the basic terminology and concepts of both metropolitan area networks and wide area networks. An introduction to cloud computing was introduced.

Chapter Ten, The Internet, delves into the details of the Internet, including TCP/IP, DNS, and the World Wide Web. The discussion on the topic of Voice over IP is included, as well as the material on MPLS, service level agreements, and convergence.

Chapter Eleven, Voice and Data Delivery Networks, provides a detailed introduction to the area of telecommunications, in particular networks that specialize in local and long distance delivery of data. The topics of basic telephone systems and frame relay were reduced to show their diminishing importance in today's technology markets, while the topics of MPLS and VPNs were increased.

Chapter Twelve, Network Security, covers the current trends in network security.

Chapter Thirteen, Network Design and Management, introduces the systems development life cycle, feasibility studies, capacity planning, and baseline studies, and shows how these concepts apply to the analysis and design of computer networks.

Teaching Tools

The following supplemental materials are available when this book is used in a classroom setting. All of the teaching tools available with this book are provided to the instructor on a single CD-ROM. Many can also be found at the Cengage Course Technology Web site (*www.cengage.com/mis/white*).

Electronic Instructor's Manual—The Instructor's Manual that accompanies this textbook includes additional instructional material to assist in class preparation, including Sample Syllabi, Chapter Outlines, Technical Notes, Lecture Notes, Quick Quizzes, Teaching Tips, Discussion Topics, and Key Terms.

ExamView®—This textbook is accompanied by ExamView, a powerful testing software package that allows instructors to create and administer printed, computer (LAN-based), and Internet exams. ExamView includes hundreds of questions that correspond to the topics covered in this text, enabling students to generate detailed study guides that include page references for further review. The computer-based and Internet testing components allow students to take exams at their computers and also save the instructor time by grading each exam automatically.

PowerPoint Presentations—This book comes with Microsoft Power-Point slides for each chapter. These are included as a teaching aid for classroom presentation, to make available to students on the network for chapter review, or to be printed for classroom distribution. Instructors can add their own slides for additional topics they introduce to the class.

Distance Learning—Cengage Course Technology offers online WebCT and Blackboard (version 5.0 and 6.0) courses for this text to provide the most complete and dynamic learning experience possible. When you add online content to one of your courses, you're adding a lot: automated tests, topic reviews, quick quizzes, and additional case projects with solutions. For more information on how to bring distance learning to your course, contact your local Cengage Course Technology sales representative.

Acknowledgments

Producing a textbook requires the skills and dedication of many people. Unfortunately, the final product displays only the author's name on the cover and not the names of those who provided countless hours of input and professional advice. I would first like to thank the people at Course Technology for being so vitally supportive and one of the best teams an author could hope to work with: Charles McCormick, Jr., Senior Acquisitions Editor; Kate Mason, Senior Product Manager; and Karunakaran Gunasekaran, Content Product Manager.

I must also thank my colleagues at DePaul University who listened to my problems, provided ideas for exercises, proofread some of the technical chapters, and provided many fresh ideas when I could think of none myself.

Finally I thank my family: my wife Kathleen, my daughter Hannah, and my son Samuel. It was your love and support (again!) that kept me going, day after day, week after week, and month after month.

Curt M. White

1

Introduction to Computer Networks and Data Communications

◆◆

MAKING PREDICTIONS is a difficult task, and predicting the future of computing is no exception. History is filled with computer-related predictions that were so inaccurate that today they are amusing. For example, consider the following predictions:

"I think there is a world market for maybe five computers." *Thomas Watson, chairman of IBM, 1943*

"I have traveled the length and breadth of this country, and talked with the best people, and I can assure you that data processing is a fad that won't last out the year." *Editor in charge of business books for Prentice Hall, 1957*

"There is no reason anyone would want a computer in their home." *Ken Olson, president and founder of Digital Equipment Corporation, 1977*

"640K ought to be enough for anybody." *Bill Gates, 1981*

"We believe the arrival of the PC's little brother [PCjr] is as significant and lasting a development in the history of computing as IBM's initial foray into microcomputing has proven to be." *PC Magazine, December 1983 (The PCjr lasted less than one year.)*

Apparently, no matter how famous you are or how influential your position, it is very easy to make very bad predictions. Nevertheless, it is hard to imagine that anyone can make a prediction worse than any of those above. Buoyed by this false sense of optimism, let us make a few forecasts of our own:

Someday before you head out the door, you will reach for your umbrella, and it will tell you what kind of weather to expect outside. A radio signal will connect the umbrella to a local weather service that will download the latest weather conditions for your convenience.

Someday you will be driving a car, and if you go faster than some predetermined speed, the car will send a text message to your parents informing them of your "driving habits."

Someday we will wear a computer—like a suit of clothes—and when we shake hands with a person, data will transfer down our skin, across the shaking hands, and into the other person's "computer."

Sometime in the not too distant future, you will be able to create a business card that when waved near a computer-will automatically have the computer perform a function such as placing an Internet telephone call or creating a data entry for the information on the card.

Someday you will have a car battery that, when the power in the battery gets too weak to start the car, will call you on your cell phone to inform you that you need a replacement.

One day you will be in a big city and place a call on your cell phone to request a taxi. The voice on the other end will simply say, "Stay right where you are. Do you see the taxi coming down the street? When it stops in front of you, hop in."

Do these predictions sound far-fetched and filled with mysterious technologies that only scientists and engineers can understand? They shouldn't, because they are not predictions. They are scenarios happening today with technologies that already exist. What's more, none of these advances would be possible today were it not for computer networks and data communications.

Objectives ▶

After reading this chapter, you should be able to:

▶ Define the basic terminology of computer networks

▶ Recognize the individual components of the big picture of computer networks

▶ Outline the basic network connections

▶ Define the term "convergence" and describe how it applies to computer networks

▶ Cite the reasons for using a network architecture and explain how they apply to current network systems

▶ List the layers of the TCP/IP protocol suite and describe the duties of each layer

▶ List the layers of the OSI model and describe the duties of each layer

▶ Compare the TCP/IP protocol suite and OSI model and list their differences and similarities

Introduction ▶

The world of computer networks and data communications is a surprisingly vast and increasingly significant field of study. Once considered primarily the domain of network specialists and technicians, computer networks now involve business managers, computer programmers, system designers, office managers, home computer users, and everyday citizens. It is virtually impossible for the average person on the street to spend 24 hours without directly or indirectly using some form of computer network.

Ask any group, "Has anyone used a computer network today?" and more than one-half of the people may answer, "Yes." Then ask the others: "How did you get to work, school, or the store today if you did not use a computer network?" Most transportation systems use extensive communication networks to monitor the flow of vehicles and trains. Expressways and highways have computerized systems for controlling traffic signals and limiting access during peak traffic times. Some major cities are placing the appropriate hardware inside city buses so that the precise location of each bus is known. This information enables the transportation systems to keep the buses evenly spaced and more punctual.

In addition, more and more people are using satellite-based GPS devices in their cars that,if you become lost while driving, will tell you precisely where your automobile is and give you directions. Similar systems can unlock your car doors if you leave your keys in the ignition and can locate your car in a crowded parking lot—beeping the horn and flashing the headlights if you cannot remember where you parked.

But even if you didn't use mass transit or the GPS device in your car today, there are many other ways to use a computer network. Businesses are able to order parts and inventory on demand and build products to customer-designed specifications electronically, without the need for paper. Online retail outlets can track every item you look at or purchase. Using this data, they can make recommendations of similar products and inform you in the future when a similar new product becomes available. Twenty-four-hour banking machines can verify the user's identity by taking the user's thumbprint.

In addition, cable television continues to expand, offering extensive programming, pay-per-view options, video recording, digital television and music, and multi-megabit connectivity to the Internet. The telephone system, the oldest and most extensive network of communicating devices, continues to become more of a computer network every day. The most recent "telephone" networks can now deliver voice, Internet, and television over a single connection. Cellular telephone systems cover virtually the entire North American continent and include systems that allow users to upload and download data to and from the Internet, send and receive images, and download streaming video such as television programs. That hand-held device you are holding can play music, make phone calls, take pictures, surf the Web, and let you play games while you wait for the next train.

Welcome to the amazing world of computer networks! Unless you have spent the last 24 hours in complete isolation, it is nearly impossible to *not* have used some form of computer networks and data communications. Because of this growing integration of computer networks and data communications into business and life, we cannot leave this area of study to technicians. All of us—particularly information systems, business, and computer science students—need to understand the basic concepts. Armed with this knowledge, we not only will be better at communicating with network specialists and engineers, but will become better students, managers, and employees.

The Language of Computer Networks

Over the years, numerous terms and definitions relating to computer networks and data communications have emerged. To gain insight into the many subfields of study, and to become familiar with the emphasis of this textbook, let us examine the more common terms and their definitions.

A computer network is an interconnection of computers and computing equipment using either wires or radio waves and can share data and computing resources. Computer networks that use radio waves are termed wireless and can involve broadcast radio, microwaves, or satellite transmissions. Networks spanning an area of several meters around an individual are called personal area networks (PANs). Personal area networks include devices such as laptop computers, personal digital assistants, and wireless connections. Networks a little larger in geographic size—spanning a room, a floor within a building, a building, or a campus—are local area networks (LANs). Networks that serve an area up to roughly 50 kilometers—approximately the area of a typical city—are called metropolitan area networks (MANs). Metropolitan area networks are high-speed networks that interconnect businesses with other businesses and the Internet. Large networks encompassing parts of states, multiple states, countries, and the world are wide area networks (WANs). Chapters Seven and Eight concentrate on local area networks, and Chapters Nine, Ten, and Eleven concentrate on metropolitan area networks and wide area networks.

The study of computer networks usually begins with the introduction of two important building blocks: data and signals. Data is information that has been translated into a form more conducive to storage, transmission, and calculation. As we shall see in Chapter Two, a signal is used to transmit data. We will define data communications as the transfer of digital or analog data using digital or analog signals. Once created, these analog and digital signals then are transmitted over conducted media or wireless media (both of which are discussed in Chapter Three). Both the data and the signal can be analog or digital, allowing for four possible combinations. Transmitting analog data by analog signals and digital data by digital signals are fairly straightforward processes—the conversion from one form to another is relatively simple. Transmitting digital data using analog signals, however, requires the digital data to be modulated onto an analog signal, which is what happens with a modem and the telephone system. Transmitting analog data using digital signals requires the data to be sampled at specific intervals and then digitized into a digital signal, which is what happens with a device called a digitizer, or codec.

Transmitting data and signals between a sender and a receiver or between a computer and a modem requires interfacing, a topic covered in Chapter Four. Because sending only one signal over a medium at one time can be an inefficient way to interface, many systems perform multiplexing and/or compression. Multiplexing is the transmission of multiple signals on one medium. For a medium to transmit multiple signals simultaneously, the signals must be altered so that they do not interfere with one another. Compression is the technique of squeezing data into a smaller package, thus reducing the amount of time (as well as storage space) needed to transmit the data. Multiplexing and compression are covered in detail in Chapter Five.

When the signals transmitted between computing devices are corrupted and errors result, error detection and error control are necessary. These topics are discussed in detail in Chapter Six.

Once upon a time, a voice network transmitted telephone signals, and a data network transmitted computer data. Eventually, however, the differences between voice networks and data networks began to disappear. Networks designed primarily for voice now carry data, and networks designed to carry data now transmit voice in real time. Many experts predict that one day no distinction will be made and that one network will efficiently and effectively carry all types of traffic. The merging of voice and data networks is termed convergence, an important topic that will be presented later in this chapter and further developed in subsequent chapters.

Computer security (covered in Chapter Twelve) is a growing concern of both professional computer support personnel and home computer users with Internet connections. Network management is the design, installation, and support of a network and its hardware and software. Chapter Thirteen discusses many of the basic concepts necessary to support properly the design and improvement of network hardware and software, as well as the more common management techniques used to support a network.

The Big Picture of Networks

If you could create one picture that tries to give an overview of a typical computer network, what might this picture include? Figure 1-1 shows such a picture, and it includes examples of local, personal, and wide area networks. Note that this picture shows two different types of local area networks (LAN 1 and LAN 2). Although a full description of the different components comprising a local area network is not necessary at this time, it is important to note that most LANs include the following hardware:

- ▶ Workstations, which are personal computers/microcomputers (desktops, laptops, net books, hand helds, etc.) where users reside
- ▶ Servers, which are the computers that store network software and shared or private user files
- ▶ Switches, which are the collection points for the wires that interconnect the workstations
- ▶ Routers, which are the connecting devices between local area networks and wide area networks

Wide area networks also can be of many types. Although many different technologies are used to support wide area networks, all wide area networks include the following components:

- ▶ Nodes, which are the computing devices that allow workstations to connect to the network and that make the decisions about where to route a piece of data
- ▶ Some type of high-speed transmission line, which runs from one node to another
- ▶ A sub-network, or cloud which consists of the nodes and transmission lines, collected into a cohesive unit

Figure 1-1
An overall view of the interconnection between different types of networks

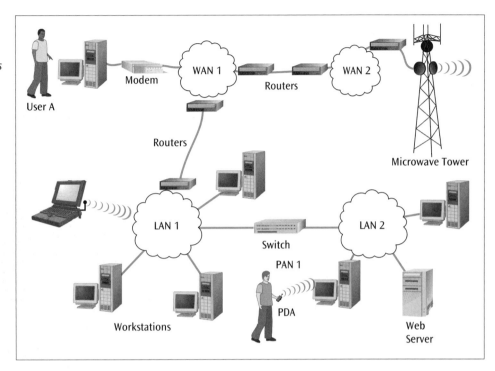

To see how the local area networks and wide area networks work together, consider User A (in the upper-left corner of Figure 1-1), who wishes to retrieve a Web page from the Web server shown in the lower-right corner. To do this, User A's computer must have both the necessary hardware and software required to communicate with the first wide area network it encounters, WAN1—namely, User A's Internet service provider. Assuming that User A's computer is connected to this wide area network through a DSL telephone line, User A needs some type of modem. Furthermore, if this wide area network is part of the Internet, User A's computer requires software that talks the talk of the Internet: TCP/IP (Transmission Control Protocol/Internet Protocol).

Notice that no direct connection exists between WAN 1, where User A resides, and LAN 2, where the Web server resides. To ensure that User A's Web page request reaches its intended receiver (the Web server), User A's software attaches the appropriate address information that WAN 1 uses to route User A's request to the router that connects WAN 1 to LAN 1. Once the request is on LAN 1, the switch-like device connecting LAN 1 and LAN 2 uses address information to pass the request to LAN 2. Additional address information then routes User A's Web page request to the Web server, whose software accepts the request.

Under normal traffic and conditions, this procedure may take only a fraction of a second. When you begin to understand all the steps involved and the great number of transformations that a simple Web page request must undergo, the fact that it takes *only* a fraction of a second to deliver is amazing.

Communications Networks—Basic Connections

The beginning of this chapter described a few of the application areas of computer networks and data communications that you encounter in everyday life. From that sampling, you can see that setting out all the different types of jobs and services that use some sort of computer network and data communications would generate an enormous list. Instead, let us examine basic network systems and their connections to see how extensive the uses of data communications and computer networks are. The basic connections that we will examine include:

- Microcomputer-to-local area network
- Microcomputer-to-Internet
- Local area network-to-local area network
- Personal area network-to-workstation
- Local area network-to-metropolitan area network
- Local area network-to-wide area network
- Wide area network-to-wide area network
- Sensor-to-local area network
- Satellite and microwave
- Cell phones
- Terminal/microcomputer-to-mainframe computer

Microcomputer-to-local area network connections

Perhaps the most common network connection today, the microcomputer-to-local area network (LAN) connection is found in virtually every business and academic environment—and even in many homes. The microcomputer—which also is commonly known as the personal computer, pc, desktop computer, laptop computer, notebook, netbook, or workstation—began to emerge in the late 1970s and early 1980s. (For the sake of consistency, we will use the older term "microcomputer" to signify any type of computer based on a microprocessor, disk drive, and memory.) The LAN, as we shall see in Chapter Seven, is an excellent tool for sharing software and peripherals. In some LANs, the data set that accompanies application software resides on a central computer called a server. Using microcomputers connected to a LAN, end users can request and download the data set, then execute the application on their computers. If users wish to print documents on a high-quality network printer, the LAN contains the network software necessary to route their print requests to the appropriate printer. If users wish to access their e-mail from the corporate e-mail server, the local area network provides a fast, stable connection between user workstations and the e-mail server. Figure 1-2 shows a diagram of this type of microcomputer-to-local area network connection.

One common form of microcomputer-to-local area network connection in the business world is the client/server system. In a client/server system, a user at a microcomputer, or client machine, issues a request for some form of data or service. This could be a request for a database record from a database server or a request to retrieve an e-mail message from an e-mail server. This request travels across the system to a server that contains a large repository of data and/or programs. The server fills the request and returns the results to the client, displaying the results on the client's monitor.

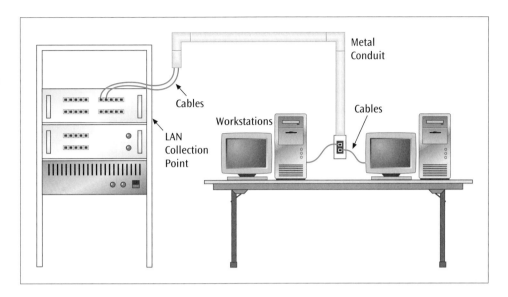

A type of microcomputer-to-local area network connection that continues to grow in popularity is the wireless connection. A user sitting at a workstation or laptop uses a wireless communication device to send and receive data to and from a wireless access point. This access point is connected to the local area network and basically serves as the "bridge" between the wireless user device and the wired network. Although this setup uses radio frequency transmissions, we still consider it a microcomputer-to-local area network connection.

Microcomputer-to-Internet connections

With the explosive growth of the Internet and the desire of users to connect to the Internet from home (either for pleasure or work-related reasons), the microcomputer-to-Internet connection continues to grow steadily. Currently, fewer than half of all home users connect to the Internet using a modem and a dial-up telephone service, which at the time provides data transfer rates of approximately 56,000 bits per second (56 kbps). (The connections do not actually achieve 56 kbps, but that is a mystery we will examine in Chapter Eleven.) The growing number of users who wish to connect at speeds higher than 56 kbps use telecommunications services such as digital subscriber line (DSL) or access the Internet through a cable modem service. All of these alternative telecommunications services will be examined in Chapter Eleven. (In comparing the various data transfer rates of services and devices, we will use the convention in which lowercase k = 1000. Also as part of the convention, lowercase b will refer to bits, while uppercase B refers to bytes.)

To communicate with the Internet using a dial-up or DSL modem, a user's computer must connect to another computer already communicating with the Internet. The easiest way to establish this connection is through the services of an Internet service provider (ISP). In this case, the user's computer requires software to communicate with the Internet. The Internet "talks" only TCP/IP, so users must use software that supports the TCP and IP protocols. Once the user's computer is talking TCP/IP, a connection to the Internet can be established. Figure 1-3 shows a typical microcomputer-to-Internet connection.

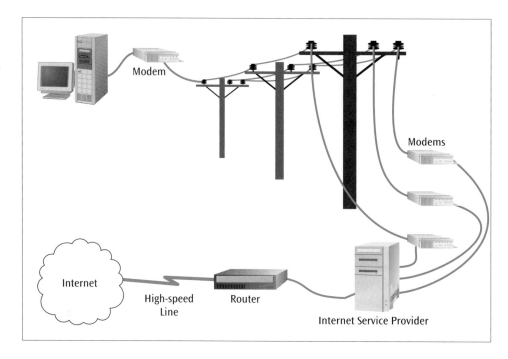

Local area network-to-local area network connections

Because the local area network is a standard in business and academic environments, it should come as no surprise that many organizations need the services of multiple local area networks and that it may be necessary for these LANs to communicate with each other. For example, a company may want the local area network that supports its research department to share an expensive color laser printer with its marketing department's local area network. Fortunately, it is possible to connect two local area networks so that they can share peripherals as well as software. The devices that usually connect two or more LANs are the switch and router.

In some cases, it may be more important to *prevent* data from flowing between local area networks than to allow data to flow from one network to another. For instance, some businesses have political reasons for supporting multiple networks—each division may want its own network to run as it wishes. Additionally, there may be security reasons for limiting traffic flow between networks; or allowing data destined for a particular network to traverse other networks simply may generate too much network traffic. Devices that connect local area networks can help manage these types of services as well. For example, the switch can filter out traffic not intended for the neighboring network, thus minimizing the overall amount of traffic flow. Figure 1-4 provides an example of two LANs connected by a switch.

Figure 1-4
Two local area networks connected by a switch

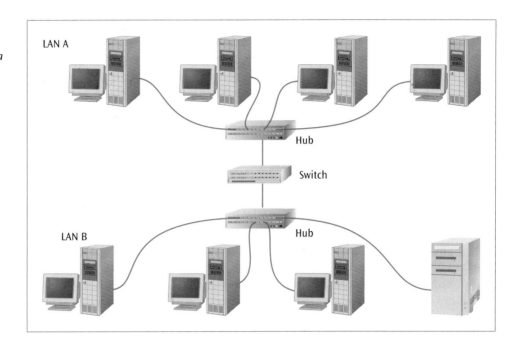

Personal area network-to-workstation connections

The personal area network was created in the late 1990s and is one of the newer forms of computer networks. Using wireless transmissions with devices such as personal digital assistants (PDAs), laptop computers, and portable music players, an individual can transfer voice, data, and music from handheld devices to other devices such as microcomputer workstations (see Figure 1-5). Likewise, a user can download data from a workstation to one of these portable devices. For example, a user may use a PDA to record notes during a meeting. Once the meeting is over, the user can transmit the notes over a wireless connection from the PDA to his or her workstation. The workstation then runs a word processor to clean up the notes, and the formatted notes are uploaded to a local area network for corporate dissemination. Another example is the hands-free Bluetooth-enabled connection that people hang on their ear so they can converse with their cell phone without placing the cell phone up to their ear.

Figure 1-5
A user transferring data from a personal digital assistant via a personal area network to a workstation attached to a local area network

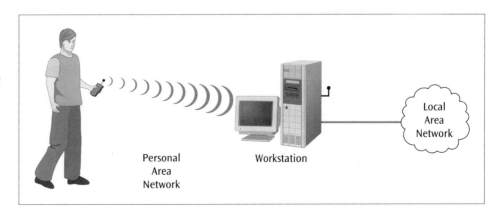

Local area network-to-metropolitan area network connections

Toward the end of the twentieth century, a new form of network appeared that interconnected businesses within a metropolitan area. Typically, this interconnection uses only fiber-optic links at extremely high speeds. These new networks are labeled metropolitan area networks. A metropolitan area network (MAN) is a high-speed network that interconnects multiple sites within a close geographic region, such as a large urban area. For example, businesses that require a high-speed connection to their Internet service providers may use a metropolitan area network for interconnection (see Figure 1-6). As we shall see in more detail in Chapter Nine, metropolitan area networks are a cross between local area networks and wide area networks. They can transfer data at fast, LAN speeds but over larger geographic regions than typically associated with a local area network.

Figure 1-6
Businesses interconnected within a large metropolitan area via a metropolitan area network

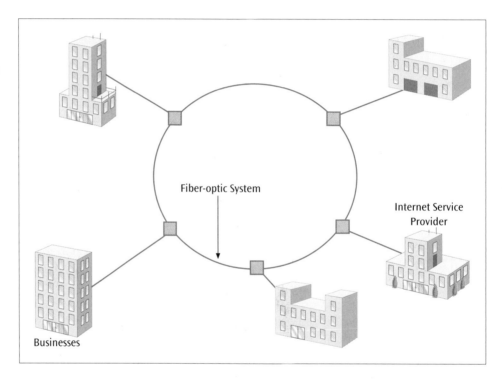

Local area network-to-wide area network connections

You have already seen that the local area network is commonly found in business and academic environments. If a user working at a microcomputer connected to a local area network wishes to access the Internet (a wide area network), the user's local area network has to have a connection to the Internet. A device called a router is employed to connect these two networks. A router converts the local area network data into wide area network data. It also performs security functions and must be properly programmed to accept or reject certain types of incoming and outgoing data packets. Figure 1-7 shows a local area network connected to a wide area network via a router.

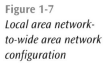

Figure 1-7
Local area network-to-wide area network configuration

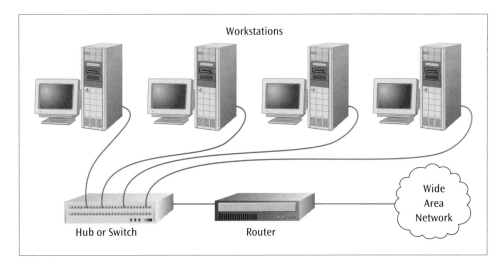

Wide area network-to-wide area network connections

The Internet is not a single network but a collection of thousands of networks. In order to travel any distance across the Internet, a data packet undoubtedly will pass through multiple wide area networks. Connecting a wide area network to a wide area network requires special devices that can route data traffic quickly and efficiently. These devices are high-speed routers. After the data packet enters the high-speed router, an address in the network layer (the IP address) is extracted, a routing decision is made, and the data packet is forwarded onto the next wide area network segment. As the data packet travels across the Internet, router after router makes a routing decision, moving the data toward its final destination. We will examine the Internet in more detail in Chapter Ten, then follow up with a discussion of several other types of wide area network technologies in Chapter Eleven.

Sensor-to-local area network connections

Another common connection found in everyday life is the sensor-to-local area network connection. In this type of connection, the action of a person or object triggers a sensor—for example, a left-turn light at a traffic intersection—that is connected to a network. In many left-turn lanes, a separate left-turn signal will appear if and only if one or more vehicles are in the left-turn lane. A sensor embedded in the roadway detects the movement of an automobile in the lane above and triggers the left-turn mechanism in the traffic signal control box at the side of the road. If this traffic signal control box is connected to a larger traffic control system, the sensor is connected to a local area network.

Another example of sensor-to-local area network connection is found within manufacturing environments. Assembly lines, robotic control devices, oven temperature controls, and chemical analysis equipment often use sensors connected to data-gathering computers that control movements and operations, sound alarms, and compute experimental or quality control results. Figure 1-8 shows a diagram of a typical sensor-to-local area network connection in a manufacturing environment.

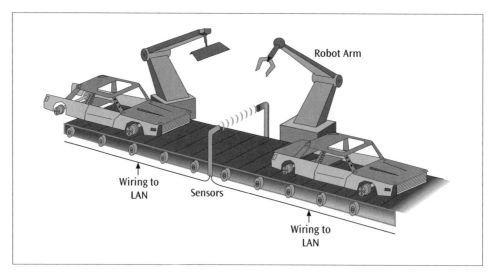

Satellite and microwave connections

Satellite and microwave connections are continuously evolving technologies used in a variety of applications. If the distance between two networks is great and running a wire between them would be difficult (if not impossible), satellite and microwave transmission systems can be an extremely effective way to connect the two networks or computer systems. Examples of these applications include digital satellite TV; meteorology; intelligence operations; mobile maritime telephony; GPS (Global Positioning System) navigation systems; wireless e-mail, paging, and worldwide mobile telephone systems; and videoconferencing. Figure 1-9 shows a diagram of a typical satellite system.

Figure 1-9
Example of a television company using a satellite system to broadcast television services into homes and businesses

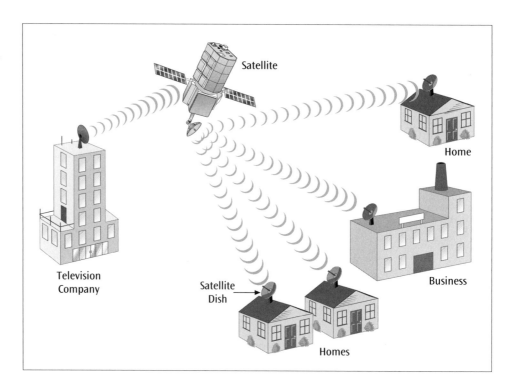

Cell phone connections

One of the most explosive areas of growth in recent years has been cell phone, or wireless telephone networks. The cell phone has almost replaced the pager, and newer wireless technologies that conduct telephone conversations with less background noise and can transmit varying amounts of data are joining the older services. Figure 1-10 shows an example of a handheld personal digital assistant (PDA) that, along with making telephone calls, can transmit and receive data. The PDA has a modem installed, which transmits the PDA's data across the wireless telephone network to the wireless telephone switching center. The switching center then transfers the PDA's data over the public telephone network or through a connection onto the Internet. Many newer handheld devices have combined data accessing capabilities with a cell phone and can transfer data over wireless telephone connections.

Figure 1-10

An example of a PDA connected to a wireless telephone system to transmit and receive data

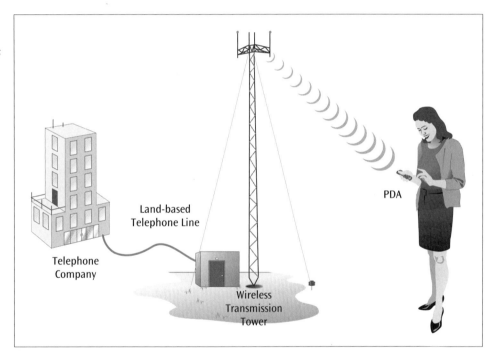

Terminal/microcomputer-to-mainframe computer connections

Today, many businesses still employ a terminal-to-mainframe connection, although the number of these systems in use is not what it used to be. During the 1960s and 1970s, the terminal-to-mainframe connection was in virtually every office, manufacturing, and academic environment. These types of systems are still being used for inquiry/response applications, interactive applications, and data-entry applications, such as you might find when applying for a new driver's license at the Department of Motor Vehicles (Figure 1-11).

Figure 1-11
Using a terminal to perform a text-based input transaction

Terminal-to-mainframe connections of the 60s and 70s used "dumb" terminals because the end user was doing relatively simple data-entry and retrieval operations and a workstation with a lot of computing power and storage was not necessary. A computer terminal was a device that was essentially a keyboard and screen with no long-term storage capabilities and little, if any, processing power. Computer terminals were used for entering data into a system, such as a mainframe computer, and then displaying results from the mainframe. Because the terminal did not possess a lot of computing power, the mainframe computer controled the sending and receiving of data to and from each terminal. This required special types of protocols (sets of rules used by communication devices), and the data was usually transmitted at relatively slow speeds, such as 9600 or 19,200 bits per second (bps).

During this period, many of the same end users who had terminals on their desks also now found a microcomputer there (and thus had very little room for anything else). In time, terminal-emulation cards were developed, which allowed a microcomputer to imitate the abilities of a computer terminal. As terminal emulation cards were added to microcomputers, terminals were removed from end users' desks, and microcomputers began to serve both functions. Now, if users wished, they could download information from the mainframe computer to their microcomputers, perform operations on the data, and then upload the information to the mainframe. Today, one rarely sees dumb computer terminals. Instead, most users use microcomputers and access the mainframe using either a terminal emulation card, a web browser and web interface, Telnet software (more on this in Chapter 10), or a thin client. A thin client workstation is similar to a microcomputer but has no hard drive storage.

Convergence

A dictionary might define "convergence" as the process of coming together toward a single point. With respect to computer networks and communications systems, this definition is fairly relevant. Over the years, the communications industry has seen and continues to see different network applications and the technologies that support them converge into a single technology capable of supporting various applications. In particular, we can define three different types of convergence: technological convergence, protocol convergence, and industrial convergence. For example, one of the earliest and most common examples of technological convergence was the use of computers and modems to transmit data over the telephone system. This was an example of voice transmission systems converging with data transmission systems and yielding one system capable of supporting both data and voice. By the 1990s, telephone systems carried more computer data than voice. At about the same time, local area networks began to transfer telephone calls. Because local area networks originally were designed for data applications, this was another example of voice and data systems converging. Now we are seeing substantial growth in the Voice over Internet Protocol (VoIP) field. VoIP involves converting voice signals to packets and then sending those packets over packet-driven networks such as local area networks and the Internet.

Today we see many more examples of technological convergence, particularly in the wireless markets. For example, it is now quite common to snap a photo using a cell phone and then transfer the image over the cell phone network to another cell phone. Shortly after the introduction of photo-enabled cell phones, cell phones also became capable of sending and receiving instant messages. Then in 2005, cell phone providers started offering services that allow a user to transmit high-speed data over a cell phone connection. These all are examples of the convergence of two different applications (for example, digital photography and cell phones in the case of photo-enabled cell phones) into a single technology. As we will see in a later chapter, many of the telephone companies that provide local and long distance telephone service have converged into fewer companies. These are examples of industrial convergence. Also in a later chapter, we will see how older network protocols have given way or merged with other protocols, thus demonstrating protocol convergence.

Throughout the rest of this book, we will examine other examples of convergence within the communications industry. In addition to introducing the technologies involved, we will also examine the effects a given convergence of technologies may have on individual users and businesses.

Network Architectures

Now that you know the different types of networks and connections, you need a framework to understand how all the various components of a network interoperate. When someone uses a computer network to perform an application, many pieces come together to assist in the operation. A **network architecture**, or communications model, places the appropriate network pieces in layers. The layers define a *model* for the functions or services that need to be performed. Each layer in the model defines what services either the hardware or software (or both) provide. The two most common architectures known today are the TCP/IP protocol suite and the Open Systems Interconnection (OSI) model. The TCP/IP protocol suite is a working model (currently used on the Internet), while the OSI model (originally designed to be a working model), has been relegated as a theoretical model. We will discuss these two architectures in more detail in the following pages. But first you should know a bit more about the components of a network and how a network architecture helps organize those components.

Consider that a typical computer network within a business contains the following components that must interact in various ways:

- ▶ Wires
- ▶ Printed circuit boards
- ▶ Wiring connectors and jacks
- ▶ Computers
- ▶ Centrally located wiring concentrators
- ▶ Disk and tape drives
- ▶ Computer applications such as word processors, e-mail programs, and accounting, marketing, and electronic commerce software
- ▶ Computer programs that support the transfer of data, check for errors when the data is transferred, allow access to the network, and protect user transactions from unauthorized viewing

This large number of network components and their possible interactions inspires two questions. First, how do all of these pieces work together harmoniously? You do not want two pieces performing the same function, or no pieces performing a necessary function. Like the elements of a well-oiled machine, all components of a computer network must work together to produce a product.

Second, does the choice of one piece depend on the choice of another piece? To make the pieces as modular as possible, you do not want the selection of one piece to constrain the choice of another piece. For example, if you create a network and originally plan to use one type of wiring but later change your mind and use a different type of wiring, will that change affect your choice of word processor? Such an interaction would seem highly unlikely. Alternately, can the choice of wiring affect the choice of the software program that checks for errors in the data sent over the wires? The answer to this question is not as obvious.

To keep the pieces of a computer network working together harmoniously and to allow modularity between the pieces, national and international organizations developed network architectures, which are cohesive layers of protocols defining a set of communication services. Consider the following non-computer example. Most organizations that produce some type of product or perform a service have a division of labor. Secretaries do the paperwork; accountants keep the books; laborers perform the manual duties; scientists design products; engineers test the products; and managers control operations. Rarely is one person capable of performing all these duties. Large software applications operate the same way. Different procedures perform different tasks, and the whole would not function without the proper operation of each of its parts. Computer network applications are no exception. As the size of the applications grows, the need for a division of labor becomes increasingly important. Computer network applications also have a similar delineation of job functions. This delineation is the network architecture. Let's examine two network architectures or models: the TCP/IP protocol suite, followed by the OSI model.

The TCP/IP protocol suite

The TCP/IP protocol suite was created by a group of computer scientists in order to support a new type of network (the ARPANET) being installed across the United States in the 1960s and 70s. The goal was to create an open architecture that would allow virtually all networks to inter-communicate. The design was based on a number of layers, in which the user would connect at the upper-most layer and would be isolated from the details of the electrical signals found at the lower layer.

The number of layers in the suite is not a static entity. In fact, some books present the TCP/IP protocol suite as four layers, while others present five. Even then, different sources use different names for each of the layers. For this book, we will define five layers, as shown in Figure 1-12: application, transport, network, network access, and physical. Note that the layers do not specify precise protocols or exact services. In other words, the TCP/IP protocol suite does not tell us, for example, what kind of wire or what kind of connector to use to connect the pieces of a network. That choice is left to the designer or implementer of the system. Instead, the suite simply says that if you specify a type of wire or a specific connector, you do that in a particular layer. In addition, each layer of the TCP/IP protocol suite provides a service for the next layer. For example, the transport layer makes sure the data received at the very end of a transmission is exactly the same as the data originally transmitted, but it relies upon the network layer to find the best path for the data to take from one point to the next within the network. With each layer performing its designated function, the layers work together to allow an application to send its data over a network of computers. Let us look at a simple e-mail application example (Figure 1-12) to understand how the layers of the TCP/IP protocol suite work together.

Figure 1-12

The five layers of the TCP/IP protocol suite

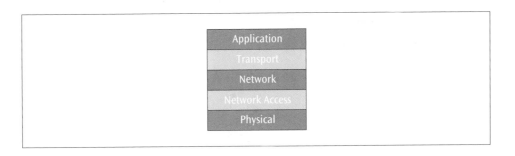

A common network application is e-mail. An e-mail program that accepts and sends the message "Andy, how about lunch? Sharon" has many steps. Using the TCP/IP protocol suite, the steps might look like the following. To begin, the e-mail "application worker" prompts the user to enter a message and specify an intended receiver. The application worker would create the appropriate data package with message contents and addresses and send it to a "transport worker," which is responsible for providing overall transport integrity. The transport worker may establish a connection with the intended receiver, monitor the flow between sender and receiver, and perform the necessary operations to recover lost data in case some data disappears or becomes unreadable.

The "network worker" would then take the data package from the transport worker and may add routing information so that the data package can find its way through the network. Next to get the data package would be the "network access worker," which would insert error-checking information and prepare the data package for transmission. The final worker would be the "physical worker," which would transmit the data package over some form of wire or through the air using radio waves.

Each worker has his own job function. Figure 1-13 shows how these workers work together to create a single package for transmission.

Figure 1-13
Network workers per-form their job duties at each layer in the model

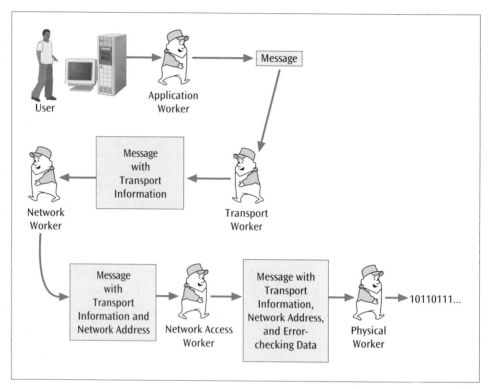

Let's examine each of the layers in more detail. The top layer of the TCP/IP protocol suite, the application layer, supports the network applications and may in some cases include additional services such as encryption or compression. The TCP/IP application layer includes several frequently used applications:

- ▶ Hypertext Transfer Protocol (HTTP) to allow Web browsers and servers to send and receive World Wide Web pages
- ▶ Simple Mail Transfer Protocol (SMTP) to allow users to send and receive electronic mail
- ▶ File Transfer Protocol (FTP) to transfer files from one computer system to another
- ▶ Telnet to allow a remote user to log in to another computer system
- ▶ Simple Network Management Protocol (SNMP) to allow the numerous elements within a computer network to be managed from a single point

The next layer in the TCP/IP protocol suite is the transport layer. The TCP/IP transport layer commonly uses the Transmission Control Protocol (TCP) to maintain an error-free end-to-end connection. To maintain this connection, TCP includes error control information in case one packet from a sequence of packets does not arrive at the final destination and packet sequencing information so that all the packets stay in the proper order. TCP is not the only possible protocol found at the TCP/IP transport layer. User Datagram Protocol (UDP) is an alternative also used, though less frequently, in the TCP/IP protocol suite.

TCP/IP's network layer, sometimes called the Internet layer, is roughly equivalent to OSI's network layer. The protocol used at this layer to transfer data within and between networks is the Internet Protocol (IP). The Internet Protocol is the software that prepares a packet (a fixed-size collection) of data so that it can move from one network to another on the Internet or within a set of corporate networks.

The next lower layer of the TCP/IP protocol suite is the network access layer. If the network layer deals with passing packets through the Internet, then the network access layer is the layer that gets the data from the user workstation to the Internet. In a majority of cases, the connection that gets the data from the user workstation to the Internet is a local area network. Thus, the network access layer prepares a data packet (called a frame at this layer) for transmission from the user workstation to a router sitting between the local area network and the Internet. This is also the last layer before the data is handed off for transmission across the medium. The network access layer is often called the data link layer.

The bottom-most layer in the TCP/IP protocol suite (or at least according to many) is the physical layer. The physical layer is the layer in which the actual transmission of data occurs. As noted earlier, this transmission can be over a physical wire, or it can be a radio signal transmitted through the air. Note that some people combine the network access layer and physical layer into one layer.

Having distinctly defined layers enables you to "pull" out one layer and insert an equivalent layer without affecting the other layers. For example, let us assume a network was designed for copper-based wire. Later, the system owners decided to replace the copper-based wire with fiber-optic cable. Even though a change is being made at the physical layer, it should not be necessary to make any changes at any other layers. In reality, however, a few relationships exist between the layers of a communication system that cannot be ignored. For example, if the physical organization of a local area network is changed, it is likely that the frame description at the data link layer also will need to be changed. (We will examine this phenomenon in Chapter Seven.) The TCP/IP protocol suite recognizes these relationships and merges many of the services of the physical and data link layers into one layer.

The OSI Model

Although the TCP/IP protocol suite is the model of choice for most installed networks, it is important to study both this architecture and the OSI model. Many books and articles, when describing a product or a protocol, often refer to the OSI model with a statement such as, "This product is compliant with OSI layer xxx." If you do not become familiar with the various layers of the OSI model and the TCP/IP protocol suite, this lack of important basic knowledge might impede your understanding of more advanced concepts in the future.

The OSI model was designed with seven layers, as shown in Figure 1-14. Note further the relationship between the five layers of the TCP/IP protocol suite and the seven layers of the OSI model. The top layer in the OSI model is the application layer, where the application using the network resides. Although many kinds of applications employ computer networks, certain ones are in widespread use. Applications such as electronic mail, file transfer systems, remote login systems, and Web browsing are so common that various standards-making organizations have created specific standards for them.

Figure 1-14
The seven layers of the OSI model compared to the five of the TCP/IP Protocol suite

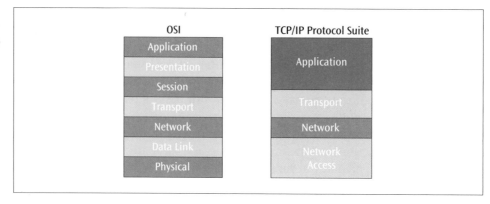

Details ▶

The Internet's Request for Comment (RFC)

Network models, like communications protocols, computer hardware, and application software, continue to evolve daily. The TCP/IP protocol suite is a good example of a large set of protocols and standards constantly being revised and improved. An Internet standard is a tested specification that is both useful and adhered to by users who work with the Internet. Let us examine the path a proposal must follow on the way to becoming an Internet standard.

All Internet standards start as an Internet draft, which is a preliminary work in progress. One or more internal Internet committees work on a draft, improving it until it is in an acceptable form. When the Internet authorities feel the draft is ready for the public, it is published as a Request for Comment (RFC), a document open to all interested parties. The RFC is assigned a number, and it enters its first phase: proposed standard. A proposed standard is a proposal that is stable, of interest to the Internet community, and fairly well understood. The specification is tested and implemented by a number of different groups, and the results are published. If the proposal passes at least two independent and interoperable implementations, the proposed standard is elevated to draft standard. If, after feedback from test implementations is taken into account, the draft standard experiences no further problems, the proposal is finally elevated to Internet standard.

If, however, the proposed standard is deemed inappropriate at any point along the way, it becomes a historic RFC and

is kept for historical perspective. (Internet standards that are replaced or superseded also become historic.) An RFC can also be categorized as experimental or informational. In these cases, the RFC in question probably was not meant to be an Internet standard, but was created either for experimental reasons or to provide information. Figure 1-15 shows the levels of progression for an RFC.

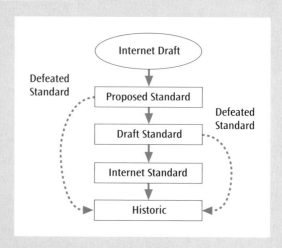

Figure 1-15 *Levels of progression as an RFC moves toward becoming a standard*

The next layer in the OSI model, the presentation layer, performs a series of miscellaneous functions necessary for presenting the data package properly to the sender or receiver. For example, the presentation layer might perform ASCII-to-non-ASCII character conversions, encryption and decryption of secure documents, and the compression of data into smaller units.

The session layer is responsible for establishing sessions between users and for token management, a service that controls which user's computer talks when during the current session by passing a software token back and forth. Additionally, the session layer establishes synchronization points, which are backup points used in case of errors or failures. For example, while transmitting a large document such as an electronic book, the session layer may insert a synchronization point at the end of each chapter. If an error occurs during transmission, both sender and receiver can back up to the last synchronization point (to the beginning of a previously transmitted chapter) and start retransmission from there. Many network applications do not include a specific session layer and do not use tokens to manage a conversation. If they do, the "token" is inserted by the application layer, or possibly the transport layer, instead of the session layer. Likewise, if network applications use synchronization points, these points often are inserted by the application layer.

It is possible to obtain a printed listing of each RFC. See the Internet Engineering Task Force's Web page at http://www.ietf.org/rfc.html for the best way to access RFCs.

The Internet is managed by the work of several committees. The topmost committee is the Internet Society (ISOC). ISOC is a nonprofit, international committee that provides support for the entire Internet standards-making process. Associated with ISOC is the Internet Architecture Board (IAB), which is the technical advisor to the ISOC. Under the IAB are two major committees: the Internet Engineering Task Force (IETF) and the Internet Research Task Force (IRTF). IETF manages the working groups that create and support functions such as Internet protocols, security, user services, operations, routing, and network management. IRTF manages the working groups that focus on the long-range goals of the Internet, such as architecture, technology, applications, and protocols.

Internet committees are not the only groups that create protocols or approve standards for computer networks, data communications, and telecommunications. Another organization that creates and approves network standards is the International Organization for Standardization (ISO), which is a multinational group composed of volunteers from the standards-making committees of various governments throughout the world. ISO is involved in developing standards in the field of information technology and created the OSI model for a network architecture.

Other standards-making organizations include:

▶ American National Standards Institute (ANSI)—A private, nonprofit organization not associated with the U.S. government, ANSI strives to support the U.S. economy and protect the interests of the public by encouraging the adoption of various standards.

▶ International Telecommunication Union-Telecommunication Standardization Sector (ITU-T)—Formerly the Consultative Committee on International Telegraphy and Telephony (CCITT), ITU-T is devoted to the research and creation of standards for telecommunications in general and telephone and data systems in particular.

▶ Institute for Electrical and Electronics Engineers (IEEE)—The largest professional engineering society in the world, IEEE strives to promote the standardization of the fields of electrical engineering, electronics, and radio. Of particular interest to us is the work IEEE has performed on standardizing local area networks.

▶ Electronic Industries Association (EIA)—Aligned with ANSI, EIA is a nonprofit organization devoted to the standardization of electronics products. Of particular interest is the work EIA has performed on standardizing the interfaces between computers and modems.

The fourth layer in the OSI model, the transport layer, ensures that the data packet that arrives at the final destination is identical to the data packet that left the originating station. By "identical" we mean there were no transmission errors, the data arrived in the same order as it was transmitted, and there was no duplication of data. Thus, we say that the transport layer performs *end-to-end* error control and *end-to-end* flow control. This means the transport layer is not in use while the data packet is hopping from point to point within the network—it is used only at the two endpoints of the connection. If the underlying network experiences problems such as reset or restart conditions, the transport layer will try to recover from the error and return the end-to-end connection to a known safe state. As we shall see, to ensure that the data arrives error-free at the final destination, the transport layer must be able to work across all kinds of networks, whether they are reliable or not.

The four layers described so far are called end-to-end layers. They are responsible for the data transmitted between the endpoints of a network connection. In other words, these layers perform their operations only at the beginning point and ending point of the network connection. The remaining three layers—the network, data link, and physical layers—are not end-to-end layers. They perform their operations at each node along the network path, not just at the endpoints. The network layer is responsible for creating, maintaining, and ending network connections. As this layer sends the package of data from node to node within a network and between multiple networks, it generates the network addressing necessary for the system to recognize the next intended receiver. To choose a path through the network, the network layer determines routing information and applies it to each packet or group of packets. The network layer also performs congestion control, which ensures that the network does not become saturated at any one point. In networks that use a broadcast distribution scheme, such as a local area network, where the transmitted data is sent to all other stations, the network layer may be very simple.

The data link layer is responsible for taking data from the network layer and transforming it into a cohesive unit called a frame. This frame contains an identifier that signals the beginning and end of the frame, as well as spaces for control information and address information. The address information identifies a particular workstation in a line of multiple workstations. In addition, the data link layer can incorporate some form of error detection software. If an error exists, the data link layer is responsible for error control, which it does by informing the sender of the error. The data link layer must also perform flow control. In a large network where the data hops from node to node as it makes its way across the network, flow control ensures that one node does not overwhelm the next node with too much data. Note that these data link operations are quite similar to the transport layer operations. The primary difference is that the transport layer performs its operations only at the endpoints, while the data link layer performs its operations at every stop (node) along the path.

The bottom layer in the OSI model—the physical layer—handles the transmission of bits over a communications channel. To perform this transmission of bits, the physical layer handles voltage levels, plug and connector dimensions, pin configurations, and other electrical and mechanical issues. The choice of wire or wireless transmission media is usually determined at the physical layer. Furthermore, because the digital or analog data is encoded or modulated onto a digital or analog signal at this point in the process, the physical layer also determines the encoding or modulation technique to be used in the network.

Logical and physical connections

An important concept to understand with regard to the layers of a communication model is the lines of communication between a sender and a receiver. Consider Figure 1-16, which shows sender and receiver using a network application designed on the TCP/IP protocol suite.

Figure 1-16
Sender and receiver communicating using the TCP/IP protocol suite

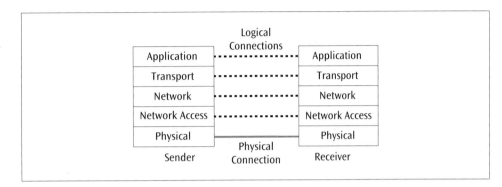

Notice the dashed lines between the sender's and receiver's application layers, transport layers, network layers, and network access layers. No data flows over these dashed lines. Each dashed line indicates a logical connection. A logical connection is a nonphysical connection between sender and receiver that allows an exchange of commands and responses. The sender's and receiver's transport layers, for example, share a set of commands used to perform transport-type functions, but the actual information or data has to be passed through the physical layers of the sender and receiver, as there is no direct connection between the two transport layers. Without a logical connection, the sender and receiver would not be able to coordinate their functions. The physical connection is the only direct connection between sender and receiver and is at the physical layer, where actual 1s and 0s—the digital content of the message—are transmitted over wires or airwaves.

For an example of logical and physical connections, consider an imaginary scenario in which the dean of arts and sciences wants to create a new joint degree with the school of business. In particular, the dean would like to create a degree that is a cross between computer science and marketing. The dean of arts and sciences could call the dean of business to create the degree, but deans are not necessarily experts at assembling all the details involved in a new degree. Instead, the dean of arts and sciences starts the process by issuing a request for a new degree from the dean of business. Before this request gets to the dean of business, however, the request must pass through several layers. First, the request goes to the chairperson of the computer science department. The chairperson will examine the request for a new degree and add the necessary information to staff the program. The chairperson will then send the request to the computer science curriculum committee, which will design several new courses. The curriculum committee will send the request to its department secretary, who will type all the memos and create a readable package. This package is then placed in the intercampus mail and sent to the marketing department in the school of business.

Once the request arrives at the marketing department, the secretary in the marketing department opens the envelope and gives all the materials to the marketing curriculum committee. The marketing curriculum committee looks at the proposed courses from the computer science curriculum committee and makes some changes and additions. Once these changes are made, the proposal is given to the chair of the marketing department who looks at the staffing needs suggested by the chair of computer science, checks the request for accuracy, and makes some changes. The chair of marketing then hands the request to the dean of business, who examines the entire document and gives approval with a few small changes. The request then works its way back down to the secretary of the marketing department, who sends it back to the secretary of computer science. The computer science secretary then sends the reply to the request up the layers until it reaches the dean of arts and sciences. Figure 1-17 shows how this request for a degree might move up and down through the layers of a university's bureaucracy.

Figure 1-17
Flow of data through the layers of bureaucracy

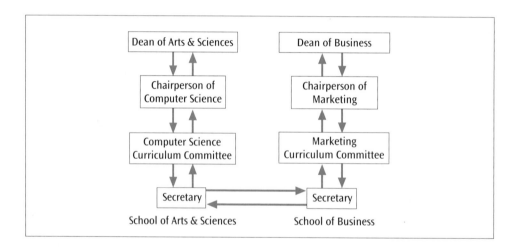

Note that the data did not flow directly between deans, nor did it flow directly between department chairpersons or curriculum committees. Instead, the data had to flow all the way down to the physical layer (in this case, the secretaries) and then back up the other side. At each layer in the process, information that may be useful to the "peer" layer on the other side was added. This example stretches the truth a little; college curriculums are not actually designed this way (the process is in fact much more complicated). Therefore, we will examine a more realistic example in which a person using a Web browser requests a Web page from somewhere on the Internet. But before we examine this more difficult scenario, let us take a look at an example of the connections that occur when a user connects his laptop to the company local area network.

Network Connections In Action ▶

Let us consider a scenario in which an employee of a company is sitting with a laptop computer at work and is accessing the corporate local area network via a wireless connection (see Figure 1-18). The employee is using a Web browser and is trying to download a Web page from the Internet. What are the network connections involved in this scenario? First, the connection between the user's wireless laptop and the corporate local area network is a micro-computer-to-local area network connection. Once the Web page request is in the corporate local area network, it may have to be transferred over multiple local area networks within the corporate system. These connections between local area networks would be local area network-to-local area network connections. To access the Internet, we need a local area network-to-wide area network connection. Or perhaps the corporate local area network connects to a metropolitan area network, in which case we would need a local area network-to-metropolitan area network connection to access the Internet. Once the employee's Web page request is on the Internet, it is difficult to tell what connections are involved. There could be more wide area network-to-wide area network interconnections, as well as a microwave or satellite connection. Once the Web page request nears its final destination, there could be another metropolitan area network, or multiple local area network connections. The return trip might take the same path or might involve new network paths and connections. Clearly, many different types of network connections are involved even in common, daily applications.

Figure 1-18
The numerous network connections involved with a user downloading a Web page at work

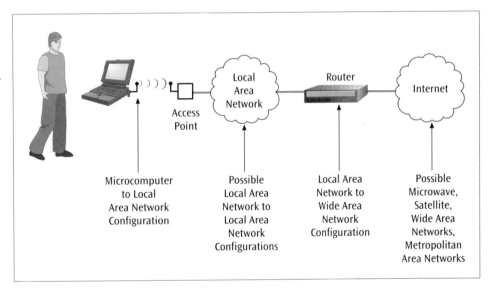

The TCP/IP Protocol Suite In Action ▶

A more detailed and more challenging example of a request for a service moving through the layers of a communications model will help make the concepts involved clearer. Consider Figure 1-19, in which a user browsing the Internet on a personal computer requests a Web page to be downloaded and then displayed on his or her screen.

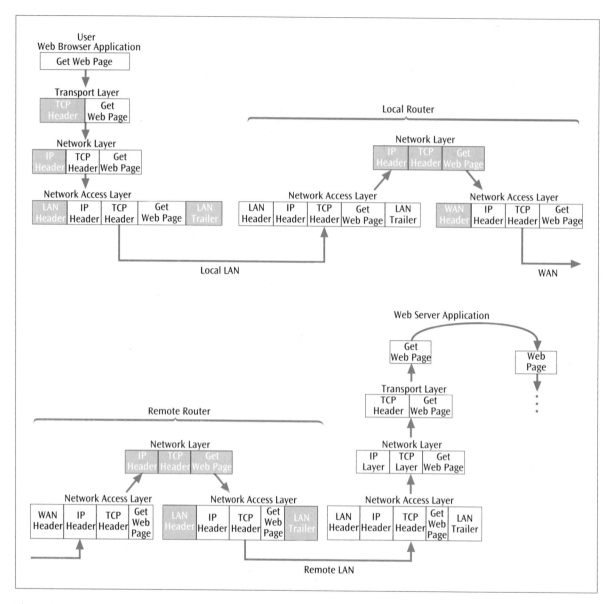

Figure 1-19

Path of a Web page request as it flows from browser to Internet server and back

Beginning in the upper-left corner of the figure, the process is initiated when the user clicks a link on the current Web page. In response, the browser software (the application) creates a Get Web Page command that is given to the browser's transport layer, TCP. TCP adds a variety of header information to be used by the TCP software on the receiving end. Added to the front of the packet, this information may be used to control the transfer of the data. This information assists with end-to-end error control and end-to-end flow control and provides the address of the receiving application (the Web server).

The enlarged packet is now sent to the network layer, where IP adds its header. The information contained within the IP header assists the IP software on the receiving end, as well as assisting the IP software at each intermediate node (router) during the data's progress through the Internet. This assistance includes providing the Internet address of the workstation that contains the requested Web page.

The packet is now given to the network access layer. Because the user's computer is connected to a local area network, the appropriate local area network headers are added. Note that sometimes in addition to headers, control information is added to the end of the data packet as trailers. One of the most important pieces of information included in the local area network header is the address of the device (the router) that connects the local area network to the wide area network (the Internet).

Eventually, the binary 1s and 0s of the data packet are transmitted across the user's local area network via the physical layer, where they encounter a router. The router is a device that serves as the gateway to the Internet. The router removes the local area network header and trailer. The information in the IP header is examined, and the router determines that the data packet must go out to the Internet. New wide area network (WAN) header information, which is necessary for the data packet to traverse the wide area network, is applied, and the binary 1s and 0s of the data packet are placed onto the wide area network.

After the data packet moves across the Internet, it will arrive at the router connected to the local area network that contains the desired Web server. This remote router removes the wide area network information, sees that the packet must be placed on the local area network, and inserts the local area network header and trailer information. The packet is placed onto the local area network, and using the address information in the LAN header, travels to the computer holding the Web server application. As the data packet moves up the layers of the Web server's computer, the LAN, IP, and TCP headers are removed. The Web server application receives the Get Web Page command, retrieves the requested Web page, and creates a new data packet with the requested information. This new data packet now moves down the layers and back through the routers to the user's network and workstation. Finally, the Web page is displayed on the user's monitor.

It is interesting to note that as a packet of data flows down through a model and passes through each layer of the system, the data packet grows in size. This growth is attributable to the fact that each layer adds more information to the original data. Some of this layer-added information is needed by the nodes and routers in the data packet's path, and some is required by the data packet's final destination. This information aids in providing services such as error detection, error control, flow control, and network addressing. The addition of control information to a packet as it moves through the layers is called encapsulation. Note also that as the packet moves up through the layers, the data packet shrinks in size. Each layer removes the header it needs to perform its job duty. Once the job duty is complete, the header information is discarded and the smaller packet is handed to the next higher layer.

❖ ❖

SUMMARY

▶ Many services and products that we use every day employ computer networks and data communications in one way or another. Telephone systems, banking systems, cable television, audio and video systems, traffic control systems, and wireless telephones are a few examples.

▶ The field of data communications and computer networks includes data networks, voice networks, wireless networks, local area networks, metropolitan area networks, wide area networks, and personal area networks.

▶ The application areas of computer networks and data communications can be understood in terms of general network connections:

  ▶ Microcomputer-to-local area network

  ▶ Microcomputer-to-Internet

  ▶ Local area network-to-local area network

  ▶ Personal area network-to-workstation

  ▶ Local area network-to-metropolitan area network

- ▶ Local area network-to-wide area network
- ▶ Wide area network-to-wide area network
- ▶ Sensor-to-local area network
- ▶ Satellite and microwave
- ▶ Cell phone
- ▶ Terminal/microcomputer-to-mainframe computer

▶ A key concept in networking these days is convergence, the phenomena in which network applications and the technologies that support them converge into a single technology capable of supporting various applications. In particular, we can define three different types of convergence: technological convergence, protocol convergence, and industrial convergence.

▶ A network architecture, or communications model, places network pieces in layers. The layers define a *model* for the functions or services that need to be performed. Each layer in the model defines the services that either hardware or software or both provide.

▶ To standardize the design of communications systems, the International Organization for Standardization (ISO) created the Open Systems Interconnection (OSI) model. There are currently no actual implementations of the OSI model. The OSI model is based on seven layers:

- ▶ The application layer is the top layer of the OSI model, where the application using the network resides.
- ▶ The presentation layer performs a series of miscellaneous functions necessary for presenting the data package properly to the sender or receiver.
- ▶ The session layer is responsible for establishing sessions between users.
- ▶ The transport layer is concerned with an error-free end-to-end flow of data.
- ▶ The network layer is responsible for creating, maintaining, and ending network connections.
- ▶ The data link layer is responsible for taking the raw data and transforming it into a cohesive unit called a frame.
- ▶ The physical layer handles the transmission of bits over a communications channel.

▶ Another network architecture (or communications model) called the TCP/IP protocol suite has become the de facto standard for network models. The TCP/IP protocol suite is also known as the Internet model and is composed of five layers:

- ▶ The application layer contains the network applications for which one uses a network and the presentation services that support that application.
- ▶ The transport layer maintains an error-free end-to-end connection.
- ▶ The network layer, or Internet layer, uses the Internet Protocol (IP) to transfer data between networks.
- ▶ The network access layer defines the frame that incorporates flow and error control.
- ▶ The physical layer is the bottom-most layer and performs the actual transfer of signals through a medium.

▶ A logical connection is a flow of ideas that occurs, without a direct physical connection, between the sender and receiver at a particular layer.

KEY TERMS

American National Standards Institute (ANSI)

client/server system

cloud

code

compression

computer network

computer terminal

convergence

data communications

data network

Electronic Industries Association (EIA)

encapsulation

File Transfer Protocol (FTP)

frame

hub

Hypertext Transfer Protocol (HTTP)

Institute for Electrical and Electronics Engineers (IEEE)

International Organization for Standardization (ISO)

International Telecommunication Union-Telecommunication Standardization Sector (ITU-T)

Internet Protocol (IP)

local area network (LAN)

logical connection

metropolitan area network (MAN)

multiplexing

network architecture

network management

node

Open Systems Interconnection (OSI) model

application layer

presentation layer

session layer

transport layer

network layer

data link layer

physical layer

personal area network (PAN)

physical connection

protocol

router

server

Simple Mail Transfer Protocol (SMTP)

Simple Network Management Protocol (SNMP)

sub-network

switch

synchronization point

TCP/IP protocol suite

application layer

transport layer

network layer

network access layer

physical layer

Telnet

token management

voice network

wide area network (WAN)

wireless

workstation

REVIEW QUESTIONS

1. What is the definition of:
   a. a computer network?
   b. data communications?
   c. telecommunications?
   d. a local area network?
   e. a personal area network?
   f. a metropolitan area network?
   g. a wide area network?
   h. network management?
   i. convergence?
2. What is the relationship between a sub-network and a node?
3. What kind of applications might use a computer terminal-to-mainframe computer connection?
4. What kind of applications might use a microcomputer-to-mainframe computer connection?
5. What "language" does a microcomputer have to use in order to interface to the Internet?
6. What kind of applications might use a sensor-to-local area network connection?
7. Why is a network architecture useful?
8. List the seven layers of the OSI model.
9. List the five layers of the TCP/IP protocol suite.
10. How do the layers of the OSI model compare with the layers of the TCP/IP protocol suite?
11. What are some of the more common applications found in the TCP/IP protocol suite?

12. What is the difference between a logical connection and a physical connection?
13. How does convergence apply to the communications industry?

EXERCISES

1. Create a list of all the actions you perform in an average day that use data communications and computer networks.

2. If you could design your own home, what kinds of labor-saving computer network or data communications devices would you incorporate?

3. Two companies are considering pooling their resources to perform a joint venture. The CEO of the first company meets with his legal team, and the legal team consults a number of middle managers in the proposed product area. Meanwhile, the CEO of the first company sends an e-mail to the CEO of the second company to offer a couple of suggestions concerning the joint venture. Does this scenario follow the OSI model? Explain your answer.

4. Using a laptop computer with a wireless connection to a company's local area network, you download a Web page from the Internet. List all the different network connections involved in this operation.

5. You are working from home using a microcomputer, a DSL modem, and a telephone connection to the Internet. Your company is connected to the Internet and has both local area networks and a mainframe computer. List all the different network connections involved in this operation.

6. You are sitting at the local coffee shop, enjoying your favorite latte. You pull out your laptop and, using the wireless network available at the coffee shop, access your e-mail. List all the different network connections involved in this operation.

7. With your new cell phone, you have just taken a snapshot of your best friend. You decide to send this snapshot to the e-mail account of a mutual friend across the country. List all the different network connections involved in this operation.

8. You are driving in a new city and have just gotten lost. Using your car's GPS system, you submit a request for driving directions from a nearby intersection to your destination. List all the different network connections involved in this operation.

9. The layers of the TCP/IP protocol suite and OSI are different. Which layers are "missing" from the TCP/IP suite? Are they really missing?

10. If the data link layer provides error checking and the transport layer provides error checking, isn't this redundant? Explain your answer.

11. Similarly, the data link layer provides flow control, and the transport layer provides flow control. Are these different forms of flow control? Explain your answer.

12. You are watching a television show in which one character is suing another. The lawyers for both parties meet and try to work out a settlement. Is there a logical or physical connection between the lawyers? What about between the two parties?

13. You want to download a file from a remote site using the File Transfer Protocol (FTP). To perform the file transfer, your computer issues a Get File command. Show the progression of messages as the Get File command moves from your computer, through routers, to the remote computer, and back.

14. What characteristics distinguish a personal area network from other types of networks?

15. Isn't a metropolitan area network just a big local area network? Explain your answer.

16. List the OSI layer that performs each of the following functions:
    a. data compression
    b. multiplexing

    c. routing
    d. definition of a signal's electrical characteristics
    e. e-mail
    f. error detection
    g. end-to-end flow control
17. For each of the functions in the previous exercise, list the TCP/IP protocol suite layer that performs that function.
18. You are sending and receiving instant messages (IM) with a friend. Is this IM session a logical connection or a physical connection? Explain your answer.

THINKING OUTSIDE THE BOX

1 You have been asked to create a new network architecture model. Will it be layered, or will its components take some other form? Show your model's layers or its new form, and describe the functions performed by each of its components.

2 Take an example from your work or school in which a person requests a service and diagram that request. Does the request pass through any layers before it reaches the intended recipient? Do logical connections as well as physical connections exist? If so, show them in the diagram.

3 This chapter listed several different types of network connections. Do any other connections exist in the real world that are not listed in the chapter? If so, what are they?

4 Describe a real-life situation that uses at least five of the network connections described in this chapter.

HANDS-ON PROJECTS

1. Recall a job you have had (or still have). Was a chain of command in place for getting tasks done? If so, draw that chain of command on paper or using a software program. How does this chain of command compare to either the OSI model or the TCP/IP protocol suite?
2. Because the TCP/IP protocol suite is not carved in stone, other books may discuss a slightly different layering. Find two other examples of the TCP/IP protocol suite that differ from this book's layering and cite the sources. How are those two suites alike, and how do they differ? How do they compare to the TCP/IP protocol suite discussed in this chapter? Write a short, concise report summarizing your findings.
3. What is the more precise form of the Get Web Page command shown in Figure 1-19? Show the form of the command, and describe the responsibility of each field.
4. What types of network applications exist at your place of employment or your college? Are local area networks involved? Wide area networks? List the various network connections. Draw a diagram or map of these applications and their connections.
5. What other network models exist or have been in existence besides the OSI model and TCP/IP protocol suite? Research this topic and write a brief description of each network model you find.
6. What are the names of some of the routing protocols currently in use on the Internet? Can you describe each protocol with a sentence or two?

2

Fundamentals of Data and Signals

◆ ◆

WE CAN'T SAY we weren't warned.The U.S. government told us years ago that someday all analog television signals would cease and they would be replaced by the more modern digital signals. Digital signals, we were told, would provide for a much better picture. Beginning in 1998, some television stations across the United States began broadcasting digital pictures and sound on a limited scale. According to the FCC, more than 1000 stations were broadcasting digital television signals by May 2003. The FCC announced that when at least 85 percent of the homes in a given area were able to accept a digital television signal, it would discontinue providing analog television broadcasting to those areas. The first planned date was set for February 18, 2009. But because the government was overwhelmed with requests for digital converter boxes, the FCC backed off on this date and set a new date of June 12, 2009. That date arrived and without too much surprise, thousands of viewers were caught off-guard and could no longer receive television signals using the older analog equipment. Many

people stood in long lines hoping to snag either a converter box or at least a coupon to later receive a converter box. Nonetheless, the digital age of television has officially begun. I think most would certainly agree that watching television with an old-fashioned antenna has certainly improved. Where there used to be fuzzy pictures with multiple ghosts we now see crystal clear pictures and often in high definition.

Nonetheless, when it comes to analog signals versus digital signals, many questions still remain:

Why are digital signals so much better than analog signals?

What other applications have been switched from analog to digital?

Do any applications remain that someday may be converted to digital?

Source: DTV.gov, downloaded on June 18, 2009.

Objectives ▶

After reading this chapter, you should be able to:

▶ Distinguish between data and signals, and cite the advantages of digital data and signals over analog data and signals

▶ Identify the three basic components of a signal

▶ Discuss the bandwidth of a signal and how it relates to data transfer speed

▶ Identify signal strength and attenuation, and how they are related

▶ Outline the basic characteristics of transmitting analog data with analog signals, digital data with digital signals, digital data with discrete analog signals, and analog data with digital signals

▶ List and draw diagrams of the basic digital encoding techniques, and explain the advantages and disadvantages of each

▶ Identify the different shift keying (modulation) techniques, and describe their advantages, disadvantages, and uses

▶ Identify the two most common digitization techniques, and describe their advantages and disadvantages

▶ Identify the different data codes and how they are used in communication systems

Introduction ▶

When the average computer user is asked to list the elements of a computer network, most will probably cite computers, cables, disk drives, modems, and other easily identifiable physical components. Many may even look beyond the obvious physical ones and cite examples of software, such as application programs and network protocols. This chapter will deal primarily with two ingredients that are even more difficult to see physically: data and signals.

Data and signals are two of the basic building blocks of any computer network. It is important to understand that the terms "data" and "signal" do not mean the same thing, and that in order for a computer network to transmit data, the data must first be converted into the appropriate signals. The one thing data and signals have in common is that both can be in either analog or digital form, which gives us four possible data-to-signal conversion combinations:

▶ Analog data-to-analog signal, which involves amplitude and frequency modulation techniques

▶ Digital data-to-digital signal, which involves encoding techniques

▶ Digital data-to-(a discrete) analog signal, which involves modulation techniques

▶ Analog data-to-digital signal, which involves digitization techniques

Each of these four combinations occurs quite frequently in computer networks, and each has unique applications and properties, which are shown in Table 2-1.

Table 2-1
Four combinations of data and signals

Data	Signal	Encoding or Conversion Technique	Common Devices	Common Systems
Analog	Analog	Amplitude modulation Frequency modulation	Radio tuner TV tuner	Telephone AM and FM radio Broadcast TV Cable TV
Digital	Digital	NRZ-L NRZI Manchester Differential Manchester Bipolar-AMI 4B/5B	Digital encoder	Local area networks Telephone systems
Digital	(Discrete) Analog	Amplitude shift keying Frequency shift keying Phase shift keying	Modem	Dial-up Internet access DSL Cable modems Digital Broadcast TV
Analog	Digital	Pulse code modulation Delta modulation	Codec	Telephone systems Music systems

Converting analog data to analog signals is fairly common. The conversion is performed by modulation techniques and is found in systems such as telephones, AM radio, FM radio, broadcast television, and cable television. Later in this chapter, we will examine how AM radio signals are created. Converting digital data to digital signals is relatively straightforward and involves numerous digital encoding techniques. We label these discrete analog signals because, despite the fact that they are fundamentally analog signals, they take on a discrete number of levels. Many people call these types of signals digital as opposed to analog, as we will see shortly. The local area network is one of the most common examples of a system that uses this type of conversion. We will examine a few representative encoding techniques and discuss their basic advantages and disadvantages. Converting digital data to (discrete) analog signals requires some form of a modem. Converting analog data to digital signals is generally called digitization. Telephone systems and music systems are two common examples of digitization. When your voice signal travels from your home and reaches a telephone company's switching center, it becomes digitized. Likewise, music and video are digitized before they can be recorded on a CD or DVD. In this chapter, two basic digitization techniques will be introduced, and their advantages and disadvantages shown. In all of this chapter's examples, data is converted to a signal by a computer or computer-related device, then transmitted over a communications medium to another computer or computer-related device, which converts the signal back to data. The originating device is the transmitter, and the destination device is the receiver.

A big question arises during the study of data and signals: Why should people interested in the business aspects of computer networks concern themselves with this level of detail? One answer to that question is that a firm understanding of the fundamentals of communication systems will provide a solid foundation for the further study of the more advanced topics of computer networks. Also, this chapter will introduce many terms that are used by network personnel. In order to be able to understand these individuals and to interact knowledgeably with them, we must spend a little time covering the basics of communication systems. Imagine you are designing a new online inventory system and you want to allow various users within the company to access this system. The network technician tells you this cannot be done because downloading one inventory record in a reasonable amount of time (X seconds) will require a connection of at least Y million bits per second—which is not possible, given the current network structure. How do you know the network technician is correct? Do you really want to just believe everything she says? The study of data and signals will also explain why almost all forms of communication, such as data, voice, music, and video, are slowly being converted from their original analog forms to the newer digital forms. What is so great about these digital forms of communications, and what do the signals that represent these forms of communication look like? We will answer these questions and more in this chapter.

Data and Signals

Information stored within computer systems and transferred over a computer network can be divided into two categories: data and signals. Data is entities that convey meaning within a computer or computer system. Common examples of data include:

- A computer file of names and addresses stored on a hard disk drive
- The bits or individual elements of a movie stored on a DVD
- The binary 1s and 0s of music stored on a CD or inside an iPod

- ▶ The dots (pixels) of a photograph that has been digitized by a digital camera and stored on a memory stick
- ▶ The digits 0 through 9, which might represent some kind of sales figures for a business

In each of these examples, some kind of information has been electronically captured and stored on some type of storage device.

If you want to transfer this data from one point to another, either via a physical wire or through radio waves, the data has to be converted into a signal. Signals are the electric or electromagnetic impulses used to encode and transmit data. Common examples of signals include:

- ▶ A transmission of a telephone conversation over a telephone line
- ▶ A live television news interview from Europe transmitted over a satellite system
- ▶ A transmission of a term paper over the printer cable between a computer and a printer
- ▶ The downloading of a Web page as it transfers over the telephone line between your Internet service provider and your home computer

In each of these examples, data, the static entity or tangible item, is transmitted over a wire or airwave in the form of a signal, which is the dynamic entity or intangible item. Some type of hardware device is necessary to convert the static data into a dynamic signal ready for transmission, and then convert the signal back to data at the receiving destination.

Before examining the basic characteristics of data and signals and the conversion from data to signal, however, let us explore the most important characteristic that data and signals share.

Analog vs. digital

Although data and signals are two different entities that have little in common, the one characteristic they do share is that they can exist in either analog or digital form. Analog data and analog signals are represented as continuous waveforms that can be at an infinite number of points between some given minimum and maximum. By convention, these minimum and maximum values are presented as voltages. Figure 2-1 shows that between the minimum value A and maximum value B, the waveform at time t can be at an infinite number of places. The most common example of analog data is the human voice. For example, when a person talks into a telephone, the receiver in the mouthpiece converts the airwaves of speech into analog pulses of electrical voltage. Music and video, when they occur in their natural states, are also analog data. Although the human voice serves as an example of analog data, an example of an analog signal is the telephone system's electronic transmission of a voice conversation. Thus, we see that analog data and signals are quite common, and many systems have incorporated them for many years.

Figure 2-1
A simple example of an analog waveform

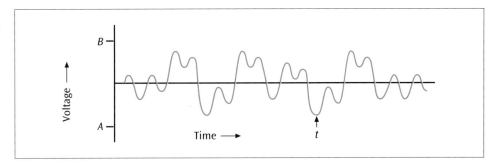

One of the primary shortcomings of analog data and analog signals is how difficult it is to separate noise from the original waveform. Noise is unwanted electrical or electromagnetic energy that degrades the quality of signals and data. Because noise is found in every type of data and transmission system, and because its effects range from a slight hiss in the background to a complete loss of data or signal, it is especially important that noise be reduced as much as possible. Unfortunately, noise itself occurs as an analog waveform, and this makes it challenging, if not extremely difficult, to separate noise from an analog waveform that represents data.

Consider the waveform in Figure 2-2, which shows the first few notes of an imaginary symphonic overture. Noise is intermixed with the music—the data. Can you tell by looking at the figure what is the data and what is the noise? Although this example may border on the extreme, it demonstrates that noise and analog data can appear similar.

Figure 2-2
The waveform of a symphonic overture with noise

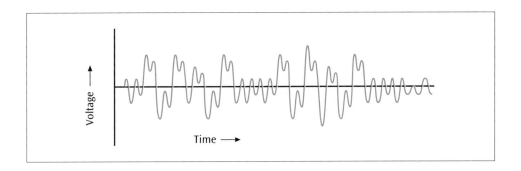

The performance of a record player provides another example of noise interfering with data. Many people have collections of albums, which produce pops, hisses, and clicks when played; albums sometimes even skip. Is it possible to create a device that filters out the pops, hisses, and clicks from a record album without ruining the original data, the music? Various devices were created during the 1960s and 1970s to perform these kinds of filtering, but only the devices that removed hiss were (relatively-speaking) successful. Filtering devices that removed the pops and clicks also tended to remove parts of the music. Filters now exist that can fairly effectively remove most forms of noise from analog recordings, but they are, interestingly, digital—not analog—devices. Even more interestingly, some people download software from the Internet that lets them insert clicks and pops into digital music to make it sound old-fashioned (in other words, as though it were being played from a record album).

Another example of noise interfering with an analog signal is the hiss you hear when you are talking on the telephone. Often the background hiss is so slight that most people do not notice it. Occasionally, however, the hiss rises to such a level that it interferes with the conversation. Yet another common example of noise interference occurs when you listen to an AM radio station during an electrical storm. The radio signal crackles with every lightning strike within the area.

Digital data and **digital signals** are discrete waveforms, rather than continuous waveforms. Between a minimum value *A* and a maximum value *B*, the digital waveform takes on only a finite number of values. In the example shown in Figure 2-3, the digital waveform takes on only two different values. In this example, the waveform is a classic example of a square wave.

Figure 2-3
A simple example of a digital waveform

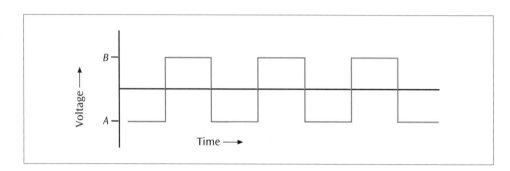

What happens when you introduce noise into digital data and digital signals? As stated earlier, noise has the properties of an analog waveform and thus can occupy an infinite range of values; digital waveforms occupy only a finite range of values. When you combine analog noise with digital waveform, it is fairly easy to separate the original digital waveform from the noise. Figure 2-4 shows a digital signal with some noise.

Figure 2-4
A digital signal with some noise introduced

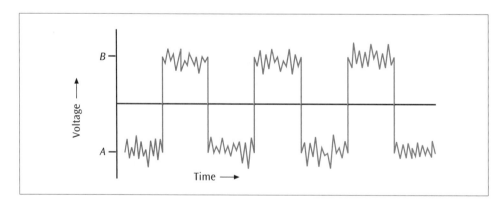

If the amount of noise remains low enough that the original digital waveform can still be interpreted, then the noise can be filtered out, thereby leaving the original waveform. In the simple example in Figure 2-4, as long as you can tell a high part of the waveform from a low part, you can still recognize the digital waveform. If, however, the noise becomes so great that it is no longer possible to distinguish a high from a low, as shown in Figure 2-5, then the noise has taken over the signal and you can no longer understand this portion of the waveform.

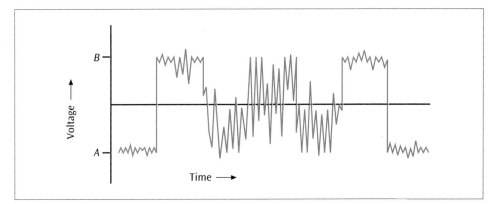

The ability to separate noise from a digital waveform is one of the great strengths of digital systems. When data is transmitted as a signal, the signal will *always* incur some level of noise. In the case of digital signals, however, it is relatively simple to pass the noisy digital signal through a filtering device that removes a significant amount of the noise and leaves the original digital signal intact.

Despite this strong advantage that digital has over analog, not all systems use digital signals to transmit data. The reason for this is that the electronic equipment used to transmit a signal through a wire or over the airwaves usually dictates the type of signals the wire can transmit. Certain electronic equipment is capable of supporting only analog signals, while other equipment can support only digital signals. Take, for example, the local area networks within your business or your house, most of which have always supported digital signals. The primary reason is that local area networks were designed for transmitting computer data, which is digital. Thus, the electronic equipment that supports the transmission of local area network signals is also digital.

Now that we have learned the primary characteristic that data and signals share (that they can exist in either analog or digital form) along with the main feature that distinguishes analog from digital (that the former exists as a continuous waveform, while the latter is discrete), let us examine the important characteristics of signals in more detail.

Fundamentals of signals

Let us begin our study of analog and digital signals by examining their three basic components: amplitude, frequency, and phase. A sine wave is used to represent an analog signal, as shown in Figure 2-6. The amplitude of a signal is the height of the wave above (or below) a given reference point. This height often denotes the voltage level of the signal (measured in volts), but it also can denote the current level of the signal (measured in amps) or the power level of the signal (measured in watts). That is, the amplitude of a signal can be expressed as volts, amps, or watts. Note that a signal can change amplitude as time progresses. In Figure 2-6, you see one signal with two different amplitudes.

Figure 2-6
*A signal with two
different amplitudes*

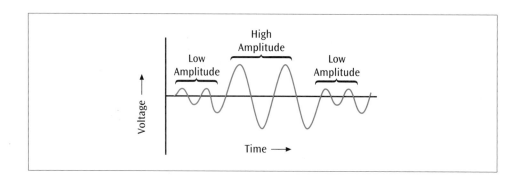

The frequency of a signal is the number of times a signal makes a complete cycle within a given time frame. The length, or time interval, of one cycle is called its period. The period can be calculated by taking the reciprocal of the frequency (1/frequency). Figure 2-7 shows three different analog signals. If the time *t* is one second, the signal in Figure 2-7(a) completes one cycle in one second. The signal in Figure 2-7(b) completes two cycles in one second. The signal in Figure 2-7(c) completes three cycles in one second. Cycles per second, or frequency, is represented by hertz (Hz). Thus, the signal in Figure 2-7(c) has a frequency of 3 Hz.

Figure 2-7
*Three signals of
(a) 1 Hz, (b) 2 Hz, and
(c) 3 Hz*

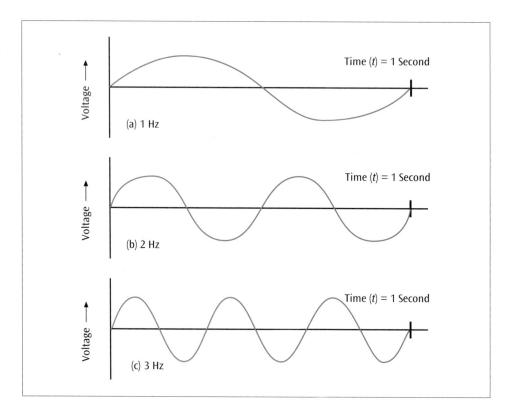

Human voice, audio, and video signals—indeed most signals—are actually composed of multiple frequencies. These multiple frequencies are what allow us to distinguish one person's voice from another's and one musical instrument from another. The frequency range of the average human voice usually goes no lower than 300 Hz and no higher than approximately 3400 Hz. Because a telephone is designed to transmit a human voice, the telephone system transmits signals in the range of 300 Hz to 3400 Hz. The piano has a wider range of frequencies than the human voice. The lowest note possible on the piano is 30 Hz, and the highest note possible is 4200 Hz.

The range of frequencies that a signal spans from minimum to maximum is called the spectrum. The spectrum of our telephone example is simply 300 Hz to 3400 Hz. The bandwidth of a signal is the absolute value of the difference between the lowest and highest frequencies. The bandwidth of a telephone system that transmits a single voice in the range of 300 Hz to 3400 Hz is 3100 Hz. Because extraneous noise degrades original signals, an electronic device usually has an effective bandwidth that is less than its bandwidth. When making communication decisions, many professionals rely more on the effective bandwidth than the bandwidth, because most situations must deal with the real-world problems of noise and interference.

The phase of a signal is the position of the waveform relative to a given moment of time, or relative to time zero. In the drawing of the simple sine wave in Figure 2-8(a), the waveform oscillates up and down in a repeating fashion. Note that the wave never makes an abrupt change but is a continuous sine wave. A phase change (or phase shift) involves jumping forward (or backward) in the waveform at a given moment of time. Jumping forward one-half of the complete cycle of the signal produces a 180-degree phase change, as seen in Figure 2-8(b). Jumping forward one-quarter of the cycle produces a 90-degree phase change, as in Figure 2-8(c). Some systems, as you will see in this chapter's "Transmitting digital data with analog signals" section, can generate signals that do a phase change of 45, 135, 225, and 315 degrees on demand.

Figure 2-8

A sine wave showing (a) no phase change, (b) a 180-degree phase change, and (c) a 90-degree phase change

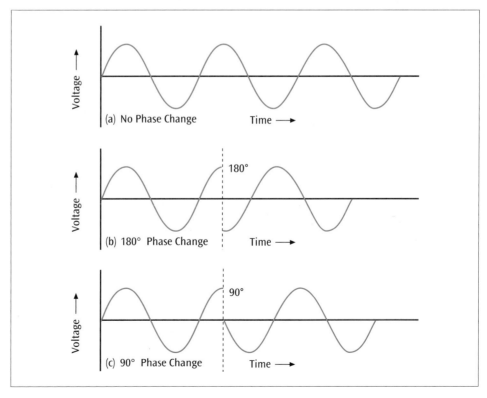

Loss of signal strength

Imagine a scenario in which you are recommending a computer network solution for a business problem. You tell the network specialists that you want to place a computer workstation at the company's reception desk so that the receptionist can handle requests for scheduling meeting rooms. The network specialist says it cannot be done because the wire connecting the workstation to the network will be too long and the signal will be too weak. Or worse, the network specialist uses computer jargon: "The signal will have too much attenuation and will drop below an acceptable threshold, and noise will take over." Is the network specialist accurate, or is he using computer jargon to dissuade you because he doesn't want to take the time to install the wire and the workstation? A little knowledge of the loss of signal strength will help in such situations.

Details ▶

Composite Signals

Almost all of the example signals shown in this chapter are simple, periodic sine waves. You do not always find simple, periodic sine waves in the real world, however. In fact, you are more likely to encounter combinations of various kinds of sines and cosines that when combined produce unique waveforms.

One of the best examples of this is how multiple sine waves can be combined to produce a square wave. Stated differently, multiple analog signals can be combined to produce a digital signal. A branch of mathematics called Fourier analysis shows that any complex, periodic waveform is a composite of simpler periodic waveforms. Consider, for example, the first two waveforms shown in Figure 2-9. The formula for the first waveform is $1 \sin(2\pi ft)$, and the formula for the second waveform is $\frac{1}{3} \sin(2\pi 3ft)$. In each formula, the number at the front (the 1 and $\frac{1}{3}$, respectively) is a value of amplitude, the term "sin" refers to the sine trigonometric function, and the terms "ft" and "3ft" refer to the frequency over a given period of time. Examining both the waveforms and the formulas shows us that, whereas the amplitude of the second waveform is $\frac{1}{3}$ as high as the amplitude of the first waveform, the frequency of the second waveform is 3 times as high as the frequency of the first waveform. The third waveform in Figure 2-9(c) is a composite, or addition, of the first two waveforms.

Note the relatively square shape of the composite waveform. Now suppose you continued to add more waveforms to this composite signal—in particular, waveforms with amplitude values of $\frac{1}{5}$, $\frac{1}{7}$, $\frac{1}{9}$, and so on (odd-valued denominators) and frequency multiplier values of 5, 7, 9, and so on. The more waveforms you added, the more the composite signal would resemble the square waveform of a digital signal. Another way to interpret this transformation is to state that

adding waveforms of higher and higher frequency—that is, of increasing bandwidth—will produce a composite that looks (and behaves) more and more like a digital signal. Interestingly, a digital waveform is, in fact, a combination of analog sine waves.

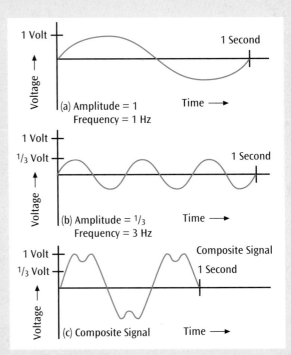

Figure 2-9 *Two simple, periodic sine waves (a) and (b) and their composite (c)*

When traveling through any type of medium, a signal always experiences some loss of its power due to friction. This loss of power, or loss of signal strength, is called attenuation. Attenuation in a medium such as copper wire is a logarithmic loss (in which a value decrease of 1 represents a tenfold decrease) and is a function of distance and the resistance within the wire. Knowing the amount of attenuation in a signal (how much power the signal lost) allows you to determine the signal strength. Decibel (dB) is a relative measure of signal loss or gain and is used to measure the logarithmic loss or gain of a signal. Amplification is the opposite of attenuation. When a signal is amplified by an amplifier, the signal gains in decibels.

Because attenuation is a logarithmic loss and the decibel is a logarithmic value, calculating the overall loss or gain of a system involves adding all the individual decibel losses and gains. Figure 2-10 shows a communication line running from point A, through point B, and ending at point C. The communication line from A to B experiences a 10 dB loss, point B has a 20 dB amplifier (that is, a 20 dB gain occurs at point B), and the communication line from B to C experiences a 15 dB loss. What is the overall gain or loss of the signal between point A and point C? To answer this question, add all dB gains and losses:

$$-10 \text{ dB} + 20 \text{ dB} + (-15 \text{ dB}) = -5 \text{ dB}$$

Figure 2-10
Example demonstrating decibel loss and gain

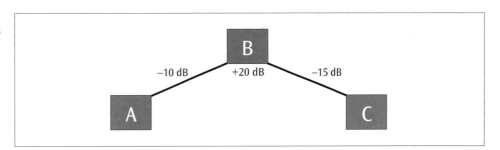

Let us return to the earlier example of the network specialist telling you that it may not be possible to install a computer workstation as planned. You now understand that signals lose strength over distance. Although you do not know how much signal would be lost, nor at what point the strength of the signal would be weaker than the noise, you can trust part of what the network specialist told you. But let us investigate a little further. If a signal loses 3 dB, for example, is this a significant loss or not?

The decibel is a relative measure of signal loss or gain and is expressed as

$$dB = 10 \log_{10} (P_2 / P_1)$$

in which P_2 and P_1 are the ending and beginning power levels, respectively, of the signal expressed in watts. If a signal starts at a transmitter with 10 watts of power and arrives at a receiver with 5 watts of power, the signal loss in dB is calculated as follows:

$$
\begin{aligned}
dB &= 10 \log_{10} (5/10) \\
&= 10 \log_{10} (0.5) \\
&= 10 (-0.3) \\
&= -3
\end{aligned}
$$

In other words, a 3 dB loss occurs between the transmitter and receiver. Because decibel is a relative measure of loss or gain, you cannot take a single power level at time *t* and compute the decibel value of that signal without having a reference or a beginning power level.

Rather than remembering this formula, let us use a shortcut. As we saw from the previous calculation, any time a signal loses half its power, a 3 dB loss occurs. If the signal drops from 10 watts to 5 watts, that is a 3 dB loss. If the signal drops from 1000 watts to 500 watts, this still is a 3 dB loss. Conversely, a signal whose strength is doubled experiences a 3 dB gain. It follows then that if a signal drops from 1000 watts to 250 watts, this is a 6 dB loss (1000 to 500 is a 3 dB loss, and 500 to 250 corresponds to another 3 dB). Now we have a little better understanding of the terminology. If the network specialist tells us a given section of wiring loses 6 dB, for example, then the signal traveling through that wire has lost three-quarters of its power!

Now that we are up to speed on the fundamentals of and differences between data and signals, let us investigate how to convert data into signals for transmission.

Converting Data into Signals

Like signals, data can be analog or digital. Often, analog signals convey analog data, and digital signals convey digital data. However, you can use analog signals to convey digital data and digital signals to convey analog data. The decision about whether to use analog or digital signals often depends on the transmission equipment and the environment in which the signals must travel. Recall that certain electronic equipment is capable of supporting only analog signals, while other types of equipment support only digital signals. For example, the telephone system was created to transmit human voice, which is analog data. Thus, the telephone system was originally designed to transmit analog signals. Although telephone wiring is capable of carrying either analog or digital signals, the electronic equipment used to amplify and remove noise from many of the lines can accept only analog signals. Therefore, to transmit digital data from a computer over these telephone lines, it is common to use analog signals. Transmitting analog data with digital signals is also fairly common. Originally, cable television companies transmitted analog television channels using analog signals. More recently, the analog television channels are converted to digital signals in order to provide clearer images and higher definition signals. As we saw in the chapter introduction, broadcast television is now transmitting using digital signals. As you can see from these examples, there are four main combinations of data and signals:

- ▹ Analog data transmitted using analog signals
- ▹ Digital data transmitted using digital signals
- ▹ Digital data transmitted using discrete analog signals
- ▹ Analog data transmitted using digital signals

Let us examine each of these in turn.

Transmitting analog data with analog signals

Of the four combinations of data and signals, the analog data-to-analog signal conversion is probably the simplest to comprehend. This is because the data is an analog waveform that is simply being transformed to another analog waveform, the signal, for transmission. The basic operation performed is modulation. Modulation is the process of sending data over a signal by varying either its amplitude, frequency, or phase. Land-line telephones, AM radio, FM radio, and pre-June 2009 broadcast television are the most common examples of analog data-to-analog signal conversion. Consider Figure 2-11, which shows AM radio as an example. The audio data generated by the radio station might appear like the first sine wave shown in the figure. To convey this analog data, the station uses a carrier wave signal, like that shown in Figure 2-11(b). In the modulation process, the original audio waveform and the carrier wave are essentially added together to produce the third waveform. Note how the dotted lines superimposed over the third waveform follow the same outline as the original audio waveform. Here, the original audio data has been modulated onto a particular carrier frequency (the frequency at which you set the dial to tune in a station) using amplitude modulation—hence, the name AM radio. Frequency modulation also can be used in similar ways to modulate analog data onto an analog signal, and it yields FM radio.

Figure 2-11
An audio waveform modulated onto a carrier frequency using amplitude modulation

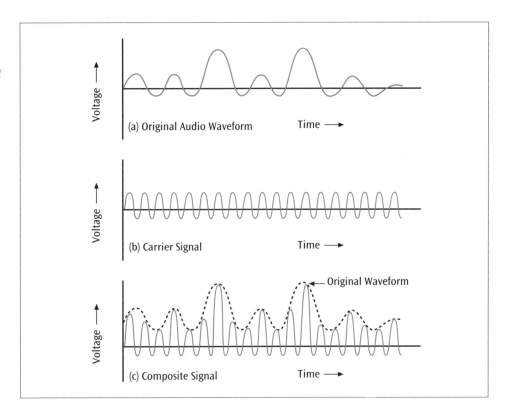

Transmitting digital data with digital signals: Digital encoding schemes

To transmit digital data using digital signals, the 1s and 0s of the digital data must be converted to the proper physical form that can be transmitted over a wire or airwave. Thus, if you wish to transmit a data value of 1, you could do this by transmitting a positive voltage on the medium. If you wish to transmit a data value of 0, you could transmit a zero voltage. You could also use the opposite scheme: a data value of 0 is positive voltage, and a data value of 1 is a zero voltage. Digital encoding schemes like this are used to convert the 0s and 1s of digital data into the appropriate transmission form. We will examine six digital encoding schemes that are representative of most digital encoding schemes: NRZ-L, NRZI, Manchester, differential Manchester, bipolar-AMI, and 4B/5B.

Nonreturn to Zero Digital Encoding Schemes

The nonreturn to zero-level (NRZ-L) digital encoding scheme transmits 1s as zero voltages and 0s as positive voltages. The NRZ-L encoding scheme is simple to generate and inexpensive to implement in hardware. Figure 2-12(a) shows an example of the NRZ-L scheme.

Figure 2-12

Examples of five digital encoding schemes

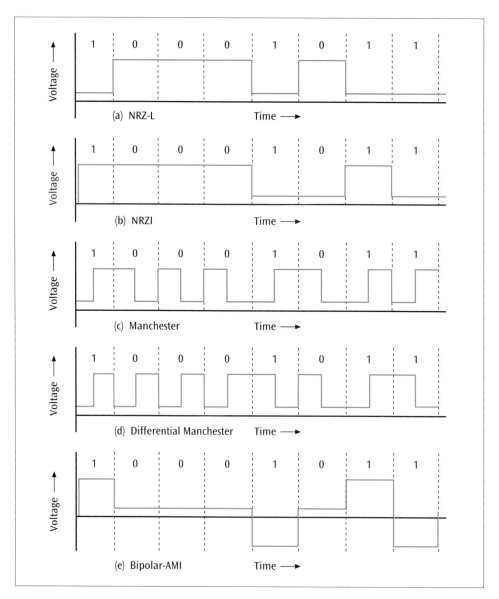

The second digital encoding scheme, shown in Figure 2-12(b), is nonreturn to zero inverted (NRZI). This encoding scheme has a voltage change at the beginning of a 1 and no voltage change at the beginning of a 0. A fundamental difference exists between NRZ-L and NRZI. With NRZ-L, the receiver has to check the voltage *level* for each bit to determine whether the bit is a 0 or a 1. With NRZI, the receiver has to check whether there is a *change at the beginning* of the bit to determine if it is a 0 or a 1. Look again at Figure 2-12 to see this difference between the two NRZ schemes.

An inherent problem with the NRZ-L and NRZI digital encoding schemes is that long sequences of 0s in the data produce a signal that never changes. Often the receiver looks for signal changes so that it can synchronize its reading of the data with the actual data pattern. If a long string of 0s is transmitted and the signal does not change, how can the receiver tell when one bit ends and the next bit begins? (Imagine how hard it would be to dance to a song that has no regular beat, or worse, no beat at all.) One potential solution is to install in the receiver an internal clock that knows when to look for each successive bit. But what if the receiver has a different clock from the one the transmitter used to generate the signals? Who is to say that these two clocks keep the same time? A more accurate system would generate a signal that has a change for each and every bit. If the receiver could count on each bit having some form of signal change, then it could stay synchronized with the incoming data stream.

Manchester Digital Encoding Schemes

The Manchester class of digital encoding schemes ensures that each bit has some type of signal change, and thus solves the synchronization problem. Shown in Figure 2-12(c), the Manchester encoding scheme has the following properties: to transmit a 1, the signal changes from low to high in the *middle* of the interval; to transmit a 0, the signal changes from high to low in the *middle* of the interval. Note that the transition is always in the middle, a 1 is a low-to-high transition, and a 0 is a high-to-low transition. Thus, if the signal is currently low and the next bit to transmit is a 0, the signal has to move from low to high at the *beginning* of the interval so that it can do the high-to-low transition in the middle. Manchester encoding is used in most local area networks for transmitting digital data over a local area network cable.

The differential Manchester digital encoding scheme, which is also used in some local area networks for transmitting digital data over a local area network cable, is similar to the Manchester scheme in that there is always a transition in the middle of the interval. But unlike the Manchester code, the direction of this transition in the middle does not differentiate between a 0 or a 1. Instead, if there is a transition at the *beginning* of the interval, then a 0 is being transmitted. If there is no transition at the beginning of the interval, then a 1 is being transmitted. Because the receiver must watch the beginning of the interval to determine the value of the bit, the differential Manchester is similar to the NRZI scheme (in this one respect). Figure 2-12(d) shows an example of differential Manchester encoding.

The Manchester schemes have an advantage over the NRZ schemes: In the Manchester schemes, there is always a transition in the middle of a bit. Thus, the receiver can expect a signal change at regular intervals and can synchronize itself with the incoming bit stream. The Manchester encoding schemes are called self-clocking, because the occurrence of a regular transition is similar to seconds ticking on a clock. As you will see in Chapter Four, it is very important for a receiver to stay synchronized with the incoming bit stream, and the Manchester codes allow a receiver to achieve this synchronization.

The big disadvantage of the Manchester schemes is that roughly half the time there will be two transitions during each bit. For example, if the differential Manchester encoding scheme is used to transmit a series of 0s, then the signal has to change at the beginning of each bit, as well as change in the middle of each bit. Thus, for each data value 0, the signal changes twice. The number of times a signal changes value per second is called the baud rate, or simply baud. In Figure 2-13, a series of binary 0s is transmitted using the differential Manchester encoding scheme. Note that the signal changes twice for each bit. After one second, the signal has changed 10 times. Therefore, the baud rate is 10. During that same time period, only 5 bits were transmitted. The data rate, measured in bits per second (bps), is 5, which in this case is one-half the baud rate. Many individuals mistakenly equate baud rate to bps (or data rate). Under some circumstances, the baud rate may equal the bps, such as in the NRZ-L or NRZI encoding schemes shown in Figure 2-12. In these, there is at most one signal change for each bit transmitted. But with schemes such as the Manchester codes, the baud rate is not equal to the bps.

Figure 2-13
Transmitting five binary 0s using differential Manchester encoding

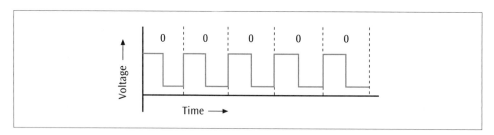

Why does it matter that some encoding schemes have a baud rate twice the bps? Because the Manchester codes have a baud rate that is twice the bps, and the NRZ-L and NRZI codes have a baud rate that is equal to the bps, hardware that generates a Manchester-encoded signal has to work twice as fast as hardware that generates a NRZ-encoded signal. If 100 million 0s per second are transmitted using differential Manchester encoding, the signal has to change 200 million times per second (as opposed to 100 million times per second with NRZ encoding). As with most things in life, you do not get something for nothing. Hardware or software that handles the Manchester encoding schemes is more elaborate and more costly than the hardware or software that handles the NRZ encoding schemes. More importantly, as we shall soon see, signals that change at a higher rate of speed are more susceptible to noise and errors.

Bipolar-AMI Encoding Scheme

The bipolar-AMI encoding scheme is unique among all the encoding schemes seen thus far because it uses three voltage levels. When a device transmits a binary 0, a zero voltage is transmitted. When the device transmits a binary 1, either a positive voltage or a negative voltage is transmitted. Which of these is transmitted depends on the binary 1 value that was last transmitted. For example, if the last binary 1 transmitted a positive voltage, then the next binary 1 will transmit a negative voltage. Likewise, if the last binary 1 transmitted a negative voltage, then the next binary 1 will transmit a positive voltage (Figure 2-12).

The bipolar scheme has two obvious disadvantages. First, as you can see in Figure 2-12(e), we have the long-string-of-0s synchronization problem again, as we had with the NRZ schemes. Second, the hardware must now be capable of generating and recognizing negative voltages as well as positive voltages. On the other hand, the primary advantage of a bipolar scheme is that when all the voltages are added together after a long transmission, there should be a total voltage of zero. That is, the positive and negative voltages essentially cancel each other out. This type of zero voltage sum can be useful in certain types of electronic systems (the question of why this is useful is beyond the scope of this text).

4B/5B Digital Encoding Scheme

The Manchester encoding schemes solve the synchronization problem but are relatively inefficient because they have a baud rate that is twice the bps. The 4B/5B scheme tries to satisfy the synchronization problem and avoid the "baud equals two times the bps" problem. The 4B/5B encoding scheme takes 4 bits of data, converts the 4 bits into a unique 5-bit sequence, and encodes the 5 bits using NRZI.

The first step the hardware performs in generating the 4B/5B code is to convert 4-bit quantities of the original data into new 5-bit quantities. Using 5 bits (or five 0s and 1s) to represent one value yields 32 potential combinations ($2^5 = 32$). Of these possibilities, only 16 combinations are used, so that no code has three or more consecutive 0s. This way, if the transmitting device transmits the 5-bit quantities using NRZI encoding, there will never be more than two 0s in a row transmitted (unless one 5-bit character ends with 00, and the next 5-bit character begins with a 0). If you never transmit more than two 0s in a row using NRZI encoding, then you will never have a long period in which there is no signal transition. Figure 2-14 shows the 4B/5B code in detail.

Figure 2-14

The 4B/5B digital encoding scheme

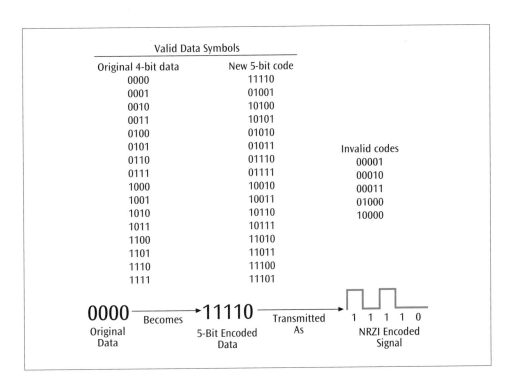

How does the 4B/5B code work? Let us say, for example, that the next 4 bits in a data stream to be transmitted are 0000, which, you can see, has a string of consecutive zeros and therefore would create a signal that does not change. Looking at the first column in Figure 2-14, we see that 4B/5B encoding replaces 0000 with 11110. Note that 11110, like all the 5-bit codes in the second column of Figure 2-14, does not have more than two consecutive zeros. Having replaced 0000 with 11110, the hardware will now transmit 11110. Because this 5-bit code is transmitted using NRZI, the baud rate equals the bps and thus is more efficient. Unfortunately, converting a 4-bit code to a 5-bit code creates a 20 percent overhead (one extra bit). Compare that to a Manchester code, in which the baud rate can be twice the bps and thus yield a 100 percent overhead. Clearly, a 20 percent overhead is better than a 100 percent overhead. Many of the newer digital encoding systems that use fiber-optic cable also use techniques that are quite similar to 4B/5B. Thus, an understanding of the simpler 4B/5B can lead to an understanding of some of the newer digital encoding techniques.

Transmitting digital data with discrete analog signals

The technique of converting digital data to an analog signal is also an example of modulation. But in this type of modulation, the analog signal takes on a discrete number of signal levels. It could be as simple as two signal levels (such as the first technique shown in the next paragraph) or something more complex as 256 levels as is used with digital television signals. The receiver then looks specifically for these unique signal levels. Thus, even though they are fundamentally analog signals, they operate with a discrete number of levels, much like a digital signal from the previous section. So to avoid confusion, we'll label them discrete analog signals. Let's examine a number of these discrete modulation techniques beginning with the simpler techniques (shift keying) and ending with the more complex techniques used for systems such as digital television signals—quadrature amplitude modulation.

Amplitude Shift Keying

The simplest modulation technique is amplitude shift keying. As shown in Figure 2-15, a data value of 1 and a data value of 0 are represented by two different amplitudes of a signal. For example, the higher amplitude could represent a 1, while the lower amplitude (or zero amplitude) could represent a 0. Note that *during* each bit period, the amplitude of the signal is constant.

Figure 2-15
Example of amplitude shift keying

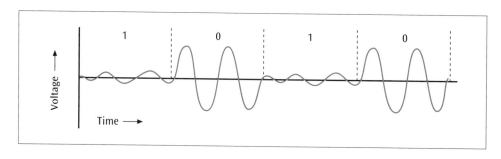

Amplitude shift keying is not restricted to two possible amplitude levels. For example, we could create an amplitude shift keying technique that incorporates four different amplitude levels, as shown in Figure 2-16. Each of the four different amplitude levels would represent 2 bits. You might recall that when counting in binary, 2 bits yield four possible combinations: 00, 01, 10, and 11. Thus, every time the signal changes (every time the amplitude changes), 2 bits are transmitted. As a result, the data rate (bps) is twice the baud rate. This is the opposite of a Manchester code in which the data rate is one-half the baud rate. A system that transmits 2 bits per signal change is more efficient than one that requires two signal changes for every bit.

Figure 2-16
Amplitude shift keying using four different amplitude levels

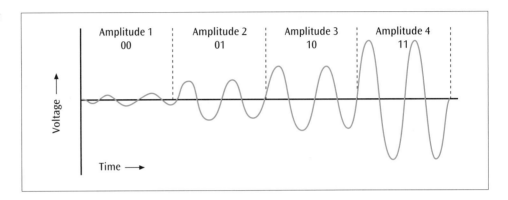

Amplitude shift keying has a weakness: It is susceptible to sudden noise impulses such as the static charges created by a lightning storm. When a signal is disrupted by a large static discharge, the signal experiences significant increases in amplitude. For this reason, and because it is difficult to accurately distinguish among more than just a few amplitude levels, amplitude shift keying is one of the least efficient encoding techniques and is not used on systems that require a high data transmission rate. When transmitting data over standard telephone lines, amplitude shift keying typically does not exceed 1200 bps.

Frequency Shift Keying

Frequency shift keying uses two different frequency ranges to represent data values of 0 and 1, as shown in Figure 2-17. For example, the lower frequency signal might represent a 1, while the higher frequency signal might represent a 0. *During each bit period, the frequency of the signal is constant.*

Figure 2-17
Simple example of frequency shift keying

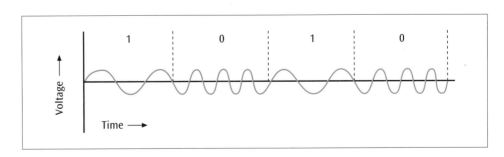

Unlike amplitude shift keying, frequency shift keying does not have a problem with sudden noise spikes that can cause loss of data. Nonetheless, frequency shift keying is not perfect. It is subject to intermodulation distortion, a phenomenon that occurs when the frequencies of two or more signals mix together and create new frequencies. Thus, like amplitude shift keying, frequency shift keying is not used on systems that require a high data rate.

Phase Shift Keying

A third modulation technique is phase shift keying. Phase shift keying represents 0s and 1s by different changes in the phase of a waveform. For example, a 0 could be no phase change, while a 1 could be a phase change of 180 degrees, as shown in Figure 2-18.

Figure 2-18

An example of simple phase shift keying of a sine wave

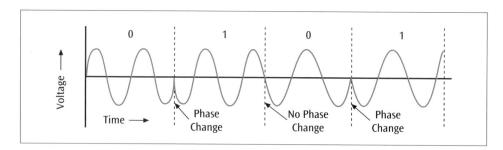

Phase changes are not affected by amplitude changes, nor are they affected by intermodulation distortions. Thus, phase shift keying is less susceptible to noise and can be used at higher frequencies. Phase shift keying is so accurate that the signal transmitter can increase efficiency by introducing multiple phase-shift angles. For example, quadrature phase shift keying incorporates four different phase angles, each of which represents 2 bits: a 45-degree phase shift represents a data value of 11, a 135-degree phase shift represents 10, a 225-degree phase shift represents 01, and a 315-degree phase shift represents 00. Figure 2-19 shows a simplified drawing of these four different phase shifts. Because each phase shift represents 2 bits, quadrature phase shift keying has double the efficiency of simple phase shift keying. With this encoding technique, one signal change equals 2 bits of information; that is, 1 baud equals 2 bps.

The efficiency of this technique can be increased even further by combining 12 different phase-shift angles with two different amplitudes. Figure 2-20(a) (known as a constellation diagram) shows 12 different phase-shift angles with 12 arcs radiating from a central point. Two different amplitudes are applied on each of four angles. Figure 2-20(b) shows a phase shift with two different amplitudes. Thus, eight phase angles have a single amplitude, and four phase angles have double amplitudes, resulting in 16 different combinations. This encoding technique is an example from a family of encoding techniques termed quadrature amplitude modulation, which is commonly employed in contemporary modems and uses each signal change to represent 4 bits (4 bits yield 16 combinations). Therefore, the bps of the data transmitted using quadrature amplitude modulation is four times the baud rate. For example, a system using a signal with a baud rate of 2400 achieves a data transfer rate of 9600 bps (4 × 2400). Interestingly, it is techniques like this that enable us to access the Internet via DSL and watch digital television broadcasts.

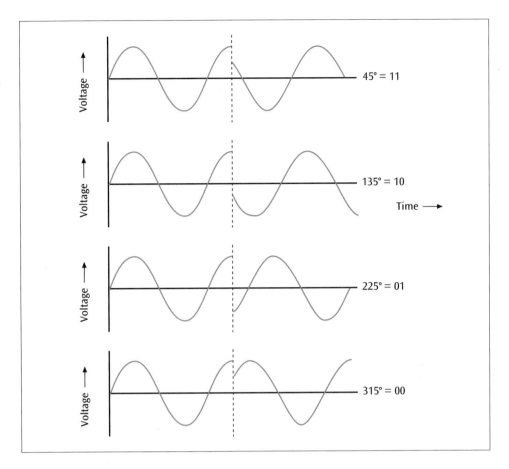

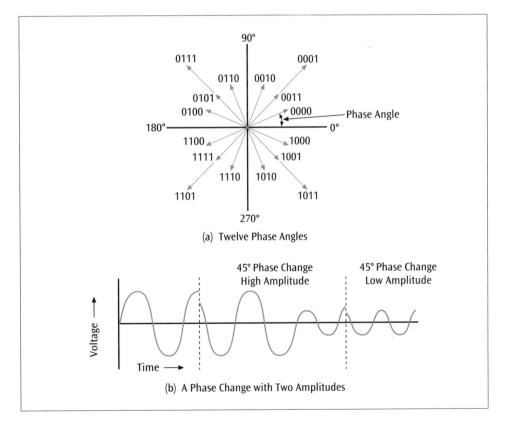

(a) Twelve Phase Angles

(b) A Phase Change with Two Amplitudes

Transmitting analog data with digital signals

It is often necessary to transmit analog data over a digital medium. For example, many scientific laboratories have testing equipment that generates test results as analog data. This analog data is converted to digital signals so that the original data can be transmitted through a computer system and eventually stored in memory or on a magnetic disk. A music recording company that creates a CD also converts analog data to digital signals. An artist performs a song that produces music, which is analog data. A device then converts this analog data to digital data so that the binary 1s and 0s of the digitized music can be stored, edited, and eventually recorded on a CD. When the CD is used, a person inserts the disc into a CD player that converts the binary 1s and 0s back to analog music. Let us look at the two techniques for converting analog data to digital signals.

Pulse Code Modulation

One encoding technique that converts analog data to a digital signal is pulse code modulation (PCM). Hardware—specifically, a codec—converts the analog data to a digital signal by tracking the analog waveform and taking "snapshots" of the analog data at fixed intervals. Taking a snapshot involves calculating the height, or voltage, of the analog waveform above a given threshold. This height, which is an analog value, is converted to an equivalent fixed-sized binary value. This binary value can then be transmitted by means of a digital encoding format. Tracking an analog waveform and converting it to pulses that represent the wave's height above (or below) a threshold is termed pulse amplitude modulation (PAM). The term "pulse code modulation" actually applies to the conversion of these individual pulses into binary values. For the sake of brevity, however, we will refer to the entire process simply as pulse code modulation.

Figure 2-21 shows an example of pulse code modulation. At time t (on the x-axis), a snapshot of the analog waveform is taken, resulting in the decimal value 14 (on the y-axis). The 14 is converted to a 5-bit binary value (such as 01110) by the codec and transmitted to a device for storage. In Figure 2-21, the y-axis is divided into 32 gradations, or quantization levels. (Note that the values on the y-axis run from 0 to 31, corresponding to 32 divisions.) Because there are 32 quantization levels, each snapshot generates a 5-bit value ($2^5 = 32$).

What happens if the snapshot value falls between 13 and 14? If it is closer to 14, we would approximate and select 14. If closer to 13, we would approximate and select 13. Either way, our approximation would introduce an error into the encoding because we did not encode the exact value of the waveform. This type of error is called quantization error, or quantization noise, and causes the regenerated analog data to differ from the original analog data.

To reduce this type of quantization error, we could have tuned the y-axis more finely by dividing it into 64 (i.e., double the number of) quantization levels. As always, we do not get something for nothing. This extra precision would have required the hardware to be more precise, and it would have generated a larger bit value for each sample (because 64 quantization levels requires a 6-bit value, or $2^6 = 64$). Continuing with the encoding of the waveform in Figure 2-21, we see that at time $2t$, the codec takes a second snapshot. The voltage of the waveform here is found to have a decimal value of 6, and so this 6 is converted to a second 5-bit binary value and stored. The encoding process continues in this way—with the codec taking snapshots, converting the voltage values (also known as PAM values) to binary form, and storing them—for the length of the waveform.

Figure 2-21

Example of taking "snapshots" of an analog waveform for conversion to a digital signal

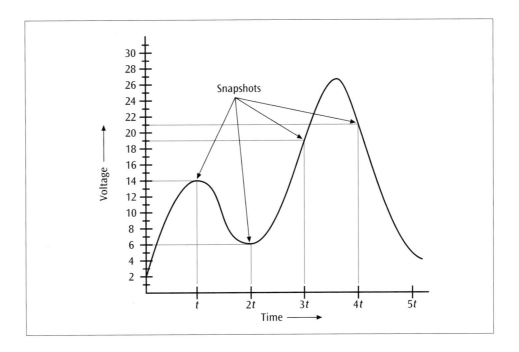

To reconstruct the original analog waveform from the stored digital values, special hardware converts each *n*-bit binary value back to decimal and generates an electric pulse of appropriate magnitude (height). With a continuous incoming stream of converted values, a waveform close to the original can be reconstructed, as shown in Figure 2-22.

Figure 2-22

Reconstruction of the analog waveform from the digital "snapshots"

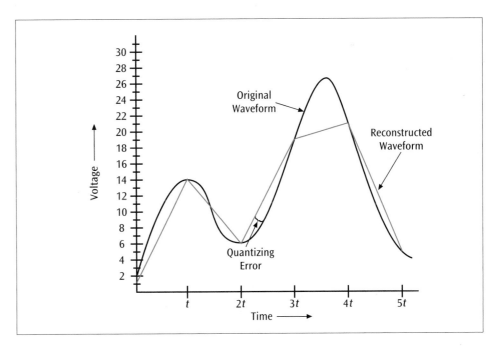

Sometimes this reconstructed waveform is not a good reproduction of the original. What can be done to increase the accuracy of the reproduced waveform? As we have already seen, we might be able to increase the number of quantization levels on the y-axis. Also, the closer the snapshots are taken to one another (the smaller the time intervals between snapshots, or the finer the resolution), the more accurate the reconstructed waveform will be. Figure 2-23 shows a reconstruction that is closer to the original analog waveform. Once again, however, you do not get something for nothing. To take the snapshots at shorter time intervals, the codec must be of high enough quality to track the incoming signal quickly and perform the necessary conversions. And the more snapshots taken per second, the more binary data generated per second. The frequency at which the snapshots are taken is called the sampling rate. If the codec takes samples at an unnecessarily high sampling rate, it will expend much energy for little gain in the resolution of the waveform's reconstruction. More often codec systems generate too few samples—use a low sampling rate—which reconstructs a waveform that is not an accurate reproduction of the original.

Figure 2-23

A more accurate reconstruction of the original waveform, using a higher sampling rate

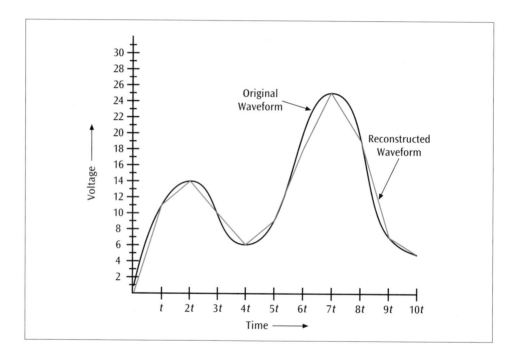

What then is the optimal balance between too high a sampling rate and too low? According to a famous communications theorem created by Nyquist, the sampling rate using pulse code modulation must be at least twice the highest frequency of the original analog waveform to ensure a reasonable reproduction. Using the telephone system as an example and assuming that the highest possible voice frequency is 3400 Hz, the sampling rate should be at least 6800 samples per second to ensure reasonable reproduction of the analog waveform. The telephone system actually allocates a 4000-Hz channel for a voice signal, and thus samples at 8000 times per second.

Delta Modulation

A second method of analog data-to-digital signal conversion is delta modulation. Figure 2-24 shows an example. With delta modulation, a codec tracks the incoming analog data by assessing up or down "steps." During each time period, the codec determines whether the waveform has risen one delta step or dropped one delta step. If the waveform rises one delta step, a 1 is transmitted. If the waveform drops one delta step, a 0 is transmitted. With this encoding technique, only 1 bit per sample is generated. Thus, the conversion from analog to digital using delta modulation is quicker than with pulse code modulation, in which each analog value is first converted to a PAM value, and then the PAM value is converted to binary.

Two problems are inherent with delta modulation. If the analog waveform rises or drops too quickly, the codec may not be able to keep up with the change, and slope overload noise results. What if a device is trying to digitize a voice or music that maintains a constant frequency and amplitude, like one person singing one note at a steady volume? Analog waveforms that do not change at all present the other problem for delta modulation. Because the codec outputs a 1 or a 0 only for a rise or a fall, respectively, a nonchanging waveform generates a pattern of 1010101010…, thus generating quantizing noise. Figure 2-24 demonstrates delta modulation and shows both slope overload noise and quantizing noise.

Figure 2-24

Example of delta modulation that is experiencing slope overload noise and quantizing noise

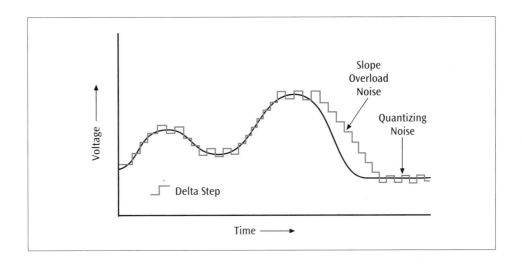

Details ▶

The Relationship between Frequency and Bits per Second

When a network application is slow, users often demand that the network specialists transmit the data faster and thus solve the problem. What many network users do not understand is that if you want to send data at a faster rate, one or two things must change: (1) the data must be transmitted with a higher frequency signal, or (2) more bits per baud must be transmitted. Furthermore, neither of these solutions will work unless the medium that transmits the signal is capable of supporting the higher frequencies. To begin to understand all these interdependencies, it is helpful to both understand the relationship between bits per second and the frequency of a signal and be able to use two simple measures—Nyquist's theorem and Shannon's theorem—to calculate the data transfer rate of a system.

An important relationship exists between the frequency of a signal and the number of bits a signal can convey per second: The greater the frequency of a signal, the higher the possible data transfer rate. The converse is also true: The higher the desired data transfer rate, the greater the needed signal frequency. You can see a direct relationship between the frequency of a signal and the transfer rate (in bits per second, or bps) of the data that a signal can carry. Consider the amplitude modulation encoding, shown twice in Figure 2-25, of the bit string 1010…. In the first part of Figure 2-25, the signal (amplitude) changes four times during a one-second period (baud rate equals 4). The frequency of this signal is 8 Hz (8 complete cycles in one second), and the data transfer rate is 4 bps. In the second part of the figure, the signal changes amplitude eight times (baud rate equals 8) during a one-second period. The frequency of the signal is 16 Hz, and the data transfer rate is 8 bps. As the frequency of the signal increases, the data transfer rate (in bps) increases.

This example is simple because it contains only two signal levels (amplitudes), one for a binary 0 and one for a binary 1. What if we had an encoding technique with four signal levels, as shown in Figure 2-26? Because there are four signal levels, each signal level can represent 2 bits. More precisely, the first

signal level can represent a binary 00, the second a 01, the third a 10, and the fourth signal level a binary 11. Now when the signal level changes, 2 bits of data will be transferred.

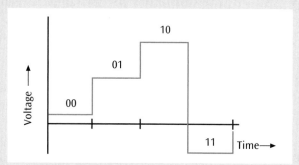

Figure 2-26   *Hypothetical signaling technique with four signal levels*

Two formulas express the direct relationship between the frequency of a signal and its data transfer rate: Nyquist's theorem and Shannon's theorem. **Nyquist's theorem** calculates the data transfer rate of a signal using its frequency and the number of signaling levels:

$$C = 2f \times \log_2 (L)$$

in which C is how fast the data can transfer over a medium in bits per second (the channel capacity), f is the frequency of the signal, and L is the number of signaling levels. For example, given a 3100-Hz signal and two signaling levels, the resulting channel capacity is 6200 bps, which results from $2 \times 3100 \times \log_2 (2) = 2 \times 3100 \times 1$. Be careful to use $\log_2$ and not $\log_{10}$. A 3100-Hz signal with four signaling levels yields 12,400 bps. Note further that the Nyquist formula does not incorporate noise, which is always present. (Shannon's formula, shown next, does.) Thus, many use the Nyquist formula not to solve for the data rate, but instead, given the data rate and frequency, to solve for the number of signal levels L.

Shannon's theorem calculates the maximum data transfer rate of an analog signal (with any number of signal levels) and incorporates noise:

$$\text{Data rate} = f \times \log_2 (1 + S/N)$$

in which the data rate is in bits per second, f is the frequency of the signal, S is the power of the signal in watts, and N is the power of the noise in watts.

Consider a 3100-Hz signal with a power level of 0.2 watts and a noise level of 0.0002 watts:

$$
\begin{aligned}
\text{Data rate} \quad &= 3100 \times \log_2 (1 + 0.2/0.0002) \\
&= 3100 \times \log_2 (1001) \\
&= 3100 \times 9.97 \\
&= 30,901 \text{ bps}
\end{aligned}
$$

(If your calculator does not have a $\log_2$ key, as most do not, you can always approximate an answer by taking the $\log_{10}$ and then dividing by 0.301.)

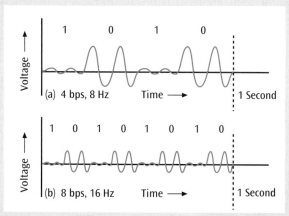

Figure 2-25   *Comparison of signal frequency with bits per second*

Data Codes

One of the most common forms of data transmitted between a transmitter and a receiver is textual data. For example, banking institutions that wish to transfer money often transmit textual information, such as account numbers, names of account owners, bank names, addresses, and the amount of money to be transferred. This textual information is transmitted as a sequence of characters. To distinguish one character from another, each character is represented by a unique binary pattern of 1s and 0s. The set of all textual characters or symbols and their corresponding binary patterns is called a data code. Three important data codes are EBCDIC, ASCII, and Unicode. Let us examine each of these in that order.

EBCDIC

The Extended Binary Coded Decimal Interchange Code, or EBCDIC, is an 8-bit code allowing 256 ($2^8 = 256$) possible combinations of textual symbols. These 256 combinations of textual symbols include all uppercase and lowercase letters, the digits 0 to 9, a large number of special symbols and punctuation marks, and a number of control characters. The control characters, such as linefeed (LF) and carriage return (CR), provide control between a processor and an input/output device. Certain control characters provide data transfer control between a computer source and computer destination. All the EBCDIC characters are shown in Figure 2-27.

Figure 2-27
The EBCDIC character code set

Bits				4→	0	0	0	0	0	0	0	0	1	1	1	1	1	1	1	1
				3→	0	0	0	0	1	1	1	1	0	0	0	0	1	1	1	1
				2→	0	0	1	1	0	0	1	1	0	0	1	1	0	0	1	1
				1→	0	1	0	1	0	1	0	1	0	1	0	1	0	1	0	1
8	7	6	5																	
0	0	0	0		NUL	SOH	STX	EXT	PF	HT	LC	DEL			SMM	VT	FF	CR	SO	SI
0	0	0	1		DLE	DC$_1$	DC$_2$	DC$_3$	RES	NL	BS	IL	CAN	EM	CC		IFS	IGS	IHS	IUS
0	0	1	0		DS	SOS	FS		BYP	LF	EOB	PRE			SM			ENQ	ACK	BEL
0	0	1	1				SYN		PN	RS	UC	EOT				DC$_4$	NAK			SUB
0	1	0	0		SP												<	(	+	|
0	1	0	1		&										!	$	.	)	:	¬
0	1	1	0		—												%	-	>	?
0	1	1	1													@			=	"
1	0	0	0			a	b	c	d	e	f	g	h	i						
1	0	0	1			j	k	l	m	n	o	p	q	r						
1	0	1	0				s	t	u	v	w	x	y	z						
1	0	1	1																	
1	1	0	0			A	B	C	D	E	F	G	H	I						
1	1	0	1			J	K	L	M	N	O	P	Q	R						
1	1	1	0				S	T	U	V	W	X	Y	Z						
1	1	1	1		0	1	2	3	4	5	6	7	8	9						

For example, if you want a computer to send the message "Transfer $1200.00" using EBCDIC, the following characters would be sent:

```
1110 0011    T
1001 1001    r
1000 0001    a
1001 0101    n
```

1010 0010	s
1000 0110	f
1000 0101	e
1001 1001	r
0100 0000	space
0101 1011	$
1111 0001	1
1111 0010	2
1111 0000	0
1111 0000	0
0101 1100	0
1111 0000	0
1111 0000	0

IBM mainframe computers are major users of the EBCDIC character set.

ASCII

The American Standard Code for Information Interchange (ASCII) is a government standard in the United States and is one of the most widely used data codes in the world. The ASCII character set exists in a few different forms, including a 7-bit version that allows for 128 (2^7 = 128) possible combinations of textual symbols, representing uppercase and lowercase letters, the digits 0 to 9, special symbols, and control characters. Because the byte, which consists of 8 bits, is a common unit of data, the 7-bit version of ASCII characters usually includes an eighth bit. This eighth bit can be used to detect transmission errors (a topic that will be discussed in Chapter Six). It can provide for 128 additional characters defined by the application using the ASCII code set, or it can simply be a binary 0. Figure 2-28 shows the ASCII character set and the corresponding 7-bit values.

Figure 2-28

The ASCII character set

		High-Order Bits (7, 6, 5)							
		000	001	010	011	100	101	110	111
	0000	NUL	DLE	SPACE	0	@	P	`	p
	0001	SOH	DC1	!	1	A	Q	a	q
	0010	STX	DC2	"	2	B	R	b	r
	0011	ETX	DC3	#	3	C	S	c	s
	0100	EOT	DC4	$	4	D	T	d	t
Low-Order Bits (4, 3, 2, 1)	0101	ENQ	NAK	%	5	E	U	e	u
	0110	ACK	SYN	&	6	F	V	f	v
	0111	BEL	ETB	'	7	G	W	g	w
	1000	BS	CAN	(	8	H	X	h	x
	1001	HT	EM	)	9	I	Y	i	y
	1010	LF	SUB	*	:	J	Z	j	z
	1011	VT	ESC	+	;	K	[	k	{
	1100	FF	FS	,	<	L	\	l		
	1101	CR	GS	-	=	M	]	m	}
	1110	SO	RS	.	>	N	^	n	~
	1111	SI	US	/	?	O	—	o	DEL

To send the message "Transfer $1200.00" using ASCII, the corresponding characters would be:

1010100	T
1110010	r
1100001	a
1101110	n
1110011	s
1100110	f
1100101	e
1110010	r
0100000	space
0100100	$
0110001	1
0110010	2
0110000	0
0110000	0
0101110	.
0110000	0
0110000	0

Unicode

One of the major problems with both EBCDIC and ASCII is that they cannot represent symbols other than those found in the English language. Further, they cannot even represent all the different types of symbols in the English language, for example many of the technical symbols used in engineering and mathematics. And what if we want to represent the other languages around the world? For this, what we need is a more powerful encoding technique—Unicode. Unicode is an encoding technique that provides a unique coding value for every character in every language, no matter what the platform. Currently, Unicode supports more than 110 different code charts (languages and symbol sets). For example, the Greek symbol β has the Unicode value of hexadecimal 03B2 (binary 0000 0011 1011 0010). Even ASCII is one of the supported code charts. Many of the large computer companies such as Apple, HP, IBM, Microsoft, Oracle, Sun, and Unisys have adopted Unicode, and many others feel that its acceptance will continue to increase with time. As the computer industry becomes more of a global market, Unicode will continue to grow in importance. Because Unicode is so large, we will not show it here. If you are interested, you can view the Unicode Web site at *www.unicode.org*.

Returning to the example of sending a textual message, if you sent "Transfer $1200.00" using Unicode, the corresponding characters would be:

0000 0000 0101 0100	T
0000 0000 0111 0010	r
0000 0000 0110 0001	a
0000 0000 0110 1110	n
0000 0000 0111 0011	s
0000 0000 0110 0110	f

```
0000 0000 0110 0101  e
0000 0000 0111 0010  r
0000 0000 0010 0000  space
0000 0000 0010 0100  $
0000 0000 0011 0001  1
0000 0000 0011 0010  2
0000 0000 0011 0000  0
0000 0000 0011 0000  0
0000 0000 0010 1110  .
0000 0000 0011 0000  0
0000 0000 0011 0000  0
```

Data and Signal Conversions In Action: Two Examples

Let us examine two typical business applications in which a variety of data and signal conversions are performed to see how analog and digital data, analog and digital signals, and data codes work together. First, consider a person at work who wants to send an e-mail to a colleague, asking about the time for the next meeting. For simplicity, let us assume the message says, "Sam, what time is the meeting with accounting? Hannah" and that it is being sent from a microcomputer connected to a local area network, which, in turn, is connected to the Internet. We will pretend this is a small business, so the connection to the Internet is over a dial-up modem (Figure 2-29).

Figure 2-29
User sending e-mail from a personal computer over a local area network and the Internet, via a modem

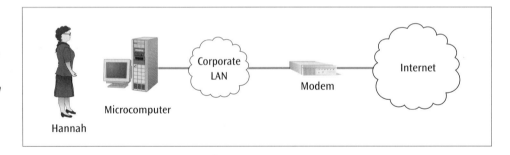

Hannah enters the message into the e-mail program and clicks the Send icon. The e-mail program prepares the e-mail message, which contains the data "Sam, what time is the meeting with accounting? Hannah" plus whatever other information is necessary for the e-mail program to send the message properly. Because this e-mail program uses ASCII, the text of this message is converted to the following:

Original message :   Sam, what time is the meeting with accounting? Hannah
ASCII string :   1010011 1100001 1101101 ...   (For brevity, only the
                           S   a   m   "Sam" portion of the message
                                                      appears here in ASCII.)

Next, the ASCII message is transmitted over a local area network (LAN) within the company. Assume this LAN uses differential Manchester encoding. The ASCII string now appears as a digital signal, as shown in Figure 2-30.

Figure 2-30

The first three letters of the message "Sam, what time is the meeting with accounting? Hannah" using differential Manchester encoding

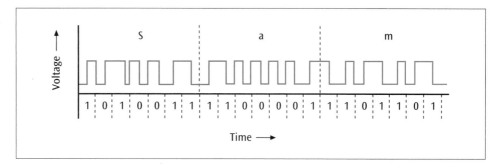

This differential Manchester encoding of the message travels over the local area network and arrives at another computer, which is connected to a modem. This computer converts the message back to an ASCII string and then transmits the ASCII string to the modem. The modem prepares the message for transmission over the Internet, using frequency modulation. For brevity, only the first 7 bits of the ASCII string (corresponding, in this case, to the "S" in Sam) are converted using simple frequency shift keying (Figure 2-31).

Figure 2-31

The frequency modulated signal for the letter "S"

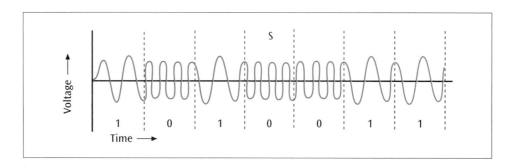

This frequency modulated signal travels over the telephone lines and arrives at the appropriate Internet gateway (the Internet service provider), which demodulates the signal into an ASCII string. From there, the ASCII string representing the original message moves out to the Internet and finally arrives at the intended receiver's computer. The process of transmitting over the Internet and delivering the message to the intended receiver's computer involves several more code conversions. Because we have not yet discussed what happens over the Internet, nor do we know what kind of connection the receiver has, this portion of the example has been omitted. Nevertheless, this relatively simple example demonstrates the number of times a conversion from data to signal to data is performed during a message transfer.

A second example involves the commonplace telephone. The telephone system in the United States is an increasingly complex marriage of traditionally analog telephone lines and modern digital technology. About the only portion of the telephone system that remains analog is the local loop—the wire that leaves your house, apartment, or business and runs to the nearest telephone switching center. Your voice, as you speak into the telephone, is analog data that is converted to an analog signal that travels over a wire to the local switching center, where it is digitized and transmitted to another switching center somewhere in the vast telephone network.

Because the human voice is analog, but a good portion of the telephone system is digital, what kind of analog-to-digital signal conversions are performed? As mentioned earlier in the chapter, the human voice occupies analog frequencies from 300 Hz to 3400 Hz and is transmitted over the telephone system with a bandwidth of 4000 Hz (4 kHz). When this 4000-Hz signal reaches the local telephone office, it is sampled at two times the greatest frequency (according to Nyquist's theorem), or 8000 samples per second. Telephone studies have shown that the human voice can be digitized using only 128 different quantization levels. Since 2^7 equals 128, each of the 8000 samples per second can be converted into a 7-bit value, yielding 8000×7, or 56,000 bits per second. If the voice signal were digitized using 256 different quantization levels, it would generate a 64,000-bps signal ($256 = 2^8$, 8×8000 yields 64,000). How are these 64,000 bits per second then sent over a wire or airwave? That depends upon the modulation technique chosen. Lower speed data streams might use frequency modulation, while higher speed streams would probably use some variation of quadrature amplitude modulation. So to summarize, we started with an analog human voice, which was digitized, and then converted back to an analog signal for transmission. Two more similar and common examples are digital cable television and digital broadcast television.

As has been stated, to send a person's voice over a telephone circuit, 128 quantization levels are adequate. But what if we want to make a recording of an artist singing a song and playing a guitar? Assuming that we wanted to create a recording of sufficiently high quality to burn on a CD, we would need many more than 128 different quantization levels. In fact, digitizing a song in order to burn it onto a CD requires thousands, maybe even tens of thousands, of quantization levels. More precisely, a music CD has a sampling rate of 44.1 kHz and uses 16-bit conversions ($2^{16} = 65,536$ quantization levels). Thus, the digitizing circuitry that converts analog music into digital form for storing on a CD is much more technically complex than the circuitry involved in making a simple telephone call.

◆ ◆

SUMMARY

▶ Data and signals are the two basic building blocks of computer networks. All data transmitted over any communications medium is either digital or analog. Data is transmitted with a signal that, like data, can be either digital or analog. The most important difference between analog and digital data and signals is that it is easier to remove noise from digital data and signals than from analog data and signals.

▶ All signals consist of three basic components: amplitude, frequency, and phase.

▶ Two important factors affecting the transfer of a signal over a medium are noise and attenuation.

▶ Because both data and signals can be either digital or analog, four basic combinations of data and signals are possible: analog data converted to an analog signal, digital data converted to a digital signal, digital data converted to a discrete analog signal, and analog data converted to a digital signal.

▶ To transmit analog data over an analog signal, the analog waveform of the data is combined with another analog waveform in a process known as modulation.

▶ Digital data carried by digital signals is represented by digital encoding formats, including the popular Manchester encoding schemes. Manchester codes always have a transition in the middle of the bit, which allows the receiver to synchronize itself with the incoming signal.

▶ For digital data to be transmitted using discrete analog signals, the digital data must first undergo a process called shift keying or modulation. The three basic techniques of shift keying are amplitude shift keying, frequency shift keying, and phase shift keying.

▶ Two common techniques for converting analog data so that it may be carried over digital signals are pulse code modulation and delta modulation. Pulse code modulation converts samples of the analog data to multiple-bit digital values. Delta modulation tracks analog data and transmits only a 1 or a 0, depending on whether the data rises or falls within the next time period.

▶ Data codes are necessary to transmit the letters, numbers, symbols, and control characters found in text data. Three important data codes are ASCII, EBCDIC, and Unicode. The EBCDIC data code uses an 8-bit code and allows for 256 different letters, digits, and special symbols. IBM mainframes use the EBCDIC code. The ASCII data code uses a 7-bit code and allows for 128 different letters, digits, and special symbols. ASCII is the most popular data code in the United States. Unicode is a 16-bit code that supports more than 110 different languages and symbol sets from around the world.

KEY TERMS

4B/5B

amplification

amplitude

amplitude shift keying

analog data

analog signals

ASCII

attenuation

bandwidth

baud rate

bipolar-AMI

bits per second (bps)

codec

data

data code

data rate

decibel (dB)

delta modulation

differential Manchester

digital data

digital signals

digitization

EBCDIC

effective bandwidth

frequency

frequency shift keying

hertz (Hz)

intermodulation distortion

Manchester

modulation

noise

nonreturn to zero inverted (NRZI)

nonreturn to zero-level (NRZ-L)

Nyquist's theorem

period

phase

phase shift keying

pulse amplitude modulation (PAM)

pulse code modulation (PCM)

quadrature amplitude modulation

quadrature phase shift keying

quantization error

quantization levels

quantization noise

sampling rate

self-clocking

Shannon's theorem

shift keying

signals

slope overload noise

spectrum

Unicode

REVIEW QUESTIONS

1. What is the difference between data and signals?
2. What are the main advantages of digital signals over analog signals?
3. What is the difference between a continuous signal and a discrete signal?
4. What are the three basic components of all signals?
5. What is the spectrum of a signal?
6. What is the bandwidth of a signal?
7. Why would analog data have to be modulated onto an analog signal?
8. How does a differential code such as the differential Manchester code differ from a nondifferential code such as the NRZs?
9. What does it mean when a signal is self-clocking?
10. What is the definition of "baud rate"?
11. How does baud rate differ from bits per second?
12. What are the three main types of shift keying?
13. What is the difference between pulse code modulation and delta modulation?
14. What is meant by the sampling rate of analog data?
15. What are the differences among EBCDIC, ASCII, and Unicode?

EXERCISES

1. What is the frequency in Hertz of a signal that repeats 80,000 times within one minute? What is its period (the length of one complete cycle)?
2. What is the bandwidth of a signal composed of frequencies from 50 Hz to 500 Hz?
3. Draw in chart form (as shown in Figure 2-12) the voltage representation of the bit pattern 11010010 for the digital encoding schemes NRZ-L, NRZI, Manchester, differential Manchester, and bipolar-AMI.
4. What is the baud rate of a digital signal that employs differential Manchester and has a data transfer rate of 2000 bps?

5. Show the equivalent 4B/5B code of the bit string 1101 1010 0011 0001 1000 1001.

6. What is the data transfer rate in bps of a signal that is encoded using phase modulation with eight different phase angles and a baud rate of 2000?

7. If quadrature amplitude modulation is used to transmit a signal with a baud rate of 8000, what is the corresponding bit rate?

8. Draw or give an example of a signal for each of the following conditions: the baud rate is equal to the bit rate, the baud rate is greater than the bit rate, and the baud rate is less than the bit rate.

9. A signal starts at point X. As it travels to point Y, it loses 8 dB. At point Y, the signal is boosted by 10 dB. As the signal travels to point Z, it loses 7 dB. What is the dB strength of the signal at point Z?

10. In the preceding problem, if the signal started at point X with a strength of 100 watts, what would be the power level of the signal at point Z?

11. Draw an example signal (similar to those shown in Figure 2-12) using NRZI in which the signal never changes for 7 bits. What does the equivalent differential Manchester encoding look like?

12. Show the equivalent analog sine-wave pattern of the bit string 00110101 using amplitude shift keying, frequency shift keying, and phase shift keying.

13. Twenty-four voice signals are to be transmitted over a single high-speed telephone line. What is the bandwidth required (in bps) if the standard analog-to-digital sampling rate is used and each sample is converted into an 8-bit value?

14. Given the analog signal shown in Figure 2-32, what are the 8-bit pulse code modulated values that will be generated at each time *t*?

Figure 2-32
Analog signal for Exercise 14

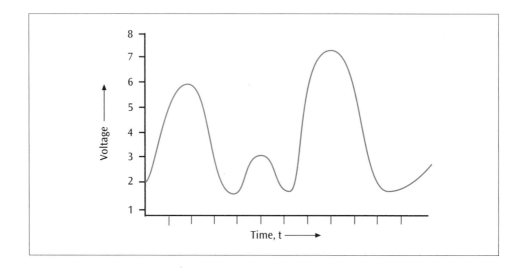

15. Using the analog signal from Exercise 14 and a delta step that is one-eighth inch long and one-eighth inch tall, what is the delta modulation output? On the drawing, point out any slope overload noise.

16. Using the EBCDIC, ASCII, and Unicode character code sets, what are the binary encodings of the message "Hello, world"?

17. You just created a pulse code modulated signal, but it is not a good representation of the original data. What can you do to improve the accuracy of the modulated signal?

18. What is the decibel loss of a signal that starts at point A with a strength of 2000 watts and ends at point B with a strength of 400 watts?

19. What is the decibel loss of signal that starts at 50 watts and experiences a 10-dB loss over a given section of wire?

20. What is the decibel loss of a signal that loses half its power during the course of transmission?

Exercises for the Details sections:

21. Using Nyquist's theorem, calculate the channel capacity C of a signal that has 16 different levels and a frequency of 20,000 Hz.

22. Using Shannon's theorem, calculate the data transfer rate given the following information:

    signal frequency = 10,000 Hz

    signal power = 5000 watts

    noise power = 230 watts

23. Using Nyquist's theorem and given a frequency of 5000 Hz and a data rate of 20,000 bps, how many signal levels (L) will be needed to convey this data?

THINKING OUTSIDE THE BOX

1 You are working for a company that has a network application for accessing a dial-up database of corporate profiles. From your computer workstation, a request for a profile travels over the corporate local area network to a modem. The modem, using a conventional telephone line, dials in to the database service. The database service is essentially a modem and a mainframe computer. Create a table (or draw a figure) that shows every time data or signals are converted to a different form in this process. For each entry in the table, show where the conversion is taking place, the form of the incoming information, and the form of the outgoing information.

2 Telephone systems are designed to transfer voice signals (4000 Hz). When a voice signal is digitized using pulse code modulation, what is the sampling rate, and how many quantization levels are used? How much data does that generate in one second? Are these the same sampling rate and quantization levels as used on a CD? Can you verify your answer?

3 If a telephone line can carry a signal with a baud rate of 6000 and we want to transmit data at 33,600 bps, how many different signal levels will be necessary? Is this how a 33,600 bps modem operates?

4 Can modems and codecs be used interchangeably? Defend your position. (The modem converts digital data to analog signals and back to digital data; the codec converts analog data to digital signals and back to analog data).

5 This chapter introduced a bipolar encoding scheme. What would be an example of a unipolar encoding scheme?

6 MegaCom is a typical company with many users, local area networks, Internet access, and so on. A user is working at home and dialing in to the corporate e-mail system. Draw a linear chart of the connection from the user's personal computer at home to the corporate e-mail server on a local area network. On this linear chart, identify each form of data and signal. Are they analog? Digital? What data/signal conversions are taking place? Where are these conversions?

HANDS-ON PROJECTS

1. Using sources from the library or the Internet, write a 2–3 page paper that describes how an iPod or a CD player performs the digital-to-analog conversion of its contents.

2. Many more digital encoding schemes exist than NRZ-L, NRZI, Manchester, differential Manchester, and bipolar-AMI. List three other encoding techniques and show an example of how each encodes.

3. What is the encoding format for the new digital high-definition television? Has the U.S. agreed upon one format, or do multiple formats exist? Are these the same formats as those used elsewhere in the world? Explain.

4. Telephone systems use a digital encoding scheme called B8ZS (pronounced "bates"), which is a variation on the bipolar-AMI encoding scheme. How does it work? Why is it used? Show an example using the binary string 01101100000000010.

5. Can you locate a Web site that shows graphically the result of adding multiple sine waves to create composite waves such as square waves or sawtooth waves? Once you locate this Web site, use the online tool to create a variety of waveforms.

6. What are the sampling rates and number of quantization levels for iPods? CD players? DVD-video players? DVD-audio players? The new Blu-Ray DVD players?

3

Conducted and Wireless Media

◆ ◆

NOT CONVINCED that cell phones have become an integral part of everyday life? Let us examine a sampling of headlines from a popular Web site that tracks wireless technology.

According to the latest research from a group called In-Stat, the cell phone market will comprise more than 2.3 billion subscribers worldwide by the year 2009 and involve an increase of 777.7 million new subscribers between 2005 and 2009.

According to a recent AP report, you can now (if you do not have change in your pocket) use your cell phone to feed a parking meter in Coral Gables, Florida. The city became the first to use CellPark, a new payment method that allows drivers to "dial in" to a meter from their cell phone: you simply enter the number assigned to your parking spot, and the meter will not expire until you call again and log off.

A cell phone ring tone that mimics the sound of a motorbike has been made into a CD and is so popular that it is outselling many of the big music hits on the British charts. This is the first time a ring tone has crossed over to the sin-gles chart and attained the number one position in England.

Now your pet can have its own cell phone. Specialized cell phones are being created that can be worn around the necks of dogs and cats. These enable pet owners to call and talk to their pet while they are at work or on vacation. The cell phones are also GPS-enabled so the owners can always know where their pets are—and locate them in case they run away.

In a poll conducted by Joel Benenson, more than 50% of 1,013 teenagers polled said they knew of someone that had used a cell phone to cheat in school.

Is your cell phone a necessary accessory when you leave the house?

Do you know which wireless technology your cell phone uses?

What are some other applications of wireless technologies besides cell phones?

Source: www.wirelessguide.org, updated July 27, 2009.

Objectives ▷

After reading this chapter, you should be able to:

▷ Outline the characteristics of twisted pair wire, including the advantages and disadvantages

▷ Outline the differences among Category 1, 2, 3, 4, 5, 5e, 6, and 7 twisted pair wire

▷ Explain when shielded twisted pair wire works better than unshielded twisted pair wire

▷ Outline the characteristics, advantages, and disadvantages of coaxial cable and fiber-optic cable

▷ Outline the characteristics of terrestrial microwave systems, including the advantages and disadvantages

▷ Outline the characteristics of satellite microwave systems, including the advantages and disadvantages as well as the differences among low-Earth-orbit, middle-Earth-orbit, geosynchronous orbit, and highly elliptical Earth orbit satellites

▷ Describe the basics of cellular telephones, including all the current generations of cellular systems

▷ Outline the characteristics of short-range transmissions, including Bluetooth

▷ Describe the characteristics, advantages, and disadvantages of broadband wireless systems and various wireless local area network transmission techniques

▷ Apply the media selection criteria of cost, speed, right-of-way, expandability and distance, environment, and security to various media in a particular application

Introduction ▶

The world of computer networks would not exist if there were no medium by which to transfer data. All communications media can be divided into two categories: (1) physical or conducted media, such as telephone lines and fiber-optic cables, and (2) radiated or wireless media, such as cellular telephones and satellite systems. Conducted media include twisted pair wire, coaxial cable, and fiber-optic cable. In addition to investigating each of these, this chapter also examines eight basic groups of wireless media used for data transfer:

- ▶ Terrestrial microwave
- ▶ Satellite transmissions
- ▶ Cellular telephone systems
- ▶ Infrared transmissions
- ▶ Broadband wireless distribution services
- ▶ Bluetooth
- ▶ Wireless local area network systems
- ▶ ZigBee short-range transmissions

The order in which the wireless topics are covered is roughly the order in which the technologies became popular.

As you read this paragraph, someone somewhere is undoubtedly designing new materials and building new equipment that is better than what currently exists. The transmission speeds and distances given in this chapter will continue to evolve. Please keep this in mind as you study the media.

The chapter will conclude with a comparison of all the media types, followed by several examples demonstrating how to select the appropriate medium for a particular application.

Conducted Media

Even though conducted media have been around as long as the telephone itself (even longer, if you include the telegraph), there have been few recent or unique additions to this technology. One exception to this is the newest member of the conducted media family: fiber-optic cable, which became widely used by the telephone companies in the 1980s and by computer network designers in the 1990s. But let us begin our discussion of the three existing types of conducted media with the oldest, simplest, and most common one: twisted pair wire.

Twisted pair wire

The term "twisted pair" is almost a misnomer, as one rarely encounters a single pair of wires. More often, twisted pair wire comes as two or more pairs of single-conductor copper wires that have been twisted around each other. Each single-conductor wire is encased within plastic insulation and cabled within one outer jacket, as shown in Figure 3-1.

Figure 3-1

Example of four-pair twisted pair wire

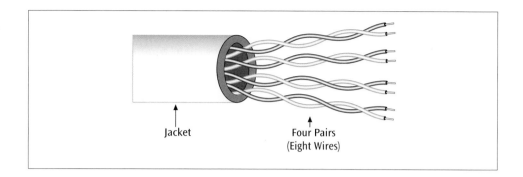

Jacket   Four Pairs (Eight Wires)

Unless someone strips back the outer jacket, you may not see the twisting of the wires, which is done to reduce the amount of interference one wire can inflict on the other, one pair of wires can inflict on another pair of wires, and an external electromagnetic source can inflict on one wire in a pair. You might recall two important laws from physics: (1) A current passing through a wire creates a magnetic field around that wire, and (2) a magnetic field passing over a wire induces a current in that wire. Therefore, a current or signal in one wire can produce an unwanted current or signal, called crosstalk, in a second wire. If the two wires run parallel to each other, as shown in Figure 3-2(a), the chance for crosstalk increases. If the two wires cross each other at perpendicular angles, as shown in Figure 3-2(b), the chance for crosstalk decreases. Although not exactly producing perpendicular angles, the twisting of two wires around each other, as shown in Figure 3-2(c), at least keeps the wires from running parallel and thus helps reduce crosstalk.

Figure 3-2

(a) Parallel wires—greater chance of crosstalk
(b) Perpendicular wires—lesser chance of crosstalk
(c) Twisted wires—crosstalk reduced because wires keep crossing each other at nearly perpendicular angles

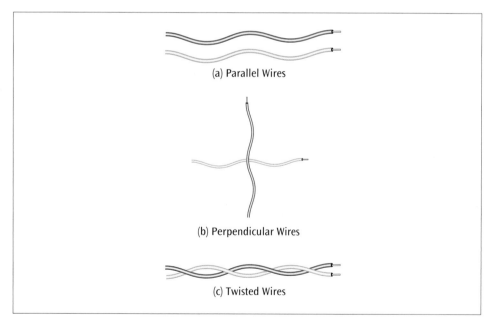

(a) Parallel Wires

(b) Perpendicular Wires

(c) Twisted Wires

You have probably experienced crosstalk many times. Remember when you were talking on the telephone and heard a conversation ever-so-faintly in the background? Your telephone connection, or circuit, was experiencing crosstalk from another telephone circuit.

As simple as twisted pair wire appears to be, it actually comes in many forms and varieties to support a wide number of applications. To help identify the numerous varieties of twisted pair wire, specifications known as Category 1-7, abbreviated as CAT 1-7 have been developed. Category 1 twisted pair is standard telephone wire and was designed to carry analog voice or data at low speeds (less than or equal to 9600 bps). Category 1 twisted pair wire, however, is not recommended for transmitting megabits of computer data. Because the wire is made from lower-quality materials and the twisting of the wire pairs is relatively minimal, Category 1 wire is susceptible to experiencing noise and signal attenuation and should not be used for high-speed data connections. Category 1 wire has been replaced (nearly out of existence) with better-quality wire. Although you might still be able to find some vendors selling Category 1 wire, it is not the wire you would want to install in a modern networked system.

Category 2 twisted pair wire is also used for telephone circuits but is a higher-quality wire than Category 1, producing less noise and signal attenuation. Category 2 twisted pair is sometimes found on T-1 and ISDN lines and in some installations of standard telephone circuits. T-1 is the designation for a digital telephone circuit that transmits voice or data at 1.544 Mbps. ISDN is a digital telephone circuit that can transmit voice or data or both from 64 kbps to 1.544 Mbps. (Chapter Eleven provides more detailed descriptions of T-1.) Once again, advances in twisted pair wire such as the use of more twists are leading to Category 2 wire being replaced with higher-quality wire, and so it is very difficult to locate anyone still selling this wire. But even if they were selling it, you would never use it for a modern network.

Category 3 twisted pair was designed to transmit 10 Mbps of data over a local area network for distances up to 100 meters (328 feet). (Note that the units typically used for specifying conducted media are metric—when necessary, the English equivalent will be provided.) Although the signal does not magically stop at 100 meters, it does weaken (attenuate), and the level of noise continues to grow such that the likelihood of the wire transmitting errors after 100 meters increases. The constraint of no more than 100 meters applies to the distance from the device that generates the signal (the source) to the device that accepts the signal (the destination). This accepting device can be either the final destination or a repeater. A repeater is a device that generates a new signal by creating an exact replica of the original signal. Thus, Category 3 twisted pair can run farther than 100 meters from its source to its final destination, as long as the signal is regenerated at least every 100 meters. Much of the Category 3 wire sold today is used for telephone circuits instead of computer network installations. There may be, however, a few older computer network installations that still use Category 3 wire. Installation of new Category 3 wire for networks is not recommended.

Category 4 twisted pair was designed to transmit 20 Mbps of data for distances up to 100 meters. It was created at a time when local area networks required a wire that could transmit data faster than the 10-Mbps speed of Category 3. Category 4 wire is rarely, if ever, sold anymore, and essentially has been replaced with newer types of twisted pair.

Category 5 twisted pair was designed to transmit 100 Mbps of data for distances up to 100 meters. (Technically speaking, Category 5 is specified for a 100-MHz signal, but because most systems transmit 100 Mbps over the 100-MHz signal, 100 MHz is equivalent to 100 Mbps.) Category 5 twisted pair has a higher number of twists per inch than the Category 1 to 4 wires, and thus introduces less noise.

Approved at the end of 1999, the specification for Category 5e twisted pair is similar to Category 5's in that this wire is also recommended for transmissions of 100 Mbps (100 MHz) for 100 meters. Many companies are producing Category 5e wire at 125 MHz for 100 meters. Although the specifications for the earlier Category 1 to 5 wires described only the individual wires, the Category 5e specification indicates exactly four pairs of wires and provides designations for the connectors on the ends of the wires, patch cords, and other possible components that connect directly with a cable. Thus, as a more detailed specification than Category 5, Category 5e can better support the higher speeds of 100-Mbps (and higher) local area networks. See the Details section "Category 5e wire and 1000 Mbps local area networks" to learn how Category 5e can support 1000 Mbps local area networks.

Category 6 twisted pair is designed to support data transmission with signals as high as 250 MHz for 100 meters. This makes Category 6 wire a good choice for 100 meter runs in local area networks with transmission speeds of 250 to 1000 Mbps. Interestingly, Category 6 twisted pair costs only pennies per foot more than Category 5e twisted pair wires. Therefore, given a choice among Category 5, 5e, or 6 twisted pair wires, you probably should install Category 6—in other words, the best-quality wire—regardless of whether or not you will be taking immediate advantage of the higher transmission speeds.

Category 7 twisted pair is the most recent addition to the twisted pair family. Category 7 wire is designed to support 600 MHz of bandwidth for 100 meters. The cable is heavily shielded—each pair of wires is shielded by a foil, and the entire cable has a shield as well. Some companies are considering using Category 7 for Gigabit and 10-Gigabit Ethernet, but currently its price is fairly high—well over $1 per foot.

Details ▶

Category 5e wire and 1000-Mbps local area networks

If Category 5e wire is designed to support 125-Mbps data transmission for 100 meters, how can it be used in 1000 Mbps (also known as Gigabit Ethernet) local area networks? The first trick is to use four pairs of Category 5e wire for the 1000-Mbps local area networks (as opposed to two pairs with 100-Mbps local area networks). With four pairs, 250 Mbps is sent over each pair. Four pairs times 250 Mbps equals 1000 Mbps. But that still does not answer how a pair of wires designed for 125-Mbps transmissions is able to send 250 Mbps. This answer involves a second trick: that Gigabit Ethernet networks use an encoding scheme called 4D-PAM5 (Pulse Amplitude Modulation). While the details of 4D-PAM5 are rather advanced and thus beyond the scope of this text, let us just say that it is a technique that employs four-dimensional (4D) data encoding coupled with a five-voltage level signal (PAM5). This combination enables Gigabit Ethernet to transmit 250 Mbps over a pair of Category 5e wires.

All of the wires described so far with the exception of Category 7 wire can be purchased as unshielded twisted pair. Unshielded twisted pair (UTP) is the most common form of twisted pair; none of the wires in this form is wrapped with a metal foil or braid. In contrast, shielded twisted pair (STP), which also is available in Category 5 through 6 (as well as numerous wire configurations), is a form in which a shield is wrapped around each wire individually, around all the wires together, or both. This shielding provides an extra layer of isolation from unwanted electromagnetic interference. Figure 3-3 shows an example of shielded twisted pair wire.

Figure 3-3
An example of shielded twisted pair

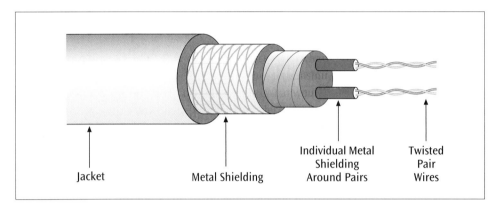

Jacket   Metal Shielding   Individual Metal   Twisted
                                                 Shielding   Pair
                                                 Around Pairs   Wires

If a twisted pair wire needs to go through walls, rooms, or buildings where there is sufficient electromagnetic interference to cause substantial noise problems, using shielded twisted pair can provide a higher level of isolation from that interference than unshielded twisted pair wire, and thus a lower level of errors. Electromagnetic interference is often generated by large motors, such as those found in heating and cooling equipment or manufacturing equipment. Even fluorescent light fixtures generate a noticeable amount of electromagnetic interference. Large sources of power can also generate damaging amounts of electromagnetic interference. Therefore, it is generally not a good idea to strap twisted pair wiring to a power line that runs through a room or through walls. Furthermore, even though Categories 5 to 6 shielded twisted pair have improved noise isolation, you cannot expect to push them past the 100-meter limit. Finally, be prepared to pay a premium for shielded twisted pair. It is not uncommon to spend an additional $1 per foot for good-quality shielded twisted pair. In contrast, Category 5, 5e, and 6 UTP often cost between $.10 and $.20 per foot.

Table 3-1 summarizes the basic characteristics of unshielded twisted pair wires. Keep in mind that for our purposes shielded twisted pair wires have basically the same data transfer rates and transmission ranges as unshielded twisted pair wires but perform better in noisy environments. Note also that the transmission distances and transfer rates appearing in Table 3-1 are not etched in stone. Noisy environments tend to shorten transmission distances and transfer rates.

Table 3-1

A summary of the characteristics of twisted pair wires

UTP Category	Typical Use	Maximum Data Transfer Rate	Maximum Transmission Range	Advantages	Disadvantages
Category 1	Telephone wire	<100 kbps	5–6 kilometers (3–4 miles)	Inexpensive, easy to install and interface	Security, noise, obsolete
Category 2	T-1, ISDN	<2 Mbps	5–6 kilometers (3–4 miles)	Same as Category 1	Security, noise, obsolete
Category 3	Telephone circuits	10 Mbps	100 m (328 ft)	Same as Category 1, with less noise	Security, noise
Category 4	LANs	20 Mbps	100 m (328 ft)	Same as Category 1, with less noise	Security, noise, obsolete
Category 5	LANs	100 Mbps (100 MHz)	100 m (328 ft)	Same as Category 1, with less noise	Security, noise
Category 5e	LANs	250 Mbps per pair (125 MHz)	100 m (328 ft)	Same as Category 5. Also includes specifications for connectors, patch cords, and other components	Security, noise
Category 6	LANs	250 Mbps per pair (250 MHz)	100 m (328 ft)	Higher rates than Category 5e, less noise	Security, noise, cost
Category 7	LANs	600 MHz	100 m (328 ft)	High data rates	Security, noise, cost

Details ▶

More Characteristics of Twisted Pair

When you are selecting a wire, you can use the Categories 5 through 7 and shielding to make several distinctions between different types of twisted pair wire. In addition to knowing these distinctions, you need to consider where your wire will be placed and what its width should be. For example, when run between the rooms within a building, will the wire be traveling within a plenum or through a riser? A plenum is the space within a building that was created by building components and designed for the movement of breathable air—for example, the space above a suspended ceiling. A plenum can also be a hidden walkway between rooms that houses heating and cooling vents, telephone lines, and other cable services. Plenum wire is designed so that, in the event of a fire, it does

not spread flame and noxious fumes. To meet these standards, the wire's jacket is made of special materials, and this, of course, significantly increases the wire's cost. In fact, plenum wire can sometimes cost twice as much as standard twisted pair wire.

If, on the other hand, your wire is going to run through a riser—a hollow metal tube that runs between walls, floors, and ceilings and encloses the individual wires—then flames and noxious fumes are not as serious an issue. In this case, standard plastic jacketing can be used. This type of twisted pair wire is typically the cable advertised in and discussed with regard to new wire installations, because a majority of these installations involve running wires through risers.

Coaxial cable

Coaxial cable, in its simplest form, is a single wire (usually copper) wrapped in a foam insulation, surrounded by a braided metal shield, then covered in a plastic jacket. The braided metal shield is very good at blocking electromagnetic signals from entering the cable and producing noise. Figure 3-4 shows a coaxial cable and its braided metal shield. Because of its good shielding properties, coaxial cable is good at carrying analog signals with a wide range of frequencies. Thus, coaxial cable can transmit large numbers of video channels, such as those found on the cable television services that are delivered into homes and businesses. Coaxial cable has also been used for long-distance telephone transmission, under rare circumstances as the cabling within a local area network, and as a connector between a computer terminal and a mainframe computer.

Figure 3-4
Example of coaxial cable showing metal braid

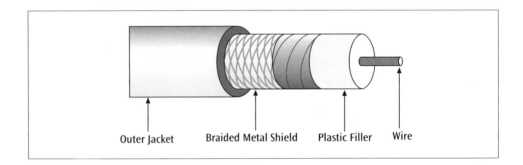

Outer Jacket   Braided Metal Shield   Plastic Filler   Wire

Two major coaxial cable technologies exist and are distinguished by the type of signal each carries: baseband and broadband. Baseband coaxial technology uses digital signaling in which the cable carries only one channel of digital data. A fairly common application for baseband coaxial used to be the interconnection of switches within a local area network. In such networks, the baseband cable would typically carry one 10- to 100-Mbps signal and require repeaters every few hundred kilometers. Currently, fiber-optic cable is replacing baseband coaxial cable as the preferred method for interconnecting LAN hubs.

Broadband coaxial technology typically transmits analog signals and is capable of supporting multiple channels of data simultaneously. Consider the coaxial cable that transmits cable television. Many cable companies offer 100 or more channels. Each channel or signal occupies a bandwidth of approximately 6 MHz. When 100 channels are transmitted together, the coaxial cable is supporting a 100×6 MHz or 600-MHz composite signal. Compared to the data capacity of twisted pair wire and baseband cable, each broadband *channel* is quite robust, as it can support the equivalent of millions of bits per second. To support such a wide range of frequencies, broadband coaxial cable systems require amplifiers (recall the amplification of analog signals from Chapter Two) approximately every three to four kilometers. Although the splitting and joining of broadband signals and cables is possible, it is a rather precise science that is best left to specialists in the field. Thus, many network administrators often hire outside experts to install and maintain broadband systems.

In addition to the two signal-based categories, coaxial cable also is available in a variety of thicknesses, with two primary physical types: thick coaxial cable and thin coaxial cable, which are both shown in Figure 3-5. Thick coaxial cable ranges in size from approximately 6 to 10 mm (1/4 to 3/8 inch) in diameter. Thin coaxial cable is approximately 4 mm (less than 1/4 inch) in diameter. Compared to thick coaxial cable, which typically carries broadband signals, thin coaxial cable has limited noise isolation and typically carries baseband signals. Thick coaxial cable has better noise immunity and is generally used for the transmission of analog data, such as single or multiple video channels. Some thick coaxial cable is so thick and so stiff that some people jokingly call it *frozen garden hose*. Thick and thin coaxial cable prices vary depending on the quality and construction of the cable, but they typically run $.20 to $1.00 per foot, and sometimes even higher.

Figure 3-5
Examples of thick coaxial cable and thin coaxial cable

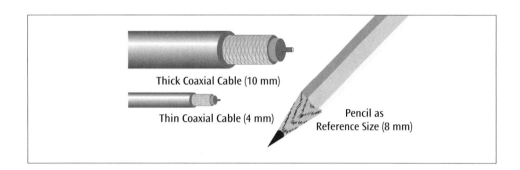

Thick Coaxial Cable (10 mm)

Thin Coaxial Cable (4 mm)

Pencil as Reference Size (8 mm)

Details ▶

More Characteristics of Coaxial Cable

An important characteristic of coaxial cable is its ohm rating. Ohm is the measure of resistance within a medium. The higher the ohm rating, the more resistance in the cable. Although resistance is not a primary concern when choosing a particular cable, the ohm value is indirectly important, because coaxial cables with certain ohm ratings work better with certain kinds of signals, and thus with certain kinds of applications. A coaxial cable's type is designated by radio guide (RG), a composite rating that accounts for many characteristics, including wire thickness, insulation thickness, electrical properties, and more. Table 3-2 summarizes the different types of coaxial cable, their ohm values, and applications.

Another characteristic of coaxial cables that is sometimes considered is whether the wire that runs down the center of the coaxial cable is single-stranded or braided. Single-stranded coaxial cable contains, as the name implies, a single wire. Braided coaxial cable is composed of many fine wires twisted around each other, acting as a single conductor. If the wire is braided, it is often less expensive and easier to bend than a single strand, which is usually thicker.

Type of Cable	Ohm Rating	Application/Comments
RG-6	75 Ohm	Cable television, satellite television, and cable modems
RG-8	50 Ohm	Older Ethernet local area networks; RG-8 is being replaced with RG-58
RG-11	75 Ohm	Broadband Ethernet local area networks and other video applications
RG-58	50 Ohm	Baseband Ethernet local area networks
RG-59	75 Ohm	Closed-circuit television; cable television (but RG-6 is better here)
RG-62	93 Ohm	Interconnection of IBM 3270 computer terminals

Table 3-2 *Common coaxial cables, ohm values, and applications*

Fiber-optic cable

All the conducted media discussed so far have one great weakness: electromagnetic interference. Electromagnetic interference is the electronic distortion that a signal passing through a metal wire experiences when a stray magnetic field passes over it. Related to this is the problem that a signal, as you learned earlier in the chapter, passing through a metal wire also generates a magnetic field and thus itself produces electromagnetic interference. Another related problem (and weakness of twisted pair wire and coaxial cable) is that it is possible for someone to wiretap these media—that is, tap into this electromagnetic interference and listen to the data traveling through the cable without being detected. Electromagnetic interference can be reduced with proper shielding, but it cannot be completely avoided unless you use fiber-optic cable. Fiber-optic cable (or optical fiber) is a thin glass cable, a little thicker than a human hair, surrounded by a plastic coating. When fiber-optic cable is packaged into an insulated cable, it is surrounded by Aramid yarn and a strong plastic jacket to protect it from bending, heat, and stress. You can see an example of fiber-optic cable in Figure 3-6.

Figure 3-6

A person holding a plain fiber-optic cable and a fiber-optic cable in an insulated jacket

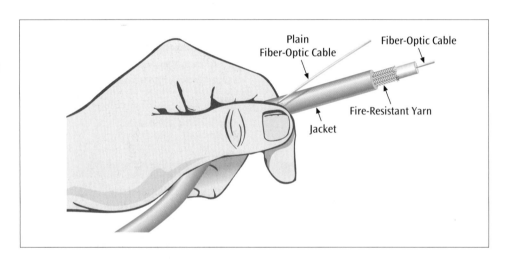

How does a thin glass cable transmit data? A light source, called a photo diode, is placed at the transmitting end and quickly switched on and off to produce light pulses. These light pulses travel down the glass cable and are detected by an optic sensor called a photo receptor on the receiving end. The light source can be either a simple and inexpensive light-emitting diode (LED), such as those found in many pocket calculators, or a more complex laser. The laser is much more expensive than the LED, and it can produce much higher data transmission rates. Fiber-optic cable is capable of transmitting data at over 100 Gbps (i.e., 100 billion bits per second!) over several kilometers. Because many common local area network installations use an LED source, however, real-world fiber-optic transmissions are effectively limited to 10 gigabits for 300 meters. (See the Details section "More Characteristics of Fiber-Optic Cable" for a discussion on LED and laser fiber-optic applications.)

In addition to providing high-speed, low-error data transmission rates, fiber-optic cable offers a number of other advantages over twisted pair wire and coaxial cable. Because fiber-optic cable passes electrically nonconducting photons through a glass medium, it is virtually impossible to wiretap. The only possible way to wiretap a fiber-optic line is to physically break into the line, an intrusion that would be noticed. Also, because fiber-optic cable cannot generate nor be disrupted by electromagnetic interference, no noise is generated from extraneous electromagnetic signals. Although fiber-optic cable still experiences noise as the light pulses bounce around inside the glass cable, this noise is significantly less than the noise generated in the metallic wire of twisted pair wires or coaxial cables. This lack of significant noise is one of the main reasons fiber-optic cable can transmit data for such long distances.

Despite these overwhelming advantages, fiber-optic cable has two small but significant disadvantages. First, due to the way the light source and photo receptor arrangement works, light pulses can travel in one direction only. Thus, to support a two-way transmission of data, two fiber-optic cables are necessary. For this reason, most fiber-optic cable is sold with at least two (if not more) individual strands of fiber bundled into a single package, as shown in Figure 3-7.

Figure 3-7
A fiber-optic cable with multiple strands of fiber

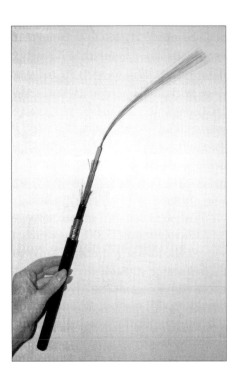

A second disadvantage of fiber-optic cable is its higher cost, but this disadvantage is slowly disappearing. For example, it is now possible to purchase bulk, general-purpose duplex (two-strand) fiber-optic cable for approximately $.50 per foot (as opposed to paying a few dollars per foot several years ago), which is close to the price of many types of coaxial cable and shielded twisted pair cable. When you consider its lower error rates and higher data transmission rates, fiber-optic cable is indeed a bargain, even compared to inexpensive twisted pair wire. Interestingly, it is not so much the fiber-optic cable itself that is expensive; the cable's high cost is due to the hardware that transmits and receives the light pulses at the ends of the fiber cable. But even this situation is changing. Starting in 1999, the prices for photo diodes and photo receptors started to drop significantly. Before 1999, it was common to use fiber-optic cable only as the backbone—the main connecting cable that runs from one end of the installation to another—of a network, and to use Category 5e or 6 twisted pair from the backbone connection up to the workstation. An illustration of a fiber-optic backbone is shown in Figure 3-8.

Figure 3-8
A fiber-optic backbone with Category 6 twisted pair running to the workstations

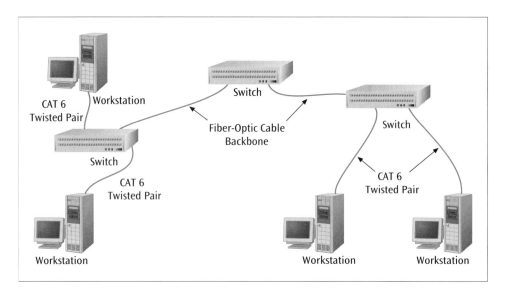

Photonic Fiber

Fiber-optic cables, as we have just seen, are solid glass wires that transmit light pulses. Unfortunately, as the length of the glass cable—and hence the distance the light must travel—grows, the light signal attenuates (scatters within the glass) due to reflection and refraction. (See the Details section "More Characteristics of Fiber-Optic Cable" for a discussion of reflection and refraction). A new type of medium, photonic fiber, has been introduced that virtually eliminates this attenuation. This glass cable is similar to fiber-optic cable in that it is as thin as a hair and transparent. The big difference, however, is that this new cable is full of holes. If you look at a cross-section of photonic fiber cable, you will see that the glass has a honeycomb pattern. The light source that is transmitted over the cable actually travels through the holes, which are just air. Because light traveling through air moves virtually unhindered, the signal does not degrade. The trick with photonic fiber is getting the cladding that surrounds the cable not to absorb the light. To resolve this problem, scientists have created a reflective surface that is even more reflective than a mirror. It is estimated that once this technology is perfected, photonic fibers will have transmission speeds and distances at least 10 times that of current fiber-optic cable.

Table 3-3 summarizes the conducted media discussed in this chapter. Category 1 and 3 twisted pair wires have been grouped together because they are commonly used for telephone systems, while Category 5 through 7 wires have been grouped together because they are typically used for local area networks. In almost all cases, maximum data rate and maximum transmission range are typical values and can be less or more, depending on environmental factors.

Details ▷

More Characteristics of Fiber-Optic Cable

When light from a source is sent through a fiber-optic cable, the light wave both bounces around inside the cable and passes through the cable to the outer protective jacket. When a light signal inside the cable bounces off the cable wall and back into the cable, this is called **reflection**. When a light signal passes from the core of the cable into the surrounding material, this is called **refraction**. Figure 3-9 demonstrates the difference between reflection and refraction.

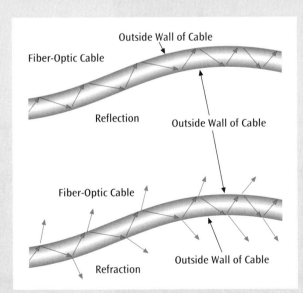

Figure 3-9 *A simple demonstration of reflection and refraction in a fiber-optic cable*

Light can be transmitted through a fiber-optic cable using two basic techniques. The first technique, called **single-mode**

transmission, requires the use of a very thin fiber-optic cable and a very focused light source, such as a laser. When a laser is fired down a narrow fiber, the light follows a tight beam, and so there is less tendency for the light wave to reflect or refract. Thus, this technique allows for a very fast signal with little signal degradation (and thus less noise) over long distances. Because lasers are used as the light source, single-mode transmission is a more expensive technique than the second fiber-optic cable signaling technique. Any application that involves a large amount of data transmitted at high speeds is a candidate for single-mode transmission.

The second signaling technique, called **multimode transmission**, uses a slightly thicker fiber cable and an unfocused light source, such as an LED. Because the light source is unfocused, the light wave experiences more refraction and reflection (i.e., noise) as it propagates through the wire. This noise results in signals that cannot travel as far or as fast as the signals generated with the single-mode technique. Correspondingly, multimode transmission is less expensive than single-mode transmission. Local area networks that employ fiber-optic cables often use multimode transmissions.

Single-mode and multimode transmission techniques use fiber-optic cable with different characteristics. The core of single-mode fiber-optic cable is 8.3 microns wide, and the material surrounding the fiber—the cladding—is 125 microns wide. Hence, single-mode fiber-optic cable is labeled **8.3/125 cable**. The core of multimode fiber-optic cable is most commonly 62.5 microns wide, and the cladding is 125 microns. Multimode fiber-optic cable is labeled **62.5/125 cable**. Other sizes of multimode fiber-optic cable include 50/125 and 100/140 microns.

Table 3-3

A summary of the characteristics of conducted media

Type of Conducted Medium	Typical Use	Maximum Data Rate	Maximum Transmission Range	Advantages	Disadvantages
Twisted pair Category 1, 3	Telephone systems	<2 Mbps	5–6 kilometers (3–4 miles)	Inexpensive, common	Noise, security, obsolete
Twisted pair Category 5, 5e, 6, 7	LANs	100–1000 Mbps	100 m (328 feet)	Inexpensive, versatile	Noise, security
Thin Coaxial Cable (baseband single channel)	LANs	10 Mbps	100 m (328 feet)	Low noise	Security
Thick Coaxial Cable (broadband multichannel)	LANs, cable TV, long-distance telephone, short-run computer system links	10–100 Mbps	5–6 kilometers (3–4 miles) (at lower data rates)	Low noise, multiple channels	Security
LED Fiber-Optic	Data, video, audio, LANs	Gbps	300 meters (approx. 1000 feet)	Secure, high capacity, low noise	Interface expensive but decreasing in cost
Laser Fiber-Optic	Data, video, audio, LANs, WANs, MANs	100s Gbps	100 kilometers (approx. 60 miles)	Secure, high capacity, very low noise	Interface expensive

Now that you are familiar with the various types of conducted media, let us turn our attention to wireless media. As we examine the various wireless technologies, let us keep an important issue in mind—the issue of right-of-way. Right-of-way is the legal capability of a business to install a wire or cable across someone's property. If a business wants to install a conducted medium between two buildings, and the business does not own the property in between the buildings, the business has to receive the right-of-way from the owner of the in-between property. As we will see in the following sections, wireless transmissions generally do not have to deal with right-of-way issues. This often provides a strong advantage for wireless media over conducted media.

Wireless Media

The introduction of this chapter lists eight separate types of wireless media. Despite the fact that each type of medium might be used by a different application, and different sets of frequencies are often assigned to each, all wireless media share the same basic technology—the transmission of data using radio waves. (Strictly speaking, in all these types of wireless technology, the actual *medium* through which the radio waves must travel is air or space. For the purposes of this discussion, however, we will expand the term "medium" to include the technology transmitting the signal.) Let us examine this growing technology and then discuss each of the eight types of wireless media, along with their basic characteristics and application areas.

Wireless transmission became popular in the 1950s with AM radio, FM radio, and television, and in 1962, transmissions were sent through the first orbiting satellite, Telstar. In the 60 or so years since wireless transmission emerged, this technology has spawned hundreds, if not thousands, of applications, some of which will be discussed in this chapter.

In wireless transmission, various types of electromagnetic waves are used to transmit signals. Radio transmissions, satellite transmissions, visible light, infrared light, X-rays, and gamma rays are all examples of electromagnetic waves or electromagnetic radiation. In general, electromagnetic radiation is energy propagated through space and, indirectly, through solid objects in the form of an advancing disturbance of electric and magnetic fields. In the particular case of, say, radio transmissions, this energy is emitted in the form of radio waves by the acceleration of free electrons, such as occurs when an electrical charge is passed through a radio antenna wire. The basic difference between various types of electromagnetic waves is their differing wavelengths, or frequencies, as shown in Figure 3-10.

Figure 3-10
Electromagnetic wave frequencies

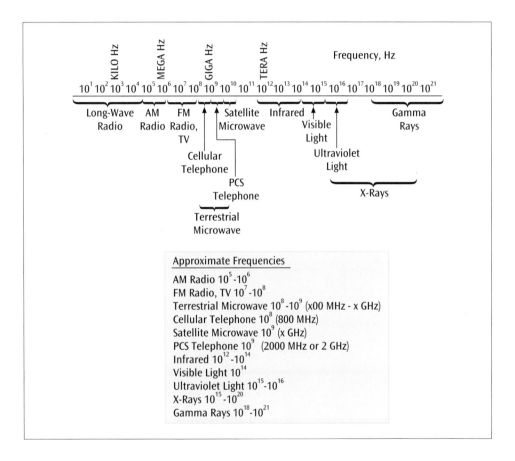

Note that all types of transmission systems such as AM radio, FM radio, television, cellular telephones, terrestrial microwaves, and satellite systems are all confined to relatively narrow bands of frequencies. The Federal Communications Commission (FCC) keeps tight control on what frequencies are used by which application. Occasionally, the FCC will assign an unused range of frequencies to a new application. At other times, the FCC will auction off unused frequencies to the highest bidder. The winner of the auction is then allowed to use those frequencies for the introduction of a particular product or service. It is important to note, however, that only so many frequencies are available to be used for applications. Thus, it is crucial that each application use its assigned frequencies as well as possible. The cellular telephone system, as we will see, provides a good example of how an application can use its assigned frequencies efficiently. Let us keep this conservative frequency allocation process in mind as we discuss eight different areas of wireless communication systems, beginning with one of the oldest—terrestrial microwave transmission.

Terrestrial microwave transmission

Terrestrial microwave transmission systems transmit tightly focused beams of radio signals from one ground-based microwave transmission antenna to another. The two most common application areas of terrestrial microwave are telephone communications and business intercommunication. Many telephone companies implement a series of antennas, placing a combination receiver and transmitter tower every 15 to 30 miles. These systems provide telephone service by spanning metropolitan as well as intrastate and interstate areas. Businesses also can use terrestrial microwave to implement telecommunications systems between corporate buildings. Maintaining such an arrangement might be less expensive in the long run than leasing a high-speed telephone line from a telephone company, which requires an ongoing monthly payment. With terrestrial microwave, once the system is purchased and installed, no telephone service fees are necessary.

Possibly the best-selling point for terrestrial microwave is its ability to transmit signals up to hundreds of millions of bits per second without the use of interconnecting wires. Figure 3-11 shows a typical microwave antenna installation.

Figure 3-11
A typical microwave tower and antenna

Microwave transmissions do not follow the curvature of the Earth, nor do they pass through solid objects, both of which limit their transmission distance. Microwave antennas use line-of-sight transmission, which means that to receive and transmit a signal, each antenna must be in sight of the next antenna (see Figure 3-12). Many microwave antennas are located on top of free-standing towers, and the typical distance between microwave towers is roughly 15 to 30 miles. The higher the tower, the farther the possible transmission distance. Thus, towers located on hills or mountains, or atop tall buildings, can transmit signals farther than 30 miles. Another factor that limits transmission distance is the number of objects that might obstruct the path of transmission signals. Buildings, hills, forests, and even heavy rain and snowfall all interfere with the transmission of microwave signals. (Assuming there is no interference, however, and that amplifiers are used on the towers to regenerate the signal, terrestrial microwave can run for an unlimited distance.) Considering these limitations, the disadvantages of terrestrial microwave can include loss of signal strength (attenuation) and interference from other signals (intermodulation), in addition to the costs of either leasing the service or installing and maintaining the antennas.

Figure 3-12
A microwave antenna on top of a free-standing tower transmitting to another antenna on the top of a building

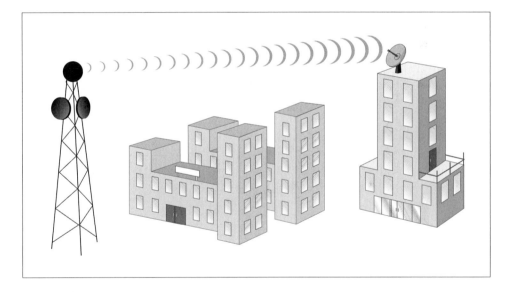

Satellite microwave transmission

Satellite microwave transmission systems are similar to terrestrial microwave systems except that the signal travels from a ground station on Earth to a satellite and back to another ground station on Earth, thus achieving much greater distances than Earth-bound line-of-sight transmissions. In fact, a satellite located at the farthest possible point from the Earth—36,000 kilometers or 22,300 miles—can receive and send signals approximately one-third the distance around the Earth. Satellite systems can also transmit a signal partway around the Earth by bouncing it from one satellite to another.

One way of categorizing satellite systems is by how far the satellite is from the Earth. The closer a satellite is to the Earth, the shorter the times required to send data to the satellite—to uplink—and receive data from the satellite—to downlink. This transmission time from ground station to satellite and back to ground station is called **propagation delay**. The disadvantage to being closer to Earth is that the satellite must continuously circle the Earth to remain in orbit. Thus, these satellites are constantly moving and eventually pass beyond the horizon, ruining the line-of-sight transmission. Satellites that are always over the same point on Earth can be used for long periods of high-speed data transfers. In contrast, because satellites that are close to Earth do not stay in the same position over the Earth, they are used with applications requiring shorter periods of data transfer, such as mobile telephone systems. As Figure 3-13 shows, satellites orbit the Earth from four possible ranges: low Earth orbit (LEO), middle Earth orbit (MEO), geosynchronous Earth orbit (GEO), and highly elliptical Earth orbit (HEO).

Figure 3-13
Earth and the four Earth orbits: LEO, MEO, GEO, and HEO

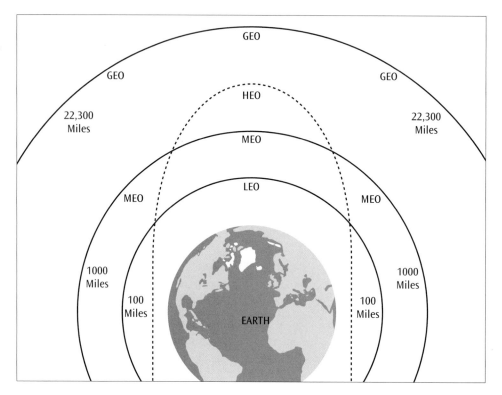

Low-Earth-orbit (LEO) satellites are closest to the Earth. They can be found as close as 100 miles from the surface and as far as 1000 miles. The number of low-Earth-orbit satellites is growing rapidly. At the end of the twentieth century, there were approximately 300 LEO satellites. By the year 2005, an estimated one thousand LEO satellites were in orbit. Low-Earth-orbit satellites are used primarily for the wireless transfer of electronic mail, worldwide mobile telephone networks, spying, remote sensing, and videoconferencing.

One of the late twentieth century's most ambitious and notorious projects was Motorola's Iridium handheld global satellite telephone and paging network. The Iridium system was originally designed to use seven layers of satellites with 11 satellites in each layer, or a total of 77 satellites. The network got its name from the element iridium, which has an atomic weight of 77. After some rethinking, it was determined that the system would also work with six layers of 11 satellites, and thus the project was scaled back to 66 satellites (the name, however, was not changed to the corresponding element dysprosium—which, apparently, did not have the same ring). Even with only 66 satellites at work, a person could, from any point on Earth, receive or place a telephone call when using an Iridium mobile telephone. Unfortunately, by the summer of 1999, the Iridium system had failed to interest enough subscribers, causing the owners of the system to sell the network of 66 satellites.

Middle-Earth-orbit (MEO) satellites can be found roughly 1000 to 3000 miles from the Earth. At the end of the twentieth century, approximately 65 MEO satellites were orbiting the Earth. Although MEO satellite systems are not growing at the same phenomenal rate as LEO systems, industry experts estimate that the number of MEO satellites in 2005 is close to 120.

Middle-Earth-orbit satellites are used primarily for global positioning system surface navigation applications. Global positioning systems are complex, but it is worthwhile to take a brief look at how they work. The global positioning system

Details ▷

Satellite Configurations

Besides being classified as LEO, MEO, GEO, and HEO, satellite systems can be categorized into three basic topologies: bulk carrier facilities, multiplexed Earth stations, and single-user Earth stations. Figure 3-14 illustrates each of these topologies.

Bulk Carrier Facilities
Figure 3-14(a) shows that in a bulk carrier facility, the satellite system and all its assigned frequencies are devoted to one user. Because a satellite is capable of transmitting large amounts of data in a very short time, and the system itself is expensive, only a very large application could economically justify the exclusive use of an entire satellite system by one user. For example, it would make sense for a telephone company to use a bulk carrier satellite system to transmit thousands of long-distance telephone calls. Typical bulk carrier systems operate in the 6/4-GHz bands (6-GHz uplink, 4-GHz downlink) and provide a 500-MHz bandwidth, which can be broken further into multiple channels of 40–50 MHz.

Multiplexed Earth Station
In a multiplexed Earth station satellite system, the ground station accepts input from multiple sources and in some fashion interweaves the data streams, either by assigning different frequencies to different signals or by allowing different signals to take turns transmitting. Figure 3-14(b) shows a diagram of how a typical multiplexed Earth station satellite system operates.

How does this type of satellite system satisfy the requests of users and assign time slots? Each user could be asked in turn if he or she has data to transmit, but because so much time could be lost by the asking process, this technique would not be economically feasible. A first-come, first-served scenario, in which each user competed with every other user, would also be an extremely inefficient design. The technique that seems to work best for assigning access to multiplexed satellite systems is a reservation system. In a reservation system, users place a reservation for future time slots. When the reserved time slot arrives, the user transmits his or her data on the system. Two types of reservation systems exist: centralized reservation and distributed reservation. In a centralized reservation system, all reservations go to a central location, and that site handles the incoming requests. In a distributed reservation system, no central site handles the reservations, but individual users come to some agreement on the order of transmission.

Single-User Earth Station
In a single-user Earth station satellite system, each user employs his or her own ground station to transmit data to the satellite. Figure 3-14(c) shows a typical single-user Earth station satellite configuration. The Very Small Aperture Terminal (VSAT) system is an example of a single-user Earth station satellite system with its own ground station and a small antenna (two to six feet across). Among all the user ground stations is one master station that is typically connected to a mainframe-like computer system. The ground stations communicate with the mainframe computer via the satellite and master station. A VSAT end user needs an indoor unit, which consists of a transceiver that interfaces the user's computer system with an outside satellite dish (the outdoor unit). This transceiver, which is small, sends signals to and receives signals from a LEO satellite via the dish. VSAT is capable of handling data, voice, and video signals over much of the Earth's surface.

(GPS) is a system of 24 satellites that were launched by the U.S. Department of Defense and are used for identifying locations on Earth. By triangulating signals from at least four GPS satellites (each of which provides the directional coordinates X, Y, Z, and time), a receiving unit can pinpoint its own current location to within a few yards anywhere on Earth. Many companies now produce hand-held and automotive GPS devices, which are accurate within a few city blocks, and major automobile manufacturers offer automobiles with built-in GPS so their customers can access driving directions and even the location of the nearest gas station while driving in their cars. These systems also enable manufacturers to determine, in case a driver is lost or in an accident, the location of the automobile anywhere in the country.

Geosynchronous-Earth-orbit (GEO) satellites are found 36,000 kilometers (22,300 miles) from the Earth and are always positioned over the same point on Earth (somewhere over the equator). Thus, two ground stations can conduct continuous transmissions from Earth to the satellite and back to Earth. Geosynchronous-Earth-orbit satellites are most commonly used for signal relays for broadcast, cable, and direct television; meteorology; government intelligence operations; and mobile maritime telephony. The primary advantage of GEO satellites is their capacity for delivering high-speed, high-quantity bulk transmissions that can cover up to one-third of the surface of the Earth. Companies that operate GEO satellites can commit all of their transmission resources to one client or can share the satellite time with multiple clients. The use of a GEO satellite system by a single client is expensive

Details ▶

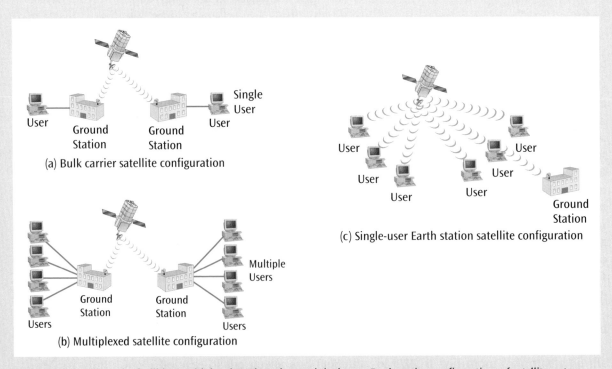

(a) Bulk carrier satellite configuration

(b) Multiplexed satellite configuration

(c) Single-user Earth station satellite configuration

Figure 3-14   *Bulk carrier facilities, multiplexed Earth station, and single-user Earth station configurations of satellite systems*

and usually involves the transfer of great amounts of data. An estimated 360 GEO satellites were in orbit by the end of 2009.

A fourth satellite system that has unique properties is the highly elliptical orbit (HEO) satellite, which is used by governments for spying (via satellite photography) and by scientific agencies for observing celestial bodies. An HEO satellite follows an elliptical pattern, as shown in Figure 3-15. When the satellite is at its perigee (closest point to the Earth), it takes photographs of the Earth. When the satellite reaches its apogee (farthest point from the Earth), it transmits the data to the ground station. At its apogee, the satellite can also photograph objects in space.

Satellites and terrestrial microwave systems can also be categorized by the frequencies with which the systems transmit, or the band of the satellite. All radio transmission systems, such as satellite, ground-based microwave, and television and radio systems transmit their signals in FCC-approved transmission bands. For example, the radio bands as defined by the ITU (International Telecommunications Union) are as follows:

Band Number	Symbol	Frequency	Common Use
4	VLF (very low frequency)	3–30 kHz	Radio navigations systems
5	LF (low frequency)	30–300 kHz	Radio beacons
6	MF (medium frequency)	300 kHz–3 MHz	AM radio
7	HF (high frequency)	3–30 MHz	CB radio
8	VHF (very high frequency)	30–300 MHz	VHF TV, FM radio
9	UHF (ultra high frequency)	300 MHz–3 GHz	UHF TV, cell phones, pagers
10	SHF (superhigh frequency)	3–30 GHz	Satellite
11	EHF (extremely high freq)	30–300 GHz	Satellite, radar systems

Many devices that transmit signals using the above radio bands transmit their signals in a broadcast manner, in which the signals propagate out from a transmission tower in all directions. Conversely, most terrestrial microwave and many satellite systems transmit their signals in a narrow line-of-sight path and are often categorized by radar bands. IEEE and NATO have designated the following radar bands:

Radar Band	Frequency	Common Use
L	~1–2 GHz	GPS, government use, GSM cell phones
S	2–4 GHz	Weather systems, digital satellite radio system
C	4–8 GHz	Commercial satellite systems
X	~7–12.5 GHz	Some communication satellites, weather
Ku	12–18 GHz	NASA, television station remotes to station
Ka	18–40 GHz	Communication satellites
V	50–75 GHz	Not heavily used
W	75–111 GHz	Misc (military, car radar systems)

Note the overlap between the two naming conventions. The SHF radio band (3–30 GHz) shares the same frequencies as the L and S radar bands (1–2 GHz and 2–4 GHz, respectively).

Figure 3-15
Diagram of a highly elliptical Earth orbit satellite

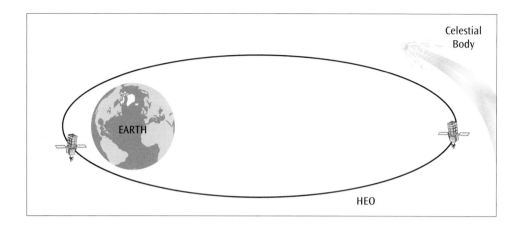

Cellular telephones

Another wireless technology that uses radio waves is the cellular telephone, or cell phone, system. Four basic generations of cellular telephone systems currently exist: first-generation analog cellular telephones; second-generation digital Personal Communications Services (PCS); a third generation (interestingly labeled generation 2.5), which saw the convergence of data signals with voice signals; and the current generation of cellular telephones. During the 1980s and 1990s, many individuals carried pagers as a means of keeping in touch with their businesses or families. At the turn of the century, as cellular telephones grew in popularity, pager use declined; and in the recent years, it might even be said that the cell phone has effectively replaced the pager. Cellular telephones have become so popular that, according to Forrester Research, a telecommunications watchdog group, as of November 2009, seventy-one percent of all households have a cell phone, and twenty percent of all households have only a cell phone and no land-line phone.

The name "cellular telephone" raises an interesting question: What does the term *cellular* mean? To answer this question, you need to examine the interactive radio network that existed in the 1990s, before cellular telephones became popular: Improved Mobile Telephone Services (IMTS). IMTS allowed only 12 concurrent users within an entire city. The reason for so few concurrent users was mentioned earlier in the chapter—the FCC makes only so many radio frequencies available for a particular application. When a user talks to another user, two channels are necessary. One channel is for one direction of transmission, and a second channel is for the opposite direction. Each channel requires a sufficient range of frequencies to carry a voice signal. To support the prospect of hundreds and thousands of simultaneous users within a metropolitan area, an extremely large range of frequencies was required. The FCC could not allocate this many frequencies to a single application, so it created an alternative by dividing the country into 700+ mobile service areas (MSAs), or markets. Each market, which usually encompasses an entire metropolitan area, is further broken into adjacent cells (as shown in the upper-left corner of Figure 3-16). Notice how the cells form a honeycomb-like pattern.

Figure 3-16
One cellular telephone market divided into cells

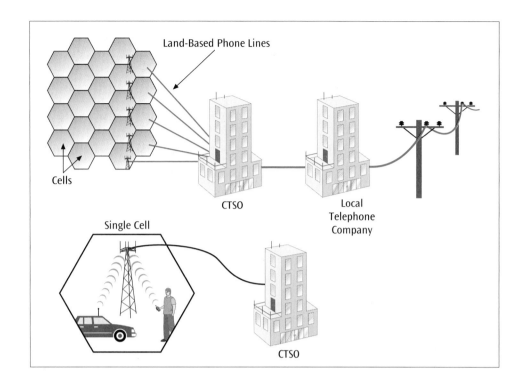

Cells can range in size from one-half mile in radius to 50 miles in radius. Located at the intersection of each cell is a low-power transmitter/receiver, which is often placed on a free-standing tower (see Figure 3-17). Not long after these towers started appearing in neighborhoods, residents began complaining about their presence. Consequently, cellular telephone companies have gotten creative at disguising their antennas on the tops and sides of buildings, inside church steeples, and even into the shapes of trees (see Figure 3-18).

A cellular telephone within a cell communicates to the cell tower, which in turn is connected to the cellular telephone switching office (CTSO) by a telephone line. The cellular telephone switching office is then connected to the local telephone system. If the cellular telephone moves from one cell to another, the cellular telephone switching office hands off the connection from one cell to another.

Because each cell uses low-power transmissions, it is not likely that a transmission within one particular cell will interfere with a transmission in another cell that is more than one or two cells away. Thus, only near-adjacent cells need to use different sets of frequencies. As a result, the frequencies used within a cell can be reused in other cells, which allows for more simultaneous connections in a market than there are available frequencies.

In each cell, at least one channel, the setup channel, is responsible for the setup and control of calls. As soon as a cellular telephone is turned on, the telephone locks onto the setup channel and transmits basic telephone identity (ID) information. The cellular telephone company accepts the telephone ID information and identifies the particular phone (and user). Now the cellular telephone network knows in which cell the user is located. Every so many seconds, the cellular telephone retransmits its ID information, just in case the telephone is actually moving.

Figure 3-17
*A cellular telephone
tower*

Figure 3-18
Cell phone towers disguised as trees

In earlier generation cell phone systems, a cellular telephone company could tell only the cell in which you were located. It could not determine exactly where in the cell you were. Shortly after 2000, the FCC asked the cellular telephone companies to devise a system in which the exact location of a cellular telephone could be determined. One of the ideas behind this feature was that it would enable emergency services to locate a cell phone user should the user request a 911 service. The FCC gave the cell phone companies until the end of 2005 to have 95% of their cell phones enabled with a GPS chip so that under emergency situations, the cell phone company could locate the cell phone and user within 100 feet. Needless to say, not all the cell phone companies could meet this requirement and thus were fined. Today, all cell phone companies can offer this service but it depends upon how modern your cell phone is.

A cellular telephone, in attempting to place a call, transfers the dialed telephone number along with any other identification, such as the cellular telephone's ID, to the CTSO via the setup channel. The user's account is checked for validity, and if the telephone bill has been paid, the CTSO assigns a channel to that connection. The cellular telephone then releases the setup channel, seizes the assigned channel, and proceeds to place the telephone call.

What happens when someone tries to call a cellular telephone? Because the cellular telephone company knows (if the telephone is turned on) which cell a cellular telephone is in, the cellular telephone company transmits the cellular telephone's ID in that one cell. When the ID is recognized by the cellular telephone, the cellular telephone tries to seize the local cell's setup channel. When the setup channel is seized, the cellular telephone sends a signal to the CTSO, the CTSO verifies the cellular ID number, a channel is assigned to the cellular telephone, and the incoming call is connected.

Currently, several cellular telephone technologies are in operation in the United States. Advanced Mobile Phone Service (AMPS) was the first-generation cellular telephone system; it covered almost all of North America and was found in more than 35 other countries. AMPS used frequency division multiplexing technology (discussed in detail in Chapter Five), which operates like television transmissions. AMPS was the cellular equivalent of the "plain old telephone system" (POTS). All U.S. cellular companies no longer offer AMPS service. Digital-Advanced Mobile Phone Service (D-AMPS) is the newer, digital equivalent of analog cellular telephone service. It uses time division multiplexing technology (also discussed in Chapter Five) in addition to frequency division multiplexing and provides greater signal clarity and security than AMPS. Because D-AMPS starts with frequency division multiplexing and then adds time division multiplexing techniques, analog cellular telephone systems can be upgraded to D-AMPS—an upgrade that increases their signal clarity, security features, the number of special services offered, and the number of available channels per cell. Most cellular providers have upgraded their AMPS cellular telephones to D-AMPS service in order to compete with newer, all digital services.

The next category of cellular telephone technology is Personal Communications Services (PCS), which does not rely on older analog techniques. These systems are considered second-generation cellular telephones. PCS cell phones were approved by the FCC in 1993, and the first PCS cellular telephone system appeared in Washington, D.C., in November 1995. Since then, three competing (and incompatible) PCS technologies have emerged. The first PCS technology uses a form of time division multiplexing called Time Division Multiple Access (TDMA) technology to divide the available user channels by time, giving each transmitting cellular telephone a brief turn to transmit. The second PCS technology uses Code Division Multiple Access (CDMA) technology, which spreads the transmission of a cellular telephone signal over a wide range of frequencies, using mathematical values. CDMA is based on spread spectrum technology, which will be introduced in Chapter Twelve. The third PCS technology is Global System for Mobile (GSM) Communications and uses a different form of time division multiple access technology.

After PCS, the next generation of cell phone service was supposed to be the third generation, but for a number of reasons (most noticeably, the fact that it represented such a nominal improvement in technology relative to that of the previous generation), the cell phone industry dubbed it the 2.5 generation. The 2.5-generation cell phones are capable of receiving and transmitting digital data between a cellular telephone and an Internet service provider. Using a cellular telephone that has a small screen, a user can download small amounts of information from the Internet. Examples of the type of data that may be downloaded include stock prices, weather, sports scores, travel directions, and other low-volume, text-based information. Many cellular telephones also have built-in cameras, and users are able to send pictures from one cellular telephone to another.

To support the 2.5-generation's more bandwidth-intensive data streams of text and images, new protocols for transmitting data at faster speeds were developed. In particular, GSM networks were converted to General Packet Radio Service (GPRS), which can transmit data at 30 kbps to 40 kbps. CDMA networks were converted to an updated form of CDMA called CDMA2000 1xRTT (one-carrier Radio Transmission Technology). This newer technology is capable of transmitting data at 50 kbps to 75 kbps.

The current generation of cell phone technology got under way in early 2005 and does actually represent the third generation. In order to support the increasing bandwidth demand for uploading and downloading text and images via the cell phone, GPRS systems are being converted to Universal Mobile Telecommunications System (UMTS) technology. UMTS is capable of supporting downstream data rates of 220 kbps to 320 kbps. 1xRTT systems are being converted to a technology called 1xEV (1 x Enhanced Version)—specifically, a version of 1xEV called Evolution Data Only (EV-DO). EV-DO systems are capable of supporting downstream data rates of 300 kbps to 500 kbps. Fourth generation systems, which might start appearing in 2010, may be based on either LTE (Long Term Evolution) technology or Wi-MAX (discussed shortly). Confused? You are not alone.

Details ▶

Channel Division Amongst Cells

To gain a better understanding of how frequencies are divided amongst the cells of a cellular phone system, let us take a look at a simple example: the first-generation cell phone system. Both AMPS and D-AMPS cellular telephone systems allocated their channels using frequency ranges within the 800–900-Megahertz (MHz) spectrum. To be more precise, the 824–849-MHz range was used for receiving signals from cellular telephones (the uplink), while the 869–894-MHz range was used for transmitting signals to cellular telephones (the downlink). Within a metropolitan area, these two bands of frequencies allowed for approximately 50 MHz in which a

signal can be transmitted. These bands of frequencies were further divided into 30-kHz sub-bands called channels. This division of the spectrum into sub-band channels was achieved through Frequency Division Multiple Access (FDMA), whereby each channel was assigned (as with television and radio) a different set of frequencies on which to transmit.

A total of 1666 channels was available for signal transmission in a metropolitan area (50 MHz divided by 30 kHz per channel yields 1666 channels). To carry on a two-way conversation on a cellular phone, two channels are required—one for the uplink and one for the downlink. With every conversation requiring two channels, 833 (1666 channels divided in half) two-way connections were available in a metropolitan area. Furthermore, the FCC allowed up to two competing carriers to offer AMPS mobile telephone service within any given metropolitan area. Thus, 416 connections per carrier (833 connections divided by 2 carriers) per metropolitan area were available for use. Finally, these 416 connections were divided among all of the cells in a metropolitan area. Thus, a metropolitan area like New York City could have two cellular telephone companies, each offering *only* 416 concurrent telephone calls for the entire city. Fortunately, there is one more crucial step. Recall that the cells in an area form a honeycomb pattern (it is a nice honeycomb pattern on paper, but not so neat in real life) and that sets of frequencies can be reused. Because only the frequencies in adjacent cells are unique, the cellular telephone companies clump seven cells together.

Thus, only seven sets of frequencies are needed; the next clump of seven cells can reuse these seven sets of frequencies. This resulted in 59 two-way connections per cell (416 divided by 7) per carrier being available in a metropolitan area. These 59 connections were further reduced by the fact that a few channels in each cell were used for call setup.

Although 59 available AMPS connections per cell does not seem like a large number compared to the total number of cellular telephone users in an area, keep the following facts in mind:

- The cell phone users within a cell are not all using their telephones concurrently.
- Cells can be as small as one-half mile across.
- Multiple service providers existed for AMPS and D-AMPS systems.

Infrared transmissions

Infrared transmission is a special form of radio transmission that uses a focused ray of light in the infrared frequency range (10^{12}–10^{14} MHz). Working much like the remote control devices used to operate television sets, this focused ray of infrared information is sent from transmitter to receiver over a line-of-sight transmission. Usually these devices are only about three to ten feet (one to three meters) apart, but infrared systems that can transmit up to one and one-half miles do exist.

Infrared transmission systems are often associated with laptop computers, handheld computers, peripheral devices such as printers and fax machines, digital cameras, and even children's handheld electronic games. Infrared transmission works well in the following activities:

- Transmitting a document from your laptop computer to a printer or a modem
- Exchanging small files such as business cards between handheld computers
- Synchronizing electronic telephone books and schedulers
- Retrieving bank records from 24-hour automatic teller machines by walking up to the machine and pointing a handheld device such as a PDA at the terminal

In each of these examples, the transmitter and receiver are within the same room or a short distance apart, and data transfer rates are typically no faster than 4 Mbps. Infrared systems that can transfer data at speeds up to 16 Mbps exist, and even higher speeds are in development.

Despite the Infrared Data Association (IrDA) leading the charge in standardizing infrared technology and in incorporating infrared into many application areas, experts are beginning to question whether infrared is going to expand much beyond current implementations. In fact, a relatively new technology that we will examine later, Bluetooth, has the potential to replace infrared for short-distance wireless connections.

Broadband wireless systems

A broadband wireless system, also known as the wireless local loop or fixed-point wireless, is one of the latest techniques for delivering Internet services into homes and businesses. These systems bypass the telephone company's local loop (the last stretch of telephone line between the telephone central office and the home or business) by transmitting voice, data, and video over very high radio frequencies. As Figure 3-19 shows, the broadband wireless service provider (broadband switching center) receives broadband transmissions from either a satellite system or a high-speed Internet connection. These broadband transmissions are then sent to the one or more local transmission antennas (base stations). Businesses and homes then receive these transmissions with a receiver dish, which converts the signals into ones appropriate for a computer or computer network.

Figure 3-19
Broadband wireless configuration in a metropolitan area

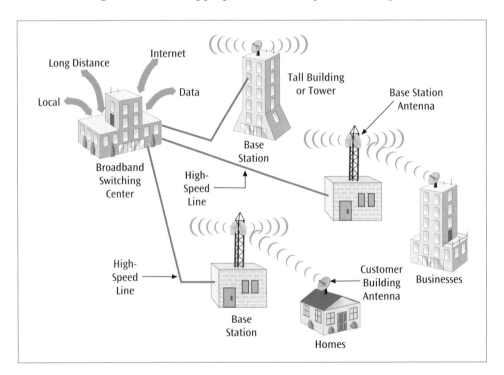

Two broadband wireless transmission technologies were introduced at the turn of the twenty-first century—Local Multipoint Distribution Service (LMDS) and Multichannel Multipoint Distribution Service (MMDS), but both seem to have been dropped by the wayside in recent years. In their place is a new technology: WiMAX. WiMAX is a broadband wireless transmission technology that is based upon a series of IEEE standards. For example, a WiMAX technology that was designed to deliver high-speed Internet access to homes and small businesses and thus compete against DSL and cable modems is called IEEE 802.16a. IEEE 802.16a operates in the 2–11-GHz spectrum, provides line-of-sight and non-line-of-sight-connections, and can transfer data up to 70 Mbps for 30 miles. Another WiMAX technology, IEEE 802.16c was designed to operate in the 10–66-GHz spectrum and is only line-of-sight. IEEE 802.16d (and sometimes called IEEE 802.16–2004) consolidates the revisions of 802.16a and 802.16c into one standard. IEEE 802.16e is a revision of the 802.16 standard that provides a high-speed connection for slowly moving devices, such as when someone is walking or driving down a residential street and using a

WiMAX-enabled cell phone. Because 802.16 was originally designed for fixed devices, however, the 802.16e revision is not an optimal standard for moving devices in general. More preferable is the newly released (in 2008) IEEE 802.20 standard, which was created specifically for high-speed mobile devices—in this case, the "high-speed" refers to both the data rate and the rate of movement the device can experience while connected. IEEE 802.20 can operate with devices moving up to 180 miles per hour, effectively performs hand-offs (the passing from one set of transmission frequencies to the next), and can transmit data in the 100s of kbps to over a million bits per second range. Some industry experts predict that the IEEE 802.20 standard may lead to the future growth of a fourth-generation cell phone standard. The 802.20 standard is one we will have to keep our eye on in the next couple of years.

Bluetooth

The Bluetooth protocol—named after the Viking crusader Harald Bluetooth, who unified Denmark and Norway in the tenth century—is a wireless technology that uses low-power, short-range radio frequencies to communicate between two or more devices. More precisely, Bluetooth uses the 2.45-GHz ISM (Industrial, Scientific, Medical) band and is typically limited to distances between 10 cm and 10 meters (corresponding to about 4 inches and 30 feet). Unlike infrared, Bluetooth is capable of transmitting through nonmetallic objects. Thus, a device that is transmitting Bluetooth signals can be carried in a pocket, purse, or briefcase. Furthermore, it is possible with Bluetooth to transfer data at reasonably high speeds. Also, an asymmetric connection that can transfer data at 57.6 kbps in one direction and 722 kbps in the opposite direction is available.

Bluetooth can also communicate among multiple devices. For example, consider an office environment with many computers, printers, fax machines, and photocopiers. With Bluetooth, each device can send signals to other devices or to a single point—for example, to indicate service instructions such as "out of paper" or "low toner." A small network like this with eight or fewer devices is called a piconet. Another term for a piconet is personal area network or PAN, which we introduced in Chapter One. Multiple piconets can be interconnected to form a scatternet.

The most interesting aspect of Bluetooth is the list of applications that benefit from such a short-range transmission technology. These applications include:

- ▶ Wireless transmission between a portable music player and a headset
- ▶ Transmissions between a personal digital assistant (PDA) and another computer
- ▶ Transmissions between peripheral devices and a computer
- ▶ Wireless transmissions between a PDA and an automobile, house, or the workplace

To appreciate the potential power of Bluetooth technology, consider the following, more descriptive examples: You can automatically synchronize all e-mail messages between your PDA and your desktop/laptop computer; as you approach your car, your PDA will tell the car to unlock its doors and change the radio to your favorite station; as you walk up to the front door of your house, your PDA will instruct your house to unlock the front door, turn on the lights, and turn on an entertainment system; and as you sit in a business meeting, your PDA/laptop will wirelessly transmit your slide presentation to a projector and your notes to each participant's PDA/laptop.

Despite large support, Bluetooth has been relatively slow in impacting the marketplace. Current Bluetooth technology is still having problems with getting multiple (more than two) devices to synchronize data with each other. The short transmission distance of 10 meters (30 feet) is also seen by many as a detriment. Although it is possible (under special conditions) to transmit Bluetooth signals up to 100 meters (328 feet), this will require more powerful batteries for the transmitting devices and could increase the problem of interference. Nonetheless, Bluetooth is a technology that certainly needs to be watched and understood.

Wireless local area networks

Even though local area networks will be discussed in detail in Chapter Seven, it may be helpful to introduce the wireless form of local area networks now, while we are discussing various types of wireless systems. The first wireless local area network standard was introduced in 1997 by IEEE and has the name IEEE 802.11. IEEE 802.11 is capable of supporting data rates up to 2 Mbps and allows wireless workstations up to roughly several hundred feet away to communicate with an access point. This access point is the connection into the wired portion of the local area network. In 1999, IEEE approved a new 11-Mbps protocol, IEEE 802.11b. This protocol is also known as wireless fidelity (Wi-Fi) and transmits data in the 2.4-GHz frequency range. Following 802.11b were two more protocols—802.11a and 802.11g. IEEE 802.11a transmits data at speeds up to 54 Mbps (a theoretical 54 Mbps but in reality about half that) using frequencies in the 5-GHz range. 802.11g also transmits data at speeds up to 54 Mbps (theoretical) but uses the same frequencies— 2.4 GHz—as 802.11b. Because 802.11b and 802.11g share the same frequency range, 802.11g is more attractive than 802.11a to those users who already have 802.11b installed and want to upgrade their system. Originally, 802.11a was called Wi-Fi5, but that term is no longer being used.

A fourth wireless LAN protocol that was recently approved at the end of 2009 is IEEE 802.11n. This standard is capable of supporting a 100-Mbps signal between wireless devices and uses multiple antennas to support multiple independent data streams. All these protocols—802.11a, 802.11b, 802.11g, and 802.11n—are now termed Wi-Fi.

Free space optics and ultra-wideband

Two additional wireless technologies worth mentioning are free space optics and ultra-wideband transmissions. Free space optics uses lasers, or, in some cases, infrared transmitting devices, to transmit data between two buildings over short distances, like across the street. Data transfer speeds with this technology can be as high as 45 Mbps, and higher speeds are possible in the future. One of the major problems, however, with free space optics is fog. Lasers lose their strength when transmitting through fog. Thus, if the fog is thick, transmission distances can be cut down to less than 50 meters (150 feet).

The second wireless medium is ultra-wideband. Ultra-wideband systems transmit data over a wide range of frequencies rather than limiting transmissions to a narrow, fixed band of frequencies. The interesting aspect about transmitting over a wide range of frequencies is that some of those frequencies are used by other sources, such as cellular phone systems. So do ultra-wideband signals interfere with signals from these other sources? Proponents for ultra-wideband claim that even though a wide range of frequencies is used, ultra-wideband transmits at power levels low enough that other sources should not be affected. Opponents of ultra-wideband argue that this is not correct—that ultra-wideband transmissions *do* affect other sources and thus should be carefully controlled. Despite this interference issue, ultra-wideband is capable of supporting speeds up to 100 Mbps, but over small distances such as those found in wireless local area networks.

ZigBee

ZigBee is a relatively new wireless technology supported by the IEEE 802.15.4 standard. It has been designed for data transmission between smaller, often embedded, devices that require low data transfer rates (20–250 KBps) and corresponding low power consumption. For example, the ZigBee Alliance states that ZigBee is ideal for applications such as home and building automation (heating, cooling, security, lighting, and smoke and CO detectors), industrial control, automatic meter reading, and medical sensing and monitoring. It operates in the industrial, scientific, and medical (ISM) radio bands and requires very small software support and very little power. In fact, power consumption is so low that some suppliers claim that their ZigBee-equipped devices will last multiple years on an original battery.

An interesting aspect of ZigBee is how devices are able to keep power consumption low. The first technique employed is mesh communications. Using mesh communications, all devices do not transmit directly to a single receiver. Instead, each device transmits its signal to the next closest ZigBee device, which in turn passes this signal on to the next device. Eventually, the destination receiver will be reached and an action will take place. Because transmission distances are typically shorter in a mesh configuration, less power is needed to transmit the signal.

Secondly, ZigBee-enabled devices do not need to constantly communicate with other devices. When not transmitting a signal to a receiver, the device can put itself to sleep. When someone or something activates a device with ZigBee, the ZigBee circuit wakes up, transmits the signal, and then goes back to sleep.

As many people confuse ZigBee with Bluetooth, it will be interesting to see if both survive. Both should because each one is targeted for a different application area—Bluetooth is best at replacing cables for short distances, while ZigBee will be good at sending low-speed signals over short to medium distances.

Table 3-4 summarizes the wireless media discussed here, including the typical use, maximum data transfer rate, maximum transmission range, advantages, and disadvantages of each.

Table 3-4

Summary of wireless media

Type of Wireless Medium	Typical Use	Maximum Data Transfer Rate	Maximum Trans-mission Range	Advantages	Disadvantages
Terrestrial Microwave	Long-haul tele-communications, building to building	100s-Mbps	20–30 miles	Reliable, high speed, high volume	Long haul, expensive to implement, line-of-sight
Satellite LEO	Communications such as e-mail, paging, worldwide mobile phone network, spying, remote sensing, video conferencing	100s-Mbps	Depends on number of satellites	High-speed transfers, very wide distance, some applications inexpensive	Some applications expensive, interference
Satellite MEO	GPS-style surface navigation systems	100s-Mbps	Depends on number of satellites	High-speed transfers, wide distance	Expensive to lease, some interference
Satellite GEO	Signal relays for cable and direct television	100s-Mbps	One-third the Earth's circumference (8000 miles)	Very long distance, high speed, and high volume	Expensive to lease, some interference
Satellite HEO	Global surveillance, scientific applications	100s-Mbps	Variable	Variability of distance	Expensive
Cellular (AMPS and D-AMPS)	Cellular telephones	19.2 kbps	Each cell: 0.5–50 mile radius, but nationwide coverage	Widespread, inexpensive applications	Noise
PCS	Cellular telephones	9.6 kbps	Each cell: 0.5–25 mile radius	Digital, low noise	Slow data rates
GPRS, 1xRTT	Cellular telephones	30–75 kbps	Each cell: 0.5–25 mile radius	Digital, low noise	Slow data rates
UMTS	Cellular telephones	320 kbps	Each cell: 0.5–25 mile radius	Digital, low noise	
EV-DO	Cellular telephones	500 kbps	Each cell: 0.5–25 mile radius	Digital, low noise	
Infrared	Short-distance data transfer	16 Mbps	1.5 miles	Fast, inexpensive, secure	Short distances, line of sight
WiMAX	Wireless Internet access	30 Mbps	30 miles	High speed	
Bluetooth	Short-distance transfer	722 kbps	30 feet (10 meters)	Universal protocol	Limited distances

Table 3-4
Summary of wireless media (continued)

Type of Wireless Medium	Typical Use	Maximum Data Transfer Rate	Maximum Transmission Range	Advantages	Disadvantages
Wireless LANs	Local area networks	100 Mbps	<328 feet (<100 meters)	Relative ease of use	Several standards
Free Space Optics	Short-distance, high-speed transfers	45 Mbps	1000s feet (100s meters)	High speed	Line of sight, affected by fog
Ultra-wideband	Short-distance, high-speed transfers	100 Mbps	<328 feet (100 meters)	High speed, not restricted to fixed frequencies	May interfere with other sources
ZigBee	Short-to-medium distance, low-speed transfers	250 KBps	Unlimited distance (mesh)	Low power	Low transfer speeds

Now that you are familiar with the categories and types of wireless and conducted media available, you need to consider other criteria in order to make a decision about which media to select.

Media Selection Criteria

When designing or updating a computer network, the selection of one type of medium over another is an important issue. Computer-network-based projects have performed poorly and possibly even failed as the result of a poor decision about the appropriate type of medium. Furthermore, it is worth noting that the purchase price and installation costs for a particular medium are often the largest costs associated with computer networks. Once the time and money have been spent installing a particular medium, a business must use the chosen medium for a number of years to recover these initial costs. In short, the choice of medium should not be taken lightly. Assuming you have the option of choosing a medium, you should consider many media selection criteria before making that final choice or choices. The principal factors you should consider in your decision include cost, right-of-way, speed, expandability and distance, environment, and security. The following discussion will consider these factors in relation to twisted pair wire, coaxial cable, fiber-optic cable, terrestrial microwave, satellite microwave, cellular systems, infrared, WiMAX, Wi-Fi, and Bluetooth.

Cost

Costs are associated with all types of media, and there are different types of costs. For example, twisted pair cable is generally less expensive than both fiber-optic cable and coaxial cable. To make a cost-effective selection decision, however, it is necessary to consider more than just the initial cost of the cable—you must also consider the cost of the supporting devices that originate and terminate the cables, the installation cost, and the price/performance ratio. For example, twisted pair wire is typically the least expensive medium to purchase. Each wire usually ends with a small modular jack similar to the jack that connects a telephone to a telephone wall jack. These modular jacks are mostly plastic and very inexpensive, costing only pennies each. Installation of twisted pair is typically straightforward, but it can be quite costly, depending upon the particular installation environment and who does the installation.

In comparison, coaxial cable is often a more expensive cable to purchase, sometimes costing more than $1 per foot. The connectors that terminate coaxial cable are mostly metal and a little more costly than twisted pair connectors. It is also a little more difficult to install coaxial cable.

Fiber-optic cable, if purchased with two conductors and in bulk, is more expensive than twisted pair wire but can be comparable in cost to coaxial cable. The connectors that terminate fiber-optic cables are, as previously stated, more expensive than those of either twisted pair or coaxial cable. More importantly, if you need to connect a fiber-optic cable to a non-fiber-optic cable or device, the cost increase is even more dramatic, because you must convert light pulses to electric signals and vice versa. Although the installation costs of the three conducted media are not always significantly different, the ones associated with fiber-optic cable are often the most expensive. Consider, however, the price/performance ratio of fiber-optic cable compared to that of both twisted pair and coaxial cable. Although fiber may be a little more expensive to purchase and install, it has the greatest transmission capabilities with the least amount of noise. Which is more important: saving money on the purchase of a medium, or having a medium that is capable of very high transmission speeds?

Terrestrial microwave systems spanning great distances and satellite systems are expensive media, considering the cost of microwave towers and satellites. Very few businesses, however, install their own towers and launch their own satellites. Instead, most companies lease time from other companies that specialize in microwave systems. Considering the lease option, and the fact that microwave systems can send fast data streams, some companies may find a microwave solution that is less expensive than a wire-based solution. Privately owned microwave systems, which are mounted on company buildings, are significantly less expensive than wide-area terrestrial microwave and satellite systems. Considering the cost of installing cables (assuming that you even can install them), short-distance microwave systems may, in certain situations, also offer a reasonable alternative to installing cable or leasing time from a service provider.

In many cases, it is not possible to install your own cabling. For example, if you have two buildings that are separated by a public street, you may not have any way to connect a cable from one building to the other. Even if there were an overhead or underground passage through which a cable could be drawn, do you own the passage? In other words, you may not have the right-of-way to install the cables. If the passage is not yours, will the owner of the passage allow you to run cables through it? And if so, what will be the cost to you? If it is not possible or feasible to install your own cables, you might consider some form of wireless transmission. Or, you might consider contacting a local telecommunications service provider (such as the local telephone company) to see what options are available (more on this in Chapter Eleven).

Each type of medium has the additional cost of maintenance. Will a certain type of wire last for x years when subjected to a particular environment? This is a difficult question to answer, but it should be asked of the company supplying the medium. Whereas it is easy to browse catalogs and learn the initial costs of a particular type of cable, it is more difficult to determine the maintenance costs two, five, or ten years down the road. Too often, overfocusing on initial cost inhibits decision makers from taking long-term maintenance cost into account and, therefore, considering better types of media.

Speed

To evaluate media properly, you need to consider two types of speed: data transmission speed and propagation speed. Data transmission speed is the number of bits per second that can be transmitted. The maximum bits per second for a particular medium depends proportionally on the effective bandwidth of that medium, the distance the data must travel, and the environment through which the medium must pass (noise). If one of the requirements of the network you are designing is a minimally acceptable data transmission speed, then the medium you choose has to support that speed. This issue might sound trivial, but it is complicated by the difficulty of predicting network growth. Although a chosen medium may support a particular level of traffic at the moment, the medium may not be able to support a future addition of new users or new applications. Thus, careful planning for future growth is necessary for proper network support (we will take a look at this issue in more detail in Chapter Thirteen). Another important issue to consider is that even though a particular technology advertises a data transmission speed, that may not be the actual throughput of data. For example, IEEE 802.11a and 802.11g wireless LANs offer data transmission speeds of 54 Mbps, but in reality the users experience data transfer rates that are roughly one-half of this value. This, as we shall see in later chapters, is caused by noise, interference, and weak signals.

Propagation speed is the speed at which a signal moves through a medium. The propagation speed of fiber-optic media is very near the speed of light, and wireless media actually do propagate at the speed of light, which is 186,000 miles per second (3×10^8 meters per second). For electrically conducted media (twisted pair and coaxial cable), propagation speed is approximately two-thirds the speed of light, or 124,000 miles per second (2×10^8 meters per second). Although these speeds seem fast enough for most applications, keep in mind that the time required to send a signal to a satellite in an outer Earth orbit and back—the propagation delay—is approximately 0.25 to 0.75 seconds, depending upon the actual distance to the satellite and the number of devices through which the signals must pass. If you are transferring data from one side of the state to the other, you might want to consider a medium with a lower propagation delay.

Expandability and distance

Certain media lend themselves more easily to expansion. Twisted pair cable is easier to expand than either coaxial cable or fiber-optic cable, and coaxial cable is easier to expand than fiber-optic cable. Coaxial cable is more difficult to expand than twisted pair because of the types of connectors on the ends of the cable. Fiber-optic connectors are even more elaborate, and joining two pieces of fiber-optic cable requires practice and the proper set of tools.

Another expandability-related consideration is that most forms of twisted pair can operate for only 100 meters (328 feet) before the signal requires regeneration.

Some forms of coaxial cable systems can run for longer lengths (miles), and fiber-optic cable can extend for many miles before regeneration of the signal is necessary.

Privately owned terrestrial microwave has a high transmission rate, but if the microwave dishes are mounted on corporate buildings, the buildings can be no farther than 20 to 30 miles apart. Furthermore, this setup will work only if the buildings are tall and no obstructions, such as other buildings, are in the way. Cellular systems are widespread and expanding continuously. New technologies such as third-generation cell phones and WiMAX show promise for expandability. Incidentally, when it comes to wireless technologies, it is important not to confuse the various application areas. Bluetooth and infrared are strictly for short distances. Wi-Fi is designed for local area networks and has a maximum range of a few thousand feet (corresponding to a few hundred meters). WiMAX is designed for high-speed Internet access at distances up to 30 miles. Third-generation cell phones such as UMTS are capable of transferring data in the 100s of kbps and will eventually span the country.

When considering expandability, do not forget the right-of-way issue. If you are trying to run a cable across land that does not belong to you, you have to obtain permission from the landowner. Sometimes that permission may not be granted, and sometimes you may get the permission but have to pay a recurring fee to the landowner.

If you expect to create a system that may expand in the future, it is worthwhile to consider using a medium that can expand at a reasonable cost. Note, however, that the expansion of a system is, many times, determined more by the design of the system and the use of the supporting electronic equipment than by the selection of a type of medium.

Environment

Another factor that must be considered in the media selection process is the environment. Many types of environments are hazardous to certain media. Industrial environments with heavy machinery produce electromagnetic radiation that can interfere with improperly shielded cables. If your cabling may be traveling through an electromagnetically noisy environment, you should consider using shielded cable or fiber-optic cable.

Wireless transmission also can be disrupted by electromagnetic noise and interference from other transmissions. Sunspots, although they do not happen often, can be disruptive to satellite transmissions. Because so many people now rely on wireless services for voice, data, and paging, newspapers often warn the public when sunspot activity is expected to be high. It is worth noting that microwave transmission and free space optics can be hampered by bad weather. Before selecting a medium, it is important to know the medium's intended environment and be aware of how this environment may influence or interfere with transmissions.

Security

If data must be secure during transmission, it is important that the medium not be easy to tap. All conducted media, except fiber-optic cable, can be wiretapped easily, meaning someone can "listen" to the electromagnetic signal traveling through the wire. Wireless communications also can be intercepted, but it is much easier to wiretap a broadcast wireless transmission such as a wireless local area network than a narrow-beam wireless system such as microwave. Fortunately, there are means of improving the data security of both conducted and wireless media. Encryption and decryption software can be used with conducted media. Spread spectrum technology can be applied to wireless communications, making them virtually impervious to interception. Encryption, decryption, and spread spectrum technology will be discussed in detail in Chapter Twelve.

Now that we have covered the different types of media and selection criteria, let us turn our attention to how conducted and wireless media work in a network.

Conducted Media In Action: Two Examples ▶

Let us consider the wiring for a local area network. Figure 3-20 shows a common situation where a microcomputer workstation connected to a local area network must first connect to a device such as a switch. A switch is a device that connects multiple workstations and passes the transmission signal from any workstation on to any other workstations (Chapters Seven and Eight examine switches in much more detail). In typical installations, it is quite unlikely that the cable leaving the workstation runs directly to the switch. Instead, the cable that leaves the back of the microcomputer first connects to a wall jack in the room where the microcomputer is located. This wall jack is a passive device, a simple connection point between two runs of cable that does not regenerate the signal on the cable.

Figure 3-20
Example of a wiring situation involving a workstation and a local area network

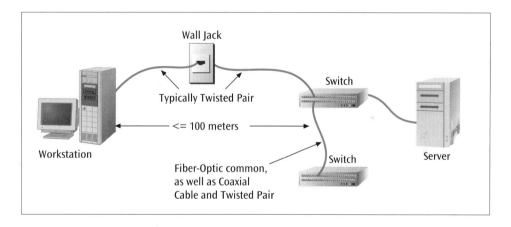

In order to select the proper medium for this connection, you must take into account two main issues: cable distance and data rate. To account for cable distance in this situation, you must consider the total distance from the back of the microcomputer to the switch. If this distance is less than 100 meters (328 feet), you can consider using twisted pair to connect the microcomputer to the switch. If the cable does not pass through a noisy environment, then you can consider using unshielded twisted pair, the least expensive and easiest cable with which to work. If you can assume that the data rate of the connection will not exceed 1000 Mbps, it should be possible to support the connection using four-pairs of Category 5e, Category 6, or Category 7.

If, however, the data rate is higher than 1000 Mbps, you may want to consider alternatives to twisted pair. Fiber-optic cable is a good choice for data transmission rates over 1000 Mbps, but it will cost more than a twisted pair connection. Also, because fiber-optic cable is a one-way cable and the data flow between a workstation and a switch is two-way, two fiber-optic cables would be necessary.

In this case, because the data rate between the microcomputer workstation and switch is 100 Mbps or less, Category 5, 5e, or 6 unshielded twisted pair would be an acceptable choice (Category 7 would be an unnecessary expense at this point). Fiber-optic cable also would be a reasonable choice, as it would be adaptable to higher transmission speeds in the future.

What about cabling the connection between the switch and the next point in the local area network (usually another switch)? If the distance between the two switches exceeds the 100 meter length or if the cable travels through a noisy environment, such as a heating and cooling mechanical room, a cable other than twisted pair should be considered. Under these circumstances, you would consider either a very good coaxial cable or, even better yet, a fiber-optic cable for connecting the two switches. Why should you consider the more expensive fiber-optic cable? The price difference between coaxial cable and fiber-optic cable is much smaller than the performance differences. Therefore, it makes sense to go with the best cabling and install fiber-optic cable.

In a second example, let us consider the scenario where a company has two buildings approximately one mile apart. The company wishes to transmit data between the two buildings on a fairly routine basis at speeds up to 100 Mbps. What would you recommend as the best type of interconnecting medium? Before we even consider any form of conducted media, we need to ask the question: Is the property between the two buildings owned by the company? Let us assume it is not (the likely case). The company might want to consider either using a form of wireless transmission, such as terrestrial microwave or free space optics (assuming there are no intervening structures), or contacting a telephone company and inquiring about whether a 100 Mbps data transmission service is available that will interconnect the two buildings.

If the property between the two buildings is company owned, the choice of a conducted medium is still difficult. Which conducting medium is capable of supporting 100 Mbps for one mile? Fiber-optic cable will meet those requirements, but how is the company going to install the cable—underground in some form of pipe or tunnel, or overhead on some type of telephone pole? Both solutions would be quite expensive, unless an infrastructure to support the new fiber-optic cables already exists. Installing wire within a building is easier than trying to install it between buildings, but neither task is simple. Because installing a medium is expensive and long-term, careful planning and decision making are essential before the project can be started.

Let us consider each of the media selection criteria as applied to this two-building solution:

▸ Cost—Fiber-optic is the most expensive conducted medium option, but is worth the cost, given the requirements of the problem. Some form of wireless might be a good alternative, along with contacting the local telephone company for its suggested solutions.

▸ Speed—Fiber-optic, twisted pair, and coaxial cable will support the necessary speed requirements, but so would microwave and free space optics.

▸ Expandability and Distance—The distance of one mile eliminates twisted pair and coaxial cable from consideration. Right-of-way is definitely an issue in this case. If you do not have right-of-way, you cannot install your own cable.

▸ Environment—Fiber-optic cable should not be affected by the environment. If a wireless solution is applied, line-of-sight and weather could be two serious impediments.

▸ Security—A fiber-optic system should be secure from wiretapping.

In conclusion, running wire between two buildings is often not possible, due to right-of-way issues or maximum distance restrictions. When conducted media are not viable, a number of wireless choices should be considered. Even if conducted media are viable, a wireless solution may prove more economical in the long run. Many times the solution to interconnecting multiple businesses involves the telephone company, as we will see in a later chapter.

Wireless Media In Action: Three Examples ▶

In your home you have a couple of computers, each in a different room. Each computer has a fairly inexpensive ink-jet printer attached to it, and only one computer has access to the Internet. You would like to hook up both computers to the Internet and purchase a good laser printer to be shared by both. For this, you will need to interconnect the computers, but pulling wires through your walls and floors does not sound like an attractive option. What about wireless? You can purchase wireless network interface cards that use one of the IEEE 802.11 protocols and create a wireless local area network. IEEE 802.11b products transmit data at speeds up to 11 Mbps and are very common, and thus would keep your costs reasonable. The newer 802.11g protocol can support higher data transfer speeds, but, being newer, it will cost a little more than 802.11b.

You might also want to consider replacing your existing devices, such as printers, modems, and scanners, with Bluetooth-enabled devices. By using Bluetooth, you will be able to reduce your reliance on running cables from workstations to peripherals and thus increase your flexibility in locating equipment. We will discuss the topic of home local area networks (often called Small Office/Home Office, or SOHO, installations) in more detail in a later chapter.

For a second example, let us consider DataMining Corporation, a large organization that has a main office in Chicago and a second office in Los Angeles. DataMining collects data from grocery stores from every purchase made by every customer. Using this data, the company extracts spending trends and sells this information to other businesses that market salable goods. The data is collected at the Chicago office and transmitted to the Los Angeles office, where it is stored and later retrieved. Thus, there is a need to transmit large amounts of data between the two sites on a continuous basis.

Currently, DataMining is leasing a telephone service between Chicago and Los Angeles, but the telephone bills are high. The company is trying to reduce costs and is considering alternatives to leased telephone services. Other forms of telephone service are available, but these will not be introduced until Chapter Eleven. Let us simply say that DataMining has examined these other forms of service and found them to be expensive—hence the need to consider other alternatives. DataMining has discovered that a number of companies can provide various levels of satellite communication services. For example, Hughes Network Systems can provide local area network internetworking, multimedia image transfers, interactive voice connections, interactive and batch data transfers, and broadcast video and data communications services. Because DataMining is primarily interested in two-way data transmissions, it is considering a two-way data communications service offered by Hughes through a VSAT satellite system. As shown in Figure 3-21, this two-way data service would require the installation of an individual Earth station satellite dish at each of DataMining's corporate locations.

Figure 3-21
*VSAT satellite solution
for DataMining
Corporation*

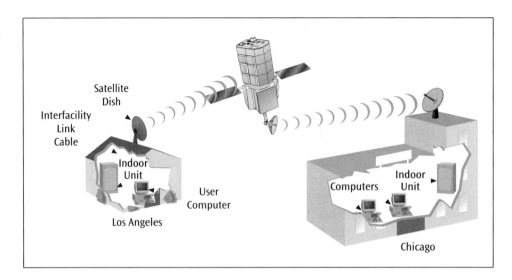

Each individual Earth station is composed of two pieces: an indoor unit and an outdoor unit. The outdoor unit consists of the satellite dish and is usually mounted on the roof of the building. The size of the satellite dish depends on the data rates used and the satellite coverage required. The outdoor unit is connected to the indoor unit via a single interfacility link cable. The indoor unit has one or more ports to which the company's data-processing equipment can be connected.

Maintenance and support for this VSAT service are provided by the satellite company on a 24-hours-per-day, 7-days-a-week basis, and include equipment configuration, system status reporting, bandwidth allocation, downloading of any necessary software, and the dispatching of field personnel if necessary.

DataMining Corporation has decided to install the VSAT satellite system. To summarize, let us consider each of the media selection criteria as applied to the VSAT solution:

▶ Cost—The VSAT system is relatively expensive, but it delivers a high data transfer rate with high reliability.

▶ Speed—The VSAT system can support the required data transfer rates of DataMining Corporation.

▶ Expandability and Distance—The satellite system can easily extend from Chicago to Los Angeles. Right-of-way is not an issue in this case.

▶ Environment—Satellite systems can be disrupted by strong electromagnetic forces, which could be a problem. If DataMining cannot tolerate any disruption of service, they might want to consider installing a backup system in case the VSAT system momentarily fails.

▶ Security—VSAT satellite systems are difficult to intercept because the transmission beam sent between ground stations and the satellite is small. Additionally, the data stream can be encrypted.

A second company, the American Insurance Company, has two offices, both located in Peoria, Illinois. The first office collects all the premium payments, and the second office contains the main data-processing equipment. American Insurance needs to transfer the collected premium information to the data-processing center on a nightly basis. The two offices are approximately two miles apart, but it is not possible to run the company's own cable over public property and private personal property. Some form of telephone system might provide a reasonable solution, but American Insurance is interested in investing in its own system and would like to avoid recurring monthly telephone charges, because these would, at some point, exceed the cost of installing its own system.

ProNet is a company that offers private terrestrial microwave systems that can transfer private voice, local area network data, video conferencing, and high-resolution images between remote sites up to 15 miles apart. Figure 3-22 shows a typical microwave communication setup as would be arranged between the two corporate offices of American Insurance.

Figure 3-22
Microwave communication between American Insurance's corporate buildings

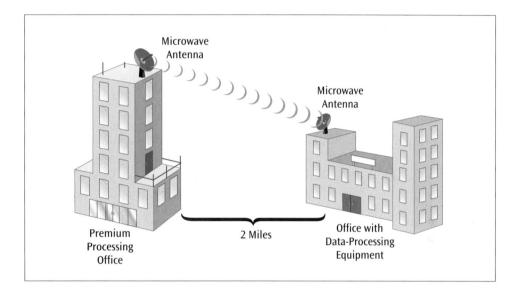

Let us consider each of the media selection criteria as applied to the ProNet terrestrial microwave solution:

▶ Cost—The ProNet system is expensive at first when the equipment is purchased, but after that American has to pay only for maintenance.

▶ Speed—The ProNet system can support the required data transfer rates of American Insurance.

▶ Expandability and Distance—The terrestrial microwave system can transmit up to 15 miles. The two corporate buildings are two miles apart. Right-of-way is not an issue in this case.

▶ Environment—Microwave systems can be disrupted by strong electromagnetic forces and inclement weather. ProNet, however, states that their system is unaffected by fog or snow and that it delivers a service reliability and availability of 99.97 percent.

▶ Security—Terrestrial microwave systems can be intercepted, but the data stream can be encrypted.

American Insurance will seriously consider using ProNet's terrestrial microwave system, as it compares favorably to other systems and satisfies the company's goals of private ownership, high transfer rates, and low recurring costs.

◆ ◆

SUMMARY

▸ All data communication media can be divided into two basic categories: (1) physical or conducted media, such as wires, and (2) radiated or wireless media, such as satellite systems.

▸ The three types of conducted media are twisted pair, coaxial cable, and fiber-optic cable.

▸ Twisted pair and coaxial cable are both metal wires and are subject to electromagnetic interference. Fiber-optic cable is a glass wire and is impervious to electromagnetic interference; therefore, it experiences a lower noise level than twisted pair and coaxial cable.

▸ Fiber-optic cable has the best transmission speeds and long-distance performance of all conducted media, because (1) it has a lower noise level, and (2) light signals do not attenuate as quickly as electric signals.

▸ Several basic groups of wireless media exist: terrestrial microwave transmissions, satellite transmissions, cellular telephone systems, infrared transmissions, WiMAX, Bluetooth, Wi-Fi, free space optics, ultra-wideband, and ZigBee.

▸ Each of the wireless technologies is designed for specific applications. Data transfer speeds, transmission distances, advantages, and disadvantages must be considered for each.

▸ When trying to select a particular medium for an application, it helps to compare the different media using these five criteria: cost, speed, expandability and distance, right-of-way, environment, and security.

KEY TERMS

1xEV (1 x Enhanced Version)
62.5/125 cable
8.3/125 cable
Advanced Mobile Phone Service (AMPS)
backbone
baseband coaxial
Bluetooth
braided coaxial cable
broadband coaxial
broadband wireless system
Category 1
Category 2
Category 3
Category 4
Category 5
Category 5e
Category 6
Category 7
Category 1–7 (CAT 1–7) twisted pair
CDMA2000 1xRTT
coaxial cable
Code Division Multiple Access (CDMA)
crosstalk

data transmission speed
Digital-Advanced Mobile Phone Service (D-AMPS)
downlink
Evolution Data Only (EV-DO)
fiber-optic cable
free space optics
General Packet Radio Service (GPRS)
geosynchronous-Earth-orbit (GEO) satellite
global positioning system (GPS)
Global System for Mobile (GSM) Communications
highly elliptical orbit (HEO) satellite
infrared transmission
line-of-sight transmission
low-Earth-orbit (LEO) satellite
media selection criteria
    cost
    speed
    expandability and distance
    environment
    security
middle-Earth-orbit (MEO) satellite

mobile service area (MSA)
multimode transmission
passive device
Personal Communications Services (PCS)
photo diode
photo receptor
photonic fiber
piconet
propagation delay
propagation speed
reflection
refraction
repeater
right-of-way
satellite microwave
scatternet
shielded twisted pair (STP)
single-mode transmission
single-stranded coaxial cable
terrestrial microwave
thick coaxial cable
thin coaxial cable
Time Division Multiple Access (TDMA)

twisted pair wire	•	unshielded twisted pair (UTP)	•	WiMAX
ultra-wideband	•	uplink	•	wireless fidelity (Wi-Fi)
Universal Mobile Telecommunications	•	Very Small Aperture Terminal (VSAT)	•	ZigBee
System (UMTS)	•		•	

REVIEW QUESTIONS

1. Why is twisted pair wire called twisted pair?
2. How does crosstalk occur in twisted pair wire?
3. For what purposes are Category 1, 2, 3, 4, 5, 5e, 6, and 7 twisted pair wire used?
4. What are the advantages and disadvantages of shielded twisted pair?
5. What is the primary advantage of coaxial cable compared to twisted pair?
6. What is the difference between baseband coaxial and broadband coaxial cable?
7. Why is fiber-optic cable immune to electromagnetic interference?
8. What are the advantages and disadvantages of fiber-optic cable?
9. What is the difference between terrestrial microwave and satellite microwave?
10. What is an average distance for transmitting terrestrial microwave?
11. What kind of objects can interfere with terrestrial microwave transmissions?
12. List a few common applications for terrestrial microwave.
13. What are the four orbit levels for satellite systems?
14. List a few common application areas for each orbit level satellite system.
15. What is the sequence of events that happens when someone places a call from a cellular telephone?
16. What is the function of a cellular telephone switching office?
17. What is the primary difference between AMPS and D-AMPS cellular systems?
18. What is the primary difference between AMPS (or D-AMPS) cellular systems and the newer PCS mobile telephones?
19. What are the differences between the 2.5-generation cell phone services such as GPRS and 1xRTT and the newer UMTS, 1xEV, and EV-DO?
20. What is meant by line-of-sight?
21. What are the WiMAX protocols used for?
22. What is the advantage of IEEE 802.20 over IEEE 802.16e?
23. Infrared transmission can be used for which type of applications?
24. What are the main advantages and disadvantages of ZigBee?
25. Broadband wireless service supports what kind of applications?
26. What are the main advantages and disadvantages of Bluetooth?
27. List three possible application areas of Bluetooth.
28. What are the different wireless local area network protocols?
29. In what situation might we use free space optics?
30. What are the different types of costs of conducted media?
31. What is the difference between data transmission speed and propagation speed?
32. What is meant by right-of-way?

EXERCISES

1. Table 3-1 shows Category 1 wire transmitting a signal for 5–6 kilometers (3–4 miles) but Category 5e for only 100 meters (328 feet). Does this mean Category 1 is the best wire for long-distance transmissions? Explain.

2. List three different examples of crosstalk that do not involve wires and electric signals. (*Hint*: Look around you.)

3. What characteristics of Category 5/5e unshielded twisted pair make it the most commonly used conducted wire?

4. Can you transmit a video signal over twisted pair wire? Explain. Be sure to consider multiple scenarios.

5. The local cable TV company is considering removing all the coaxial cable and replacing it with fiber-optic cable. List the advantages and disadvantages of this plan.

6. The local cable TV company has changed its mind. It is now going to replace all the existing coaxial cable with unshielded twisted pair. List the advantages and disadvantages of this plan.

7. Rank the following five media examples—twisted pair, coaxial cable, fiber-optic cable, microwave, and satellite—in order from highest data transmission speed to lowest data transmission speed.

8. Using the same five media examples from the previous exercise, rank them in order from most noisy transmission to least noisy transmission.

9. Using the same five media examples from the previous exercises, rank them in order from most secure transmission to least secure transmission.

10. Terrestrial microwave is a line-of-sight transmission. What sorts of objects are tall enough to interfere with terrestrial microwave?

11. Your company has two offices located approximately one mile apart. Data needs to be transferred between the two offices at speeds up to 100 Mbps. List as many solutions as possible for interconnecting the two buildings. Is each solution technically feasible? Financially feasible? Politically feasible? Defend your position.

12. Given that a satellite signal travels at the speed of light, exactly how long does it take for a signal to go from the Earth to a satellite in geosynchronous orbit and back to Earth? Show the calculations.

13. How long does it take a signal to reach a satellite in low Earth orbit? Show the calculations.

14. You are walking down the street, and your cellular telephone rings. What was the sequence of events that allowed a person with a conventional telephone to call you on your cellular telephone?

15. Which of the wireless technologies can transmit through solid objects? Which wireless technologies cannot?

16. You are talking on your cellular telephone as you pass from one cell to another. Will your cellular telephone use the same set of frequencies in the new cell as it was using in the previous cell? Explain.

17. Why do cellular telephone systems need only seven sets of frequencies in a metropolitan area?

18. What is one potentially serious problem with using your personal digital assistant and Bluetooth to unlock doors wirelessly? Explain.

19. A company in your community is starting to offer a WiMAX service for Internet access. The company promises 2 Mbps downloads. If the company predicts that this new service will attract 2000 customers, what is the bandwidth necessary to support this service?

20. A T-1 service offered by voice and data communications companies is capable of supporting 1.5 Mbps of continuous data transfer over a high-quality telephone wire. What are the advantages and disadvantages of such a service when compared to services such as WAP, Bluetooth, and terrestrial microwave?

21. You are considering replacing your terrestrial microwave transmission system with a free space optics system. What are the advantages and disadvantages of doing this?

THINKING OUTSIDE THE BOX

1 You have been asked to recommend a type of wiring for a manufacturing building. The size of the building is 200 meters by 600 meters (corresponding to about 650 feet by 2000 feet), and it houses large, heavy machinery. Approximately 50 devices in this building must be connected to a computer system. Each device transmits data at 2 Mbps and sends a small packet of data every two to three seconds. Due to various building-related factors, any cable used for this network has to be suspended high overhead in a hard-to-reach location. What type of cable would you recommend? Use the media selection criteria introduced in this chapter to arrive at your answer.

2 GJ Enterprises has hired you as a productivity consultant. Currently, the company employs six people who routinely exchange information via flash drives. All GJ's employees are in the same brick building, but not in the same room. They will be sending word-processing documents and small spreadsheets as well as e-mail to each other. GJ wants the least expensive network wiring solution with only the minimal required hardware support. What medium would you recommend, and why?

3 You are the technology guru for an interstate trucking company. You need to maintain constant contact with your fleet of trucks. Which wireless technologies will enable you to do this?

4 You need to connect two buildings across a public road, and it is not feasible to use a direct cable connection. Under what circumstances would you use a microwave link? A radio link? An infrared laser link? What about the other wireless alternatives?

5 For many years, you have had a computer at home for all members of the family to use. Recently, however, you added a second computer. You want to connect the two computers so they can share data and a high-quality printer. Unfortunately, one computer is on the main floor, and the second one is upstairs. How are you going to connect the two computers? List as many possible solutions as you can, with the advantages and disadvantages of each. For example, should you use a wired or wireless medium? If wired, what kind of wires? Where will the wires go? Can you use existing wiring? If you go wireless, what technology could you use?

6 You are sitting at your desk at work, using your laptop computer. The boss calls an emergency meeting for you and several coworkers and asks everyone to bring his or her laptop computer. When you get to the meeting room, the boss wants to download an important file from his laptop to all your coworkers' laptops. List three possible media solutions that will support this download, along with their advantages and disadvantages.

HANDS-ON PROJECTS

1. Collect and label samples of as many kinds of conducted media as possible, and display them neatly on a piece of cardboard.

2. Using the Internet, locate the technical specifications of two different types of cables.

3. How many different types of conducted media are in place in your business or school? How are they used? Draw a rough diagram that shows the approximate locations and types of wire.

4. Using any sources possible, investigate a company that can offer a microwave service in your area. Report on what kinds of applications can be supported, what equipment will be necessary, where the equipment will be located, and what services this company offers.

5. Using any sources possible, investigate a company that can offer VSAT satellite service in your area. Report on what kinds of applications can be supported, what equipment will be necessary, where the equipment will be located, and what services this company offers.

6. Does anyone in your area offer a WiMAX service? Is yes, write a one-page summary that includes the service's main features.

7. How many different cellular telephone companies offer services within your area? Are the services D-AMPS or PCS? If PCS, are they CDMA, TDMA, or GSM? Are they GPRS or CDMA2000 1xRTT? Are they UMTS or EV-DO? Do the companies have estimates of the number of current subscribers? Do they know how many cells are in your market?

8. Ask your local cable television company how they receive their television signals. If it is a satellite service, is the service LEO, MEO, or GEO? What is the frequency range of the signals they receive? Do they receive signals from multiple sources? Do they have a backup plan if one of their services is disabled?

9. Visit the FCC's Web site (*www.fcc.gov*) and report what frequencies are currently being auctioned.

10. Does a Category 8 twisted pair exist yet? Does a need for such a wire exist? Use either paper sources or the Internet to find the answer.

11. Using an outside source, such as the Internet or the library, determine the typical height of a terrestrial microwave tower. If the tower's height is raised by 10 meters, how much farther will the tower be able to transmit?

12. Using a laptop computer with a wireless LAN card installed, locate the wireless LANs on your campus, and then create a map of these networks.

4

Making Connections

◆ ◆

CONNECTING PERIPHERAL DEVICES to a computer has never been an easy task. The interface between a computer and a peripheral is complex and contains many layers of hardware and software. Experts in the field have been working for years on simplifying the interconnection process, and the Universal Serial Bus is leading the charge as one of the best contenders for a universal interface standard.

Unfortunately, new interface standards rarely work as intended on the first try. One of the most classic examples occurred during the Comdex Spring 98 Computer Show, at which Microsoft Corporation CEO Bill Gates and an associate, in an attempt to demonstrate the then soon-to-be-released Windows 98 and its Universal Serial Bus interface, tried to connect a page scanner to a computer while the computer was turned on. Despite the fact that the Universal

Serial Bus and Windows 98 were designed to accept peripherals automatically whenever they are plugged in, Gates' computer did not accept the device and crashed. To the delight of the audience, Gates quickly responded, "That must be why we're not shipping Windows 98 yet."

What is involved in connecting a computer to other devices?

Is interfacing so difficult that even the specialists have trouble?

Do we really need to know what is happening at the interface level?

Source: CNN Interactive, downloaded from *www.cnn.com/ TECH/computing/9804/20/gates.comdex/* on August 8, 2005.

Objectives ▷

After reading this chapter, you should be able to:

▶ List the four components of all interface standards

▶ Discuss the basic operations of the USB interface standard

▶ Recognize the difference between half-duplex and full-duplex connections

▶ Cite the advantages of FireWire, SCSI, iSCSI, InfiniBand, and Fibre Channel interface standards

▶ Outline the characteristics of asynchronous, synchronous, and isochronous Data Link interfaces

▶ Identify the operating characteristics of terminal-to-mainframe connections and why they are unique compared to other types of computer connections

Introduction ▷

A computer would be of no use if we could not connect it to anything. Imagine, if you will, a computer with no monitor for viewing output and no keyboard for entering data. Many people also feel that their computers would not be worth much if there were no way to connect the computer to a printer, or if they could not connect it to a DSL modem to surf the Internet or access a remote computer system. Many people in the corporate world depend almost exclusively on an interconnection between their computer and a company's local area network. Through this interconnection they are able to access corporate databases, e-mail, the Internet, and other software applications. Aware of these sorts of consumer expectations, computer and computer-related manufacturers are constantly creating new devices to interconnect to computers. These peripheral devices include music playback devices (such as iPods), document scanners, digital cameras, and video cameras, among others.

Connecting a peripheral device to a computer can be a challenging task. Various levels of hardware and software have to agree completely before the computer can "talk" to the device, or vice versa. Questions such as the following need to be resolved: Is the connector on the end of the cable coming from the device compatible with the plug on the back of the computer? Will the electrical properties of the two devices be compatible? Even if the answer to these questions is yes, another question remains: Will the computer and device "speak the same language"? Connecting computers to other devices has many pitfalls and obstacles.

To better understand the interconnection between a computer and a peripheral device, you must first become familiar with the concept of interfacing. Considered primarily a physical layer activity, interfacing is a complex, relatively technical process that varies greatly, depending upon the type of device, the computer, and the connection desired between the device and computer. We will examine the four basic components of an interface—electrical, mechanical, functional, and procedural—and then introduce several of the more common interface standards, such as EIA-232F, Universal Serial Bus, FireWire, SCSI, iSCSI, InfiniBand, and Fibre Channel.

Connecting a computer to a device requires more than just resolving connections at the physical layer, however. It is also necessary to define the packaging of the data as it gets transferred between computer and device. In general, the basic configuration of this packaging is determined by Data Link layer standards. Therefore, we will examine three popular Data Link layer configurations: asynchronous connections, synchronous connections, and isochronous connections.

Finally, we will examine the connection between a terminal and a mainframe computer. Because a terminal is a relatively unintelligent device, a mainframe computer creates a unique dialog called polling, which prompts the terminal to see if it has data to submit to the mainframe.

But first let us start with the basics of connecting a computer to other devices.

Interfacing a Computer to Peripheral Devices

Most people would agree that a computer is a fantastic tool. This tool is capable of performing a myriad of operations primarily because of two important facts: a computer is programmable, and it can connect to a wide range of input/output devices, or peripherals. The connection to a peripheral is often called the interface, and the process of providing all the proper interconnections between a computer and a peripheral is called interfacing. You cannot discuss interfacing without discussing standards. We will be discussing many different types of connections, and each connection will have many possible layers of interface standards. Interfacing a device to a computer is considered primarily a physical layer activity, because it deals directly with analog signals, digital signals, and hardware components.

Let us begin our discussion of interfacing by exploring the general characteristics of interface standards, and then examining the particular features of two very popular interface standards (EIA-232F and Universal Serial Bus) and of some newer high-speed interfaces.

Characteristics of interface standards

Many years ago, computer and peripheral manufacturers and users realized that if one company made a computer and another company made a peripheral device, the odds of the two being able to "talk" to one another were slim. Thus, various organizations set about creating a standard interface between devices such as computers and modems. Because there were so many different transmission and interface environments, however, one standard alone did not suffice. As a result, hundreds of standards have been created. Despite this, all interface standards have two basic characteristics: They have been created and approved by an acceptable standards-making organization, and they can consist of one to four components, each of which will be discussed shortly.

The primary organizations involved in making standards are:

- ▶ International Telecommunication Union (ITU), formerly the Consultative Committee on International Telegraphy and Telephony (CCITT)
- ▶ Electronic Industries Association (EIA)
- ▶ Institute for Electrical and Electronics Engineers (IEEE)
- ▶ International Organization for Standardization (ISO)
- ▶ American National Standards Institute (ANSI)

Often individual companies are in such a hurry to rush a product to market that they will create a new product that incorporates a nonstandard interface protocol. While there is a definite marketing advantage to being the first to offer a new technology, there is also a considerable disadvantage to using an interface protocol that has not yet been approved by one of these standards-making organizations. For example, shortly after your company introduces its new product, one of the standards-making organizations may create a new protocol that performs the same function as your nonstandard protocol, and this may lead to your company's product becoming obsolete. In the quickly changing world of computer technology, creating a product that conforms to an approved interface standard is difficult but highly recommended. On occasion, a company will create a protocol that, while not an official standard, becomes so popular that other companies start using it. In this case, the protocol is considered a de facto standard. For example, Microsoft's operating system for personal computers is not an official standard. Nonetheless, ninety-some percent of all personal computers use the Microsoft operating system, thereby making the Windows desktop operating system a de facto standard.

The second basic characteristic of an interface standard is its composition. An interface standard can consist of four parts, or components, all of which reside at the physical layer: the electrical component, the mechanical component, the functional component, and the procedural component. All of the standards currently in existence address one or more of these components. The electrical component deals with voltages, line capacitance, and other electrical issues. The electrical components of interface standards are primarily the responsibility of a technician, and therefore we will not discuss them in great detail. The mechanical component deals with items such as the connector or plug description. Questions typically addressed by the mechanical component include: What is the size and shape of a connector? How many pins are found on the connector? What is the pin arrangement? The functional component describes the function of each pin (which is referred to as a circuit when you also take into account the signal that travels through the pin and wire) used in a particular interface. The procedural component describes how the particular circuits are used to perform an operation. While the functional component of an interface standard may, for example, describe two circuits, such as Request to Send and Clear to Send, the procedural component would describe how these two circuits are used so that the computer can transfer data to the peripheral, and vice versa.

An early interface standard

The USB interface is currently the most popular form of personal computer interface and will probably remain that way for some time. But it wasn't always that way. Let's briefly introduce one of the earlier interface standards just to give us a reference point for how far things have progressed. The RS-232 interface, which was created in 1962, is a classic example of one of the early interfaces. Because of its popularity, RS-232 has evolved—instead of becoming obsolete—over the years, and its current incarnation is named EIA-232F (also known as TIA-232-F). More precisely, EIA-232F is an interface standard for connecting a computer or terminal (or DTE) to a voice-grade modem (or DCE) for use on analog public telecommunications systems. (In modem interfacing terminology, the computer (or terminal) end of an interface is referred to as data terminating equipment (DTE), while the modem is referred to as the data communicating equipment (DCE).)

The EIA-232F interface standard is actually a composite of several other standards: the ITU V.28 standard, which defines EIA-232F's electrical component; the ISO 2110 standard, which defines the mechanical component; and the ITU V.24 standard, which defines the functional and procedural components. (For a more complete description of these supporting standards, please visit the author's Web site at http://facweb.cs.depaul.edu/cwhite.)

It is worth noting that an EIA-232F interface is what we call a full-duplex connection. A full-duplex connection is one in which both sender and receiver may transmit at the same time. This is possible with EIA-232F because there is one wire for transmitting and a separate wire for receiving. Some systems, for various reasons, allow only one side or the other (that is, either the sender or the receiver) to transmit at one time. This type of connection, as we will see shortly, is an example of a half-duplex connection.

Despite the longevity of the EIA-232F standard, it appears that other interface standards have essentially replaced it. Nonetheless, EIA-232F is a classic example of an interface standard, and its study provides an interesting window into the inner-workings of the communication between a computer and a peripheral. Now let us take a look at the interface standard that is most likely to replace EIA-232F.

Universal Serial Bus (USB)

The Universal Serial Bus (USB) is a modern standard for interconnecting many types of peripheral devices to computers. More precisely, USB is a digital interface that uses a standardized connector (plug) for all serial and parallel type devices. Because USB provides a digital interface, it is not necessary to convert the digital signals of the microcomputer to analog signals for transfer over a connection. As you may recall from Chapter Two, systems that undergo digital-to-analog or analog-to-digital conversions usually have more noise in their signals as a result of the conversion. USB avoids the introduction of such noise. Furthermore, USB is a relatively thin, space-saving cable to which devices can be added and removed while the computer and peripheral are active—a feature that makes USB *hot pluggable*. The idea behind hot plugging is that the peripheral can simply be plugged in and turned on, and that the computer should dynamically recognize the device and establish the interface. In other words, the casing of the computer does not have to be opened, nor do any software or hardware switches have to be set. When using peripherals designed with a USB connector, it is also possible to connect one USB peripheral to another. This technique of connecting a device to each subsequent device (instead of back to the computer) is known as daisy-chaining. Another unique feature of USB is that it is possible for the USB cable to provide the electrical power required to operate the peripheral. With this option, it is not necessary to find multiple electrical outlets (one for each peripheral). Finally, data transfer over a USB cable is bidirectional, but only one device—the computer or the peripheral—may transmit at one time. This makes USB another example of a half-duplex connection.

An early disadvantage of USB, at least when compared to other high-speed interfaces (such as FireWire, which will be discussed shortly), was its relatively slow speed. USB version 1.1 has a maximum transfer speed of 12 Mbps, significantly slower than FireWire's 400 Mbps. Fortunately, USB version 2.0 has a maximum transfer speed of 480 Mbps and is backwards compatible with the earlier 1.1 version, allowing devices with the newer interface to connect with the older interface (but at the 12-Mbps speed). More precisely, USB 2.0 can support low-speed (10 to 100 Kbps) devices such as keyboards, mice, and game peripherals; full-speed (500 Kbps to 10 Mbps) devices such as telephone circuits, audio, and compressed video; and high-speed devices (speeds greater than 10 Mbps) such as video, imaging devices, and broadband. The most recent version—USB 3.0—is rated at 4.8 Gbps (10 times faster than USB 2.0) and was released in November of 2008.

As we have just learned, interface standards consist of up to four components. With the USB standard, the electrical and functional components support the transfer of power and of the signal over a four-wire cable. Two of these four wires, VBUS and GND (ground), carry a 5-volt signal that can be used to power the device. The other two wires, D+ and D–, carry the data and signaling information.

The mechanical component of USB strictly specifies the exact dimensions of the interface's connectors and cabling. Four types of USB connectors are specified— a connector A, a connector B, a mini-connector A, and a mini-connector B. As you can see from Figure 4-1, both connectors A and B have four pins, one for each of the four wires in the electrical component, while the mini-connectors have five pins. The fifth pin in the mini-connectors is called the signal pin and is often simply connected to either the VBUS or GND pins. Although four different types of connectors exist, the A connector appears to be the most commonly used.

Figure 4-1

The four types of USB connectors

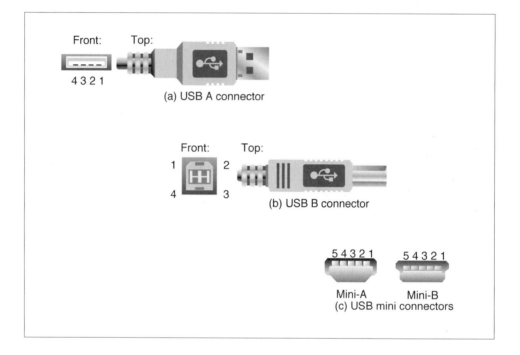

The procedural component of USB is probably the most involved of the four components. To understand how it works, you first need to become familiar with two terms: bus and polling. A bus is simply a high-speed connection to which multiple devices can attach, and polling (which is described in more detail in this chapter's section on terminal-to-mainframe computer connections) is a process in which a computer asks a peripheral if it has any data to transmit to the computer. The USB is a polled bus in which the host controller (the USB interface to the host computer) initiates all data transfers. The USB bus can recognize when a USB device has been attached to a USB port or to a USB hub (a device that is like an extension cord and can provide multiple USB ports). It can also recognize when that same device has been removed. In addition, the USB bus can support four basic types of data transfers: control transfers, which are used to configure a peripheral device at the time of attachment; bulk data transfers, which are used to support large and bursty (that is, produced in bursts) quantities of data; interrupt data transfers, which are used for timely but reliable delivery of data; and isochronous data transfers, which, as we will see a little later in this chapter, are connections that require continuous and real-time transfers of data, such as audio and video streams.

Because of the power and flexibility of USB, it will more than likely be the most commonly used interface of the future. If this happens, we will see another example of convergence—convergence toward a single interface standard that is capable of both supporting a wide variety of devices at a wide range of transfer speeds and automatically recognizing when a device is attached and acquiring the appropriate drivers.

Other interface standards

In addition to USB, other interface standards have been created over the years to provide high-speed connections to various types of peripheral devices. As we shall soon learn, some of these standards, such as FireWire, perform much like USB and, given the considerable strengths of USB, may someday be replaced by USB, while others will coexist with USB because they were designed to support different forms of interface. Let us continue with our discussion of interface standards by examining five more protocols designed to serve as high-speed interfaces between computers and peripheral devices: FireWire, SCSI, iSCSI, InfiniBand, and Fibre Channel.

FireWire

Conceived by Apple Computer in the mid-90s, and then standardized by IEEE as standard number 1394, FireWire is a type of interconnection between peripheral devices (such as wireless modems and high-speed digital video cameras) and a microcomputer. FireWire is an easy-to-use, flexible, and low-cost digital interface that is capable of supporting transfer speeds of up to 400 Mbps. (A newer version of FireWire supporting 3.2 Gbps was approved in 2007.) Because FireWire, like USB, provides a digital interface, it is not necessary to convert the digital signals of the microcomputer to analog signals for transfer over a connection. Thus, FireWire, again like USB, avoids introducing this type of noise into the signal. Furthermore, FireWire is also similar to USB in that it is hot pluggable.

FireWire supports two types of data connections: asynchronous and isochronous. Both of these data connections will be discussed in detail later in the chapter; but in brief terms, an asynchronous connection supports the more traditional peripheral devices such as modems and printers, and an isochronous connection provides guaranteed data transport at a predetermined rate, which is essential for multimedia applications. Multimedia applications are almost unique in that they require an uninterrupted transport of time-critical data and just-in-time delivery. FireWire is a good choice for interfacing digital consumer electronics and audio and visual peripherals, such as digital video cameras and digital cameras.

SCSI and iSCSI

SCSI, which stands for Small Computer System Interface and is pronounced "skuzzy," is a technique for interfacing a computer to high-speed devices such as hard disk drives, tape drives, CDs, and DVDs. Whereas USB and FireWire were designed for use with devices that may be added and removed and thus are hot pluggable, SCSI was designed to support devices of a more permanent nature. Consequently, SCSI is a *systems* interface—not just an interface technique for hard disk drives (as some people believe)—and correspondingly, SCSI optimizes the interaction between the input/output device and the central processor of a computer.

To take advantage of the higher-speed interface of SCSI, you need to install a SCSI adapter in your computer. Once this adapter is installed, you can connect up to seven different SCSI devices. To interconnect multiple SCSI devices to one

SCSI adapter, each additional device is connected to the previous SCSI device, or daisy-chained. The SCSI interface has been around for some time (since 1986) and has gone through a number of modifications. Thus, a range of SCSI techniques, from Fast SCSI and UltraSCSI to Narrow SCSI and Wide SCSI, exists. Each of these techniques differs by the transmission speed it supports and the number of bits transferred at a given moment (which is designated as though it were the "width" of the cable). For example, Ultra160 SCSI is designed for data transfers of up to 160 M *bytes*/second, which, you will notice, is almost three times faster than USB2.0's 480 Mbps. Despite some of the advantages of SCSI, many feel that SCSI will eventually give way to USB.

Another, newer variation on the SCSI interface is the iSCSI. iSCSI, or Internet SCSI, is a technique for interfacing disk storage to a computer via the Internet. Thus, if you have a large amount of disk storage located somewhere else on the Internet, you can connect your computer to that storage using the Internet and the iSCSI interface standard. Essentially what happens here is that SCSI commands are encapsulated in TCP/IP packets and sent over the Internet, like any typical Internet command such as sending e-mail or requesting a Web page.

InfiniBand and Fibre Channel

InfiniBand and Fibre Channel are two modern protocols used to interface a computer to input/output devices over a high-speed connection. More precisely, Infini-Band is a serial connection or bus that can carry multiple channels of data at the same time. It can support data transfer speeds of 2.5 billion bits (2.5 gigabits) per second (single data rate), 5 gigabits per second (double data rate), and 10 gigabits per second (quad data rate); and address (interconnect) thousands of devices, using both copper wire and fiber-optic cables. Rather than being a single shared bus, InfiniBand is a network of high-speed links and switches. Thus, traffic being passed between a computer and its input/output devices is actually moving across a high-speed network.

Fibre Channel is similar to InfiniBand in that it too is a serial, high-speed network that connects a computer to multiple input/output devices. Fibre Channel also supports data transfer rates up to billions of bits per second, but it can support the interconnection of up to 126 devices only.

Clearly, interfaces have come a long way from the early EIA-232 days. If this trend toward simple but sophisticated interfaces such as USB, FireWire, SCSI, and Fibre Channel continues (as it undoubtedly will), the topic of interfacing using EIA-232F may one day become a history lesson. Now let us turn our attention from the physical layer properties of a computer to peripheral connection and concentrate on the Data Link properties of a connection.

Data Link Connections

As we have seen, interface standards such as EIA-232F, USB, and FireWire consist of four components: electrical, mechanical, functional, and procedural. Because these four components define the physical connection between a computer and a peripheral, they reside at the physical layer of the Internet model. But there is more to creating a connection than just defining the various physical components. In order to transmit data successfully between two points on a network, such as between a DTE and a DCE or between a network sender and a network receiver, we also need to define the Data Link connections. If we once again relate this to the Internet model, you will note that the definition of the Data Link connection is performed at the Data Link, or network access layer.

To appreciate the issues involved in defining a Data Link connection, first assume that the physical layer connections are already defined by some protocol such as EIA-232F or USB. Now, given that the sender and receiver are using the same physical layer protocol, the next questions that must be resolved are as follows: What is the basic form of the data frame that is passed between sender and receiver? Is the data transmitted in single-byte blocks, or does the connection create a larger, multiple-byte block? The former data transmission option is an example of an asynchronous connection, while the latter is a synchronous connection. Does the data have to be delivered at a constant rate, such as would be required for a video camera connection? If so, this is an example of an isochronous connection. Can the connection transmit data in both directions at the same time, or in only one direction at a time? This, as we have learned, is the difference between a full-duplex connection and a half-duplex connection. Questions such as these define the type of connection found at the Data Link layer.

While examining the Data Link connection, recall the duties of the Data Link layer from the Internet model—two of which are to create a frame of data for transmission between sender and receiver and to provide some way of checking for errors during transmission. Keep these duties in mind as we examine three different types of Data Link connections.

Asynchronous connections

An asynchronous connection is one of the simplest examples of a Data Link protocol and is found primarily in microcomputer-to-modem and terminal-to-modem connections. In an asynchronous connection, a single character, or byte of data, is the unit of transfer between the sender and receiver. The sender prepares a data character for transmission, transmits that character, then begins preparing the next data character for transmission. An indefinite amount of time may elapse between the transmission of one data character and the transmission of the next character.

To prepare a data character for transmission, a few extra bits of information are added to the data bits of the character to create a frame, or small packet of data. A start bit, which is always a logic 0, is added to the beginning of the character and informs the receiver that an incoming data frame is arriving. The start bit also allows the receiver to synchronize itself to the character. At the end of the data character, one or two stop bits, which are always logic 1s, are added to signal the end of the frame. (Although there is usually only one stop bit, some systems still allow a choice of one or two.) The start and stop bits have, in essence, provided a beginning and ending frame around the data. Finally, a single parity bit, which is inserted between the data bits and the stop bit, may be added to the data. This parity bit (covered in more detail in Chapter Six) may be indicating either even parity or odd parity, and it performs an error check on only the data bits. This error check is achieved by adding a 0 or 1 such that an even or odd number of 1s is maintained. Figure 4-2 shows an example of the character A (in ASCII) with one start bit, one stop bit, and an even parity bit added.

Figure 4-2
*Example of the
character A with one
start bit, one stop bit,
and even parity*

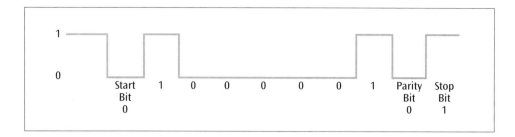

Because each character has its own start, stop, and parity bits, the transmission of multiple characters, such as HELLO, is possible. Figure 4-3 demonstrates the transmission of HELLO.

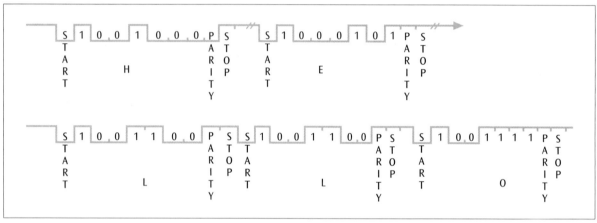

Figure 4-3
*Example of the
character string HELLO
with included start,
stop, and parity bits*

An asynchronous connection has advantages and disadvantages. On the positive side, generation of the start, stop, and parity bits is simple and requires little hardware or software. On the negative side, an asynchronous connection has one disadvantage in particular that cannot be overlooked. Given that seven data bits (ASCII character code set) are often combined with one start bit, one stop bit, and one parity bit, the resulting transmitted character contains three check bits and seven data bits, for a 3:7 ratio. In this scenario of 10 total check and data bits, 3 out of 10—or 30 percent—of the bits are used as check bits. This ratio of check bits to data bits is not very efficient for high amounts of data transfer and, therefore, results in slow data transfers.

Interestingly, the term *asynchronous connection* is misleading to beginners because the protocol actually does, despite the word *asynchronous*, maintain synchronization with the incoming data stream. Recall from Chapter Two the importance of a receiver staying synchronized with the incoming data stream, especially if the data stream contains a long sequence of unchanging values. The Manchester codes were designed to help with this problem, but they were not yet created when asynchronous connections were developed. Thus, asynchronous connections incorporate their own methods for keeping the receiver synchronized with the incoming data stream. How do they do it? Two key features of asynchronous connections help them maintain synchronization:

▶ Frame size—Because each frame in an asynchronous connection is one character plus a few check bits, the receiver will receive only a small amount of information at one time. It should not, therefore, be difficult for the receiver to stay synchronized for that short of a period.

▶ Start bit—When the receiver recognizes the start bit, the synchronization begins. Because only approximately 8 or 9 bits follow, there will not be a long sequence of unchanging values.

During the early years of the microcomputer industry, the simplicity (and thus low cost) of an asynchronous connection lent itself nicely to the hobbyist segment of the market. But the business segment of the market needed a Data Link connection that was more efficient and more powerful. Businesses needed an efficient, higher-speed connection—a synchronous connection.

Synchronous connections

The second type of Data Link connection (with a less misleading name) is the synchronous connection. With a synchronous connection, the unit of transmission is a sequence of characters. This sequence of characters may be thousands of characters in size. Similar to the way start, stop, and parity bits frame the data bits in an asynchronous connection, a start sequence (flag), a control byte, an address, a checksum, and an end sequence (flag) frame the data bits in a synchronous connection, as shown in Figure 4-4.

Figure 4-4

Block diagram of the parts of a generic synchronous connection

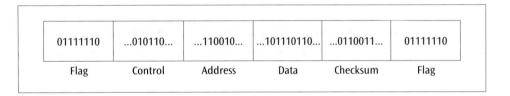

01111110	...010110...	...110010...	...101110110...	...0110011...	01111110
Flag	Control	Address	Data	Checksum	Flag

The starting and ending sequences of the synchronous connection are called flags and are each typically 8 bits (a byte) in length. Following the start sequence flag is usually one or more bytes of control information. This control information provides information about the enclosed data or provides status information pertaining to the sender or receiver or both. For example, a unique bit in the control information byte may be set to 1, indicating that the enclosed data is high-priority. Often the control byte contains addressing information that indicates where the data is coming from, or for whom it is intended. The address field indicates the destination of the frame, where the frame came from, or both. Following the data is almost always some form of error-checking sequence, such as the cyclic checksum. Cyclic checksum (explained in detail in Chapter Six) is a more advanced error-checking technique than parity checking and is used in many modern implementations of computer networks. After the error-checking sequence is the end sequence flag.

How does a synchronous connection keep the sender and receiver synchronized? This question is especially important because it is possible to send thousands of characters in a single package. Three ways are used to maintain synchronization in synchronous connections:

1. Send a synchronizing clock signal over a separate line that runs parallel to the data stream. As the data arrives on one line, a clock signal arrives on a second line. The receiver can use this clock signal to stay synchronized with the incoming data.

2. If transmitting a digital signal, use a Manchester code. Because a Manchester code always has a signal transition in the middle of each bit, the receiver can anticipate this signal transition and read the incoming data stream with no errors. A Manchester-encoded digital signal is an example of a self-clocking signal.

3. If transmitting an analog signal, use the properties of the analog signal itself for self-clocking. For example, an analog signal with a periodic phase change can provide the necessary synchronization.

Because of their higher efficiency, synchronous connections have almost completely replaced asynchronous connections. But there is still one more type of connection worth introducing—the isochronous connection.

Isochronous connections

An isochronous connection is a special kind of Data Link connection used to support various types of real-time applications. Examples of real-time applications usually include streaming voice, video, and music. A real-time application is unique in that its data must be delivered to a computer at just the right speed. If the data is delivered too slowly, then the music will distort, or the video will break up. If the data is delivered too fast, the receiving computer may not be able to buffer the data, which may result in the data being lost.

As we learned earlier in this chapter, both USB and FireWire can support isochronous connections. Before any data transfer can begin, however, the proper isochronous resources must be allocated within the connection. To perform this allocation, the sender and receiver exchange a number of initial packets, which determine, among other things, what channel will be used to transfer the data and the bandwidth that is necessary. Another issue resolved between sender and receiver is that error checking should be disabled on both ends. Error checking is not performed in real-time data transfers, because the time needed to ask the sender to retransmit along with the time it takes to retransmit the data will cause the data to arrive too late to be delivered to the viewer in real time. The main characteristic of a real-time data transfer is exactly that—the data must be delivered in real time (as or nearly as soon as it is generated). Asking the sender to retransmit a few bytes of erroneous data compromises the timeliness of real-time delivery.

Terminal-to-Mainframe Computer Connections

One type of connection that is based upon synchronous and asynchronous connections is the terminal-to-mainframe computer connection. The terminal-to-mainframe configuration was introduced in Chapter One as one of the original but still commonly found network configurations. Because terminals (or microcomputer workstations acting as terminals) possess little processing power compared to microcomputer workstations, the mainframe computer has to take control and perform all the data transfer operations. As you might imagine, the operations performed by the mainframe computer depend upon the type of physical connection that exists between a terminal and mainframe. A direct connection between a terminal and a mainframe computer, as shown in Figure 4-5(a), is a point-to-point connection. A single wire runs between the two devices, and no other terminals or computers share this connection. When multiple terminals share one direct connection to a mainframe, as shown in Figure 4-5(b), the connection is called multipoint. A multipoint connection is a single wire with the mainframe connected on one end and multiple terminals connected on the other end.

Figure 4-5
Point-to-point and multipoint connections of terminals and mainframe computer

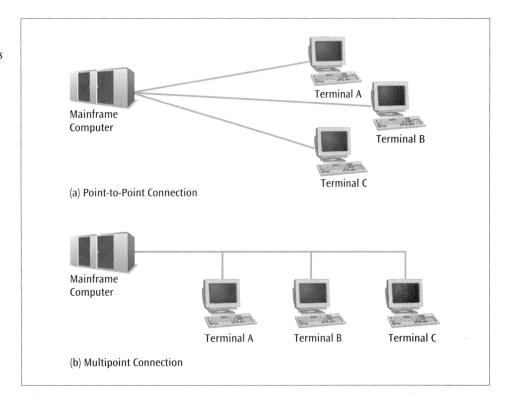

(a) Point-to-Point Connection

(b) Multipoint Connection

Once again, if multiple devices are to share a single line, something special must be done so that more than one device does not try to transmit at the same time. A technique called polling, which allows only one terminal to transmit at one time, successfully controls multiple terminals that share a connection to a mainframe computer. (As you read on, you should be aware that the concept of polling is not unique to terminals and mainframes. We will revisit this concept again in Chapter Five, when we examine multiplexing, and once again in Chapter Seven, when we introduce local area networks.) Polling originated in the early years of computing when terminals were relatively dumb devices, incapable of performing many operations beyond data entry and display. During this period, a mainframe computer was called the primary, and each terminal was called a secondary. In the polling technique, a terminal, or secondary, would transmit data only when prompted. Roll-call polling is the polling method in which the mainframe computer (primary) polls each terminal (secondary), one at a time, in round-robin fashion. If terminals A, B, and C shared one connection, as in Figure 4-6, the primary would begin by polling terminal A. If A had data to send to the host, it would do so. When A was done transmitting, the primary would poll terminal B. If B had nothing to send, it would inform the primary accordingly, and the primary would poll terminal C. When terminal C was finished, the primary would return to terminal A and continue the polling process.

Figure 4-6
Terminals A, B, and C being polled by a primary

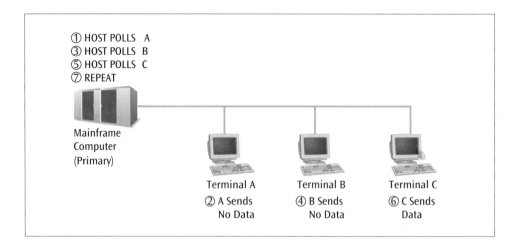

Note that only one device is talking at a time. Or more precisely, only one side of the connection is talking at a time. This is true despite the fact that the cable is capable of transmitting data in both directions. This is another example of a half-duplex connection.

An alternative to roll-call polling is hub polling. A primary that performs hub polling polls only the first terminal, which then passes the poll to the second terminal, and each successive terminal passes the poll along. For example, after being polled by the primary, terminal A, when it is finished responding, passes the poll to terminal B. When terminal B is finished transmitting, it passes the poll to terminal C. In this scenario, the primary does not need to poll each terminal separately. The process of the primary sending a poll to a terminal and waiting for a response takes time. When large amounts of data are being transmitted, this time may be significant.

When the primary wishes to send data to a terminal, it uses a process called selection. In selection, the primary creates a packet of data with the address of the intended terminal and transmits the packet. Only that specific terminal recognizes the address and accepts the incoming data. A primary can also use selection to broadcast data to all terminals.

If control simplicity is your primary goal, point-to-point connection of terminals is clearly superior to multipoint connections. With point-to-point connections, polling is not necessary because there is only one terminal per line. In multipoint connections, the terminal must possess the software necessary to support polling. Another disadvantage of multipoint connections in which several terminals share one connection is that each terminal has to wait while another terminal transmits. On the other hand, although point-to-point connections make more efficient use of transmission time, they also require more expensive hardware—that is, more cabling is required when each terminal has a direct connection to the primary.

Making Computer Connections In Action ▶

The laptop computer that your company ordered for you has just arrived. While it is pretty easy just to open the top of the computer and turn it on, you notice that there are quite a number of connectors around the four sides of the unit. In fact, there are no less than twelve places where something can plug into your new computer. Let's take a look at these connections and see what they might be used for.

The first connection is the easiest—it is where the power cord plugs in. All consumer electronics operate on DC (direct current) power. In order to keep the laptop small, the hardware that converts the AC (alternating current) from the outlet in the wall to DC is usually contained in the power cord, and not in the laptop computer. The power cord obviously keeps the batteries charged and the unit running when not on battery power.

The next three connectors are fairly simple—they are the USB connectors. You recognize them as Type A and know that you will probably be using these connectors more than anything else. The next two connectors are small and relatively square—these are the RJ-11 and RJ-45 connectors. The RJ-11 connector is used to connect the laptop to a telephone line via the built-in modem, and the RJ-45 is used to connect the laptop to an Ethernet local area network via a built-in Ethernet network interface card. The fairly small, almost square connector with the printing 1394 next to it is the Firewire connector. The two small round connectors are for plugging in a microphone and a set of headphones or external speakers.

Next we see a fairly large rectangular slot on the side of the unit. This connector is used for plugging in a PC Card/SmartCard. The PC Card, which used to be called the PCMCIA Card, is an older interface standard that allows devices such as memory cards, modems, network cards, and hard disk to be added to a laptop computer. While the PC Card has been all but replaced by the USB interface, it does have the advantage of sliding into the laptop, thus being mostly out of sight and out of the way. A SmartCard connector allows a user to plug in a SmartCard with metallic connectors (there are also SmartCards that don't use connectors but use wireless signals to communicate with other devices). SmartCards are about the size of a credit card and have embedded processing capabilities. They are often used for providing security for the laptop.

Next to the PC Card slot is a small rectangular connector about the size of a USB connector, but with a D over it. This is a DisplayPort connector. DisplayPort is a connector similar to DVI and HDMI used with televisions and other video devices and allows you connect your laptop to an audio/video device. Another fairly skinny and rectangular connector (a little wider than the DisplayPort connector) is the 7-in-1 Media Card reader. This connector allows you to plug in media cards such as SD, SD HC, xD, xD Type H, MMC, MS, and MS-PRO. These type of media cards are often found on digital cameras and digital video cameras and are used to store images and videos. Thus, if you want to copy your digital photos from your camera to your computer, you can use this slot.

Finally, there is a rectangular connector with fifteen small holes. This is an external monitor connector and can be used to connect your laptop to another video device such as an overhead projector.

Now you should have a basic understanding of the different types of connectors and connections available on a typical laptop/notebook computer. Although we did not examine each type of connection in detail, you should understand that many different types of connections are possible, and each connection has many possible layers of interface standards. Clearly, the USB connector (or any one connector) has not replaced all the other forms of interfacing just yet.

❖ ❖

SUMMARY

▶ The connection between a computer and a peripheral is often called the interface, and the process of providing all the proper interconnections between a computer and a peripheral is called interfacing.

▶ Interface standards have two basic characteristics: they have been created and approved by an acceptable standards-making organization, and they can consist of one to four components: electrical, mechanical, functional, and procedural.

▶ A DTE is a data terminating device such as a computer, and a DCE is a data circuit-terminating device such as a modem.

▶ Over the years, a number of interface standards have been developed. Two that are worthy of additional study are EIA-232F and the Universal Serial Bus. EIA-232F was one of the earliest standards and has been highly popular for years, while Universal Serial Bus is currently the most popular interface standard. Both standards provide mechanical, electrical, functional, and procedural specifications.

▶ Transmission systems that are half-duplex can transmit data in both directions, but in only one direction at a time. Full-duplex systems can transmit data in both directions at the same time.

▶ Other peripheral interfacing standards that provide power, flexibility, higher speeds, and ease-of-installation include FireWire, SCSI, iSCSI, InfiniBand, and Fibre Channel.

▶ Interfacing a device to a computer is considered a physical layer activity and thus primarily resides at the physical layer. But a Data Link connection is also required when data is transmitted between two points on a network. Three common Data Link connections are asynchronous connections, synchronous connections, and isochronous connections.

▶ Asynchronous connections use single-character frames and start and stop bits to establish the beginning and ending points of the frame.

▶ Synchronous connections use multiple-character frames, sometimes consisting of thousands of characters. A synchronizing clock signal or some kind of self-clocking must be used with synchronous transmission to enable the receiver to remain synchronized during this large frame.

▶ Isochronous connections provide real-time connections between computers and peripherals and require a fairly involved dialog to support the connection.

▶ A connection between a computer terminal and a mainframe computer that is dedicated to one terminal is called a point-to-point connection. A shared connection between more than one computer terminal and a mainframe computer is called a multipoint connection. Mainframe computers use polling techniques such as roll-call polling and hub polling to support multipoint connections.

KEY TERMS

asynchronous connection

data communicating equipment (DCE)

data terminating equipment (DTE)

de facto standard

EIA-232F

electrical component

Fibre Channel

FireWire

frame

full-duplex connection

functional component

half-duplex connection

hub polling

InfiniBand

interchange circuit

interfacing

iSCSI (Internet SCSI)

isochronous connection

mechanical component

multipoint connection

parallel port

parity bit

point-to-point connection

polling

primary

procedural component

roll-call polling

RS-232

SCSI (Small Computer System Interface)

secondary

selection

serial port

start bit

stop bit

synchronous connection

Universal Serial Bus (USB)

REVIEW QUESTIONS

1. What is a DTE, and what is a DCE?
2. What are the four components of an interface?
3. What are the advantages of USB over EIA-232F and the other types of interface standards?
4. FireWire and USB are standards to interconnect what to what?
5. What are the advantages of FireWire?
6. When might one use SCSI to interconnect a peripheral?
7. What is the difference between SCSI and iSCSI?
8. Are InfiniBand and Fibre Channel likely to be used on your home computer?
9. What are the primary differences among asynchronous, synchronous, and isochronous connections?
10. In asynchronous connections, what additional bits are added to a character to prepare it for transfer?
11. In asynchronous connections, how many characters are placed into one frame?
12. What are the advantages and disadvantages of asynchronous communication?
13. What is the basic block diagram of a synchronous frame?
14. What are the advantages and disadvantages of synchronous communication?
15. What is the difference between half-duplex and full-duplex communications?
16. What is the difference between a point-to-point connection and a multipoint connection?
17. How does a mainframe computer ask a terminal to send it data?

EXERCISES

1. What are the advantages, if any, of the older EIA 232F interface over the newer USB interface?
2. What is the major advantage of the FireWire interface over the Universal Serial Bus 1.1 interface?
3. If I have a device that has a Universal Serial Bus 2.0 interface, but my computer only has a Universal Serial Bus 1.1 connector, is my device going to work? Explain why or why not.
4. Create a table that compares the advantages and disadvantages of the Universal Serial Bus to those of the EIA-232F interface.
5. Show the sequence of start, data, and stop bits that is generated during asynchronous transmission of the character string LUNCH.
6. List two examples not mentioned in the book for each of the following connections: half-duplex and full-duplex.
7. Terminals A, B, and C are connected to a mainframe computer. Only terminal C has data to transmit. Show the sequence of messages sent between the mainframe and the three terminals using roll-call polling.
8. Using the same scenario as the previous problem, show the sequence of messages exchanged using hub polling.
9. Suppose you want to send 1000 7-bit characters of data. How many total bits will you transmit using asynchronous transmission? How many bits total will you transmit using synchronous transmission? Assume that all 1000 characters will fit within one synchronous transmission frame.

10. List two features of the asynchronous connection that allow the receiver to stay in sync with the incoming data stream.

11. How does the receiver in a synchronous connection stay in sync with the incoming data stream?

12. What types of devices are best served with an isochronous connection?

13. In what type of situation might hub polling be preferable to roll-call polling?

14. A company has a very powerful computer, and it wants to connect the computer to a large number of high-speed disk storage devices. Which protocol(s) introduced in this chapter would provide a good interface for this scenario?

THINKING OUTSIDE THE BOX

1 You are designing an application at work that transmits data records to another building within the same city. The data records are 500 bytes in length, and your application will send one record every 0.5 seconds. Is it more efficient to use a synchronous connection or an asynchronous connection? What speed transmission line is necessary to support either type of connection? Show all your work.

2 You are using a computer system that has 20 terminals connected to a mainframe computer using roll-call polling. Each terminal is polled once a second by the mainframe and sends a 200-byte record every 10 seconds. The boss says you should remove the terminals, install computer workstations, and replace the polling with asynchronous connections. Which is more efficient: keeping the terminals and polling, or using workstations and asynchronous connections?

3 How does the isochronous connection compare to the asynchronous and synchronous connections? Compare the applications and efficiencies of all three.

4 A computer manufacturer wants to streamline its products by producing a computer with a single Universal Serial Bus connector and no other connectors. Is this a good idea? Is it even possible? Explain.

5 Continuing a problem from the previous chapter (see question #5 in Chapter Three's "Thinking Outside the Box" section), you have a computer on the main floor of your house and another computer in a second-floor bedroom. You want to interconnect the two computers so that they share a high-speed Internet connection and a high-quality printer. What type of medium will you use to interconnect these two computers to each other and to a printer and Internet access device, and what type of interface/connector will each use?

HANDS-ON PROJECTS

1. One interface standard is called X.21. How does it compare to USB? What signals in addition to T, C, R, and I are used in the X.21 interface standard?

2. The asynchronous protocol, as it first appeared, described an option for 1, 1.5, or 2 stop bits. Why would anyone need or want to use 1.5 or 2 stop bits?

3. Create a list of different products that use a Universal Serial Bus interface. What are some of the more "creative" products?

4. What is the latest status of the FireWire standard? Is it going to succumb to the USB standard? Explain your response.

5

Making Connections Efficient: Multiplexing and Compression

◆ ◆

MANY PEOPLE NOW HAVE a portable music playback system such as Apple Computer's iPod. The 20-GB-sized iPod is capable of storing up to 5000 songs. How much storage is necessary to hold this quantity of music? If we consider that a typical song taken from a compact disc is composed of approximately 32 million bytes (assuming that an average song length is 3 minutes, that the music is sampled 44,100 times per second, and that each sample is 16 bits, in both left and right channels), then storing 5000 songs of 32 million bytes each would require 160 billion bytes. How is it possible to squeeze 5000 songs (160 billion bytes) into a storage space of little more than 20 billion bytes? The answer is through compression. While there are many types of compression techniques, the basic objective underlying them is the same—to squeeze as much data as possible into a limited amount of storage space.

Music compression is not the only type of data that can be compressed. iPods can also compress speech, thus allowing users to record messages and memos to themselves or for later transmission to another person. Likewise, music videos, television shows, and movies can be compressed, stored, and then replayed using various versions of the iPod as well as other various portable music and video players. Clearly, the iPod would not be the device it is today without a compression technique.

Does anything get lost when we compress data into a smaller form?

Are there multiple forms of compression?

Do certain compression techniques work better with certain types of applications?

Source:www.Apple.com.

Objectives ▶

After reading this chapter, you should be able to:

▶ Describe frequency division multiplexing and list its applications, advantages, and disadvantages

▶ Describe synchronous time division multiplexing and list its applications, advantages, and disadvantages

▶ Outline the basic multiplexing characteristics of T-1 and SONET/SDH telephone systems

▶ Describe statistical time division multiplexing and list its applications, advantages, and disadvantages

▶ Cite the main characteristics of wavelength division multiplexing and its advantages and disadvantages

▶ Describe the basic characteristics of discrete multitone

▶ Cite the main characteristics of code division multiplexing and its advantages and disadvantages

▶ Apply a multiplexing technique to a typical business situation

▶ Describe the difference between lossy and lossless compression

▶ Describe the basic operation of run-length, JPEG, and MP3 compression

Under the simplest conditions, a medium can carry only one signal at any moment in time. For example, the twisted pair cable that connects a keyboard to a microcomputer carries a single digital signal. Likewise, the Category 6 twisted pair wire that connects a microcomputer to a local area network carries only one digital signal at a time. Many times, however, we want a medium to carry multiple signals at the same time. When watching television, for example, we want to receive multiple television channels in case we do not like the program on the channel we are currently watching. We have the same expectations of broadcast radio. Additionally, when you walk or drive around town and see many people all talking on cellular telephones, you are witnessing the simultaneous transmission of multiple cell phone signals over the wireless medium. The technique of transmitting multiple signals over a single medium is multiplexing.

For multiple signals to share one medium, the medium must be "divided" somehow to give each signal a portion of the total bandwidth. Presently, a medium can be divided in three basic ways: a division of frequencies, a division of time, and a division of transmission codes. Regardless of which kind of division is performed, multiplexing can make a communications link, or connection, more efficient by combining the signals from multiple sources. We will examine the three ways a medium can be divided by describing in detail the multiplexing technique that corresponds to each division, and then follow with a discussion that compares the advantages and disadvantages of all the techniques.

Another way to make a connection between two devices more efficient is to compress the data that transfers over the connection. If a file is compressed to one half its normal size, it will take one-half the time or one-half the bandwidth to transfer that file. This compressed file will also take up less storage space, which is clearly another benefit. As we shall see, a number of compression techniques are currently used in communication (and entertainment) systems. Some of these compression techniques are capable of returning an exact copy of the original data (lossless), while others are not (lossy). But let us start first with multiplexing.

Frequency Division Multiplexing

Frequency division multiplexing is the oldest multiplexing technique and is used in many fields of communications, including broadcast television and radio, cable television, and cellular telephones. It is also one of the simplest multiplexing techniques. Frequency division multiplexing (FDM) is the assignment of nonoverlapping frequency ranges to each "user" of a medium. A user may be a television station that transmits its television channel through the airwaves (the medium) into homes and businesses. A user might also be the cellular telephone transmitting signals over the medium on which you are talking, or it could be a computer terminal sending data over a wire to a mainframe computer. To allow multiple users to share a single medium, FDM assigns each user a separate channel. A channel is an assigned set of frequencies that is used to transmit the user's signal. In frequency division multiplexing, this signal is analog.

Many examples of frequency division multiplexing can be found in business and everyday life. Cable television is still one of the more commonly found applications of frequency division multiplexing. As shown in Table 5-1, each cable television channel is assigned a unique range of frequencies by the Federal Communications Commission, and these frequency assignments are fixed, or static. Note from Table 5-1 that the frequencies of the various channels do not overlap. The television set, cable television box, or a digital video recorder contains a tuner, or channel selector. The tuner separates one channel from the next and presents each as an individual data stream to you, the viewer.

Table 5-1
Assignment of frequencies for cable television channels

	Channel	Frequency in MHz
Low-Band VHF and Cable	2	55–60
	3	61–66
	4	67–72
	5	77–82
	6	83–88
Mid-Band Cable	95	91–96
	96	97–102
	97	103–108
	98	109–114
	99	115–120
	14	121–126
	15	127–132
	16	133–138
	17	139–144
	18	145–150
	19	151–156
	20	157–162
	21	163–168
	22	169–174
High-Band VHF and Cable	7	175–180
	8	181–186
	9	187–192
	10	193–198
	11	199–204
	12	205–210
	13	211–216

Other common examples of frequency division multiplexing are cellular telephone systems. These systems divide the bandwidth that is available to them into multiple channels. Thus, the telephone connection of one user is assigned one set of frequencies for transmission, while the telephone connection of a second user is assigned a second set of frequencies. As explained in Chapter Three, first-generation cellular telephone systems allocated channels using frequency ranges within the

800 to 900 megahertz (MHz) spectrum. To be more precise, the 824 to 849 MHz range was used for receiving signals from cellular telephones (the uplink), while the 869 to 894 MHz range was used for transmitting to cellular telephones (the downlink). To enable a two-way conversation to occur, two channels were assigned to each telephone connection. The signals coming into the cellular telephone came in on one 30-kHz band (in the 869 to 894 MHz range), while the signals leaving the cellular telephone went out on a different 30-kHz band (in the 824 to 849 MHz range). While later generation cellphones may use different frequency ranges, the multiplexing concepts are similar. Cellular telephones are an example of dynamically assigned channels. When a user enters a telephone number and presses the Send button, the cellular network assigns this connection a range of frequencies based on current network availability. As you might expect, the dynamic assignment of frequencies can be less wasteful than the static assignment of frequencies, which is found in terminal-to-mainframe computer multiplexed systems and television systems.

In general, the device that accepts input from one or more users is called the multiplexor. The device attached to the receiving end of the medium and splits off each signal to deliver it to the appropriate receiver is called the second multiplexor, or demultiplexor. In all frequency division multiplexing systems, the multiplexor accepts input from the user(s), converts the data streams to analog signals using either fixed or dynamically assigned frequencies, and transmits the combined analog signals over a medium that has a wide enough bandwidth to support the total range of all the assigned frequencies. The demultiplexor then accepts the combined analog signals, separates out one or more of the individual analog signals, and delivers these to the appropriate user(s). Figure 5-1 shows a simplified diagram of frequency division multiplexing.

Figure 5-1
Simplified example of frequency division multiplexing

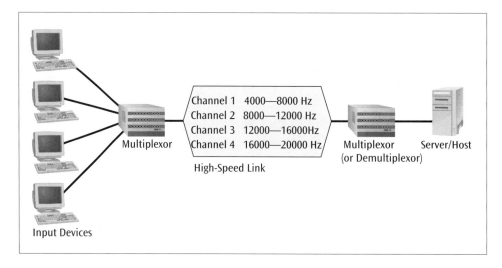

To keep one signal from interfering with another signal, a set of unused frequencies called a guard band is usually inserted between the two signals, to provide a form of insulation. These guard bands take up frequencies that might be used for other data channels, thus introducing a certain level of wastefulness. This wastefulness is much like that produced in static assignment systems when a user that has been assigned to a channel does not transmit data and, therefore, is considered to be an inefficiency in the FDM technique. In an effort to improve upon these deficiencies, another form of multiplexing—time division multiplexing—was developed.

Time Division Multiplexing

Frequency division multiplexing takes the available bandwidth on a medium and divides the frequencies among multiple channels, or users. Essentially, this division enables multiple users to transmit at the same time. In contrast, time division multiplexing (TDM) allows only one user at a time to transmit, and the sharing of the medium is accomplished by dividing available transmission *time* among users. Here, a user uses the entire bandwidth of the channel, but only for a brief moment.

How does time division multiplexing work? Suppose an instructor in a classroom poses a controversial question to students. In response, a number of hands shoot up, and the instructor calls on each student, one at a time. It is the instructor's responsibility to make sure that only one student talks at any given moment, so that each individual's response is heard. In a relatively crude way, the instructor is a time division multiplexor, giving each user (student) a moment in time to transmit data (express an opinion to the rest of the class). In a similar fashion, a time division multiplexor calls on one input device after another, giving each device a turn at transmitting its data over a high-speed line. Suppose two users, A and B, wish to transmit data over a shared medium to a distant computer. We can create a rather simple time division multiplexing scheme by allowing user A to transmit during the first second, then user B during the following second, followed again by user A during the third second, and so on. Since time division multiplexing was introduced (in the 1960s), it has split into two roughly parallel but separate technologies: *synchronous* time division multiplexing and *statistical* time division multiplexing.

Synchronous time division multiplexing

Synchronous time division multiplexing (Sync TDM) gives each incoming source signal a turn to be transmitted, proceeding through the sources in round-robin fashion. Given *n* inputs, a synchronous time division multiplexor accepts one piece of data, such as a byte, from the first device, transmits it over a high-speed link, accepts one byte from the second device, transmits it over the high-speed link, and continues this process until a byte is accepted from the *n*th device. After the *n*th device's first byte is transmitted, the multiplexor returns to the first device and continues in round-robin fashion. Alternately, rather than accepting a byte at a time from each source, the multiplexor may accept single bits as the unit input from each device. Figure 5-2 shows an output stream produced by a synchronous time division multiplexor.

Figure 5-2
*Sample output stream
generated by a
synchronous time
division multiplexor*

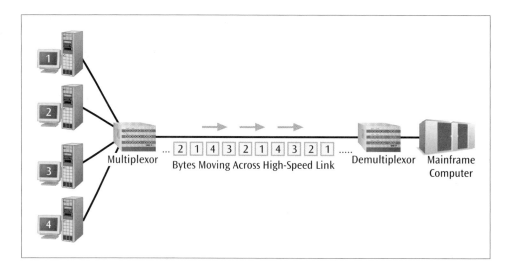

Note that the demultiplexor on the receiving end of the high-speed link must disassemble the incoming byte stream and deliver each byte to the appropriate destination. Because the high-speed output data stream generated by the multiplexor does not contain addressing information for individual bytes, a precise order must be maintained—this will allow the demultiplexor to disassemble and deliver the bytes to the respective owners in the same sequence as the bytes were input.

Under normal circumstances, the synchronous time division multiplexor maintains a simple round-robin sampling order of the input devices, as depicted in Figure 5-2. What would happen if one input device sent data at a much faster rate than any of the others? An extensive buffer (such as a large section of random access memory) could hold the data from the faster device, but this buffer would provide only a temporary solution to the problem. A better solution is to sample the faster source multiple times during one round-robin pass. Figure 5-3 demonstrates how the input from device A is sampled twice for every one sample from the other input devices. As long as the demultiplexor understands this arrangement and this arrangement does not change dynamically, in theory, no problems should occur. In reality, however, one additional condition must be met. This sampling technique will work only if the faster device is two, three, or four—an integer multiple—times faster than the other devices. If device A is, say, two and one-half times faster than the other devices, this technique will not work. In that case, device A's input stream would have to be padded with additional "unusable" bytes to make its input stream seem a full three times faster than that of the other devices.

Figure 5-3
*A synchronous time
division multiplexing
system that samples
device A twice as fast
as the other devices*

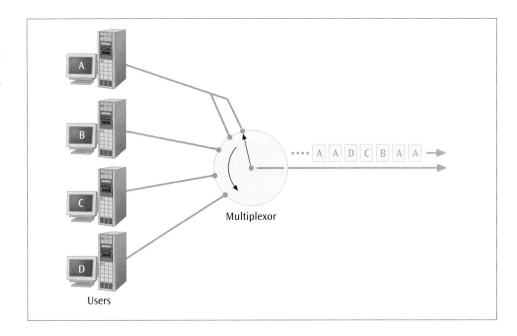

What happens if a device has nothing to transmit? In this case, the multiplexor must still allocate a slot for that device in the high-speed output stream, but that time slot will, in essence, be empty. Because each time slot is statically fixed in synchronous time division multiplexing, the multiplexor cannot take advantage of the empty slot and reassign busy devices to it. If, for example, only one device is transmitting, the multiplexor must still be going about sampling each input device (Figure 5-4). In addition, the high-speed link that connects the two multiplexors must always be capable of carrying the total of all possible incoming signals, even when none of the input sources is transmitting data.

Figure 5-4
*Multiplexor
transmission
stream with only
one input device
transmitting data*

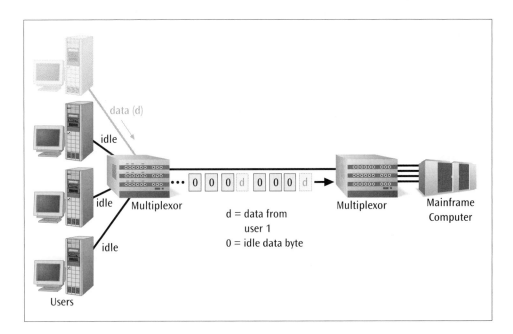

As with a simple connection between one sending device and one receiving device, maintaining synchronization across a multiplexed link is important. To maintain synchronization between sending multiplexor and receiving demultiplexor, the data from the input sources is often packed into a simple frame, and synchronization bits are added somewhere within the frame (Figure 5-5). Depending on the TDM technology used, anywhere from one bit to several bits can be added to a frame to provide synchronization. The synchronization bits act in a fashion similar to differential Manchester's constantly changing signal—they provide a constantly reappearing bit sequence that the receiver can anticipate and onto which it can lock.

Figure 5-5
Transmitted frame with added synchronization bits

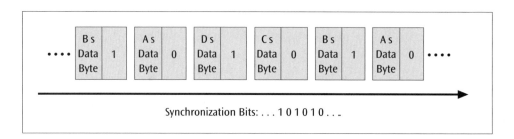

Synchronization Bits: . . . 1 0 1 0 1 0 . . .

Two types of synchronous time division multiplexing that are popular today are T-1 multiplexing and SONET/SDH. Although the details of T-1 and SONET/SDH can be highly technical, a brief examination of each technology will show how it multiplexes multiple channels of information together into a single stream of data.

T-1 Multiplexing

In the 1960s, AT&T created a service known as T-1, which multiplexed digital data and digitized voice onto a high-speed telephone line with a data rate of 1.544 megabits per second. The T-1's original purpose was to provide a high-speed connection between AT&T's switching centers. When businesses learned of this high-speed service, they began to request it to connect their computer and voice communications systems to the telephone network. In 1984, AT&T finally began offering this service to business customers.

Details ▶

T-1 Multiplexing

T-1 communications lines are a popular technology for connecting businesses to high-speed sources such as Internet service providers and other wide area networks. Because T-1 multiplexing is a classic example of synchronous time division multiplexing, it merits further examination.

A T-1 telecommunications line uses a multiplexing technique termed **DS-1 signaling**, which provides for the multiplexing of up to 24 separate channels at a total speed of 1.544 Mbps. How does the T-1 line achieve the unique transmission speed of 1.544 Mbps? To answer this question, let us consider an example in which the T-1 line supports the maximum 24 voice channels.

Because the average human voice occupies a relatively narrow range of frequencies (approximately 300 to 3400 Hz), it is fairly simple to digitize voice. In fact, an analog-to-digital converter needs only 128 different quantization levels to achieve a fair digital representation of the human voice. Because 128 equals 2^7, each pulse code modulated voice sample can fit into a 7-bit value. Two hundred and fifty-six quantization levels would allow for an even more precise representation of the human voice. Because $256 = 2^8$, and 8 bits equals 1 byte, the telephone system uses 256 quantization levels to digitize the human voice. (If you need a refresher on this material, revisit pulse code modulation in Chapter Two.)

In T-1 multiplexing, the frames of the T-1 multiplexor's output stream are divided into 24 separate digitized voice/data channels of 64 Kbps each (Figure 5-6). Users who wish to use all 24 channels are using a full T-1, while other users who need to use only part of the 24 channels may request a fractional T-1. The T-1 multiplexed stream is a continuous repetition of frames. Each frame consists of 1 byte from each of the 24 channels (users) plus 1 synchronization bit. Thus, data from the first user is followed by the data from the second user, and so on, until data from the 24th user is once again followed by data from the first user. If one of the 24 input sources has no data to transmit, the space within the frame is still allocated to that input source. The input data from a maximum of 24 devices is assigned to fixed intervals. Each device can transmit only during that fixed interval. If a device has no significant data to transmit, the time slot is still assigned to that device, and data such as blanks or zeros are transmitted. The T-1 system is a classic application of synchronous time division multiplexing.

Figure 5-6
T-1 multiplexed data stream

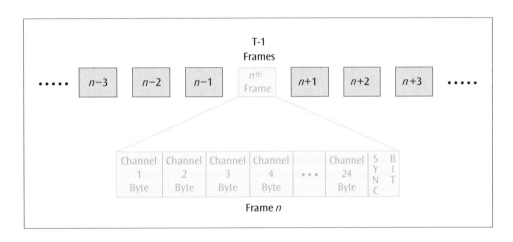

Recall that to create an accurate digital representation of an analog signal, you need to sample the analog signal at a rate that is twice the highest frequency. Given that the telephone company assigns a 4000 Hz channel to carry the voice signal, you need, when digitizing voice, to sample the analog voice signal 8000 times per second. Recall also, from Figure 5-6, the T-1 frame sequence. Because each T-1 frame contains 1 byte of voice data for 24 different channels, the system needs 8000 frames per second to maintain 24 simultaneous voice channels. Because each frame is 193 bits in length (24 channels × 8 bits per channel + 1 control bit = 193 bits), 8000 frames per second is multiplied by 193 bits per frame, which yields a rate of 1.544 Mbps.

T-1 can be used to transfer data as well as voice. If data is being transmitted, the 8-bit byte for each channel is broken into 7 bits of data and 1 bit of control information. Seven data bits per frame × 8000 frames per second = 56,000 bits per second per channel. Thus, when used for data, each of the 24 T-1 channels is capable of supporting a 56-Kbps connection.

SONET/SDH Multiplexing

Synchronous Optical Network (SONET) and Synchronous Digital Hierarchy (SDH) are powerful standards for multiplexing data streams over a single medium. SONET (developed in the United States by ANSI) and SDH (developed in Europe by ITU-T) are two almost identical standards for the high-bandwidth transmission of a wide range of data types over fiber-optic cable. SONET and SDH have two features that are of particular interest in the context of multiplexing. First, they are both synchronous multiplexing techniques. A single clock controls the timing of all transmission and equipment across an entire SONET (or SDH) network. Using only a single clock to time all data transmissions yields a higher level of synchronization, because the system does not have to deal with two or more clocks having slightly different times. This high level of synchronization is necessary to achieve the high level of precision required when data is being transmitted at hundreds and thousands of megabits per second.

Second, SONET and SDH are able to multiplex varying speed streams of data onto one fiber connection. SONET defines a hierarchy of signaling levels, or data transmission rates, called synchronous transport signals (STS). Each STS level supports a particular data rate, as shown in Table 5-2, and is supported by a physical specification called an optical carrier (OC). Note that the data rate of OC-3 is exactly three times the rate of OC-1; this relationship carries through the entire table of values. SONET is designed with this data rate relationship so that multiplexing signals is relatively straightforward. For example, it is relatively simple to multiplex three STS-1 signals into one STS-3 signal. Likewise, four STS-12 signals can be multiplexed into one STS-48 signal. The STS multiplexor in a SONET network can accept electrical signals from copper-based media, convert those electrical signals into light pulses, and then multiplex the various sources onto one high-speed stream.

Table 5-2

STS signaling levels, corresponding OC levels, and data rates

STS Level	OC Specification	Data Rate (in Mbps)
STS-1	OC-1	51.84
STS-3	OC-3	155.52
STS-9	OC-9	466.56
STS-12	OC-12	622.08
STS-18	OC-18	933.12
STS-24	OC-24	1244.16
STS-36	OC-36	1866.24
STS-48	OC-48	2488.32
STS-96	OC-96	4976.64
STS-192	OC-192	9953.28

Each SONET frame contains the data being transmitted plus a number of control bits, which are scattered throughout the frame. Figure 5-7 shows the frame layout for the STS-1 signaling level. The STS-1 signaling level supports 8000 frames per second, and each frame contains 810 bytes (6480 bits). Multiplying 8000 frames per second by 6480 bits per frame yields 51,840,000 bits per second, which is the OC-1 data rate. The other STS signaling levels are similar except for the layout of data and the placement and quantity of control bits.

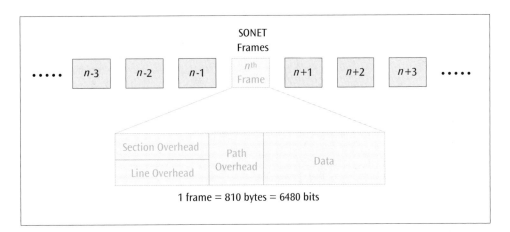

Figure 5-7
SONET STS-1 frame layout

SONET and SDH are used in numerous applications in which very high data transfer rates over fiber-optic lines are necessary. For example, two common users of SONET are the telephone company and companies that provide an Internet backbone service. Both telephone companies and Internet backbone providers have very high-speed transmission lines that span parts of the country and must transmit hundreds and thousands of millions of bits per second over long distances. Installing fiber-optic lines that support SONET transmission technology is one of the best ways to meet the demands of such challenging applications.

Statistical time division multiplexing

As you have seen in the preceding discussions, both frequency division multiplexing and synchronous time division multiplexing can waste unused transmission space. One solution to this problem is statistical time division multiplexing. Sometimes called asynchronous time division multiplexing, statistical time division multiplexing (Stat TDM) transmits data only from active users and does not transmit empty time slots. To transmit data only from active users, the multiplexor creates a more complex frame that contains data only from those input sources that have something to send. For example, consider the following simplified scenario. If four stations, A, B, C, and D, are connected to a statistical multiplexor, but only stations A and C are currently transmitting, the statistical multiplexor transmits only the data from stations A and C, as shown in Figure 5-8. Note that at any moment, the number of stations transmitting can change from two to zero, one, three, or four. If that happens, the statistical multiplexor must create a new frame containing data from the currently transmitting stations.

Figure 5-8
*Two stations out of
four transmitting via a
statistical multiplexor*

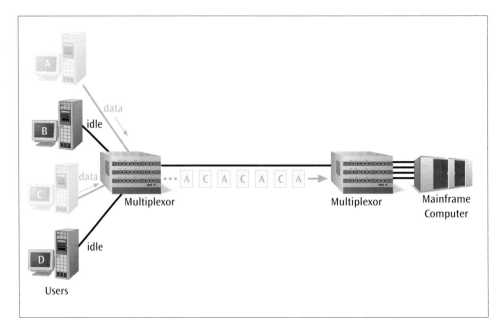

Because only two of the four stations are transmitting, how does the demultiplexor on the receiving end recognize the correct recipients of the data? Some type of address must be included with each byte of data, to identify who sent the data and for whom it is intended (Figure 5-9). The address can be as simple as a binary number that uniquely identifies the station that is transmitting. For example, if the multiplexor is connected to four stations, then the addresses can simply be 0, 1, 2, and 3 for stations A, B, C, and D. In binary, the values would be 00, 01, 10, and 11, respectively.

Figure 5-9
*Sample address and
data in a statistical
multiplexor
output stream*

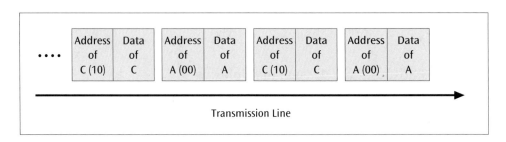

If the multiplexor transmits more than one byte of data at a time from each source, then an alternate form of address and data is required. To transmit pieces of data of variable sizes, a length field defining the length of the data block is included along with the address and data. This packet of *address/length/data/ address/length/data* is shown in Figure 5-10.

Figure 5-10
Packets of address, length, and data fields in a statistical multiplexor output stream

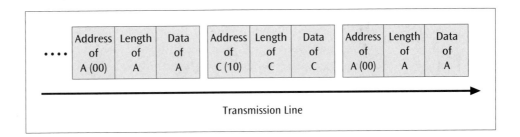

Finally, the sequence of *address/length/data/address/length/data...* is packaged into a larger unit by the statistical multiplexor. This larger unit, shown in Figure 5-11, is a more realistic example than Figure 5-10 and looks much like the frame that is transmitted using a synchronous connection. The flags at the beginning and end delimit the beginning and end of the frame. The control field provides information that is used by the sending and receiving multiplexors to control the flow of data between them. Last, the frame check sequence (FCS) provides information that the receiving multiplexor can use to detect transmission errors within the frame.

Figure 5-11
Frame layout for the information packet transferred between statistical multiplexors

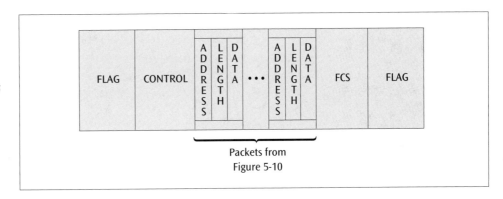

Wavelength Division Multiplexing

Although frequency division and time division are two very common multiplexing techniques, another multiplexing technique—wavelength division multiplexing—emerged a number of years ago and has since become a powerful alternative. When transmission systems employing fiber-optic cable were first installed (in the 1980s), the explosive growth of the Internet and other data transmission networks had not even been imagined. Now that the twenty-first century has begun, it is painfully obvious that early growth forecasts were gross underestimates. With Internet access growing by more than 100 percent per year and individuals requesting multiple telephone lines for use with faxes and modems, video transmissions, and teleconferencing, a single fiber-optic line (a pair for full-duplex operation) transmitting billions of bits per second is simply no longer sufficient. This inability of a single fiber-optic line to meet users' needs is called fiber exhaust. For many years, technology specialists saw few ways to resolve fiber exhaust other than by installing additional fiber lines, sometimes at great expense. Now there appears to be an attractive solution that takes advantage of currently installed fiber-optic lines—wavelength division multiplexing.

Wavelength division multiplexing (WDM) multiplexes multiple data streams onto a single fiber-optic line. It is, in essence, a frequency division multiplexing technique that assigns input sources to separate sets of frequencies. Wave division multiplexing uses different wavelength (frequency) lasers to transmit multiple signals at the same time over a single medium. The wavelength of each differently colored laser is called the lambda. Thus, WDM supports multiple lambdas.

The technique assigns a uniquely colored laser to each input source and combines the multiple optical signals of the input sources so that they can be amplified as a group and transported over a single fiber. It is interesting to note that because of the properties of the signals and glass fiber, plus the nature of light itself, each signal carried on the fiber can be transmitted at a different rate from the other signals. This means that a single fiber-optic line can support simultaneous transmission speeds such as 51.84 Mbps, 155.52 Mbps, 622.08 Mbps, and 2.488 Gbps (which, incidentally, are multiples of T-1 speeds and are defined as OC-1, OC-3, OC-12, and OC-48, the optical carrier specifications for high-speed fiber-optic lines). In addition, a single fiber-optic line can support a number of different transmission formats such as SONET, Asynchronous Transfer Mode (ATM), and others, in various combinations (see Figure 5-12).

Figure 5-12
Fiber-optic line using wavelength division multiplexing and supporting multiple-speed transmissions

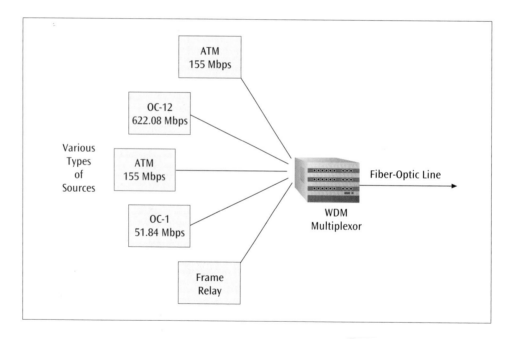

Wavelength division multiplexing is also scalable. As the demands on a system and its applications grow, it is possible to add additional wavelengths, or lambdas, onto the fiber, thus further multiplying the overall capacity of the original fiber-optic system. While most systems support fewer than 100 lambdas, some of the ultra high-priced systems can handle more than 100 lambdas. When WDM can support a large number of lambdas, it is often called dense wavelength division multiplexing (DWDM). This additional power does not come without a price tag, however. Dense wavelength division multiplexing is an expensive way to transmit signals from multiple devices due to the high number of differently colored lasers required in one unit. One less expensive variation on dense wavelength division multiplexing is coarse wavelength division multiplexing. Coarse wavelength division multiplexing (CWDM)

is a less expensive technology because it is designed for short-distance connections and has only a few lambdas, with a greater space between lambdas. Because the wavelengths are farther apart and not packed as closely together as they are in DWDM, the lasers used for coarse wavelength division multiplexing can be less expensive and do not require extensive cooling. Despite its cost and complexity, many technology experts predict that wavelength division multiplexing will remain a powerful technology.

While wavelength division multiplexing is a powerful technology that is relatively expensive and uncommon, the next type of multiplexing that we will examine—discrete multitone—is no less impressive. In addition to being quite common and inexpensive, discrete multitone is the technology behind the popular digital subscriber line (DSL) system.

Discrete Multitone

Discrete multitone (DMT) is a multiplexing technique commonly found in digital subscriber line (DSL) systems. DSL, as we have already seen, is a technology that allows a high-speed data signal to traverse a standard copper-based telephone line. We have also seen that the highest transmission speed we can achieve with a standard dial-up telephone line is 56 Kbps. DSL, however, is capable of achieving speeds into the millions of bits per second. How is this possible? The answer is the multiplexing technique DMT. DMT essentially combines hundreds of different signals, or subchannels, into one stream; unlike the previously discussed multiplexing techniques, however, DMT is designed such that all these subchannels are destined for a single user.

Details ▶

Additional Multiplexing Techniques

A number of new multiplexing techniques have appeared in the last several years, all of which are interesting and might have great promise. Three of these multiplexing techniques are Optical Spatial Division Multiplexing (OSDM), Orthogonal Frequency Division Multiplexing (OFDM), and Optical Time Division Multiplexing (OTDM). The first, Optical Spatial Division Multiplexing, allows for the multiplexing of "bursty" traffic (that is, traffic that comes in bursts and is produced by numerous voice and Internet data sources) onto an optical transmission technology that has not supported this kind of traffic well in the past. An example of one such technology is SONET. Because most, if not all, telephone companies use SONET somewhere in their high-speed backbone networks, the use of OSDM creates systems that can carry more traffic and perhaps even provide it at a lower cost.

A second multiplexing technique, Orthogonal Frequency Division Multiplexing, is a discrete multitone technology (used in DSL systems) that combines multiple signals of different frequencies into a single, more complex signal. Before the multiple signals are combined, each is individually phase-modulated. The phase-modulated signals are then combined to create a compact, high-speed data stream. OFDM is used in applications such as wireless local area networks, digital television, digital radio, and home AC power-line transmissions.

The third multiplexing technique, Optical Time Division Multiplexing, is similar to wavelength division multiplexing in that fiber-optic cables are used extensively. But where wavelength division multiplexing is a form of frequency division multiplexing, OTDM (as its name implies) is a form of time division multiplexing. An OTDM multiplexor combines the data from each input source into a high-speed time multiplexed stream. In the better systems, all input and output streams are optical, and the data, instead of changing to electrical form, remains in optical form throughout the multiplexing and demultiplexing phases. These all-optical systems are extremely fast (with speeds in the terabits-per-second range) and hold great promise for future applications.

The real power of DMT is the fact that each of the subchannels can perform its own quadrature amplitude modulation (QAM). (Recall from Chapter Two that a common example of QAM is one that involves a four-bit code in which eight phase angles have a single amplitude, and four phase angles have double amplitudes.) For example, one form of DMT supports 256 subchannels, each of which is capable of a 60-Kbps QAM modulated stream (Figure 5-13). Thus, 256 × 60 Kbps yields a 15.36-million-bps system. Unfortunately, because of noise, not all 256 subchannels can transmit at a full 60-Kbps rate. Those subchannels experiencing noise will modify their modulation technique and drop back to a slower speed. Thus, DSL systems that transmit data in the hundreds of thousands of bits per second are more the norm.

Figure 5-13
256 quadrature amplitude modulated streams combined into one DMT signal for DSL service

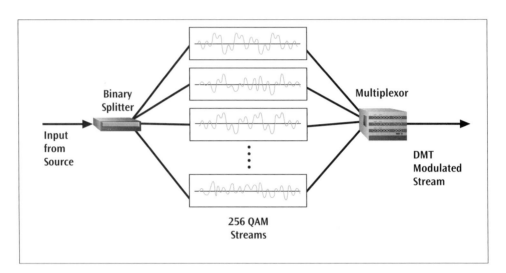

DMT is a fascinating technology that has been developed in the quest to increase data transmission speeds for the average consumer. Imagine one technology that can support 256 independently modulated streams, many of them transmitting at different speeds! Let us turn our attention to another multiplexing technique that is also enhancing the capabilities of existing technology: code division multiplexing.

Code Division Multiplexing

Also known as code division multiple access, **code division multiplexing (CDM)** is a relatively new technology that has been used extensively by both the military and cellular telephone companies. Whereas other multiplexing techniques differentiate one user from another by either assigning frequency ranges or interleaving bit sequences in time, code division multiplexing allows multiple users to share a common set of frequencies by assigning a unique digital code to each user.

More precisely, code division multiplexing is based upon a class of modulation techniques known as spread spectrum technology. Spread spectrum technology will be discussed in more detail in Chapter Eleven, but in brief terms, it is a technique used in the communications industry for modulating a signal into a new signal that is more secure and thus more resistant to wire-tapping. This technology falls into two categories—frequency hopping and direct sequence. Code division multiplexing uses direct sequence spread spectrum technology, a technique that spreads the transmission of a signal over a wide range of frequencies, using mathematical values. As the original data is input into a direct sequence modulator, each binary 1 and 0 is replaced with a larger, unique bit sequence. For example, each device in a cell phone market that uses code division multiplexing to transmit its signal is assigned its own bit sequence. When the bit sequences arrive at the destination station, the code division multiplexor is capable of telling one mobile device's bit sequence from another's. In actual cell phone systems, code division multiplexing is only used during the transmission from the mobile telephone office to the cell phones, not during transmission from the cell phones to the mobile telephone office. This is due to the synchronization problems inherent in code division multiplexing. Nonetheless, to keep the example simple, we will pretend that the cell phones are transmitting to the mobile telephone office using code division multiplexing.

Despite the fact that this is a fairly complex procedure, code division multiplexing is one of the more fascinating technologies in data communications, and it merits a little closer examination. Let us create an example using three mobile users: A, B, and C. (To aid in the understanding of this technology, this example simplifies the more technical concepts involved.) Suppose mobile user A has been assigned the binary code 11110000, mobile user B the code 10101010, and mobile user C the code 00110011. These binary codes are called the chip spreading codes. In the real world, these codes are 64 bits in length. To keep our example simple, we will use 8-bit codes. If mobile user A wishes to transmit a binary 1, it transmits instead its code—11110000. If mobile user A wishes to transmit a binary 0, it transmits the inverse of its code—00001111. Actually, the mobile user transmits a series of positive and negative voltages—a positive voltage for a 1 and a negative voltage for a 0. Now, for example, let us say mobile user A transmits a binary 1, mobile user B transmits a binary 0, and mobile user C transmits a binary 1. The following is actually transmitted:

Mobile user A sends a binary 1 (11110000), or ++++−−−−

Mobile user B sends a binary 0 (01010101), or −+−+−+−+

Mobile user C sends a binary 1 (00110011), or −−++−−++

The receiver receives all three signals at the same time and adds the voltages as shown below:

+	+	+	+	−	−	−	−
−	+	−	+	−	+	−	+
−	−	+	+	−	−	+	+
Sums: −1	+1	+1	+3	−3	−1	−1	+1

Then, to determine what each mobile user transmitted, the receiver multiplies the sums by the original code of each mobile user, expressed as + and − values, then takes the sum of those products:

Sums:	−1	+1	+1	+3	−3	−1	−1	+1
Mobile user A's code:	+1	+1	+1	+1	−1	−1	−1	−1
Products:	−1	+1	+1	+3	+3	+1	+1	−1
Sum of Products:	+8							

Because the Sum of Products is greater than or equal to +8 ($\geq +8$) in this 8-bit example, the value transmitted must have been a binary 1. In the real world, with the 64-bit system, the Sum of Products would have to be greater than or equal to

+64 ($\geq$ +64). If the Sum of Products were $\leq$ –8 (or $\leq$ –64 using real codes), the value transmitted would have been a binary 0.

The same procedure would be performed to determine mobile user B's transmitted value:

Sums:	–1	+1	+1	+3	–3	–1	–1	+1
Mobile user B's code:	+1	–1	+1	–1	+1	–1	+1	–1
Products:	–1	–1	+1	–3	–3	+1	–1	–1
Sum of Products:	–8							

Because the Sum of Products is $\leq$ –8, the value transmitted must have been a binary 0.

Now that we have examined how the various multiplexing techniques work, let us compare their advantages and disadvantages.

Comparison of Multiplexing Techniques

Frequency division multiplexing suffers from two major disadvantages. The first disadvantage is found in computer-based systems that multiplex multiple channels over a single medium. Because the frequencies are usually statically assigned, devices that do not have anything to transmit are still assigned frequencies, and thus bandwidth is wasted.

The second disadvantage of frequency division multiplexing is due to the fact that the technique uses analog signals, and analog signals are more susceptible to noise disruption than digital signals. Nonetheless, many different types of applications (such as television and radio) use frequency division multiplexing because of its simplicity, and the technique is probably going to be with us for a long time.

Synchronous time division multiplexing is also relatively straightforward, but as in frequency division multiplexing, input devices that have nothing to transmit can waste transmission space. The big advantage of synchronous TDM over frequency division multiplexing is the lower noise due to the use of digital signals during transmission. Even though T-1s use synchronous TDM, and T-1s are not going to go away over night, synchronous TDM is slowly being replaced with systems such as Ethernet, as we will see in a later chapter. Statistical TDM is one variation of TDM that transmits data only from those input devices that have data to transmit. Thus, statistical TDM wastes less bandwidth on the transmission link.

Statistical multiplexors have another very good advantage over synchronous time division multiplexors. Although both types of time division multiplexing can transmit data over a high-speed link, statistical time division multiplexing does not require as high-speed a line as synchronous time division multiplexing does. Statistical time division multiplexing assumes that all devices do not transmit at the same time; therefore, it does not require a high-speed link that is the total of *all* the incoming data streams. Another consequence of this assumption is that the output line capacity coming from the statistical multiplexor can be less than the output line capacity from the synchronous multiplexor, which also allows for a slower-speed link between multiplexors. This slower-speed link usually translates into lower costs.

One disadvantage of statistical multiplexors is their increased level of complexity. Synchronous TDM simply accepts the data from each attached device and transmits that data in an unending cycle. The statistical multiplexor must collect and buffer data from active attached devices and, after creating a frame with necessary control information, transmit that frame to the receiving multiplexor. Although this slightly higher level of complexity translates into higher initial costs, those costs are usually offset by the statistical TDM's ability to use a smaller-capacity interconnecting line.

Statistical time division multiplexing is a good choice for connecting a number of lower-speed devices that do not transmit data on a continuous basis to a remote computer system. Examples of these systems include data-entry systems, point-of-sale systems, and many other commercial applications in which users enter data at computer terminals.

Wavelength division multiplexing is a good technique for transmitting multiple concurrent signals over a fiber-optic line. Wavelength division multiplexing is also scalable. As the demands on a system and its applications grow, more wavelengths, or lambdas, can be added onto the fiber, thus further multiplying the overall capacity of the original fiber-optic system. Wavelength division multiplexing systems that use a large number of lambdas are termed *dense wavelength division multiplexing*, while those systems that use only a few lambdas are termed *coarse wavelength division multiplexing*. While wavelength division multiplexing can be a costly alternative, it may be less expensive than trying to install additional fiber-optic lines.

Discrete multitone technology is a unique form of multiplexing in that all the subchannels multiplexed together are intended for one user. Thus, discrete multitone does not directly compare with the other multiplexing techniques, in which each subchannel or channel is destined for a different user. However, discrete multitone is a complex technology and can suffer greatly from too much noise.

Finally, code division multiplexing, while using a fairly wide bandwidth of frequencies and a complex technology, is scalable like WDM and can produce system capacities that are 8 to 10 times those of frequency division multiplexing systems.

The advantages and disadvantages of each multiplexing technique are summarized in Table 5-3.

Table 5-3
Advantages and disadvantages of multiplexing techniques

Multiplexing Technique	Advantages	Disadvantages
Frequency Division Multiplexing	Simple Popular with radio, TV, cable TV All the receivers, such as cellular telephones, do not need to be at the same location	Noise problems due to analog signals Wastes bandwidth Limited by frequency ranges
Synchronous Time Division Multiplexing	Digital signals Relatively simple Commonly used with T-1, SONET	Wastes bandwidth
Statistical Time Division Multiplexing	More efficient use of bandwidth Frame can contain control and error information Packets can be of varying size	More complex than synchronous time division multiplexing
Wavelength Division Multiplexing	Very high capacities over fiber Signals can have varying speeds Scalable	Cost Complexity
Discrete Multitone	Capable of high transmission speeds	Complexity, noise problems
Code Division Multiplexing	Large capacities Scalable	Complexity Primarily a wireless technology

So far, with multiplexing, we have examined how multiple data streams can be variously combined to maximize the number of them that can be transmitted through different types of media, thus yielding a more efficient connection. Let us now examine another technique that can maximize the amount of data that can be transmitted at a time or stored in a given space—that is, the process known as compression.

Compression—Lossless versus Lossy

As we have already seen, compression is the process of taking data and somehow packing more of it into the same space, whether this is in the form of a storage device such as a hard drive or iPod, or a medium such as a fiber-optic line. When data is compressed for transmission, it transfers more quickly, because there is actually less of it, and this can result in a more efficient connection. Correspondingly, in terms of storage capacity, compression also allows for more data to be stored in the same amount of memory or disk space. The basic way to perform compression is to look for some common pattern in the data and replace each data pattern with a symbol or symbols that will consume less space during transmission or storage. For example, if a document contains a large number of occurrences of the word *snow*, the sender might want to replace the word *snow* with a symbol such as a percent sign, %. After the data is transmitted, the receiver then replaces the symbol % with the original word, *snow*. This replacement immediately raises two questions: How does the receiver know to replace the symbol % with *snow*? What happens if a percent sign (%) actually appears in the document as a percent sign? We certainly do not want the receiver replacing valid percent signs with the word *snow*. As we look at real examples of compression, you will see how these questions, and more like them, are addressed.

Before we examine some actual compression techniques, however, we should divide the compression process into two categories. If a compression technique compresses data and then decompresses it back into the original data, then it is referred to as a lossless technique. With a lossless compression technique, no data is lost due to compression. If a compression technique does lose some of the data as a result of the compression process, then it is referred to as a lossy compression technique. Consider as an example a bank that wishes to compress all of its customer accounts in order to increase its computer system's data storage space. Given the disaster that would ensue if the customer accounts were to lose data due to compression, the bank would obviously want to use a lossless compression technique to perform this task. On the other hand, suppose you wanted to copy a song from a compact disc to an iPod. To do this, you would first need to compress the song. During the compression process, if some of the data got lost, you might not even notice the loss, especially if the compression algorithm were designed to intentionally "lose" only those sounds which most human ears are not likely to detect. Because certain ranges of audio and video data cannot be detected easily, lossy compression algorithms are often used to compress music and video files, and thus are commonly incorporated into technological devices such as portable digital music players. To investigate the process of compression in more detail, let us start by examining the lossless techniques.

Lossless compression

One of the more common and simpler examples of lossless compression is run-length encoding. This technique replaces any repetitions of the same bit or byte that occur in a sequence of data with a single occurrence of the bit/byte and a run count, or simply with a run count. For example, this technique works at the binary level by counting either long strings (or runs) of binary 0s or long strings of binary 1s. Let us consider the following data string, which is composed predominantly of binary 0s:

00000100000000011000000000000000100001100000000000000000001000000

A compression technique based on run-length encoding would compress the 0s by first counting the "runs" of 0s—that is, it would start by counting the 0s until a binary 1 is encountered. If there are no 0s between a pair of 1s, then that pair would be considered a run that contains zero 0s. Performing this on our data string, we find the following runs:

5	9	0	15	4	0	20	6

Thus, in the first run, we encountered five 0s, while the second run had nine 0s. The third run had zero 0s because a 1 immediately followed a 1. In the next run, we encountered fifteen 0s, followed by a run of four 0s, zero 0s, twenty 0s, and finally six 0s.

The next step in this compression technique would be to convert each of the decimal values (5, 9, 0, 15, and so on) into 4-bit binary values, or nibbles. The only unique rule to follow during this conversion comes into play when you encounter a decimal value of 15 or greater. Because the largest decimal number that a 4-bit binary nibble can represent is 15 (which corresponds to four binary 1s—1111), you must convert a run that has a decimal value that is greater than 15 into multiple 4-bit nibbles. For example, a run of 20 would be converted into 1111 0101, in which the first nibble is the value 15, and the second nibble is the value 5. A caveat to this rule is that if you are converting the value of 15 itself, then you would also create two nibbles: 1111 followed by 0000. The reason for this is simply to be consistent—so that whenever a binary nibble of 1111 (or 15) is encountered, the following nibble (0000, which corresponds to a decimal value of 0) is added to that nibble.

Thus, converting the above runs of 5, 9, 0, 15, 4, 0, 20, and 6, would produce the following nibbles:

0101 1001 0000 1111 0000 0100 0000 1111 0101 0110

In this example, note that the original bit string, which consisted of 68 bits, is compressed to 40 bits—a reduction of 42%—and that no data has been lost (hence, the name lossless). One disadvantage of this technique is that it is worthwhile only if the original data consists predominantly of binary 0s. As we will see a little later in this chapter, run-length encoding is used in compressing video images (due to the presence of many zero values), as well as compressing other documents that have repeated characters.

A second technique that can be used to compress data when a lossless compression is necessary is the Lempel-Ziv technique. This technique is quite popular and is used by programs such as pkzip, WinZip, gzip, UNIX compress, and Microsoft compress. While the actual algorithm is fairly complex, it is possible to get a basic understanding of how the algorithm works. As the string to be transmitted is processed, the sender of the data creates a "dictionary" of character strings and associated codes. This set of codes is transmitted and the receiver then re-creates the dictionary and the original data string as the data codes are received.

The Lempel-Ziv algorithm can be fairly effective in compressing data. Studies have shown that computer program files can be reduced to 44 percent of the original size, text files can be reduced to 64 percent of the original size, and image files can be reduced to 88 percent of the original size.

It is also possible to compress music (and audio) files and not lose any of the musical content. While most portable music players use a lossy compression scheme (such as MP3), more users are turning towards lossless music compression in order to preserve a more-exact copy of their analog recordings. Likewise, most commercial users who digitize and compress analog recordings do not want to lose any of the original music. To this end, a number of lossless audio compression schemes are available. These include FLAC (Free Lossless Audio Codec), MPEG-4 ALS (Audio Lossless Coding), TTA, WavPak, Apple Lossless Encoder (ALE), and Monkey's Audio. While some of these schemes are proprietary, FLAC, TTA, and WavPak are free and/or open source. And most, if not all, can compress audio sources by at least 50%.

Lossy compression

All of the compression techniques described thus far have been examples of lossless compression. Lossless compression is necessary when the nature of the data is such that it is important that no data be lost during the compression and decompression stages. Like program, text, and image files, video images and higher-quality audio files (as we have just seen) can also be compressed using lossless compression, but the percentage of reduction is usually not as significant. This is due to the nature of the data in video and audio files—there is not one symbol or set of symbols that occur frequently enough to produce a reasonable level of compression. For example, if you take some video and digitize it, you will produce a long stream of binary 1s and 0s. To compress this stream, you can choose to perform a lossless run-length encoding on either the 1s or the 0s. Unfortunately, however, because this type of data is dynamic, there will probably not be enough repeating runs of either bit to produce a reasonable compression. Thus, you may want to consider some other compression techniques.

Music and video have other properties, however, that can be exploited in order to perform an effective compression. Let us consider music first. When one is listening to music, if two sounds play at the same time, the ear hears the louder one and usually ignores the softer one. Also, the human ear can hear sounds only within a certain range, which for an average person is 20 Hz to 20 kHz (20,000 Hz). Consequently, there are sounds, usually occurring at the extremes of the normal hearing range, that the human ear cannot hear well or even at all. Audio engineers take advantage of these (and other) facts to compress music through techniques called perceptual noise shaping, or perceptual encoding. If the perceptual encoding is performed well, the compressed version of an audio stream sounds fairly close to the uncompressed version (that is, almost CD-quality) even though some of the original data has been removed.

MP3, which is an abbreviation for MPEG (Moving Picture Experts Group) Audio Layer-3, is a common form of audio compression. (The Moving Picture Experts Group has also developed compression standards for HDTV broadcasts, digital satellite systems (DSS), and DVD movies.) After employing these perceptual encoding tricks, the MP3 encoder produces a data stream that has a much slower data rate than that of conventional CD-quality music. While a CD player is designed to reproduce music that has been encoded with 44,100 samples per second, which generates a data stream of 705,600 bits per second (44,100 samples per second times 16 bits per sample) or about 706 Kbps, an MP3 encoder typically reproduces a data stream of 128 Kbps to 192 Kbps. This kind of reduction in data leads to a 10 to 1 compression ratio for a typical song. Thus, the compression process reduces both the amount of data as well as the data transfer rate of the music within the music-generating device.

Video files can also be compressed by removing small details from the image that, in this case, the average human eye will not notice are missing. **JPEG**, which stands for Joint Photographic Experts Group, is a technique that is very commonly used to compress video images. The process of converting an image to JPEG format involves three phases: discrete cosine transformation, quantization, and run-length encoding. To perform the discrete cosine transformation, the image is broken into multiple 8 by 8 blocks of pixels, where each pixel represents either a single dot of color in a color image or a single shade of black and white in a black and white image. Each 8 by 8 block (corresponding to 64 pixels) is then subjected to a fairly common mathematical routine called the discrete cosine transformation. Essentially, what this transformation does is produce a new 8 by 8 block of values. These values, however, are now called spatial frequencies, which are cosine calculations of how much each pixel value changes as a function of its position in the block. Rather than deal with the mathematics of this process, let us examine two fairly simple examples. If we have an image with fairly uniform color changes over the area of the image—in other words, not a lot of fine details—then one of its 8 by 8 blocks of pixels might look something like the following block, where each decimal value represents a particular level of color:

15	18	21	24	28	32	36	40
19	22	25	28	32	36	40	44
22	25	28	32	36	40	44	48
26	29	32	35	39	43	47	51
30	34	38	42	46	51	56	61
34	38	42	46	51	56	61	66
38	42	46	51	56	61	66	72
43	48	53	58	63	68	74	80

After applying the discrete cosine transformation to these pixels, we would then have a set of spatial frequencies such as the following:

628	−123	12	−8	0	−2	0	−1
−185	23	−5	0	0	0	0	0
10	0	0	0	0	0	0	0
0	0	0	0	0	0	0	0
3	0	0	0	0	0	0	0
−1	0	0	0	0	0	0	0
0	0	0	0	0	0	0	0
0	0	0	0	0	0	0	0

Note the many zero entries, and that the nonzero entries are clustered toward the upper-left corner of the block. This is because of the discrete cosine calculations, which in essence depict the difference between one pixel's color relative to that of a neighbor's pixel rather than the absolute value of a particular pixel's color. The other reason for the clustering is that this image, as noted earlier, is one with fairly uniform color—that is, not a lot of color variation—and thus there is little change as you move away from the upper-left corner.

Suppose, however, that we have an image that has lots of fine detail. It will have an 8 by 8 block of pixels that has widely different values and may look like the following:

120	80	110	65	90	142	56	100
40	136	93	188	90	210	220	56
95	89	134	74	170	180	45	100
9	110	145	93	221	194	83	110
65	202	90	18	164	90	155	43
93	111	39	221	33	37	40	129
55	122	52	166	93	54	13	100
29	92	153	197	84	197	84	83

After applying the discrete cosine transformation to the pixels of this image, we would then have a set of spatial frequencies such as the following:

652	32	−40	54	−18	129	−33	84
111	−33	53	9	122	−43	65	100
−22	101	94	−32	23	104	76	101
88	33	211	2	−32	143	43	14
132	−32	43	0	122	−48	54	110
54	11	133	27	56	154	13	−94
−54	−69	10	109	65	0	27	−33
199	−18	99	98	22	−43	8	32

Notice that few zero entries are in this block of spatial frequencies. Let us continue with the conversion process by focusing on this block that corresponds to the image with lots of fine detail.

The second phase in the conversion of an image to a JPEG file is the quantization phase. The object of this phase is to try to generate more zero entries in the 8 by 8 block. To do this, we need to divide each value in the block by some predetermined number and disregard the remainders. For example, if the pixel block contains a spatial frequency with the value 9, we would divide this by 10 to get the result of 0. But we do not want to divide all 64 spatial frequencies by the same value, because the values in the upper-left corner of the block have more importance (due to the discrete cosine transformation operation). So let us divide the block of spatial frequencies with a block of values in which the upper-left corner of values are closer to 1—and thus will serve to reproduce the original number in a division. An example of such a block is as follows:

1	4	7	10	13	16	19	22
4	7	10	13	16	19	22	25
7	10	13	16	19	22	25	28
10	13	16	19	22	25	28	31
13	16	19	22	25	28	31	33
16	19	22	25	28	31	33	36
19	22	25	28	31	34	37	40
22	25	28	31	34	37	40	43

Now when we divide the block of spatial frequencies with this block of weighted values, we should produce a new block of values with more zero entries, as shown here:

652	8	−5	5	−1	8	0	3
27	−4	5	0	7	−2	2	4
−3	10	7	2	1	4	3	3
8	2	13	0	−1	5	1	0
10	−2	2	0	4	−1	1	3
3	0	6	1	2	4	0	−2
−2	−3	0	3	2	0	0	0
9	0	3	3	0	−1	0	0

A question you should ask at this point is, if we perform 64 divisions and toss out the remainders, would we not be losing something from the original image? The answer is yes, we will. But we hope to select an optimal set of values such that we do not lose too much of the original image. In other words, in striving to maximize the number of zeros in each block (so that we may successfully perform the run-length encoding of the final phase), we allow the data—i.e., the image—to change a little bit, but hopefully not so much that the human eye might detect gross differences between the original file and the one that has been compressed and decompressed.

Finally, the third phase of the JPEG compression technique is to take the matrix of quantized values and perform run-length encoding on the zeros. But the trick here is that you do not run-length encode the zeros by simply going up and down the rows of the 8 by 8 block. Instead, we take advantage of the fact that we would achieve longer runs of zeros if we encode on a diagonal, as is shown in Figure 5-14.

Figure 5-14
Run-length encoding of a JPEG image

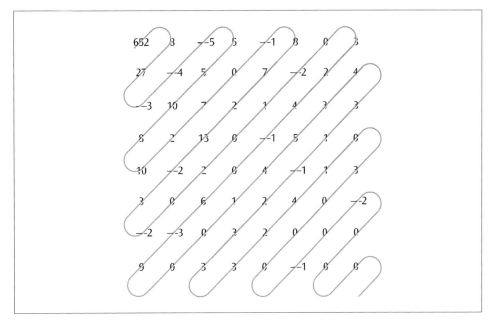

What about moving video images, such as those you encounter when watching digital television or a DVD? Does this type of data have a unique characteristic that we can exploit in order to do compression? As it turns out, it does. A video is actually a series of images. When these images, or frames, are shown in rapid succession, it appears as if the objects in the images are moving. In order to make it seem as though the movement (of characters, objects, scenery) in a movie is fluid, a movie projection device or television displays these images or frames at a rate of approximately 30 frames per second. But there is an interesting aspect of these frames—which you might have noticed if you ever tried to create a cartoon by drawing images on multiple sheets of paper and then flipping through the pages. Unless there is a complete scene change, one image looks very similar to the next. In the context of compression, the question to consider is: If successive images are very similar, why transfer the full image of each frame? Why not just transfer the *difference* between the two frames? This sort of transfer is an example of differential encoding. MPEG-1 and MPEG-2—or simply MPEG—are common examples of this form of compression.

Recall that a video device displays multiple (typically 30) frames per second. In order to save space, not all of those 30 frames are complete images. MPEG actually creates a complete frame of information, followed by several partial, or difference frames, followed by a complete frame. More precisely, the following frames are created:

I   B   B   P   B   B   I   B   B   P   ...

where the I-frame is a complete frame, the P-frame is the difference from the previous I-frame (and is created using motion-compensated prediction), and the B-frames are the difference frames, which contain the smaller differences between the I-frame and the P-frame and are inserted between I and P frames to smooth out the motion.

Because MPEG is computationally complex, processor chips have been designed, such as Intel's MMX Technology, specifically for the compression and decompression of MPEG images.

Business Multiplexing In Action ▶

XYZ Corporation has two buildings, A and B, separated by a distance of 300 meters, or roughly 1000 feet (see Figure 5-15). A 3-inch diameter tunnel runs underground between the two buildings. Building B contains 66 text-based terminals (also known as thin-client workstations as they don't have a hard drive) that need to be connected to a mainframe computer in Building A. The thin-client workstations transmit relatively low volumes of data equivalent to approximately 9600 bits per second. What are some good ways of connecting the thin-client workstations in Building B to the mainframe computer in Building A? Are there any that maximize the throughput through the connection?

Figure 5-15
Buildings A and B and the 3-inch-diameter tunnel connecting the buildings

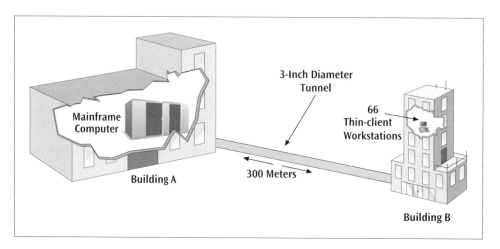

Considering the technologies that have been introduced in the text thus far, there are four possible scenarios for connecting the terminals and mainframe computer:

1. Connect each thin-client workstation to the mainframe computer using separate point-to-point lines. Each line will be some form of conducted medium.
2. Collect all the thin-client workstation outputs, and use microwave transmissions, Free-Space Optics, or WiMAX to send the data to the mainframe computer.
3. Collect all the thin-client workstation outputs using multiplexing, and send the data to the mainframe computer over a conducted-medium line.
4. Connect all the thin-client workstations to a local area network.

Let us examine the pros and cons of each solution.

The first solution of connecting each thin-client workstation to the mainframe computer using a point-to-point conducted medium has some advantages but also some serious drawbacks. The distance of 300 meters poses an immediate problem. When transmitting data in millions of bits per second, twisted pair typically has a maximum distance of 100 meters. XYZ Corporation's data is not being transmitted at millions of bits per second, but at 9600 bits per second instead. At this slower rate, we may be able to successfully transmit across a distance farther than 100 meters, but doing so may not be a good idea. Electromagnetic noise is always a potential problem, and we may discover after the installation of the wires that there is too much noise. A more noise-resistant medium, such as coaxial cable or fiber-optic cable, might be a reasonable option. But 66 coaxial cables (one for each thin-client workstation) will probably not fit in a 3-inch-diameter tunnel. Fiber-optic cable has roughly the same dimensions as coaxial cable and a much higher cost if you factor in the 66 pairs of optical devices that would be needed at the ends of the cables. Even if we could run 66 wires of some medium through the tunnel, what if management decides, a month after the installation, to add 10 more thin-client workstations to Building B? Ten additional cables are not likely to fit through the tunnel, and if they were to fit, pulling 10 more cables through would be time-consuming and expensive.

The main advantages of the first solution include the lower cost of using a relatively inexpensive conducted medium and the fact that multiple point-to-point lines eliminate the need for additional services such as polling or multiplexing.

The second solution, transmitting the data using microwave signals, Free-Space Optics, or WiMAX is interesting. All three technologies are very fast, and private ownership of the equipment can be attractive. The following concerns, however, are worth investigating:

> ► Is the line-of-sight between Building A and Building B obstructed by trees or other buildings? If there is an obstruction, microwave and free space optics will not work.
> ► What is the cost of installing a microwave, free space optic, or WiMAX system between the two buildings? If the cost is high, there may be a more reasonable alternative.
> ► The system would still need some kind of device that collects data from the 66 thin-client workstations and prepares a single data stream for transmission. Will the system handle this collection, or will we need something like a multiplexor?

Thus, microwave and free space optics are possible solutions if there is a clear line-of-sight between the two buildings and the associated costs are not too high. WiMAX, however, is capable of penetrating buildings and does not pose the same line-of-sight problems. Unfortunately, WiMAX is too new a technology currently for XYZ Corporation to consider.

The third solution—to install multiplexors at each end of the tunnel and connect the multiplexors with some type of high-speed medium—also requires some forethought and investigation. The following issues are worth considering:

▶ Can one pair of multiplexors handle 66 thin-client workstations? If not, we may have to install two pairs of multiplexors.

▶ What does a pair of multiplexors cost? Will this cost be so high that we are forced to consider other alternatives?

▶ What kind of medium could we use to connect the multiplexors? How many wires would we need to run? Fiber-optic cable or even coaxial cable would be a good choice. Even if the system required multiple strands of fiber or coaxial cable, they would fit within the 3-inch-diameter tunnel, because much fewer than 66 sets of cables would be necessary.

▶ Is the multiplexor solution scalable? Can the system expand to include additional terminals in the future? In the worst-case scenario, we would have to add an additional pair of multiplexors and another cable in the tunnel. We could plan ahead and pull several strands of fiber or coaxial cable through the tunnel so that we are prepared for future expansion.

Finally, the fourth solution is also worth considering. Is there someway that we can connect all the thin-client workstations to a local area network? Then as we saw in the first chapter, we could install the appropriate software on each thin-client that would allow the workstation to act as a terminal responding to polls from the mainframe. Unfortunately, we have not yet covered local area networks, so examination of this solution will have to wait.

In conclusion, it appears at this time that a multiplexing scheme provides the most efficient use of a small number of cables running through the small tunnel. If a high-quality cable such as fiber-optic wire is used, it will minimize noise intrusion and allow for the greatest amount of future growth. The microwave/free space optic solution is also attractive, but may cost more than a pair of multiplexors and connecting cables. WiMAX or local area networks might be very interesting solutions—ones we will have to keep an eye on in the near future.

◆ ◆

SUMMARY

▶ For multiple signals to share a single medium, the medium must be divided into multiple channels. The three basic techniques for dividing a medium into multiple channels are: a division of frequencies, a division of time, and a division of transmission codes.

▶ Frequency division multiplexing involves assigning nonoverlapping frequency ranges to different signals. Frequency division multiplexing uses analog signals, while time division multiplexing uses digital signals.

▶ Time division multiplexing of a medium involves dividing the available transmission time on a medium among the users. Time division multiplexing has two basic forms: synchronous time division multiplexing and statistical time division multiplexing.

▶ Synchronous time division multiplexing accepts input from a fixed number of devices and transmits their data in an unending repetitive pattern. T-1 and SONET/SDH telephone systems are common examples of systems that use synchronous time division multiplexing. The static assignment of input devices to particular frequencies or time slots can be wasteful if the input devices are not constantly transmitting data.

▶ Statistical time division multiplexing accepts input from a set of devices that have data to transmit, creates a frame with data and control information, and transmits that frame. Input devices that do not have data to send are not included in the frame.

▶ Wavelength division multiplexing involves fiber-optic systems and the transfer of multiple streams of data over a single fiber using multiple, colored laser transmitters. Wavelength division multiplexing systems can be dense or coarse.

▶ Discrete multitone is a technology used in DSL systems. Multiple subchannels, each supporting a form of quadrature amplitude modulation, are multiplexed together to provide one data stream for a user.

▶ Code division multiplexing allows multiple users to share the same set of frequencies by assigning a unique digital code to each user.

▶ Compression is a process that compacts data into a smaller package. When stored, compressed data saves space; when transmitted, it results in shorter transmission times.

▶ Two basic forms of compression exist: lossless, in which no data is lost during the compression and decompression stages; and lossy, in which some of the original data is lost.

▶ Two popular forms of lossless compression include run-length encoding and the Lempel-Ziv compression technique. A number of lossless audio compression schemes also exist.

▶ Lossy compression is the basis of a number of compression techniques, including MP3 for audio, JPEG for still images, and MPEG for moving video.

KEY TERMS

channel
chip spreading codes
coarse wavelength division
   multiplexing (CWDM)
code division multiplexing
   (CDM)
compression
demultiplexor
dense wavelength division
   multiplexing (DWDM)
discrete multitone (DMT)
DS-1 signaling
fiber exhaust
frequency division
   multiplexing (FDM)

guard band
JPEG
lambda
lossless compression
lossy compression
multiplexing
multiplexor
MP3
MPEG
perceptual encoding
run-length encoding
statistical time division
   multiplexing (Stat TDM)

Synchronous Digital
   Hierarchy (SDH)
Synchronous Optical
   Network (SONET)
synchronous time division
   multiplexing (Sync TDM)
synchronous transport
   signals (STS)
time division
   multiplexing (TDM)
T-1 multiplexing
wavelength division
   multiplexing (WDM)

REVIEW QUESTIONS

1. List three common examples of frequency division multiplexing.

2. Frequency division multiplexing is associated with what type of signals?

3. In what order does synchronous time division multiplexing sample each of the incoming signals?

4. What would happen if a synchronous time division multiplexor sampled the incoming signals out of order?

5. How does a synchronous time division multiplexor stay synchronized with the demultiplexor on the receiving end?

6. How many separate channels does a T-1 multiplexor combine into one stream?

7. How are a T-1 and SONET similar?

8. What are the main differences between statistical time division multiplexing and synchronous time division multiplexing?

9. If a statistical multiplexor is connected to 20 devices, does it require a high-speed output line that is equivalent to the sum of the 20 transmission streams? Defend your response.

10. Why is addressing of the individual data streams necessary for statistical multiplexing?

11. What type of medium is required to support wavelength division multiplexing?

12. How many different wavelengths can dense wavelength division multiplexing place onto one connection?

13. What is the difference between dense wavelength division multiplexing and coarse wavelength division multiplexing?

14. How is discrete multitone different from the other multiplexing techniques? How is it similar?

15. How does code division multiplexing distinguish one signal from another?

16. What are the two basic forms of compression?

17. Run-length encoding can be used to compress what kind(s) of data?

18. What are the three phases of JPEG compression?

EXERCISES

1. Compared to the other multiplexing techniques, state two advantages and two disadvantages of each of the following:
   a. frequency division multiplexing
   b. synchronous time division multiplexing
   c. statistical time division multiplexing
   d. wavelength division multiplexing

2. A benefit of frequency division multiplexing and code division multiplexing is that all the receivers do not have to be at the same location. Explain the consequences of this benefit and give an example.

3. Twenty-four *voice signals* are to be multiplexed and transmitted over twisted pair. What is the total bandwidth required if frequency division multiplexing is used?

4. Twenty voice signals are to be multiplexed and transmitted over twisted pair. What is the bandwidth required (in bps) if synchronous time division multiplexing is used, along with the standard analog-to-digital sampling rate, and each sample is converted into an 8-bit value?

5. If only four computers are transmitting digital data over a T-1 line, what is the maximum possible data rate for each computer?

6. What is the purpose of the synchronization bit in a T-1 frame? Why is it necessary?

7. Ten computer workstations are connected to a synchronous time division multiplexor. Each workstation transmits at 128 Kbps. At any point in time, 40 percent of the workstations are not transmitting. What is the minimum necessary speed of the line leaving the multiplexor? Will the answer be different if we use a statistical multiplexor instead? Explain your reasoning.

8. When data is transmitted using a statistical multiplexor, the individual units of data must have some form of address that tells the receiver the identity of the intended recipient of each piece of data. Instead of assigning absolute addresses to each piece of data, is it possible to incorporate relative addressing? If so, explain the benefits.

9. The telephone company has a fiber-optic line with time division multiplexing that runs from the United States to England and lies on the ocean floor. This fiber-optic line has reached capacity. What alternatives can the telephone company consider to increase capacity?

10. A discrete multitone system is using a modulation technique on its subchannels, each of which generates a 64-kbps stream. Assuming ideal conditions (no noise), what is the maximum data rate of the discrete multitone system?

11. The cell phone company in town uses code division multiplexing to transmit signals between its cell phones and the cell towers. You are using your cell phone while standing next to someone using her cell phone. How does the system distinguish the two signals?

12. Mobile user A is using code division multiplexing and has been assigned a binary code of 00001111. Mobile user B, also using code division multiplexing, has been assigned a binary code of 01010101. Mobile user A transmits a 1, while Mobile user B transmits a 0. Show the sum of products that results and your calculations.

13. How many frames per second does a T-1 and SONET transmit? Why this number?

14. Why is wavelength division multiplexing more like frequency division multiplexing and less like time division multiplexing?

15. Which of the multiplexing techniques can be used on both conducted media and wireless media, which on only conducted media, and which on only wireless media?

16. In theory, code division multiplexing can have 2^{64} different signals in the same area. In reality, this is not possible. Why not? Show an example.

17. Is the form of DSL that a company uses different from the form of DSL to which a home user subscribes? Explain.

18. If data has a large number of one type of symbol, which type of compression would be the most effective?

19. Given the following bit string, show the run-length encoding that would result: 000000010000011000000000000000000010000001110000000000

20. Can you compress a set of bank statements using JPEG compression? Explain.

21. MP3, JPEG, and MPEG all rely on what characteristic in the data in order to perform compression?

THINKING OUTSIDE THE BOX

1 A company has two buildings that are 50 meters (roughly 50 yards) apart. Between the buildings is private land owned by the company. A large walk-through tunnel connects the two buildings. In one building is a collection of 30 computer workstations; in the other building is a mainframe computer. What is the best way to connect the workstations to the mainframe computer? Explain your reasoning and all the possible solutions you considered.

2 A company has two buildings that are 100 meters (100 yards) apart. Between the buildings is public land with no access tunnel. In one building is a collection of 30 computer workstations; in the other building is a mainframe computer. What is the best way to connect the workstations to the mainframe computer? Explain your reasoning and all the possible solutions you considered.

3 What is the relationship, if one exists, between synchronous time division multiplexing and the synchronous and asynchronous connections described in Chapter Four?

4 Compare and contrast the older multiplexing techniques such as frequency division and time division multiplexing with the newer techniques such as discrete multitone and orthogonal frequency division multiplexing. What appears to be the trend in these newer protocols?

5 You are receiving high-speed Internet access from a DSL provider. You also live down the street from a radio station's AM broadcast antenna. Is your DSL affected by this antenna? Explain.

6 Consider a VGA screen that has 640 × 800 pixels per screen. Further assume that each pixel is 24 bits (8 for red, 8 for blue, and 8 for green). If a movie video presents 30 frames (images) per second, how many bytes will a two-hour movie require for storage? How many bytes can a standard DVD hold? What then must be the compression ratio?

HANDS-ON PROJECTS

1. Locate advertising material that lists the maximum number of devices a frequency or time division multiplexor can handle. Is this number consistent with what was presented in the chapter?

2. Digital broadcast television will someday replace conventional analog television. What form of multiplexing is used to broadcast digital television signals?

3. Broadcast radio is one of the last forms of entertainment to go digital. Find the latest material describing the current state of digital broadcast radio, and write a two- or three-page report that includes the type of multiplexing envisioned and the impact digital radio will have on the current radio market.

4. The FCC created a set of frequencies for walkie-talkie radios. This set is called the family radio service and allows two radios to transmit up to a several-mile distance. What kind of multiplexing is used with these radios? How many concurrent channels are allowed? Is there a technology newer than family radio service? If so, describe its characteristics.

5. The local loop of the telephone circuit that enters your house uses multiplexing so that the people on the two ends of the connection can talk at the same time (if they wish). What kind of multiplexing is used? State the details of the multiplexing technique.

6. Numerous forms of MPEG compression (such as MPEG-1, MPEG-2, etc.) exist. List each of the forms with an accompanying sentence describing what type of data for which that compression form is designed.

7. What other compression schemes exist besides those listed in this chapter? What are the uses of each compression scheme?

6

Errors, Error Detection, and Error Control

◆ ◆

DURING THE SUMMER OF 2003, the planet Mars came within 34,646,418 miles of Earth. This was the closest Mars had been to Earth in 59,619 years. Even at this distance, however, a radio signal would take 3 minutes and 6 seconds to travel between the planets. Now picture a probe on Mars that can traverse the Martian landscape and make maneuvers such as going forward, turning right, taking a picture, and digging a sample. The probe can either make its own decisions about what to do next, or—more likely, for safety reasons—send a picture back to Earth and then await instructions from scientists. If it takes 3 minutes and 6 seconds for a signal to propagate to Earth, several minutes for the scientists to decide what the probe should do, and then 3 minutes and 6 seconds for the signal to propagate back to Mars, the Martian probe is not likely to complete tasks very quickly. Now let us throw in one more wrinkle. Let us suppose that during that 34.6-million-mile

data transfer, the signal becomes corrupted and the receiving side has to ask the transmitting side to resend because it could not read the data. In this case, making the probe perform a simple operation could take the scientists hours. Given this, future versions of the Martian probe should probably be designed to make their own decisions rather than wasting so much time sending information back and forth between Earth and Mars. In the meantime, however, the best way to resolve such time delays might be to improve the way the transmitted signal is sent.

Is noise this much of a problem when transmitting signals?

Isn't there a way to send a signal such that the entire signal does not have to be re-sent in the event of an error?

Is it possible for an error to occur during transmission and go undetected?

Objectives ▶

After reading this chapter, you should be able to:

▶ Identify the different types of noise commonly found in computer networks

▶ Specify the different error-prevention techniques, and be able to apply an error-prevention technique to a type of noise

▶ Compare the different error-detection techniques in terms of efficiency and efficacy

▶ Perform simple parity and longitudinal parity calculations, and enumerate their strengths and weaknesses

▶ Cite the advantages of cyclic redundancy checksum, and specify what types of errors cyclic redundancy checksum will detect

▶ Cite the advantages of arithmetic checksum, and specify what types of applications use arithmetic checksum

▶ Differentiate between the basic forms of error control, and describe the circumstances under which each may be used

▶ Follow an example of a Hamming self-correcting code

Introduction

The process of transmitting data over a medium often works according to Murphy's Law: If something can go wrong, it probably will. Even if all possible error-reducing measures are applied before and during data transmission, something will invariably alter the form of the original data. If this alteration is serious enough, the original data becomes corrupt, and the receiver does not receive the data that was originally transmitted. Even with the highest-quality fiber-optic cable, noise eventually creeps in and begins to disrupt data transmission. Thus, despite one's best efforts to control noise, some noise is inevitable. When the ratio of noise power to signal power becomes such that the noise overwhelms the signal, errors occur. It is at this point that error-detection techniques become valuable tools.

Given that noise is inevitable and errors happen, something needs to be done to detect error conditions. This chapter examines some of the more common error-detection methods and compares them in terms of efficiency and efficacy.

Before you begin to learn about error-detection techniques, it is vital that you understand the different forms of noise that commonly occur during data transmission. Having a better understanding of the different types of noise and what causes them to occur will enable you to apply noise-reduction techniques to communication systems, and thus limit the amount of noise before it reaches the threshold at which errors occur.

Once an error has been detected, what action should a receiver take? There are three error control options: ignore the error, return an error message to the transmitter, or correct the error without help from the transmitter. Although ignoring the error seems like an irresponsible position, it has merit and is worth examining. The second option—return an error message to the transmitter so that the transmitter can resend the original data—is the most common error control action. The third option—correcting the error without asking for additional help from the transmitter—may sound like the ideal situation, but it is difficult to support and requires a significant amount of overhead.

How do error detection and error control fit into the TCP/IP suite introduced in Chapter One? Most people associate error detection with the data link/network access layer. When the data link layer creates a frame, it usually inserts an error-checking code after the data field. When the frame arrives at the next station, this error-checking code is extracted and the frame is checked for accuracy. But the data link layer is not the only layer that performs some type of error detection. The transport layer also includes an error-detection scheme. When the transport packet arrives at the final destination (and only the final destination), the receiver may extract an error-checking code from the transport header and perform error detection. Also, some network layer protocols, such as the Internet Protocol (IP), include an error-detection code in the network layer header. In the case of IP, however, the error detection is performed on only the IP header, and not the data field. Many applications also perform some type of error check such as detecting lost packets from a sequence of transmitted packets. For the moment, we will concentrate on the error-detection and error-control details covered in the data link and transport layers. Note, however, that all the topics discussed so far in this book, with the possible exception of asynchronous, synchronous, and isochronous connections, have been physical layer activities. To understand how error detection and control fit into the many layers of a computer network, it is important to recall the basic concept behind the TCP/IP protocol suite and the OSI model: the activities of one layer should not be affected by the activities of another layer. Thus, the selection of an error-detection scheme or error-control technique is an issue separate from the type of medium selected or the choice of multiplexing technique. Any of the error-detection and control schemes presented in this chapter can be applied to any type of communication system.

Noise and Errors

As you might expect, a number of errors can occur during data transmission. From a simple blip to a massive outage, transmitted data—both analog and digital—is susceptible to many types of noise and errors. Copper-based media have traditionally been plagued by many types of interference and noise. Satellite, microwave, and radio networks are also prone to interference and crosstalk. Even the near-perfect fiber-optic cables can introduce errors into a transmission system, although the probability of this happening is less than with the other types of media. Let us examine several of the major types of noise that occur in transmission systems.

White noise

White noise, which is also called thermal noise or Gaussian noise, is a relatively continuous type of noise and is much like the static you hear when a radio is being tuned between two stations. It is always present to some degree in transmission media and electronic devices and is dependent on the temperature of the medium. As the temperature increases, the level of noise increases due to the increased activity of the electrons in the medium. Because white noise is relatively continuous, it can be reduced significantly but never completely eliminated. White noise is the type of interference that makes an analog or digital signal become fuzzy (Figure 6-1).

Reducing white noise from a digital signal is relatively straightforward if the signal is passed through a signal regenerator before the noise completely overwhelms the original signal. Reducing white noise from an analog signal is also possible and involves passing the noisy analog signal through an appropriate set of filters, which (one hopes) leaves nothing but the original signal.

Impulse noise

Impulse noise, or noise spike, is a noncontinuous noise and one of the most difficult errors to detect, because it can occur randomly. The difficulty comes in separating the noise from the signal. Typically, the noise is an analog burst of energy. If the impulse spike interferes with an analog signal, removing it without affecting the original signal can be difficult. Recall the example from Chapter Two with regard to scratched record albums. With albums, the impulse spikes correspond to the loud pops and clicks that are produced when an album is played that can interfere with some people's enjoyment of the music For a second example of impulse noise, consider what happens when you listen to AM radio during a thunderstorm. The lightning strikes in the area cause severe static on the radio—so severe that you cannot hear the normal radio transmissions.

Figure 6-1
White noise as it interferes with a digital signal

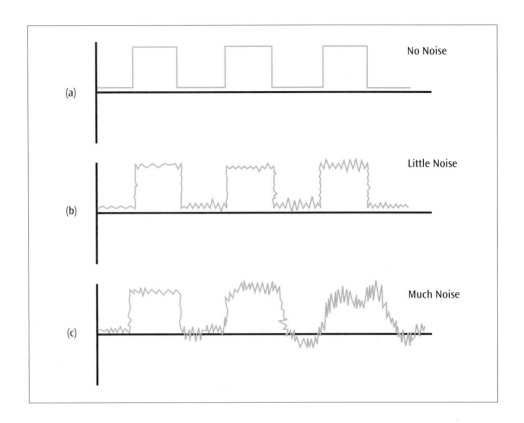

Figure 6-2
The effect of impulse noise on a digital signal

If impulse noise interferes with a digital signal, often the original digital signal can be recognized and recovered. When the noise completely obliterates the digital signal, the original signal cannot be recovered (see Figure 6-2).

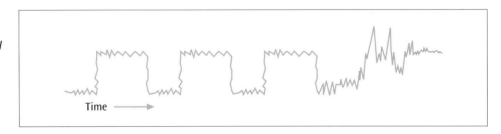

Noise is a problem for both analog and digital signals, of course, but with digital signals, transmission speed can affect whether or not noise is significant. In fact, sometimes the influence of transmission speed is quite dramatic and can be easily demonstrated. Figure 6-3 shows a digital signal transmission at relatively slow speed and at relatively high speed. Notice in the figure that when transmission speed is slower, you can still determine the value of a signal, but when transmission speed increases, you can no longer determine whether the signal is a 0 or a 1.

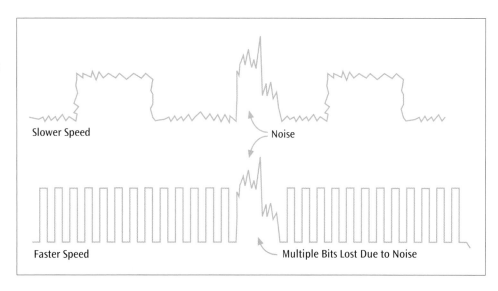

Figure 6-3
Transmission speed and its relationship to noise in a digital signal

Slower Speed

Noise

Faster Speed

Multiple Bits Lost Due to Noise

Crosstalk

Crosstalk is an unwanted coupling between two different signal paths. This unwanted coupling can be electrical, as might occur between two sets of twisted pair wire (as in a phone line), or it can be electromagnetic (as when unwanted signals are picked up by microwave antennas). Telephone signal crosstalk was a more common problem 20 to 30 years ago, before telephone companies used fiber-optic cables and other well-shielded wires. When crosstalk occurs during a phone conversation, you can hear another telephone conversation in the background (Figure 6-4). High humidity and wet weather can cause an increase in electrical crosstalk over a telephone system. Even though crosstalk is relatively continuous, it can be reduced with proper precautions and hardware, as you will see shortly.

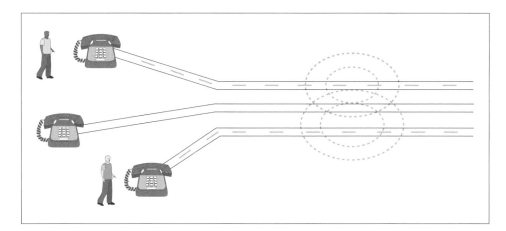

Figure 6-4
Three telephone circuits experiencing crosstalk

Echo

Echo is the reflective feedback of a transmitted signal as the signal moves through a medium. Much like the way a voice will echo in an empty room, a signal can hit the end of a cable, bounce back through the wire, and interfere with the original signal. This error occurs most often at junctions where wires are connected or at the open end of a coaxial cable. Figure 6-5 demonstrates a signal bouncing back from the end of a cable and creating an echo. To minimize the effect of echo, a device called an echo suppressor can be attached to a line. An echo suppressor is essentially a filter that allows the signal to pass in one direction only. For local area networks that use coaxial cable, a small filter is usually placed on the open end of each wire to absorb any incoming signals.

Figure 6-5
A signal bouncing back at the end of a cable and causing echo

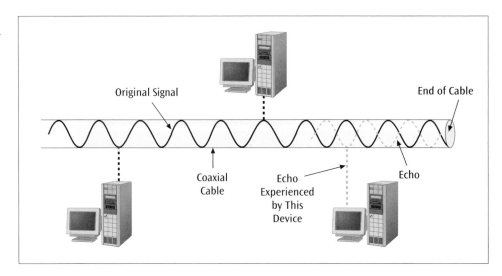

Jitter

Jitter is the result of small timing irregularities that become magnified during the transmission of digital signals as the signals are passed from one device to another. To put it another way, when a digital signal is being transmitted, the rises and falls of the signal can start to shift, or become blurry, and thus produce jitter. If unchecked, jitter can cause video devices to flicker, audio transmissions to click and break up, and transmitted computer data to arrive with errors. If jitter becomes too great, correcting it can require the transmitting devices to slow down their transmission rates, which in turn limits overall system performance. Figure 6-6 shows a simplified example of a digital signal experiencing jitter.

Figure 6-6
Original digital signal and digital signal with jitter

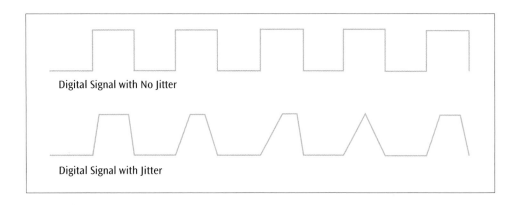

Causes of jitter can include electromagnetic interference, crosstalk, passing the signal through too many repeaters, and the use of lower-quality equipment. Possible solutions to the jitter problem involve installing proper shielding, which can reduce or eliminate electromagnetic interference and crosstalk, and limiting the number of times a signal is repeated.

Attenuation

Attenuation is the continuous loss of a signal's strength as it travels though a medium. It is not necessarily a form of error, but can indirectly lead to an increase in errors affecting the transmitted signal. As you learned in Chapter Two, attenuation can be eliminated with the use of amplifiers for analog systems or repeaters for digital systems.

Error Prevention

Because there are so many forms of noise and errors, and because the presence of one form of noise or another in a system is virtually a given, every data transmission system must take precautions to reduce noise and the possibility of errors. An unfortunate side effect of noise during a transmission is that the transmitting station has to slow down its transmission rate. For this reason, when a modem first makes a connection with another modem, the two modems participate in fallback negotiation. This means that if the transmitting modem sends data and the data arrives garbled, the receiving modem may ask the transmitting modem to fall back to a slower transmission speed. This slowdown creates a signal in which the bit duration of each 0 and 1 is longer, or in the case of quadrature amplitude modulation, a constellation diagram in which there are fewer discrete levels or targets, thus giving the receiver a better chance of distinguishing one value from the next, even in the presence of noise. If you can reduce the possibility of noise before it happens, however, the transmitting station may not have to slow down its transmission stream.

You can prevent the occurrence of many types of transmission errors by applying proper error-prevention techniques, including those listed below:

- ▶ Install wiring with the proper shielding to reduce electromagnetic interference and crosstalk.
- ▶ Be aware that many different types of wireless applications share the same wireless frequencies. Even some non-wireless devices generate signals that can interfere with wireless applications. For example, microwave ovens can interfere with wireless LAN signals.
- ▶ Replace older equipment with more modern, digital equipment; although initially expensive, this technique is often the most cost-effective way to minimize transmission errors in the long run.
- ▶ Use the proper number of digital repeaters and analog amplifiers to increase signal strength, thus decreasing the probability of errors.
- ▶ Observe the stated capacities of a medium, and, to reduce the possibility of errors, avoid pushing transmission speeds beyond their recommended limits. For example, recall from Chapter Two that twisted pair Category 5e/6 cable should not be longer than the recommended 100-meter (300-foot) distance when it is transmitting at 100 Mbps.

Reducing the number of devices, decreasing the length of cable runs, and reducing the transmission speed of the data may also be effective ways to reduce the possibility of errors. Although choices like these are not always desirable, sometimes they are the most reasonable alternatives available.

Table 6-1 lists the different types of errors that can arise and includes one or more possible error-prevention techniques for each.

Table 6-1
Summary of errors and error-prevention techniques

Type of Error	Error-Prevention Technique
White noise	Install special filters for analog signals; implement digital signal regeneration for digital signals
Impulse noise	Install special filters for analog signals; implement digital signal processing for digital signals
Crosstalk	Install proper shielding on cables
Echo	Install proper termination of cables
Jitter	Use better-quality electronic circuitry, use fewer repeaters, slow the transmission speed
Attenuation*	Install device that amplifies analog signals; implement digital signal regeneration of digital signals

* Not a type of error, but indirectly affects error

Do not be lured into thinking that simply because various error-prevention techniques have been applied, errors will not happen. Appropriate error-detection methods still need to be implemented. Let us examine the main error-detection techniques next.

Error Detection

Despite one's best attempts at prevention, errors still occur. Because most data transferred over a communication line is important, it is usually necessary to apply an error-detection technique to the received data to ensure that no errors were introduced into the data during transmission. If an error is detected, a typical response is to perform some type of request for transmission.

Error detection can be performed in several places within a communications model. One of the most common places is the data link layer. When a device creates a frame of data at the data link layer, it inserts some type of error-detection code. When the frame arrives at the next device in the transmission sequence, the receiver extracts the error-detection code and applies it to the data frame. Then the data frame is reconstructed and sent to the next device in the transmission sequence. Some protocols perform an error detection routine at the final destination. As we saw in Chapter One, TCP performs error detection at the end points of the connection.

Regardless of where the error detection is applied, all systems still recognize the importance of checking for transmission errors. The error-detection techniques themselves can be relatively simple or relatively elaborate. As you might expect, simple techniques do not provide the same degree of error checking as the more elaborate schemes. For example, the simplest error-detection technique is simple parity, which adds a single bit to a character of data, but catches the fewest number of errors. At the other end of the spectrum is the most elaborate and most effective technique available today—cyclic redundancy checksum. Not only is cyclic redundancy checksum more complex than simple parity, typically adding 8 to 32 check bits of error-detection code to a block of data, it is the most effective error-detection technique ever devised. Let us examine four error-detection techniques and evaluate the strengths and weaknesses of each.

Parity checks

The most basic error-detection techniques are parity checks, which are commonly used with asynchronous connections. Although there are various forms of single-character parity checking, one fact remains constant: Parity checks let too many errors slip through undetected. For this reason alone, parity checks are rarely, if ever, used on any kind of serious data transmissions. Despite this, two forms of parity checks—simple parity and longitudinal parity—do still exist and are worth examining.

Simple Parity

Simple parity (occasionally known as vertical redundancy check) is the easiest error-detection method to incorporate into a transmission system; it comes in two basic forms: even parity and odd parity. The basic concept of parity checking is that a bit is added to a string of bits to create either even parity or odd parity. With even parity, the 0 or 1 added to the string produces an even number of binary 1s. With odd parity, the 0 or 1 added to the string produces an odd number of binary 1s. If the 7-bit ASCII character set is used, a parity bit is added as the eighth bit. Suppose, for example, that the character "k"—which is 1101011 in binary—is transmitted and even parity is being applied. In this case, a parity bit of 1 would be added to the end of the bit stream, as follows: 11010111. There is now an even number (six) of 1s. (If odd parity were used, a 0 would be added at the end, resulting in 11010110.)

Now, if a transmission error causes one of the bits to be flipped (the value is erroneously interpreted as a 0 instead of a 1, or vice versa), the error can be detected if the receiver understands that it needs to check for even parity. Returning to the example of the character "k" sent with even parity, if you send 11010111 but 01010111 is received, the receiver will count the 1s, see that there is an odd number, and know there is an error.

What happens if 11010111 with even parity is sent and two bits are corrupted? For example, 00010111 is received. Will an error be detected? The answer is no, an error will not be detected, because the number of 1s is still even. Simple parity can detect only an odd number of erroneous bits per character. Is it possible for more than one bit in a character to be altered as a result of transmission error? Yes. Isolated single-bit errors occur 50 to 60 percent of the time. Error bursts, in which two erroneous bits are separated by fewer than 10 uncorrupted bits, occur 10 to 20 percent of the time.

Note that when the 7-bit ASCII character set is used, a parity bit is added for every 7 bits of data, resulting in a 1:7 ratio of parity bits to data bits. Thus, simple parity produces relatively high ratios of check bits to data bits, while achieving only mediocre (50 percent) error-detection results.

Longitudinal Parity

Longitudinal parity, sometimes called longitudinal redundancy check or horizontal parity, tries to solve the main weakness of simple parity—that all even numbers of errors are not detected. To provide this extra level of protection, longitudinal parity needs to use additional parity check bits, as you will see shortly. The first step of this parity scheme involves grouping individual characters together in a block, as shown in Table 6-2. Each character (also called a row) in the block has its own parity bit. In addition, after a certain number of characters are sent, a row of parity bits, or a block character check, is also sent. Each parity bit in this last row is a parity check for all the bits in the column above it. If one bit is altered in Row 1, the parity bit at the end of Row 1 signals an error. In addition, the parity bit for the corresponding column also signals an error. If two bits in Row 1 are flipped, the Row 1 parity check will not signal an error, but two column parity checks will signal errors. This is how longitudinal parity is able to detect more errors than simple parity. Note, however, that if two bits are flipped in Row 1 and two bits are flipped in Row 2, and the errors occur in the same column, no errors will be detected. This scenario, which is shown in Table 6-3, is a limitation of longitudinal parity.

Table 6-2

Simple example of longitudinal parity

	Data							Parity
Row 1	1	1	0	1	0	1	1	1
Row 2	1	1	1	1	1	1	1	1
Row 3	0	1	0	1	0	1	0	1
Row 4	0	0	1	1	0	0	1	1
Parity Row	0	1	0	0	1	1	1	0

Table 6-3

The second and third bits in Rows 1 and 2 have errors, but longitudinal parity does not detect the errors

	Data							Parity
Row 1	1	1 0	0 1	1	0	1	1	1
Row 2	1	1 0	1 0	1	1	1	1	1
Row 3	0	1	0	1	0	1	0	1
Row 4	0	0	1	1	0	0	1	1
Parity Row	0	1	0	0	1	1	1	0

Although longitudinal parity provides an extra level of protection by using a double parity check, this method, like simple parity, also introduces a high number of check bits relative to data bits, with only slightly better than mediocre error-detection results. If n characters in a block are transmitted, the ratio of check bits to data bits is n+8:7n. In other words, to transmit a 20-character block of data, for example, a simple parity bit needs to be added to each of the 20 characters, plus a full 8-bit block character check is added at the end, producing a ratio of check bits to data bits that is 28:140, or 1:5.

Arithmetic checksum

Many higher-level protocols used on the Internet (such as TCP and IP) use a form of error detection in which the characters to be transmitted are "summed" together. This sum is then added to the end of the message and the message is transmitted to the receiving end. The receiver accepts the transmitted message and performs the same summing operation and essentially compares its sum with the sum that was generated by the transmitter. If the two sums agree, then no error occurred during the transmission. If the two sums do not agree, the receiver informs the transmitter that an error has occurred. Since the sum is generated by performing relatively simple arithmetic, this technique is often called arithmetic checksum.

More precisely, let us consider the following example. Suppose we want to transmit the message "This is cool." In ASCII (from Chapter Two), that message would appear in binary as: 1010100 1101000 1101001 1110011 0100000 1101001 1110011 0100000 1100011 1101111 1101111 1101100 0101110. (Do not forget the blanks between the words and the period at the end of the sentence.)

TCP and IP actually add these values in binary to create a binary sum. But binary addition of so many operands is pretty messy. So that we do not have to add all these binary values, let us convert the binary values to their decimal form. In case you do not know binary, we will do it for you. The first binary value—1010100—is the value 84 in decimal. 1101000 equals 104. The next binary value 1101001 equals 105; 1110011 equals 115; 0100000 equals 32; 1101001 equals 105; 1110011 equals 115; 0100000 equals 32; 1100011 equals 99; 1101111 equals 111; 1101111 again is 111; 1101100 equals 108; and 0101110 equals 46. If we add this column of values, we will get the following:

```
        84
       104
       105
       115
        32
       105
       115
        32
        99
       111
       108
   +    46
      1056
```

The sum 1056 is then added to the outgoing message and sent to the receiver. The receiver will take the same characters, add their ASCII values, and if there were no errors during transmission, should get the same sum of 1056. Once again, the calculations in TCP and IP are performed in binary with a little more complexity, but hopefully you get the idea.

Worth noting is that the arithmetic checksum is relatively easy to compute and performs a fairly good job of error detection. Clearly, if noise messes up a bit or two during transmission, the receiver is more than likely not going to get the same sum. One can imagine that if the planets and stars align just right, it would be possible to "lower" the value of one character while "raising" the value of a second character just perfectly so that the sum comes out exactly the same. But the odds of that occurring are small. Nonetheless, it leaves us wondering if there is another error detection method that has odds so small that almost no error can escape detection. Indeed there is—the cyclic redundancy checksum.

Cyclic redundancy checksum

Unlike the simple parity and longitudinal parity techniques of error detection, which produce high ratios of check bits to data bits with only mediocre error-detection results, the cyclic redundancy checksum (CRC), or cyclic checksum, method typically adds 8 to 32 check bits to potentially large data packets and yields an error detection capability approaching 100 percent.

The CRC error-detection method treats the packet of data to be transmitted (the message) as a large polynomial. The rightmost bit of the data becomes the x^0 term, the next data bit to the left is the x^1 term, and so on. When a bit in the message is 1, the corresponding polynomial term is included. Thus, the data 101001101 would be equivalent to the polynomial:

x^8		$+ x^6$			$+ x^3$	$+ x^2$		$+ x^0$
1	0	1	0	0	1	1	0	1

(Because any value raised to the 0th power is 1, the x^0 term is always written as a 1.) The transmitter takes this message polynomial and, using polynomial arithmetic, divides it by a given generating polynomial, and produces a quotient and a remainder. The quotient is discarded, but the remainder (in bit form) is appended to the end of the original message polynomial, and this combined unit is transmitted over the medium. When the data plus remainder arrive at the destination, the same generating polynomial is used to detect an error. A generating polynomial is an industry-approved bit string used to create the cyclic checksum remainder. Some common generating polynomials in widespread use include:

- ▸ CRC-12: $x^{12} + x^{11} + x^3 + x^2 + x + 1$
- ▸ CRC-16: $x^{16} + x^{15} + x^2 + 1$
- ▸ CRC-CCITT: $x^{16} + x^{15} + x^5 + 1$
- ▸ CRC-32: $x^{32} + x^{26} + x^{23} + x^{22} + x^{16} + x^{12} + x^{11} + x^{10} + x^8 + x^7 + x^5 + x^4 + x^2 + x + 1$
- ▸ Asynchronous Transfer Mode CRC: $x^8 + x^2 + x + 1$

The receiver divides the incoming data (the original message polynomial plus the remainder) by the exact same generating polynomial used by the transmitter. If no errors were introduced during data transmission, the division should produce a remainder of zero. If an error was introduced during transmission, the arriving original message polynomial plus the remainder will not divide evenly by the generating polynomial and will produce a nonzero remainder, signaling an error condition.

In real life, the transmitter and receiver do not perform polynomial division with software. Instead, hardware designed into an integrated circuit can perform the calculation much more quickly.

The CRC method is almost foolproof. Table 6-4 summarizes the performance of the CRC technique. In cases where the size of the error burst is less than r + 1, where r is the degree of the generating polynomial, error detection is 100 percent. For example, suppose the CRC-CCITT is used, and so the degree, or highest power, of the polynomial is 16. In this case, if the error burst is less than r + 1 or 17 bits in length, CRC will detect it. Only in cases where the error burst is greater than or equal to r + 1 bits in length is there a chance that CRC may not detect the error. The chance or probability of an error burst of size r + 1 being detected is $1 - (½)^{(r-1)}$. Assuming again that r = 16, $1 - (½)^{(16-1)}$ equals 1 − 0.0000305, which equals 0.999969. Thus, the probability of a large error being detected is very close to 1.0 (100 percent).

Table 6-4
Error-detection performance of cyclic redundancy checksum

Type of Error	Error Detection Performance
Single-bit errors	100 percent
Double-bit errors	100 percent, as long as the generating polynomial has at least three 1s (they all do)
Odd number of bits in error	100 percent, as long as the generating polynomial contains a factor $x + 1$ (they all do)
An error burst of length < r + 1	100 percent
An error burst of length = r + 1	probability = $1 - (½)^{(r-1)}$ (very near 100%)
An error burst of length > r + 1	probability = $1 - (½)^{r}$ (very near 100%)

Details ▶

Cyclic Redundancy Checksum Calculation

Hardware performs the process of polynomial division very quickly. The basic hardware used to perform the calculation of CRC is a simple register that shifts all data bits to the left every time a new bit is entered. The unique feature of this shift register is that the leftmost bit in the register feeds back around at select points. At these points, the value of this fed-back bit is exclusive-ORed with the bits shifting left in the register (for more details about the exclusive OR operation, please see the student online companion that accompanies this book). Figure 6-7 shows a schematic of a shift register used for CRC. You can see from the figure that where there is a term in the generating polynomial, there is an exclusive OR (indicated by a plus sign in a circle $\oplus$) between two successive shift boxes. As a data bit enters the shift register (is shifted in from the right), all bits shift one position to the left. Before a bit shifts left, if there is an exclusive OR to shift through, the leftmost bit currently stored in the shift register wraps around and is exclusive-ORed with the shifting bit. Table 6-5 shows an example of CRC generation using the message 1010011010 and the generating polynomial $x^5 + x^4 + x^2 + 1$, which is not a standard generating polynomial but one created for this simplified example.

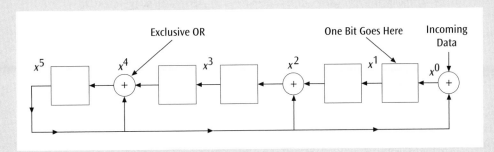

Figure 6-7 *Hardware shift register used for CRC generation*

The cyclic redundancy checksum is one of the few cases in the field of computer science in which you almost get *more* than you paid for. In contrast to parity checking, cyclic redundancy checksum detects almost 100 percent of all errors. Recall that parity checking, depending on whether it is simple parity or longitudinal parity, can detect only between 50 percent and approximately 80 percent of errors. You can perform manual parity calculations quite quickly, but the hardware methods of cyclic redundancy checksum are also fairly quick. As you saw earlier, parity schemes require a high number of check bits per data bits. In contrast, cyclic redundancy checksum requires that a remainder-sized number of check bits (either 8, 16, or 32—as you can see from the list of generating polynomials) be added to a message. The message itself can be hundreds to thousands of bits in length. Therefore, the number of check bits per data bits in cyclic redundancy can be relatively low.

Cyclic redundancy checksum is a very powerful error-detection technique and should be seriously considered for all data transmission systems. Indeed, all local area networks use CRC techniques (CRC-32 is found in Ethernet LANs) and many wide area network protocols incorporate a cyclic checksum.

Now that we understand the basic error-detection techniques, let us look at what happens once an error is detected.

	0	⊕	0		0	⊕	0		0	⊕	**Incoming Data**	↓
	0		0		0		0		1	|	1	
	0		0		0		1		0	|	0	
	0		0		1		0		1	|	1	
	0		1		0		1		0	|	0	
	1		0		1		0		0	|	0	
	1		1		1		0		0	|	1	
	0		1		1		0		0	|	1	
	1		1		0		0		0	|	0	
	0		0		1		0		0	|	1	
	0		1		0		0		0	|	0	
	1		0		0		0		0	|	0	
	1		0		1		0		1	|	0	
	1		1		1		1		1	|	0	
	0		1		0		1		1	|	0	
Remainder	1		0		1		1		0	|	0	

} *r* 0s

Table 6-5 *Example of CRC generation using shift register method*

Error Control

Once an error in the received data transmission stream is detected, what is the receiver going to do about it? The action that the receiver takes is called error control, which essentially involves taking one of three actions:

- ► Do nothing.
- ► Return a message to the transmitter asking it to resend the data packet that was in error.
- ► Correct the error without retransmission.

Let us examine each of these options in more detail.

Do nothing

The first error-control option—doing nothing—hardly seems like an option at all. Yet doing nothing for error control is becoming a mode of operation for some newer wide area network transmission techniques. For example, frame relay, which has only been in existence since 1994 and is offered by telephone companies to transfer data over wide areas, supports the "do nothing" approach to error control. If a data frame arrives at a frame relay switch and an error is detected after the cyclic checksum is performed, the frame is simply discarded. The rationale behind this action is twofold. Frame relay networks are created primarily of fiber-optic cable. Because fiber-optic cable is the medium least prone to generating errors, it is assumed that the rate of errors is low and that error control is unnecessary. If a frame is in error and is discarded, frame relay assumes that either the transport layer or the higher-layer application using frame relay to transmit data will keep track of the frames and will notice that a frame has been discarded. It would then be the responsibility of the higher layer to request that the dropped frame be retransmitted. Consider the example in which a company has a database application that sends database records across the country between two corporate locations. The database application (at the application layer) is using frame relay at the data link layer to transfer the actual records. If a record or part of a record is dropped by frame relay because of a transmission error, frame relay does not inform the application. Instead, the database application has to keep track of all records sent and received, and if one record does not arrive at the destination, the database application has to ask for a retransmission.

Return a message

The second option—sending a message back to the transmitter—is probably the most common form of error control. Returning a message was also one of the first error-control techniques developed and is closely associated with a particular flow control technique. Recall from Chapter One that flow control is a process that keeps a transmitter from sending too much data to a receiver, thus overflowing the receiver's buffer. Over the years, two basic versions of return-a-message error control have emerged: Stop-and-wait and sliding window. Let us look at the Stop-and-wait error control first.

Stop-and-Wait Error Control

Stop-and-wait error control is a technique usually associated with the Stop-and-wait flow control protocol. This protocol and its error-control technique are the oldest, simplest, and thus most restrictive. A workstation (Station A) transmits one packet of data to another workstation (Station B), then stops and waits for a reply from Station B. Four things can happen at this point. First, if the packet of data arrives without error, Station B responds with a positive acknowledgment, such as ACK. When Station A receives an ACK, it transmits the next data packet. Second, if the data arrives with an error, Station B responds with a negative acknowledgment, such as NAK or REJ (for reject). If Station A receives a NAK, it resends the previous data packet. Figure 6-8 shows an example of these transactions.

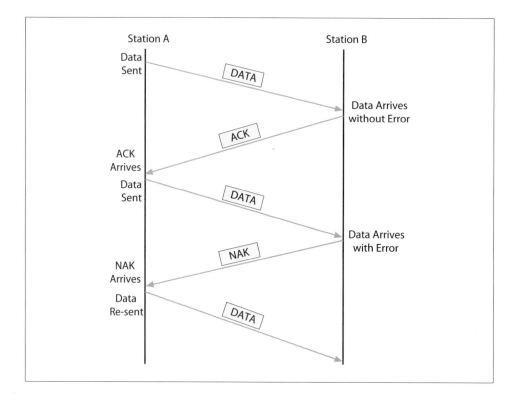

Third, a packet arrives at Station B uncorrupted, Station B transmits an ACK, but the ACK is lost or corrupted. Because Station A must wait for some form of acknowledgment, it will not be able to transmit any more packets. After a certain amount of time (called a **timeout**), Station A resends the last packet. But now if this packet arrives uncorrupted at Station B, Station B will not know that it is the same packet as the last one received. To avoid this confusion, the packets are numbered 0, 1, 0, 1, and so on. If Station A sends packet 0, and the ACK for packet 0 is lost, Station A will resend packet 0. Station B will notice two packet 0s in a row (the original and a duplicate) and deduce that the ACK from the first packet 0 was lost.

Fourth, Station A sends a packet, but the packet is lost. Because the packet did not arrive at Station B, Station B will not return an ACK. Because Station A does not receive an ACK, it will time out and resend the previous packet. For example, Station A sends packet 1, times out, and resends packet 1. If this packet 1 arrives at Station B, Station B responds with an ACK. How does Station A know whether the ACK is acknowledging the first packet or the second? To avoid confusion, the ACKs are numbered, just like the packets. In contrast to packets (which are numbered 0, 1, 0, 1, and so on), however, ACKs are numbered 1, 0, 1, 0, and so on. If Station B receives packet 0, it responds with ACK1. The ACK1 tells Station A that the *next* packet expected is packet 1. Because packet 0 just arrived, packet 1 is expected next.

One of the most serious drawbacks to the simple Stop-and-wait error control is its high degree of inefficiency. Stop-and-wait error control is a half-duplex protocol, which means that only one station can transmit at one time. The time the transmitting station wastes waiting for an acknowledgment could be better spent transmitting additional packets. More efficient techniques than Stop-and-wait are available. One of these protocols is the sliding window technique.

Sliding Window Error Control

Sliding window error control is based on the sliding window protocol, which is a flow control scheme that allows a station to transmit a number of data packets at one time before receiving some form of acknowledgment. Sliding window protocols have been around since the 1970s, a time when computer networks had two important limitations. First, line speeds and processing power were much lower than they are today. For this reason, it was important that a station transmitting data did not send data too quickly and overwhelm the receiving station. Second, memory was more expensive, and so network devices had limited buffer space in which to store incoming and outgoing data packets. Because of these limitations, standard sliding window protocols set their maximum window size to seven packets. A station that had a maximum window size of 7 (as some of the early systems did) could transmit only seven data packets at one time before it had to stop and wait for an acknowledgment. Because a window size of 7 was small, *extended* sliding window protocols were soon created that could support 127 packets. Today, the TCP protocol used on the Internet can dynamically adjust its window size into the thousands for optimum performance. For simplicity, the following examples will consider the standard protocol with a maximum window size of 7.

To follow the flow of data in a sliding window protocol with window size 7, packets are assigned numbers 0, 1, 2, 3, 4, 5, 6, and 7. Once packet number 7 is transmitted, the number sequencing starts back over at 0. Even though the packets are numbered 0 through 7, which corresponds to eight different packets, only seven data packets can be outstanding (unacknowledged) at one time. (The reasoning for this will become apparent shortly.) Because a maximum of seven data packets can be outstanding at one time, two data packets of the same number (for example, both numbered 4) can never be transmitted at the same time. If a sender has a maximum window size of 7 and transmits four packets, it can still transmit three more packets before it has to wait for an acknowledgment. If the receiver acknowledges all four packets before the sender transmits any more data, the sender's window size once again returns to 7.

Consider a scenario in which the sender transmits five packets and stops. The receiver receives the five packets and acknowledges them. Before the acknowledgment is received, the sender can send two more packets, because the window size is 7. On receipt of the acknowledgment, the sender can send an additional seven packets before it has to stop again.

The acknowledgment that a receiver transmits to the sender is also numbered. In the sliding window protocol, acknowledgments always contain a value equal to the number of the *next expected* packet. For example, if the sender, as shown in Figure 6-9, transmits three packets numbered 0, 1, and 2, and the receiver wishes to acknowledge all of them, the receiver will return an acknowledgment (ACK) with value 3, because packet 3 is the next packet the receiver expects.

Figure 6-9
Example of sliding window

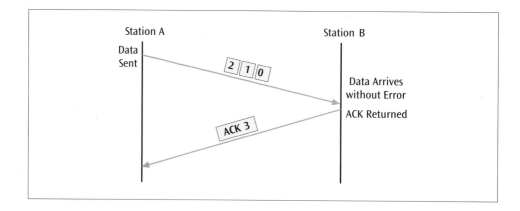

Let us go back now and consider what would happen if the protocol allowed eight packets to be sent at one time. Assume the sender sends packets numbered 0 through 7. The receiver receives all of them and acknowledges all by sending an acknowledgement numbered 0 (the next packet expected). But what if none of the packets arrived at the receiver? The receiver would not respond with a positive acknowledgment, and the sender would hear nothing. If after a short waiting period (a timeout), the sender asked the receiver for the number of the next packet expected, the receiver would answer with 0. The sender would never know if that meant all the packets were received or none of the packets was received. This potential confusion could lead a sender to resend all the packets when it is not necessary, or to resend none of the packets when the receiver is trying to indicate an error.

Now let us add error control to the sliding window protocol. Essentially four things can happen to a packet when it is transmitted:

- ▸ the packet arrives without error
- ▸ the packet is lost (never arrives)
- ▸ the packet is corrupted (arrives but has a cyclic checksum error)
- ▸ the packet is delayed (if the packet is delayed long enough, a duplicate packet may be transmitted, resulting in two copies of the same packet)

A sliding window protocol with error control must be able to account for each of these four possibilities. As you read on, it might help to remember an important distinction. A sliding window protocol's function is simply to inform the transmitter what piece of data is expected next. The function of a sliding window protocol with error control is to further specify what will occur if something goes wrong during a sliding window operation.

While we examine the four possible things that can go wrong during transmission, it is also important to note a difference in the way different sliding window protocols number data. Older sliding window protocols, such as the High-level Data Link Control protocol, number the transmitted packets, each of which may contain hundreds of bytes of data. Thus, if a transmitter sends four packets, they might be numbered 0, 1, 2, and 3, respectively. If something goes wrong with a packet, the receiver will request that packet n be transmitted again. In contrast, newer sliding window protocols such as the TCP protocol numbers the individual bytes. In this case, if a transmitter sends a packet with 400 bytes of data, the bytes may, for example, be numbered 8001 to 8400. If something goes wrong with the packet, the receiver will indicate that it needs bytes 8001 to 8400 to be transmitted again. Let us look at some examples that illustrate the four basic error-control scenarios possible with sliding window protocols. For the first scenario, you will see a packet-numbering example, but in general, we will concentrate more on examples of the byte-numbering scheme of the TCP protocol, as it is the more popular of the protocols.

In the first scenario (shown in Figure 6-10), one or more packets, numbered individually, are transmitted, and all arrive without error. More specifically, Station A transmits four packets numbered 2, 3, 4, and 5, and Station B receives them and sends an ACK 6 acknowledging all four. Notice that Station B is also telling Station A what packet it expects next (packet 6). Station A responds by sending five more packets numbered 6, 7, 0, 1, and 2. Station B acknowledges all the packets by returning an ACK 3.

Figure 6-10

Normal transfer of data between two stations with numbering of the packets

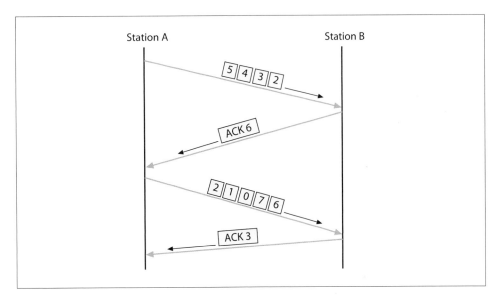

If the sliding window protocol numbers bytes instead of packets, we might have an example such as that shown in Figure 6-11. Station A transmits one packet with bytes 0–400, followed by a second packet with bytes 401–800. Station B receives both packets and acknowledges all the bytes. Note once again that the ACK tells Station A the next byte that Station B expects (801).

Figure 6-11

Normal transfer of data between two stations with numbering of the bytes

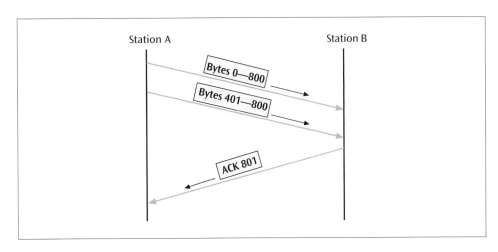

An interesting question arises: Does a receiver have to acknowledge the data every time something is received? Or can a receiver wait a while to see if something more is coming in before it sends an acknowledgment? In the world of TCP/IP, receiving stations follow a handful of rules for resolving this question. The first rule (shown in Figure 6-12) is that if a receiver just received some data and wishes to send data back to the sender, then the receiver should include an ACK with the data it is about to send. This is called piggybacking, and it saves the receiver from sending a separate ACK message. The second rule is that if the receiver does not have any data to return to the sender, and the receiver has just acknowledged the receipt of a previously sent packet of data, then the receiver must wait 500 milliseconds to see if another packet arrives. If a second packet arrives before the 500 milliseconds expire, however, then the receiver must immediately send an ACK. Lastly, the third rule states that if the receiver is waiting for a second packet to arrive, and the 500 milliseconds expire, then the receiver does not wait for a second packet and instead issues an ACK immediately.

Figure 6-12

Three examples of returning an acknowledgment (ACK)

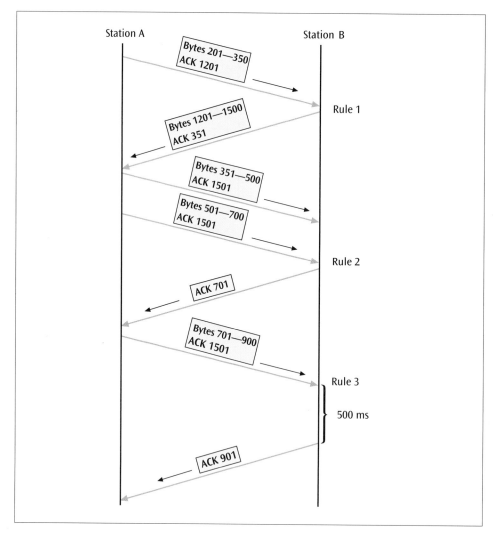

What happens when a packet is lost? Figure 6-13 illustrates the situation in which Station A transmits a sequence of packets and the second one is lost in the network. When the receiver, Station B, sees the third packet out of sequence, it returns an ACK with the sequence number of what it was expecting (byte 2401). Station A sees that something is wrong and retransmits the second packet. A similar result would occur if the second packet arrived, but with a CRC error. In both cases the packet is considered "lost."

Figure 6-13
A lost packet and Station B's response

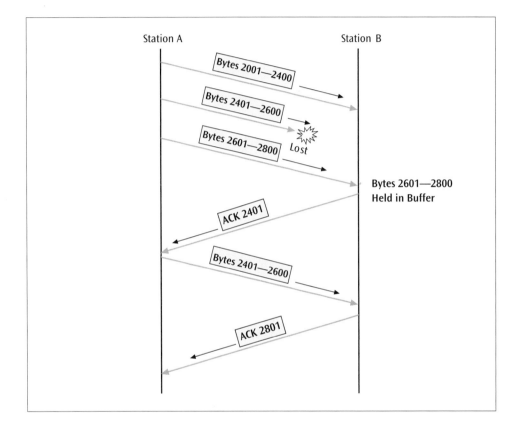

What happens if a packet is delayed, or a duplicate packet arrives at the destination? If the packet is delayed enough that it comes in out of order, the receiver also treats this as a lost packet and sends an ACK with the appropriate value. When the delayed packet or a duplicate packet arrives, the destination will see a packet with a sequence number less than the previously acknowledged bytes and simply discard the duplicate.

Finally, what happens if an acknowledgment is lost? Two possible scenarios exist. If a lost acknowledgment command is followed shortly by another acknowledgment command that does not get lost, no problem should occur, because the acknowledgments are cumulative (the second acknowledgment will have an equal or later packet number). If an acknowledgment command is lost and is not followed by any subsequent acknowledgment commands, the transmitting station will eventually time out and treat the previous packet as a lost packet and retransmit it (Figure 6-14).

Figure 6-14
A lost acknowledgment and the retransmission of a packet

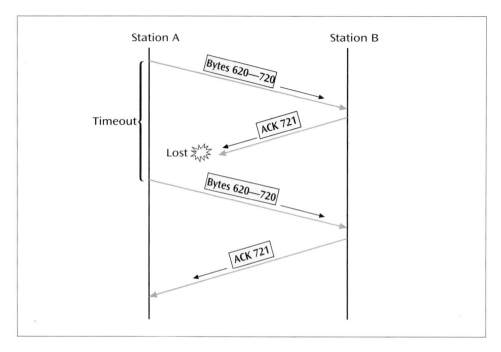

Correct the error

The beginning of this section on error control listed three actions a receiver can take if an error packet is deemed corrupted: do nothing, return an error message, or correct the error. Correcting the error seems like a reasonable solution. The data packet has been received, and error-detection logic has determined that an error has occurred. Why not simply correct the error and continue processing? Unfortunately, correcting an error is not that simple. For a receiver to be able to fix an error—in a process called forward error correction—redundant information must be present so that the receiver knows which bit or bits are in error and what their original values were. For example, if you were given the data 0110110 and informed that a parity check had detected an error, could you determine which bit or bits were corrupted? No, you would not have enough information.

To see the full extent of the problem, consider what would happen if you transmitted three identical copies of each bit. For example, you transmitted 0110110 as 000 111 111 000 111 111 000. Now, let us corrupt one bit: 000 111 111 001 111 111 000. Can you determine which bit was corrupted? If you assume that only one bit has been corrupted, you can apply what is known as the majority rules principle and determine that the error bit is the final 1 in the fourth group, 001. Note, however, that even in this simple example, forward error correction entailed transmitting *three times* the original amount of data, and it provided only a small level of error correction. This level of overhead limits the application of forward error correction.

A more useful type of forward error correction is a Hamming code. A Hamming code is a specially designed code in which special check bits have been added to data bits such that, if an error occurs during transmission, the receiver may be able to correct the error using the included check and data bits. (To learn more of the technical intricacies of a Hamming code, see the Details box titled "Forward Error Correction and Hamming Distance.") For example, let us say we want to transmit an 8-bit character, for example, the character 01010101 seen in Figure 6-15. Let us number the bits of this character b12, b11, b10, b9, b7, b6, b5, and b3. (We will number the bits from right to left, leaving spaces for the soon-to-be-added check bits.) Now add to these data bits the following check bits: c8, c4, c2, and c1, where c8 generates a simple even parity for bits b12, b11, b10, and b9. The check bit c4 will generate a simple even parity for bits b12, b7, b6, and b5. Check bit c2 will generate a simple even parity for bits b11, b10, b7, b6, and b3. Finally, c1 will generate a simple even parity for bits b11, b9, b7, b5, and b3. Note that each check bit here is checking different sequences of data bits.

Figure 6-15
Hamming code check bits generated from the data 01010101

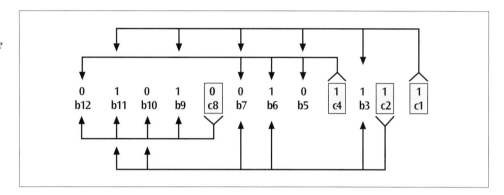

Details ▶

Forward Error Correction and Hamming Distance

For a data code such as ASCII to perform forward error correction, redundant bits must be added to the original data bits. These redundant bits allow a receiver to look at the received data and, if there is an error, recover the original data using a consensus of received bits. For a simple example, let us transmit three identical copies of a single bit (majority operation). Thus, to send a 1, 111 will be transmitted. Next, consider what would happen if the 3 bits received have the values 101. In forward error correction, the receiver would assume that the 0 bit should be a 1, because the majority of bits are 1. To understand how redundant bits are created, you need to examine the Hamming distance of a code, which is the smallest number of bits by which character codes differ. The Hamming distance is a characteristic of a code. To create a self-correcting, or forward error correcting code, you must create a code that has the appropriate Hamming distance.

In the ASCII character set, the letter B in binary is 1000010, and the letter C is 1000011. The difference between B and C is one bit—the rightmost bit. If you compare all the ASCII characters, you will find that some pairs of characters differ by one bit, and some differ by two or more bits. Because the Hamming distance of a code is based on the *smallest* number of bits by which character codes differ, the ASCII character set has a Hamming distance of 1. Unfortunately, if a character set has a Hamming distance of 1, it is

not possible to detect errors, nor to correct them. Ask yourself the following question: If a receiver accepts the character 1000010, how does it know for sure that this is the letter B and not the letter C with a 1-bit error?

When a parity bit is assigned to ASCII, the Hamming distance becomes 2. Because the rightmost bit is the parity bit, the character B, assuming even parity, becomes 10000100, and the character C becomes 10000111. Now, the last two bits of B have to change from 00 to 11 for the character B to become the character C, and the difference between the two characters is two bits. Now that the Hamming distance is 2, it is possible to detect single-bit errors, but you still cannot *correct* errors. Also, if the character B is transmitted, but one bit is flipped by error, a parity check error occurs; however, you still cannot tell what the intended character was supposed to be. For example, the character B with even parity added is 10000100. If one bit is altered, such as the second bit, the binary value is now 11000100. This character would cause a parity check, but what was the original character? Any one of the bits could have changed, thus allowing for many possible original characters. You can correct single-bit errors and detect double-bit errors when the Hamming distance of the character set is at least 3. Achieving a Hamming distance of 3 requires an even higher level of redundancy, and consequently, cost.

Let us take a closer look at how each of the Hamming code check bits in Figure 6-15 works. Note that c_8 "covers" bits b_{12}, b_{11}, b_{10}, and b_9, which are 0101. If we generate an even parity bit based on those four bits, we would generate a 0 (there are an even number of 1s). Thus, c_8 equals 0. c_4 covers b_{12}, b_7, b_6, and b_5, which are 0010, so c_4 equals 1. c_2 covers b_{11}, b_{10}, b_7, b_6, and b_3, which are 10011, so c_2 equals 1. c_1 covers b_{11}, b_9, b_7, b_5, and b_3, which are 11001, so c_1 equals 1. Consequently, if we have the data 01010101, we would generate the check bits 0111, as shown in Figure 6-15. This 12-bit character is now transmitted to the receiver. The receiver accepts the bits and performs the four parity checks on the check bits c_8, c_4, c_2, and c_1. If nothing happened to the 12-bit character during transmission, all four parity checks should result in no error. But what would happen if one of the bits is corrupted and somehow ends up the opposite value? For example, what if bit b_9 is corrupted? With the corrupted b_9, we would now have the string 010000101111. The receiver would perform the four parity checks, but this time there would be parity errors. More precisely, because c_8 checks b_{12}, b_{11}, b_{10}, b_9, and c_8 (01000), there would be a parity error—as you can see, there are an odd number of 1s in the data string, but the check bit is returning a 0. c_4 checks b_{12}, b_7, b_6, b_5, and c_4 (00101), and thus would produce no parity error. c_2 checks b_{11}, b_{10}, b_7, b_6, b_3, and c_2 (100111), and would produce no parity error. c_1 checks bits b_{11}, b_9, b_7, b_5, b_3, and c_1 (100011), which would result in a parity error. Notice that if we examine just the check bits and denote a 1 if there is a parity error and a 0 if there is no parity error, we would get 1001 (c_8 error, c_4 no error, c_2 no error, c_1 error). 1001 is binary for 9, telling us that the bit in error is in the ninth position.

Despite the additional costs of using forward error correction, are there applications that would benefit from the use of this technology? Two major groups of applications can in fact reap a benefit: digital television transmissions and applications that send data over very long distances. Digital television signals use advanced forms of forward error correction called Reed-Solomon codes and Trellis encoding. Note that it is not possible to ask for retransmission of a TV signal if there is an error. It would be too late! The data needs to be delivered in real time, thus the need for forward error correction.

In the second example, if data has to be sent over a long distance, it is costly time-wise and money-wise to retransmit a packet that arrived in error. For example, the time required for NASA to send a message to a Martian probe is several minutes. If the data arrives garbled, it will be another several minutes before the negative acknowledgment is received, and another several minutes before the data can be retransmitted. If a large number of data packets arrive garbled, transmitting data to Mars could be a very long, tedious process.

Error Detection In Action ▶

You are working for a finance company that has offices in two locations on opposite sides of a large metropolitan area. The company has many computers located at both locations, and the computers in one building constantly transmit data to the computers in the second building. At the moment, a leased telephone service transfers all the data between buildings at rates of approximately 128,000 bits per second. You have been asked to investigate the selection of an error-detection scheme. Should you select the simpler parity check or use the more elaborate cyclic redundancy checksum? Let us consider two factors as we make our decision: the probability of an error going undetected, and the extra amount of error-detection bits that are transmitted between the two locations.

Considering first the probability of an error going undetected, recall that simple parity detects only odd numbers of bit errors. Even numbers of bit errors will go undetected. Thus, only 50 percent of all errors will be caught, or 50 percent of all transmitted frames with an error will go undetected. How often is a frame transmitted with an error? Let us assume a fairly conservative error rate of one frame in 10,000, or in other words, an error rate of 10^{-4}. If an error occurs in only one frame in 10,000, and the error is not caught half the time, then an error slips through once in every 20,000 frames. Because the frame that includes a parity bit is typically 8 bits in length, an error will go undetected once in every 160,000 bits. This means that if we are transmitting data at 128,000 bits per second, then an undetected error would occur almost every second and a half.

Using a cyclic redundancy checksum, the worst-case scenario is an error burst greater than $r + 1$ in length, which has a probability of not being detected equal to $(\frac{1}{2})^r$. Given a generating polynomial of 16 bits, $(\frac{1}{2})^{16}$ equals 0.000015258. Once again assuming that a frame with an error occurs once in every 10,000 frames (a probability of 0.0001), 0.00001528×0.0001 equals 0.000000001528. Taking the reciprocal, this means that 654,450,261 frames will be transmitted before an error slips through. If we assume that a typical frame size using a cyclic redundancy checksum is 1000 bits, 6.5445×10^{11} bits will be transmitted before an error slips through. Transmitting at 128,000 bits per second, 5,112,892.7 seconds will pass before an error occurs. This roughly translates into 59 days of continuous (24/7) transmission. Compare that to having an error go undetected almost every second and a half, using parity.

Now let us consider how many extra error-detection bits must be added to each transmission. For parity, let us assume a standard 7-bit ASCII character. For each ASCII character, one parity bit is added. Thus, one out of every 8 bits transmitted is a check bit. For cyclic redundancy checksum, the remainder is size r. Once again assuming a checksum of size 16 bits and a frame size of 1000 bits, 16 out of every 1000 bits (or one out of every 62.5 bits) is a check bit. That is almost eight times better than parity.

In conclusion, cyclic redundancy checksum adds many fewer check bits to each frame and delivers a substantially increased error-detection rate. There is no doubt that, given these two factors, cyclic redundancy checksum is superior to parity.

◆ ◆

SUMMARY

▸ Noise is always present in computer networks, and if the noise level is too high, errors will be introduced during the transmission of data. The types of noise include white noise, impulse noise, crosstalk, echo, jitter, and attenuation. Only impulse noise is considered a noncontinuous noise, while the other forms of noise are continuous.

▸ Among the techniques for reducing noise are proper shielding of cables, recognizing possible wireless signal conflicts, using modern digital equipment, using digital repeaters and analog amplifiers, and observing the stated capacities of media. Other reasonable options for reducing the possibility of errors include reducing the number of devices in a transmission stream, decreasing the length of cable runs, and reducing data transmission speed.

▸ Three basic forms of error detection are parity, arithmetic checksum, and cyclic redundancy checksum. Simple parity adds one additional bit to every character and is a very simple error-detection scheme that suffers from low error-detection rates and a relatively high ratio of check bits to data bits. Longitudinal parity adds an entire character of check bits to a block of data and improves error detection, but still suffers from inadequate error-detection rates and a relatively high ratio of check bits to data bits.

▶ Arithmetic checksum is an error-detection scheme used with the Internet Protocol and produces an arithmetic sum of all the characters transmitted within a message.

▶ Cyclic redundancy checksum is a superior error-detection scheme with almost 100 percent capability of recognizing corrupted data packets. Calculation of the checksum remainder is fairly quick when performed by hardware, and it adds relatively few check bits to potentially large data packets.

▶ Once an error has been detected, there are three possible options: do nothing, return an error message, and correct the error. The "doing nothing" option is used by some of the newer transmission technologies, such as frame relay. Frame relay assumes that fiber-optic lines will be used, which significantly reduces the chance for errors. If an error does occur, a higher-layer protocol will note the frame error and will perform some type of error control. The option of returning an error message to the transmitter is the most common response to an error and involves using Stop-and-wait protocols and sliding window protocols.

▶ A Stop-and-wait protocol allows only one packet to be sent at a time. Before another packet can be sent, the sender has to receive a positive acknowledgment. A sliding window protocol allows multiple packets to be sent at one time. A receiver can acknowledge multiple packets with a single acknowledgment.

▶ Error correction is a possibility if the transmitted data contains enough redundant information so that the receiver can properly correct the error without asking the transmitter for additional information. This form of error control requires a high amount of overhead and is used only in special applications in which the retransmission of data is not desirable.

KEY TERMS

arithmetic checksum	forward error correction	piggybacking
attenuation	generating polynomial	simple parity
crosstalk	Hamming code	sliding window protocol
cyclic redundancy	Hamming distance	Stop-and-wait error control
checksum (CRC)	impulse noise	timeout
echo	jitter	white noise
error control	longitudinal parity	
even parity	odd parity	

REVIEW QUESTIONS

1. What is white noise and how does it affect a signal?
2. What is impulse noise and why is it the most disruptive?
3. What is crosstalk and how does it affect a signal?
4. What is echo and how does it affect data transmission?
5. What is jitter and why is it a digital signal problem?
6. Which types of noise introduced in this chapter are continuous, and which are noncontinuous?
7. Will proper shielding of a medium increase or decrease the chance of errors? Explain your reasoning.

8. What is the difference between even parity and odd parity?

9. What is the ratio of check bits to data bits for simple parity?

10. What is the ratio of check bits to data bits for longitudinal parity?

11. What types of errors will simple parity not detect?

12. What types of errors will longitudinal parity not detect?

13. With arithmetic checksum, what is being added?

14. What is a generating polynomial?

15. What types of errors will cyclic checksum not detect?

16. Frame relay practices which form of error control?

17. How many packets can be sent at one time using Stop-and-wait error control?

18. What is the function of an ACK in Stop-and-wait error control?

19. What is the function of a NAK in Stop-and-wait error control?

20. In communications systems, what does timeout mean?

21. Why were the window sizes in early sliding window systems so small?

22. What are the two different ways to number sequences of data?

23. What condition must be met for error correction to be performed?

EXERCISES

1. Which type of noise is the most difficult to remove from an analog signal? Why?

2. Which type of noise is the most difficult to remove from a digital signal? Why?

3. Explain the relationship between twisted pair wires and crosstalk.

4. Which type of cable is most susceptible to echo?

5. Given the character 0110101, what bit will be added to support even parity?

6. Given the character 1010010, what bit will be added to support odd parity?

7. Generate the parity bits and longitudinal parity bits for even parity for the characters 0101010, 0011010, 0011110, 1111110, and 0000110.

8. Given the message "Hello, goodbye", show the decimal arithmetic checksum that will be generated.

9. List the types of errors that can escape a cyclic redundancy checksum system.

10. List the types of errors that cannot escape a cyclic redundancy checksum system.

11. Why do all the CRC generating polynomials end with a 1?

12. In a Stop-and-wait error control system, Station A sends packet 0, and it is lost. What happens next?

13. In a Stop-and-wait error control system, Station A sends packet 0, it arrives without error, and an ACK is returned, but the ACK is lost. What happens next?

14. In a sliding window error control system in which each packet is numbered, Station A sends packets 4, 5, 6, and 7. Station B receives them and wants to acknowledge all of them. What does Station B send back to Station A?

15. In a sliding window error control system in which each byte is numbered, Station A sends a packet with bytes numbered 801 to 900. If the packet arrives with no errors, how does Station B respond?

16. In a sliding window error control system, Station A sends a packet with bytes 501–700, followed immediately by a packet with bytes 701–900. Create a diagram of this error control scenario, and show the response(s) that Station B will send if there are no errors.

17. In a sliding window error control system, Station A sends a packet with bytes 501–700, followed immediately by a packet with bytes 701–900. Create a diagram of this error control scenario, and show the response(s) that Station B will send if the second packet is lost in the network.

18. In a sliding window error control system, Station A sends a packet with bytes 501–700, followed immediately by a packet with bytes 701–900. Create a diagram of this error control scenario, and show the response(s) that Station B will send if both packets arrive but there is a checksum error in the second packet.

19. In a sliding window error control system, Station A sends three packets with bytes 0–100, 101–200, and 201–300, respectively. The second packet with bytes 101–200 is held up somewhere in the network long enough that the third packet arrives before the second one. Create a diagram of this error control scenario, and show the response(s) that Station B will send. Now assume that five seconds after Station B responds, the second packet shows up. What does Station B do now?

20. Assume a sliding window protocol has a 3-bit field to hold the window size. A 3-bit value gives eight combinations, from 0 to 7. Why then can a transmitter send out a maximum of only seven packets at one time and not eight? Show an example.

21. If a 7-bit sliding window size is used, how many packets can be sent before the transmitter has to stop and wait for an acknowledgment?

22. Is Stop-and-wait error control a half-duplex protocol or a full-duplex protocol? Explain your response.

23. Devise a code set for the digits 0 to 9 that has a Hamming distance of 2.

24. A system is going to transmit the byte 10010100. Show the four check bits c_8, c_4, c_2 and c_1 that will be added to this byte.

25. The 12-bit string 010111110010 with embedded Hamming code bits (c_8, c_4, c_2 and c_1) has just arrived. Is there an error? If so, which bit is in error?

26. Construct a 12-bit Hamming code for the characters A and 3.

THINKING OUTSIDE THE BOX

1 Your company is transmitting 500-character (byte) records at a rate of 400,000 bits per second. You have been asked to determine the most efficient way to transmit these records. If a Stop-and-wait error control system is used, how long will it take to transmit 1000 records? If a sliding window error control is used with a window size of 127, how long will it take to transmit 1000 records? If, on the average, one record in every 200 records is garbled, how long will it take to transmit 1000 records? What are your conclusions?

2 Your boss has heard a lot about cyclic redundancy checksum but is not convinced that it is much better than simple parity. You quote some of the numbers and probabilities of error detection, but they mean little. If you could give a more concrete example of how good cyclic redundancy checksum is, you might win the boss over. Recall the probabilities given in Table 6-4. If your company transmits a continuous stream of data at 128,000 bits per second, how much time will pass before cyclic redundancy checksum lets an error slip through?

3 Given today's networking technology, 3-bit and 7-bit sliding window sizes seem very restrictive. What would be a more reasonable size today? Explain your reasoning.

4 It was stated in this chapter that some newer protocols are adopting the do nothing form of error control. Does this seem like a reasonable choice? If so, can you create an example in which this would not be a reasonable choice?

HANDS-ON PROJECTS

1. What other error-detection techniques are available? How do they compare to parity and cyclic redundancy checksum?

2. What is intermodulation distortion? What sorts of signals are susceptible to this form of distortion?

3. Find an example of another Hamming code and describe how it works.

4. Describe two situations in which error-free transmission is crucial to communications.

5. Given the data 10001010010 and the CRC generating polynomial 10011, show the remainder that is generated. You may use either the long-division method or the shift register method.

6. Using the programming language of your choice, write a program that inputs a character string and calculates the cyclic checksum remainder using a given generating polynomial.

7. Are there any CRC generating polynomials other than those listed in this chapter?

7

Local Area Networks: The Basics

◆◆◆

THE AUTOMOBILES OF TODAY have, according to *Network World* magazine, more computing power than the rocket that was used in the Apollo moon landing in the 1970s. Although surprising, in a way this is not hard to believe. Today's automobiles can record driving transactions such as braking and acceleration; regulate air and fuel mixtures; control antilock brakes, antiskid controls, and suspension systems; provide separate front and rear audio and video entertainment systems; and inform the driver via built-in video when the car is about to back over a kid's bicycle.

In addition, manufacturers are designing vehicles with remote monitoring services that can assist the owner in scheduling preventive maintenance. New satellite radio services can be used to download entertainment systems and real-time news. Bluetooth-enabled systems are currently used to provide built-in, hands-free cellular phone service but in the future will also provide voice-activated automobile controls. Also, suppose your car is stopped alongside the road on a dark night. There are plans to create collision-avoidance systems that will communicate between your car and other vehicles on the road in order to prevent collisions.

At the heart of these communications systems will be one or more local area networks. In fact, local area networks will eventually take over many of the functions within a car. Lower-speed local area networks will be used to control lights, fans, and other slow-reacting systems. Higher-speed networks will be used to control antilock brakes, antiskid control, and the corresponding engine controls. Still other networks can be used to perform critical functions such as airbag deployment during a collision. Given all these existing and possible applications, the automobile is quickly becoming a network on wheels.

Are local area networks so vital to everyday life that they will become standard options in automobiles?

If your work, your home, and your automobile all depended on local area networks, what might be next?

Source: "Networks Drive Car of the Future" by Carolyn Duffy Marsan, *Network World*, May 23, 2005.

Objectives ▶

After reading this chapter, you should be able to:

▶ State the definition of a local area network

▶ List the primary function, activities, and application areas of a local area network

▶ Cite the advantages and disadvantages of local area networks

▶ Identify the physical and logical layouts of local area networks

▶ Cite the characteristics of wireless local area networks and their medium access control protocols

▶ Specify the different medium access control techniques

▶ Recognize the different IEEE 802 frame formats

▶ Describe the common local area network systems

Introduction ▶

A local area network (LAN) is a communications network that interconnects a variety of data communications devices within a small geographic area and transmits data at high data transfer rates. Several points in this definition merit a closer look. The phrase "data communications devices" covers computers such as personal computers, computer workstations, and mainframe computers, as well as peripheral devices such as disk drives, printers, and modems. Data communications devices could also include items such as motion, smoke, and heat sensors; fire alarms; ventilation systems; and motor speed controls. These latter devices are often found in businesses and manufacturing environments where assembly lines and robots are commonly used.

The next piece of the definition, "within a small geographic area" usually implies that a local area network can be as small as one room, or can extend over multiple rooms, over multiple floors within a building, and even over multiple buildings within a single campus-area. The most common geographic areas, however, are a room or multiple rooms within a single building.

Lastly, the final phrase of the definition states that local area networks are capable of transmitting data at "high data transfer rates". While early local area networks transmitted data at only 10 million bits per second, the newest local area networks can transmit data at 10 billion bits per second and higher.

Perhaps the strongest advantage of a local area network is its capability of allowing users to share hardware and software resources. For example, suppose the network version of a popular database program is purchased and installed on a local area network. The files that contain all the database information are stored in a central location such as a network server. When any user of the local area network wishes to access records from that database program, the records can be retrieved from the server and then transmitted over the local area network to the user's workstation, where they can be displayed. Similarly, a high-quality printer can be installed on the network so that all users can share access to this relatively expensive peripheral.

Since the local area network first appeared in the 1970s, its use has become widespread in commercial and academic environments. In fact, it would be difficult to imagine a collection of computer workstations within a computing environment that did not employ some form of local area network. Many individual computer users are beginning to install local area networks at home to interconnect two or more computers. Just as in office environments, one of the driving forces behind installing a local area network in a home is the capability of sharing peripherals such as high-quality printers and high-speed connections to the Internet. To better understand this phenomenon, it is necessary to examine several "layers" of local area network technology. This chapter begins by discussing the primary function of a local area network as well as its advantages and disadvantages. Next, the basic physical (hardware) layouts or topologies of the most commonly found local area network are discussed, followed by a survey of the medium access control protocols (software) that allow a workstation to transmit data on the network. We will then examine the most common local area network products, such as the various Ethernet versions. Chapter Eight will introduce the software that operates on local area networks, including the ever-so-important network operating system. Before we begin examining their basic configurations, let us discuss the main function and the advantages and disadvantages of local area networks.

Primary Function of Local Area Networks

To better understand the capabilities of local area networks, let us examine their primary function and some typical activities and application areas. The majority of users expect a local area network to provide access to hardware and software resources that will allow them to perform one or more of the following activities in an office, academic, or manufacturing environment: file serving, database and application serving, print serving, Internet accessing, e-mailing, video and music transfers, process control and monitoring, and distributed processing.

A local area network performs file serving when it is connected to a workstation with a large storage disk drive that acts as a central storage repository, or file server. For example, when the local area network offers access to a high-level application such as a commercial project management application, the network stores the project management software (or a portion of it) on the file server and transfers a copy of it to the appropriate workstation on demand. By keeping all of the application on the server—or more likely, part of it on the server and part of it on the client workstation—the network can control access to the software and can reduce the amount of disk storage required on each user's workstation for this application. As a second example, suppose two or more users wish to share a data set. In this case, the data set, like the application software, would be stored on the file server, while the network provided access to those users who had the appropriate permissions.

A local area network can also provide access to one or more high-quality printers. The local area network software called a print server provides workstations with the authorization to access a particular printer, accepts and queues print jobs, prints cover sheets, and allows users access to the job queue for routine administrative functions.

Most local area networks provide the service of sending and receiving e-mail. This e-mail service can operate both within the local area network and between the local area network and other networks, such as the Internet. Stored somewhere on the network is a database of e-mail messages, both old and new. When users log in to access their e-mail, their messages are stored and retrieved from the e-mail server.

A local area network can interface with other local area networks, wide area networks (such as the Internet), and mainframe computers. Thus, a local area network is often the glue that holds together many different types of computer systems and networks. A company can use a local area network's interfacing ability to enable its employees to interact with people external to the company, such as customers and suppliers. For example, if employees wished to send purchase orders to vendors, they could enter transactions on their workstations. These transactions would travel across the company's local area network, which would be connected to a wide area network. The suppliers would eventually receive the orders by being connected to this wide area network through their own local area network.

Figure 7-1 shows typical interconnections between a local area network and other entities. It is common to interconnect one local area network to another local area network via a device such as a switch. Equally common is the interconnection of a local area network to a wide area network via a router. A local area network can also be connected to a mainframe computer to enable the two entities to share each other's resources.

Figure 7-1
A local area network interconnecting another local area network, the Internet, and a mainframe computer

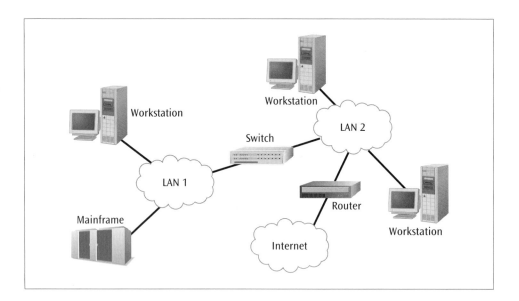

Most, if not all, of the local area networks today provide the capabilities of transferring video images and video streams. For example, a local area network could allow a user to transfer high-resolution graphic images, transfer video streams, and perform teleconferencing between two or more users.

In manufacturing and industrial environments, local area networks are often used to monitor manufacturing events and report and control their occurrence. The local area network provides process control and monitoring. An automobile assembly line that uses sensors to monitor partially completed automobiles and control robots for assembly is an excellent example of a local area network performing process control functions.

Depending on the type of network and the choice of network operating system, a local area network may support distributed processing, in which a task is subdivided and sent to remote workstations on the network for execution. Oftentimes, these remote workstations are idle; thus, the distributed processing task amounts to the "stealing" of CPU time from other machines (and is often called grid computing). The results of these remote executions are then returned to the originating workstation for dissemination or further processing. By delegating tasks to those computers that are most capable of handling specific chores, the distribution of tasks or parts of tasks can lead to an increase in execution speed.

In addition to performing these common activities, a local area network can be an effective tool in many application areas. One of the most common application areas is an office environment. A local area network in an office can provide word processing, spreadsheet operations, database functions, e-mail access, Internet access, electronic appointment scheduling, and graphic image creation capabilities over a wide variety of platforms and to a large number of workstations. Completed documents can be routed to high-quality printers to produce letterheads, graphically designed newsletters, and formal documents.

A second common application area for a local area network is an academic environment. In a laboratory setting, for example, a local area network can provide students with access to the tools necessary to complete homework assignments, send e-mail, and utilize the Internet. In a classroom setting, a local area network can enable professors to deliver tutorials and lessons with high-quality graphics and sound to students. Multiple workstations can be used to provide students with instruction at their own pace, while the instructor monitors and records each student's progress at every workstation.

A third common application area for a local area network is manufacturing. In fact, modern assembly lines operate exclusively under the control of local area networks. As products move down the assembly line, sensors control position; robots perform mundane, exacting, or dangerous operations; and product subassemblies are inventoried and ordered. The modern automobile assembly line is a technological tour de force, incorporating numerous local area networks and mainframe computers.

Now that we are familiar with the more common activities and applications of local area networks, let us examine some advantages and disadvantages.

Advantages and Disadvantages of Local Area Networks

One of the biggest advantages of local area networks is their ability to share resources in an economical and efficient manner. Shared hardware resources can include high-quality printers, tape-backup systems, plotters, CD jukeboxes, mass storage systems, and other hardware devices. On the software end, local area networks allow the sharing of commercial applications, in-house applications, and data sets with one or all user workstations. Also, with respect to communications, each workstation in a local area network can send and receive messages to and from other workstations and networks. This intercommunication allows users to send e-mail, access Web pages, send print jobs, and retrieve database records. (An interesting side effect of local area networks is that an individual workstation can survive a network failure if the workstation does not rely on software or hardware found on other workstations or on the server.) An additional advantage is that component evolution can be independent of system evolution, and vice versa. For example, if new workstations are desired, it is possible to replace older workstations with newer ones with few, if any, changes to the network itself. Likewise, if one or more network components become obsolete, it is possible to upgrade the network component without replacing or radically altering individual workstations.

Under some conditions, local area networks allow equipment from different manufacturers to be mixed on the same network. For example, it is possible to create a local area network that incorporates IBM-type personal computers with

Apple microcomputers. Two other advantages of local area networks are high transfer rates and low error rates. Local area networks typically have data transfer rates from 10 million bits per second to 10 billion bits per second. Because of these rates, documents can be transferred across a local area network quickly and with confidence. Finally, because local area networks can be purchased outright, the entire network and all workstations and devices can be privately owned and maintained. Thus, a company can offer its desired services using the hardware and software it deems best for employees.

Interestingly, however, some companies are beginning to view equipment purchases as a disadvantage. Supporting an entire corporation with the proper computing resources is expensive. It does not help that as a computer reaches its first birthday, there is a newer, faster, and less expensive computer waiting to be purchased. Thus, some companies lease local area network equipment or hire a third party to support their networks.

In addition, local area networks have a number of disadvantages. For one, local area network hardware, the operating systems, and the software that runs on the network can be expensive. The components of LANs that require significant funding include the network server, the network operating system, the network cabling system including switches and routers, the network-based applications, network security, and support and maintenance. Despite the fact that a local area network can support many types of hardware and software, the different types of hardware and software may not be able to interoperate. For example, even if a local area network supports two different types of database systems, users may not be able to share data between the two database systems. Another disadvantage is the potential for purchasing software with the incorrect user license. For example, it is almost always illegal to purchase a single-user copy of software and then install it on a local area network for multiple use. To avoid using software illegally, companies must be aware of the special licensing agreements associated with local area networks.

An important disadvantage that has often been overlooked in the past is that the management and control of the local area network requires many hours of dedication and service. A manager, or network administrator, of a local area network should be properly trained and should not assume that the network can support itself with only a few hours of attention per week. Therefore, a local area network requires specialized staff and knowledge and the right diagnostic hardware and software. Unfortunately, many hours of this support time are very often spent fighting viruses and other network security issues (Chapter Twelve is devoted entirely to network security issues).

Finally, a local area network is only as strong as its weakest link. For example, a network may suffer terribly if the file server cannot adequately serve all the requests from users of the network. Upon upgrading a server, a company may discover that the cabling is no longer capable of supporting the higher traffic. Upon upgrading the cabling, it may become apparent that the network operating system is no longer capable of performing the necessary functions. Upgrades to part of a network can cause ripple effects throughout the network, and the cycle of upgrades usually continues until it is once again time to upgrade the server.

Considering all the advantages and disadvantages associated with local area networks, it should not be surprising that the decision to incorporate a LAN into an existing environment requires much planning, training, support, and money. Let us now look more closely at how the workstations in a local area network are interconnected to serve the activities and applications discussed so far.

The First Local Area Network—The Bus/Tree

The **bus/tree local area network**, often simply called the **bus LAN**, was the first physical design when LANs became commercially available in the late 1970s, and it essentially consisted of a simple cable, or bus, to which all devices attached. Since the 1970s, the use of a bus as a local area network configuration has diminished to the point of extinction. It is interesting to note, however, that cable television signals are still delivered by a network bus. Thus, understanding the bus/tree network is still important. As shown in Figure 7-2, the bus is simply a linear coaxial cable into which multiple devices or workstations tap.

Figure 7-2
Simple diagram of a local area network bus topology

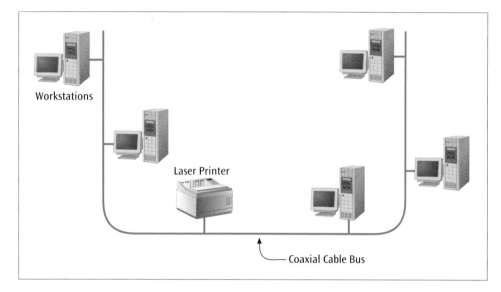

When a device transmits on the bus, all other attached devices receive the transmission. Connecting to the cable requires a simple device called a **tap** (Figure 7-3). This tap is a **passive device**, as it does not alter the signal and does not require electricity to operate. On the workstation end of the cable is a network interface card. The **network interface card (NIC)** is an electronic device, typically in the form of a computer circuit board, that performs the necessary signal conversions and protocol operations that allow the workstation to send and receive data on the network.

Figure 7-3
Tap used to interconnect a workstation and a LAN cable

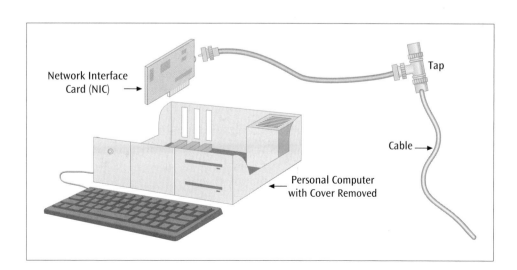

Two different signaling technologies can be used with a bus network: baseband signaling and broadband signaling. (Recall that baseband signaling and broadband signaling were introduced in Chapter Three during the discussion on coaxial cable.) Baseband signaling typically uses a single digital signal (such as Manchester encoding) to transmit data over the bus. This single digital signal uses the entire spectrum of the cable; therefore, only one signal at a time can be transmitted on the cable. All workstations must be aware that another workstation is transmitting, so they do not attempt to transmit and thereby inadvertently destroy the signal of the first transmitter. Allowing only one workstation access to the medium at one time is the responsibility of the medium access control protocol, which will be discussed in detail a little later in the chapter.

Another characteristic of baseband technology worth noting is that baseband transmission is bidirectional, which means that when the signal is transmitted from a given workstation, the signal propagates away from the source in both directions on the cable (Figure 7-4).

Figure 7-4
Bidirectional propagation of a baseband signal

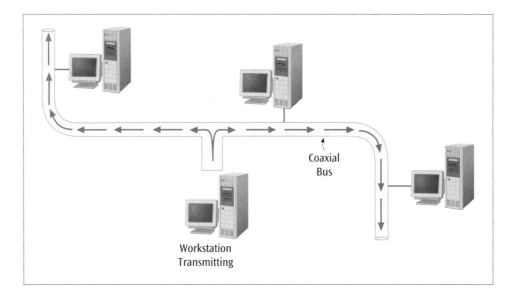

The second type of signaling technology used on the bus local area network is broadband technology (recall that only cable television systems still use broadband signaling). Broadband technology uses analog signaling in the form of frequency division multiplexing to divide the available medium into multiple channels. Each channel is capable of carrying a single stream of video, audio, or data.

It is also possible to split and join broadband cables and signals to create configurations more complex than a single linear bus. These more complex bus topologies consisting of multiple interconnected cable segments are called trees. Figure 7-5 shows an example of a tree network.

Figure 7-5
Simple example of a broadband tree topology

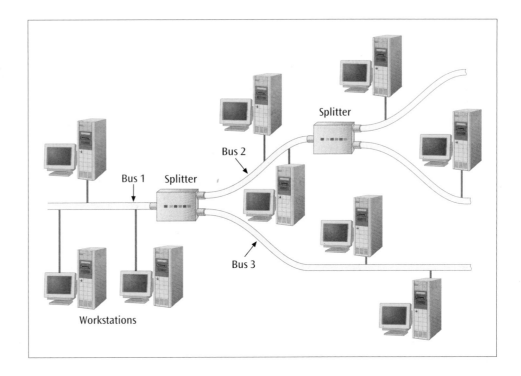

All bus networks—whether broadband or baseband—share a major disadvantage. In general, it is difficult to add a new workstation if no tap currently exists. Because there is no tap, the cable has to be cut, and a tap has to be inserted. Cutting the cable and inserting a tap disrupts the traffic on the network and is a somewhat messy job. The best way to avoid this is to anticipate where workstations will be and have the installation team install all the necessary taps in advance. As you might expect, however, predicting the exact number and location of taps is virtually impossible. With the introduction of newer technologies, bus-based local area networks have lost popularity to the point that relatively few bus-based LANs exist today. The only bus still regularly used, as we have already seen, is the one that delivers the video and data signals of cable television. One reason for this is that, if you recall, coaxial cable is a good medium for transmitting the high-frequency signals of cable television.

Let us examine the technology that replaced the local area network bus: the star-wired bus.

A More Modern LAN

The most popular configuration for a local area network today is the **star-wired bus LAN**. Today's modern star-wired bus network acts like a bus but looks like a star. To be a little more precise, the network *logically* acts as a bus, but it *physically* looks like a star. The **logical design** of a network determines how the data moves around the network from workstation to workstation. The **physical design** refers to the pattern formed by the locations of the elements of the network, as it would appear if drawn on a sheet of paper. Let us explore the details of this important distinction further.

In the first star-wired bus networks, all workstations connected to a central device such as the hub, as seen in Figure 7-6. The hub is a relatively nonintelligent device that simply and immediately retransmits the data it receives from any workstation out to *all* other workstations (or devices) connected to the hub. All workstations hear the transmitted data, because there is only a single transmission channel, and all workstations are using this one channel to send and receive. Sending data to all workstations and devices generates a lot of traffic but keeps the operation very simple, because there is no routing to any particular workstation. Thus, with regard to its logical design, the star-wired bus is acting as a bus: When a workstation transmits, all workstations (or devices) immediately receive the data. The network's physical design, however, is a star, because all the devices are connected to the hub and radiate outward in a star-like (as opposed to linear) pattern.

Figure 7-6
Simple example of a star-wired bus local area network

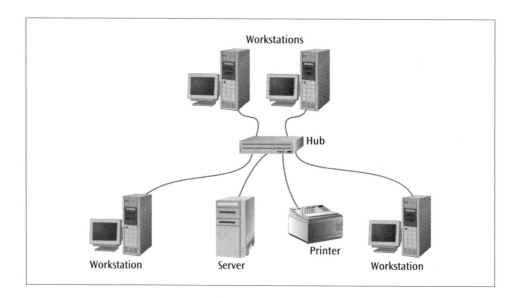

The hub at the center of star-wired bus network comes in a variety of designs. They can contain anywhere from two to hundreds of connections, or ports, as they are called. If, for example, you have a hub with 24 ports, and more are desired, it is fairly simple to either interconnect two or more hubs, or purchase a larger hub. Figure 7-7 demonstrates that to interconnect two hubs, you simply run a cable from a special connector on the front or rear of the first hub to a special connector on the front or rear of the second hub. Many hubs support multiple types of media—twisted pair, coaxial cable, and fiber-optic cable—for this interhub connection.

Figure 7-7
Interconnection of three hubs in a star-wired bus local area network

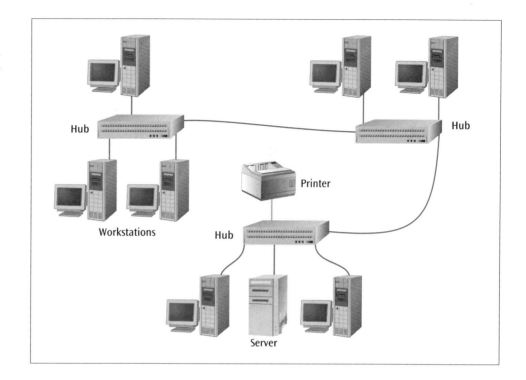

Twisted pair cabling has become the preferred medium for star-wired bus LANs, while fiber-optic cable is typically used as a connector between multiple hubs. The connectors on the ends of the twisted pair cables are simple-to-use modular RJ-45 connectors. The RJ-45 connector is very similar to, but a little bit wider than, the modular connector that connects a telephone to the wall jack (an RJ-11 connector). (You will notice this when you stick an RJ-11 connector into an RJ-45 jack. It sort of fits, but it will not work.) Twisted pair cable and modular connectors have made it much simpler to add workstations to a star-wired bus than to a coaxial-cabled bus.

The many advantages of a star-wired bus LAN include simple installation and maintenance, low-cost components (such as hubs and twisted pair wiring), and high volume of compatible products due to major market share. Perhaps the only disadvantage of a star-wired bus design is the amount of traffic its hub(s) must handle. When two or more hubs are interconnected and a workstation transmits data, *all* the workstations connected to *all* the hubs receive the data. This is an example of a shared network. All of the devices on the network are sharing the one bandwidth. As has been noted, the hub is a relatively nonintelligent device. It does not filter out any data frames, and it does not perform any routing. This, at it turns out, has become a major problem with hub-based LANs and has led to the almost-replacement of the hub. But before we explore the hub's replacement, let's take a look at the software that operates over hub-based LANs.

A medium access control protocol is the software that allows a device to place data onto a hub-based local area network (as well as other networks that require their workstations to compete for access to the network). Until several years ago, the bottom line with all medium access control protocols was this: Because a local area network is essentially a single bus using baseband technology, it is imperative that only one workstation at a time be allowed to transmit its data onto the network. This imperative, as we will see shortly, has changed with the introduction of switches (in place of hubs) and full-duplex connections. For the moment, however, we will concentrate on one workstation transmitting at a time.

The two basic categories of medium access control protocols for local area networks are:

- ▶ Contention-based protocols, such as carrier sense multiple access with collision detection
- ▶ Round-robin protocols, such as token passing

Let us examine only the first protocol, as it was the predominant form of medium access control for local area networks for many years. For a discussion of a round-robin protocol, please visit the author's Web site http://facweb.cs.depaul.edu/cwhite.

Contention-based protocols

A contention-based protocol is basically a first-come, first-served protocol—the first station to recognize that no other station is transmitting data and place its data onto the medium is the first station to transmit. The most popular contention-based protocol is carrier sense multiple access with collision detection (CSMA/CD). The CSMA/CD medium access control protocol was found almost exclusively on star-wired bus and bus local area networks and was for many years the most widely used medium access control protocol.

The name of this protocol is so long that it almost explains itself. With the CSMA/CD protocol, only one workstation at a time can transmit, and because of this, the CSMA/CD protocol is basically a half-duplex protocol. A workstation listens to the medium—that is, senses for a carrier on the medium—to learn whether any other workstation is transmitting. If another workstation is transmitting, the workstation wanting to transmit will wait and try again to transmit. The amount of time the workstation waits depends on the particular type of CSMA/CD protocol used (for more information, see the Details section titled "CSMA/CD Persistence Algorithms"). If no other workstation is actively transmitting, the workstation transmits its data onto the medium. The CSMA/CD access protocol is analogous to human beings carrying on a conversation at the dinner table. If no one is talking, someone can speak. If someone is talking, everyone else hears this and waits. If two human beings start talking at the same time, they both stop immediately (or at least polite people do) and wait a certain amount of time before trying again.

In most situations, the data being sent by a workstation is intended for one other workstation, but all the workstations on the CSMA/CD protocol-based network receive the data. (Once again, we will see that this is no longer true with the switched local area networks.) Only the intended workstation (the workstation with the intended address) will do something with the data. All the other workstations will discard the frame of data.

As the data is being transmitted, the sending workstation continues to listen to the medium, listening to its own transmission. Under normal conditions, the workstation should just hear its own data being transmitted. If the workstation hears garbage, however, it assumes a collision has occurred. A collision occurs when two or more workstations listen to the medium at the same moment, hear nothing, and then transmit their data at the same moment.

Actually, the two workstations do not need to begin transmission at exactly the same moment for a collision to occur. Consider a situation in which two workstations are at opposite ends of a bus. A signal propagates from one end of the bus to the other end in time *n*. A workstation will not hear a collision until its data has, on average, traveled halfway down the bus, collided with the other workstation's signal, and then propagated back down the bus to the first station (Figure 7-8). This interval, during which the signals propagate down the bus and back, is the collision window. During this collision window, a workstation might not hear a transmission, falsely assume that no one is transmitting, and then transmit its data.

Figure 7-8
Two workstations at opposite ends of a bus experiencing a collision

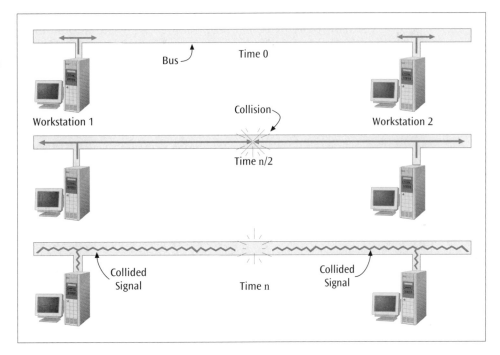

If the network is experiencing a small amount of traffic, the chances for collision are small. The chance for a collision increases dramatically when the network is under a heavy load and many workstations are trying to access it simultaneously. Studies have shown that as the traffic on a CSMA/CD network increases, the rate of collisions increases, which further degrades the service of the network. If a workstation detects a collision, it will immediately stop its transmission, wait some random amount of time, and try again. If another collision occurs, the workstation will wait once more. Because of these collisions, busy CSMA/CD networks rarely exceed 40 percent throughput. In other words, busy CSMA/CD networks waste 60 percent of their time dealing with collisions and other overhead. Many users find this rate unacceptable and consider modifications to standard CSMA/CD, such as using switches instead of hubs.

Because the number of times a workstation will have to wait is unknown, it is not possible to determine exactly when a workstation will be allowed to transmit its data without collision. Thus, CSMA/CD is a nondeterministic protocol. A nondeterministic protocol is one in which you cannot calculate the time at which a workstation will transmit. If your application must have its workstations transmit data at known times, you might want to consider a medium access control protocol other than this basic form of CSMA/CD.

What can we do to decrease the number of collisions? We need to replace our hubs with switches.

Switches

A hub is a simple device that requires virtually no overhead to operate. But it is also inefficient. When a network is experiencing a high level of traffic, a hub compounds the problem by taking any incoming frame and retransmitting it out to all connections. In contrast, the switch uses addresses and processing power to direct a frame out a particular port, thus reducing the amount of traffic on the network.

Switches, like hubs, can be used to interconnect multiple workstations on a single LAN or to interconnect multiple LANs. Whether interconnecting multiple local area networks or multiple workstations, the switch has one primary function—to direct the data frame to only the addressed receiver. Thus, the switch needs to know where all the devices are so that it can send the data out the appropriate link. It does not send the frame out all links, as the hub does (unless the data frame is a broadcast message). Thus, the switch acts as a filter. A filter examines the destination address of a frame and forwards the frame appropriately, depending on some address information stored within the switch. The switch will greatly reduce the amount of traffic on the interconnected networks by dropping frames that do not have to be forwarded.

Let us examine this filtering function a little more closely (Figure 7-9). As a frame of data moves across the first local area network and enters the switch, the switch examines the source and destination addresses stored within the frame. These frame addresses are assigned to the network interface card (NIC) when the NIC is manufactured. (All companies that produce NICs have agreed to use a formula that ensures that every NIC in the world has a unique NIC address.) The switch, using some form of internal logic, determines if a data frame's destination address belongs to a workstation on the current network. If it does, the switch does nothing more with the frame, because it is already on the appropriate network. If the destination address is not an address on the current network, the switch passes the frame on to the next local area network, assuming that the frame is intended for a station on that network.

Figure 7-9
A switch interconnecting two local area networks

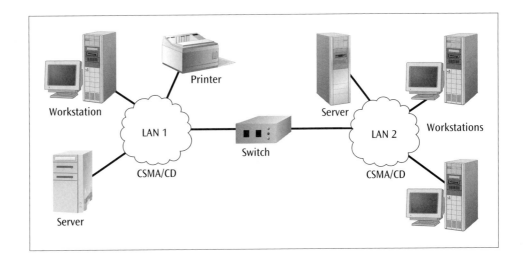

How does the switch know what addresses are on which networks? Did a technician sit down and type the address of every NIC on each interconnected network? Not likely. Most switches are transparent, which means they learn by themselves. Upon installation, the switch begins observing the addresses of the frames in transmission on the current network and creates an internal port table to be used for making future forwarding decisions. The switch creates the internal port table by using a form of backward learning—that is, by observing the location from which a frame has come. If a frame is on the current network, the switch assumes that the frame originated from somewhere on that network. The switch takes the source address from the frame and places it into an internal table. After watching traffic for a while, the switch has a table of workstation addresses for that network. If a frame arrives at the switch with a destination address that does not match any address in the table, the switch assumes the frame is intended for a workstation on some other network and passes the frame on to the next network.

For an example of how the transparent switch learns, examine Figure 7-10 and the following scenario. This particular switch has two ports, one for LAN A and the second for LAN B. When the switch is first activated, its internal port tables, one for Port A and one for Port B, are empty. Figure 7-11(a) shows the two tables as being initially empty. Now suppose Workstation 1 transmits a frame intended for Workstation 4. Because these networks are broadcast networks, the frame goes to all devices on the network, including the switch. The switch extracts Workstation 1's address and puts it in Port A's table. It has just learned that Workstation 1 is on LAN A, as shown in Figure 7-11(b). The switch still does not know the address of Workstation 4, however. Even though Workstation 4 is also on LAN A, the switch does not know this fact, because there is no entry from an *incoming* frame in the Port A table for Workstation 4. Consequently, the switch unnecessarily forwards the frame out Port B onto LAN B.

Figure 7-10
*A switch
interconnecting two
local area networks
has two internal port
tables*

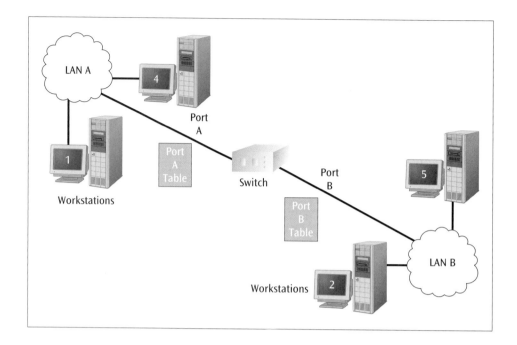

Figure 7-10
*A switch
interconnecting two
local area networks
has two internal port
tables*

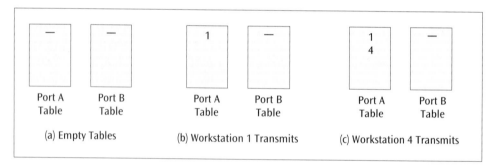

Figure 7-11
*The two internal port
tables and their new
entries*

Now suppose Workstation 4 returns a frame to Workstation 1. The switch extracts the address of Workstation 4 and places it in the Port A table, as shown in Figure 7-11(c). The frame is destined for Workstation 1, and the switch sees that there is an entry for Workstation 1 in Port A's table. Now, the switch knows Workstation 1 is on LAN A and *does not* forward the frame on to LAN B. In addition, if Workstation 1 sends another frame to Workstation 4, the switch will see that Workstation 4 is on LAN A (because of the entry in the Port A table) and *will not* forward the frame on to LAN B.

If Workstation 1 sends a frame to Workstation 5, the switch will not recognize the address of Workstation 5, because there is no entry in Port A's table, and it will forward the frame on to LAN B. The switch will perform the same learning function for LAN B and update Port B's table accordingly. Thus, the switch learns where workstations are and then uses that information for future forwarding decisions.

The above example showed a switch with two ports, one for each network. In reality, switches can have many ports. For example, you can have a LAN with fifty workstations, each one connected to a port on a switch. To support each of these ports efficiently, the main hardware of the switch—called the backplane—has to be fast enough to support the aggregate or total bandwidth of all the ports. For example, if a switch has eight 100-Mbps ports, the backplane has to support a total of 800 Mbps. This backplane is similar to a bus inside a microcomputer. It allows you to plug in one or more printed circuit cards. Each circuit card supports one port, or connection, to a workstation or other device. If the circuit cards are hot swappable, it is possible to insert and remove cards while the power to the unit is still on. This capability allows for quick and easy maintenance of the switch. As traffic enters on each port, a table is updated to reflect the source address of the received frame. Later, when a frame is to be transmitted to another workstation, this table of forwarding addresses is consulted and the frame is sent out the optimal port.

Switches can significantly decrease interconnection traffic and increase the throughput of the interconnected networks or segments, without requiring additional cabling or rearranging of the network devices. As you may recall, hub-based local area networks experience collisions. By reducing the number of unnecessarily transmitted packets, a switch can cause the number of collisions to decline. As the number of collisions declines, the overall throughput of the network should increase.

Another important advantage of a switch is that it is designed to perform much faster than a hub; this is especially true of switches that use cut-through architecture. In a cut-through architecture, the data frame begins to exit the switch almost as soon as it begins to enter the switch. In other words, a cut-through switch does not store a data frame and then forward it. In contrast, a store-and-forward device holds the entire frame for a small amount of time while various fields of the frame are examined, a procedure that diminishes the overall network throughput. This cut-through capability allows a switch to pass data frames very quickly, thus improving the overall network throughput. This major disadvantage of cut-through architecture is the potential for the device to forward faulty frames. For example, if a frame has been corrupted, a store-and-forward device will input the frame, perform a cyclic checksum, detect the error, and perform some form of error control. A cut-through device, however, is so fast that it begins forwarding the frame before the cyclic checksum field can be calculated. If there is a cyclic checksum error, it is too late to do anything about it. The frame has already been transmitted. If too many corrupted frames are passed around the network, network integrity suffers.

The switch has one physical similarity to the hub. If you decide to install a switch into a local area network, it is often as simple as unplugging a hub and plugging a switch in the hub's place. Logically, however, the switch is not like the hub. The switch can examine frame addresses and, based on the contents of an address, direct the frame out the one appropriate path. Whereas the hub simply blasts a copy of the frame out all connections, the switch uses intelligence to determine the one best connection for outgoing transmission.

Depending on user requirements, a switch can interconnect two different types of CSMA/CD network segments: shared segments and dedicated segments. In shared segment networks, as shown in Figure 7-12, a switch may be connected to a hub (or several hubs), which then connects multiple workstations. Because the workstations are first connected to a hub, they all share the one channel, or bandwidth, of the hub, which limits the transfer speeds of individual stations.

Figure 7-12
*Workstations
connected to a shared
segment local area
network*

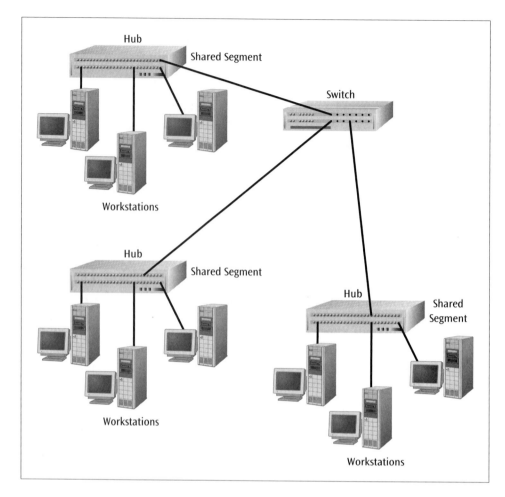

In dedicated segment networks, as shown in Figure 7-13, a switch may be directly connected to one or more workstations. Each workstation then has a private or dedicated connection. This dedicated connection increases the bandwidth for each workstation over what the bandwidth would be if the workstation were connected to the hub. Dedicated segments are useful for more powerful workstations with high communication demands. The workstations that are connected to the hub do not have dedicated connections to the network. They must share the bandwidth with the other workstations connected to that hub. The switch essentially treats all the workstations connected to the hub as a single connection.

Figure 7-13

Workstations connected to a dedicated segment local area network

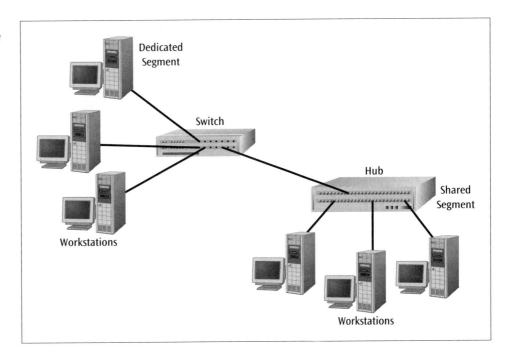

Isolating traffic patterns and providing multiple access

Whether shared or dedicated segments are involved, the primary goal of a switch is to isolate a particular pattern of traffic from other patterns of traffic or from the remainder of the network. Consider a situation in which two servers, along with a number of workstations, are connected to a switch (Figure 7-14). If Workstation A wishes to transmit to Server 1, the switch forwards the data packet/frame directly to Server 1 and to nowhere else on the network. Workstations B, C, and D do not receive a data frame from Workstation A. Furthermore, Workstation A can transmit to Server 1 at the same time that Workstation B transmits to Server 2. Finally, the switch can accommodate a high degree of intercommunication between the two servers without sending the data to any workstations on the network. Because many local area networks have a high degree of inter-server communication, this use of a switch can effectively reduce overall network traffic.

Figure 7-14

A switch with two servers allowing simultaneous access to each server

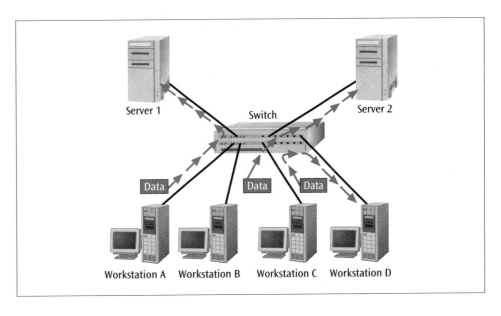

If access to a server is slow because a great number of transmissions are entering the server through its one cable and network interface card, another network interface card can be added to the server, and this new connection can be attached to an available port on the switch (see Figure 7-15). This scenario allows two workstations to transmit to the server simultaneously—thus providing more bandwidth in accessing the server—and can be especially effective if the server has multiple processors or multi-core processors installed.

Figure 7-15

A server with two NICs and two connections to a switch

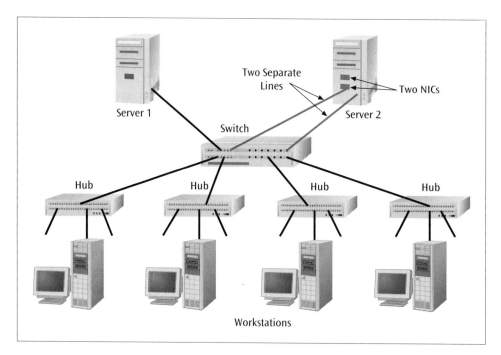

Switches can be used in combination with routers to further isolate traffic segments in a local area network. In Figure 7-16, notice that Server 1 is located near Workstations A and B and is isolated from the rest of the network via a pair of routers. If Workstations A and B routinely access Server 1, this traffic will be isolated from everyone else on the larger network. On the other side of the routers, Workstations C, D, and E typically access Servers 2, 3, and 4 through a switch. Thus, their traffic is isolated from each other and from the rest of the network.

Figure 7-16

A pair of routers (or remote bridges) and switch combination designed to isolate network traffic

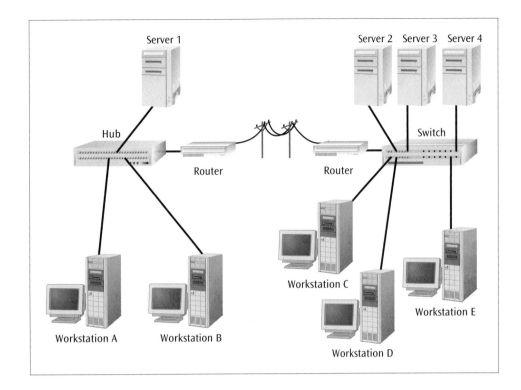

Full-duplex switches

One of the primary disadvantages of a CSMA/CD local area network with hubs is the presence of collisions. Because a workstation cannot transmit data if it hears another workstation already transmitting, a CSMA/CD network is a half-duplex system. Recall that a half-duplex system allows both sender and receiver to talk, but not at the same time. The bandwidth of a CSMA/CD network would double if it were a full-duplex system in which both sender and receiver could talk simultaneously. In addition, there would be no collisions on an individual segment, which would simplify the CSMA/CD algorithm. The full-duplex switch allows for a CSMA/CD network to simultaneously transmit and receive data to and from a workstation.

How can a workstation on a CSMA/CD network send and receive signals at the same time? Use two pairs of wire—transmit the send signal on one pair of wires and the receive signal on the other pair of wires. Figure 7-17 shows the upstream signal traversing one pair of wires and the downstream signal traversing the other set.

Figure 7-17

Full-duplex connection of workstations to a LAN switch

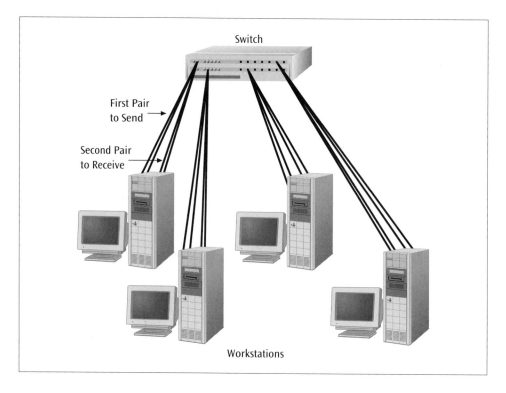

What is needed to support a full-duplex connection on a local area network? Essentially, three conditions have to be met. First, the NIC in the workstation has to be capable of supporting, and then configured to support, a full-duplex connection. Likewise, the switch has to be configured for a full-duplex connection. Finally, the cable connecting the workstation's NIC to the switch must also be able to support a full-duplex connection. Both 10 Mbps Ethernet and 100 Mbps Ethernet require two pairs (four wires) between the workstation and the switch to support a full-duplex connection. Because of its much higher transmission speed, 1000 Mbps Ethernet needs four pairs of wire. Because the connection between a workstation and a switch is not very long (usually less than 100 meters), and the wire used is typically a twisted pair, it should not be very difficult to implement a full-duplex connection. In fact, if you are installing a new LAN as well as installing its wiring, making sure that each cable has a standard set of eight wires (using an RJ-45 connector) is a simple process. Even if you do not use all eight wires now, you will have the wiring in place for future upgrades.

It is clear that the switch is a powerful tool for segmenting a local area network and thus reducing its overall amount of traffic. As a result of the reduction of traffic, a LAN that employs switches is a more efficient network, with improved network performance. Additionally, because the network's overall throughput improves, its response time when a user requests a service is faster. The full-duplex switch has become so common in many networks, that collisions are eliminated and the network software really doesn't have to listen before transmitting—thus CSMA/CD has essentially become MA (multiple access).

Virtual LANs

One of the more interesting applications for a dedicated segment network and a switch is creating a virtual LAN. A virtual LAN, or VLAN, is a logical subgroup within a local area network created via switches and software rather than by manually moving wiring from one network device to another. For example, if a company wishes to create a workgroup of employees to work on a new project, network support personnel can create a VLAN for that workgroup. Even though the employees and their actual computer workstations may be scattered throughout the building, LAN switches and VLAN software can be used to create a "network within a network." Because these workstations are connected to a dedicated segment network, they will experience fewer (if any) collisions when they transmit to other workstations within the group in the new virtual network, and their response time will be faster. Additionally, because only those workstations that are members of the VLAN will receive frames, bandwidth is optimized, and the security of the network is enhanced.

The standard IEEE 802.1Q was designed to allow multiple devices to intercommunicate and work together to create a virtual LAN. Instead of sending a technician to a wiring closet to move a workstation cable from one switch to another, an 802.1Q-compliant switch can be remotely configured by a network administrator. Thus, a workstation can be moved from one VLAN to another by a simple software change. Most modern switches can have an IP address assigned to them, allowing a network administrator to effect a VLAN change via the Internet.

Popular Local Area Network Systems

Within the last twenty years there have been roughly four popular local area network systems: Ethernet, IBM Token Ring, Fiber Distributed Data Interface, and wireless Ethernet. IBM Token Ring and Fiber Distributed Data Interface have pretty much disappeared from the LAN scene, so we will not discuss them at this time. If you are interested in learning more about either of these, consult the author's Web site. For now, let us examine the two remaining local area networks: wired Ethernet and wireless Ethernet.

Wired Ethernet

Ethernet was the first commercially available local area network system and remains, without a doubt, the most popular LAN system today. The wired version of Ethernet is based primarily on the star-wired bus topology and uses essentially the CSMA/CD medium access protocol. Because Ethernet is so popular and has been around the longest, it has evolved into a number of different forms. To avoid mass mayhem, the IEEE created a set of individual standards specifically for Ethernet or CSMA/CD local area networks, all under the category of 802.3. Let us examine the different 802.3 protocols in a little more detail. For your reference, the 802.3 standards to be discussed are summarized in Table 7-1.

The original 802.3 standards include 10Base5, 10Base2, 1Base5, and 10BaseT. The 10Base5 standard was one of the first Ethernet standards approved. The term "Base" is an abbreviation for baseband signals using a Manchester encoding. Recall that baseband signals are digital signals. Because there is no multiplexing of digital signals on any baseband LANs, there is only one channel of information on the network. The 10 of 10Base5 represents a 10-Mbps transmission speed, and the 5 represents a 500-meter maximum cable segment length. (Note that these standards, like those associated with conducted media, are based on the Metric system; where applicable, the English equivalent will be provided.) 10Base2 (nicknamed Cheapernet) was designed to allow for a less-expensive network by using less-expensive components. The **10Base2** network can transmit 10-Mbps digital signals over coaxial cable, but only for a maximum of 200 meters (the value 2 in 10Base2). 1Base5 was a system designed for twisted pair wiring, but with only a 1-Mbps data transfer rate for 500 meters. Because of technological advances, **10Base5**, 10Base2, and 1Base5 are extinct. A 10-Mbps standard that was extremely popular was 10BaseT. A 10-BaseT system transmits 10-Mbps baseband (digital) signals over twisted pair for a maximum of 100 meters per segment length. Many businesses, schools, and homes were using **10BaseT** (and then 100BaseT) as their local area network.

One of the most common standards for broadband (analog) Ethernet was the **10Broad36** specification. Using coaxial cable to transmit analog signals, 10Broad36 transmitted data at 10 Mbps for a maximum segment distance of 3600 meters. Note the much longer distance due to the use of coaxial cable and analog signals. And because broadband signals can support multiple channels, 10Broad36 delivered many concurrent channels of data, each supporting a 10-Mbps transmission stream. Too many network specialists found the coaxial cable and analog signals too difficult to work with. According to some industry reports, no one is installing any new 10Broad36 systems. Apparently, multiple concurrent channels of data were not enough to save 10Broad36.

When 10-Mbps Ethernet was first available, it was a fast protocol for many types of applications. As in most computer-based technologies, however, it was not fast enough for very long. In response to the demand for faster Ethernet systems, IEEE created the 100-Mbps Ethernet 802.3u protocol. The 100-Mbps Ethernet standards described in this paragraph are called **Fast Ethernet** to distinguish them from the 10-Mbps standards. 100BaseTX was designed to support 100-Mbps baseband signals using two pairs of Category 5 unshielded twisted pair. Like its 10BaseT counterpart, **100BaseTX** was designed for 100-meter segments. It is similar to 10BaseT systems that use twisted pair wiring and hubs with multiple workstation connections. **100BaseT4** was created to support older-category wire. Thus it can operate over Category 3 or 4 twisted pair wire, as well as Category 5/5e/6 unshielded twisted pair. It also transmits 100-Mbps baseband signals for a maximum of 100 meters, but is now extinct. Finally, **100BaseFX** was the standard created for fiber-optic systems. It can support 100-Mbps baseband signals using two strands of fiber but for much greater distances—1000 meters.

The next set of Ethernet standards to be developed was based on 1000-Mbps transmission speeds, or 1 gigabit (1 billion bits) per second. These standards define the **Gigabit Ethernet (IEEE 802.3z)**, which has become one of the hotter technologies for high-speed local area networks. The first gigabit standard—**1000BaseSX**—supports the interconnection of relatively close clusters of workstations and other devices using multimode fiber-optic cables. **1000BaseLX** is designed for longer-distance cabling within a building and uses either single-mode fiber-optic cables or multimode fiber. **1000BaseCX** is designed for short-distance (0.1 to 25 meters) jumper cables using balanced copper wire. A more recent standard, termed simply **1000BaseT**, is capable of using either the Category 5e or Category 6 cable specification.

More precisely, 1000BaseT incorporates advanced multilevel signaling to transmit data over four pairs of CAT 5e/CAT 6 cable.

A more recent Ethernet standard is 10 Gbps Ethernet. The **10 Gbps Ethernet** standard (transferring at a rate of 10 billion bits per second) was approved by IEEE in July 2002. This standard is also known as IEEE 802.3ae. The original 10 Gbps standard has already morphed into multiple secondary standards. Most of these secondary standards involve fiber-optic cable as the medium for both short and medium distances. However, a couple new copper-based standards are emerging. One such standard involves common CAT 6 twisted pair but only for 55 meters. A second copper standard may use a new variation on CAT 6 called CAT 6a and allow transmission distances up to 100 meters. Yet a third proposal involved using four pairs of twin-axial cabling. Of course, do not leave out a couple more proposals: 40 Gbps and 100 Gbps Ethernet, which are currently in draft stages.

Table 7-1 summarizes the various Ethernet standards just introduced. The table includes maximum transmission speed, signal type (either baseband, which corresponds to a digital signal, or broadband, which corresponds to analog), cable type, and maximum segment length without a repeater.

Table 7-1
Summary of Ethernet standards

Ethernet Standard	Maximum Transmission Speed	Signal Type	Cable Type	Maximum Segment Length
10BaseT	10 Mbps	Baseband	Twisted pair	100 meters
100BaseTX	100 Mbps	Baseband	2-pair Category 5 or higher unshielded twisted pair	100 meters
100BaseFX	100 Mbps	Baseband	Fiber optic	1000 meters
1000BaseSX	1000 Mbps	Baseband	Fiber optic	300 meters
1000BaseLX	1000 Mbps	Baseband	Fiber optic	100 meters
1000BaseCX	1000 Mbps	Baseband	Specialized balanced copper	25 meters
1000BaseT	1000 Mbps	Baseband	Twisted pair—four pairs	100 meters
10GBase-fiber	10 Gbps	Baseband	Fiber optic	various lengths
10GBase-T	10 Gbps	Baseband	Cat 6	55–100 meters
10GBase-CX4	10 Gbps	Baseband	Twin axial	~30 meters

One of the recent improvements to Ethernet is **power over Ethernet (PoE)** (IEEE standard 802.3af). Suppose you want to place a network interface card (NIC) in a device, but do not want to or cannot connect the device to a electrical source. For example, you want to install a surveillance camera that transfers its signal first over Ethernet and then over the Internet. Normally, you would install the camera, then install both an Ethernet connection and an electrical connection. But with PoE, you can send electrical power over the Ethernet connection, which can be used to power the camera. Although this sounds promising, one of the drawbacks is the capability to provide the Ethernet hub or switch with enough power so that the power can then be distributed over Ethernet lines to the various devices.

Wireless Ethernet

A local area network that is not based primarily on physical wiring but uses wireless transmissions between workstations is a wireless LAN, or wireless Ethernet. By attaching a transmitter/receiver to a special network interface card on a workstation or laptop, and similar hardware on a device called an access point, it is possible to transmit data between a workstation and network server at speeds into the millions of bits per second. Plus, the workstation can be located anywhere within the acceptable transmission range. This acceptable range varies with the wireless technology used, but typically falls between a few feet and 800 feet. You could argue that a wireless LAN is essentially a star topology, because the wireless workstations typically radiate around and transmit data to the access point.

Note that most wireless local area networks are actually combinations of wireless and wired technologies. The wireless portion consists of wireless workstations and laptops/notebooks/netbooks, and access points. On the other end of these access points (whose details will be examined shortly) is the wired local area network. The wired portion contains the usual Category 5e/6 wiring, hubs, switches, routers, and servers.

Clearly, one of the strongest advantages of a wireless LAN is that no cabling is necessary for the user device to communicate with the network. This makes a wireless LAN a perfect solution for many different applications. Consider an environment in which it is simply not possible to run cabling, such as in the middle of a warehouse, or on the floor of a stock exchange. Wireless LANs would also work well in historic buildings, or buildings with thick concrete or marble walls, where drilling holes through walls, ceilings, or floors is undesirable or difficult. Many business offices have incorporated wireless local area networks for other reasons as well. Suppose an employee is sitting in her cubicle working on her laptop over a wireless connection. Suddenly, the employee is called into a meeting. She picks up her laptop, walks into the meeting room, and continues to work over the wireless connection. Likewise, many if not most college and high-school campuses have set up wireless LANs, so that students can access network operations while sitting in class, working in the library, or enjoying a beautiful day on the quad.

In order to create a wireless local area network, a few basic components are necessary. The first component is the user device (also called the wireless station), such as a laptop computer, workstation, or a handheld device. The user device has a special NIC that receives and transmits the wireless signals. The second component is the wired local area network. This is the conventional network component that supports standard workstations, servers, and medium access control protocols. A vast majority of wireless networks are connected to wired local area networks. The third component is the access point, or wireless router, which is the component that communicates with the wireless user device. The access point is essentially the interface device between the wireless user device and the wired local area network. The access point also acts as a switch/bridge and supports a medium access control protocol.

Wireless local area networks are typically found in three basic configurations. The first is the single-cell wireless LAN (Figure 7-18). At the center of the cell is the access point, which is connected to the wired LAN. All user devices communicate with this one access point and compete for the same set of frequencies. The wireless LAN standards call this cell a **Basic Service Set (BSS)**.

Figure 7-18
A single-cell wireless LAN configuration

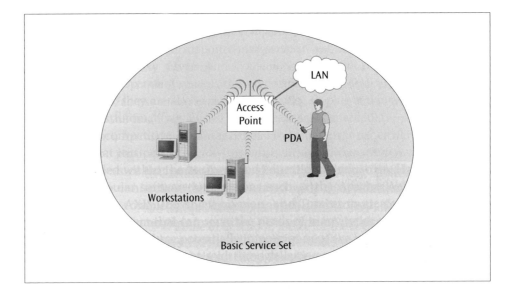

The second type of wireless LAN configuration is the multiple-cell layout (Figure 7-19). In this configuration, multiple cells are supported by multiple access points, as in a cellular telephone network. User devices communicate with the nearest access point and may move from one cell to another. Another way in which this configuration is similar to a mobile telephone cellular network is that each cell uses a different set of frequencies for communication between the user device and the access point. The wireless LAN term for a collection of multiple Basic Service Sets is **Extended Service Set (ESS)**.

Figure 7-19
A multiple-cell wireless LAN configuration

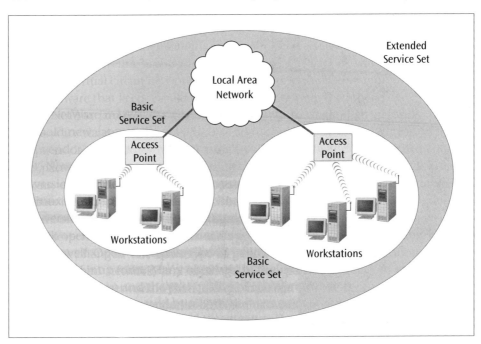

The third wireless LAN configuration is the peer-to-peer, or ad hoc, layout (Figure 7-20). With this configuration, there is no access point at the center of a cell. Each user device communicates *directly* with the other user devices. A configuration like this may be found in a business meeting in which all user devices are transmitting and sharing information at the same time.

Figure 7-20
Ad hoc configuration for a wireless LAN

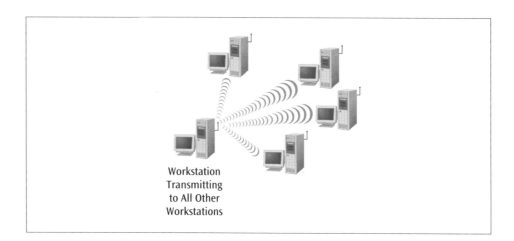

Workstation
Transmitting
to All Other
Workstations

When wireless LANs first appeared, organizations were slow to accept them. To promote acceptance, the **IEEE 802 suite of protocols** was created to support the many different types of wireless local area networks in existence. As part of this effort, the approval of the IEEE 802.11 wireless standard in June 1997 greatly helped with standardization of wireless networks and sped their growth and acceptance. Basically, the 802.11 specification defined three different types of physical layer connections to match price, performance, and operations to a particular application. The first type of physical layer defines infrared transmissions. Infrared, as you might recall from Chapter Three, is line-of-sight-based and cannot pass through solid walls. Transmission rates for infrared wireless range from 1 to 2 Mbps. The second type of physical layer defines a technology that can transmit a secure signal (spread spectrum) and can transmit data at rates up to 2 Mbps in quiet environments and 1 Mbps in noisy environments for distances up to 800 feet. The third type of physical layer is also spread spectrum but uses frequency hopping and can transmit data at 2 Mbps for 300 feet.

As we saw in Chapter Three, since the introduction of IEEE 802.11, other IEEE wireless LAN protocols have emerged. **IEEE 802.11b**, which was ratified in September 1999, can transmit data at a theoretical rate of 11 Mbps (due to noise, the actual rate is roughly one-half the theoretical) using 2.4-GHz signals. Many users who had found 802.11's data rate of 2 Mbps too slow were excited to see 802.11b products hit the market. Another name for 802.11b (as well as the other 802.11 wireless LAN standards) is Wi-Fi (wireless fidelity). **IEEE 802.11g**, introduced in 2002, transmits data at a theoretical rate of 54 Mbps (again, actual rates are about one half) using the same 2.4-GHz frequencies used in 802.11b. Also introduced in 2002 was **IEEE 802.11a**, which is capable of supporting a theoretical rate of 54-Mbps transmissions (again actual rates are about one half) using the 5-GHz frequency range.

The most recent addition to the wireless LAN market is the IEEE 802.11n standard. **802.11n** has a theoretical maximum data rate of 600 Mbps with actual data rates of roughly 100 to 145 Mbps. In order to support such high speeds, this new standard uses a technology called multiple input multiple output. Multiple input multiple output (MIMO) is a technique in which both the mobile device and the access point have multiple, smart antennae that help to reduce signal interference and reflections.

As if all the IEEE protocols were not enough to remember, yet one more protocol for wireless LANs is worth mentioning. HiperLAN/2, a European standard, is also capable of transmitting data at a theoretical 54 Mbps using the 5-GHz frequency range.

The newer wireless protocols do come with some disadvantages. The high 5-GHz frequency ranges used in IEEE 802.11a and HiperLAN/2 require so much power that laptops and portables have trouble delivering these signals for a useful amount of time. Second, the transmission distance for the higher frequency ranges is shorter than for the lower frequency ranges. A current (and rough) rule of thumb places the maximum transmission distance of 802.11b (2.4 GHz) at approximately 250 to 300 feet, and that of 802.11a (5 GHz) at approximately 90 feet. Thus, a company would need roughly three times as many access points for the 5-GHz range as it would for the 2.4-GHz range, which will increase the company's installation and maintenance costs. To further complicate matters, the shorter wavelength of 5 GHz also has trouble going through walls, floors, and furniture. This too may influence a company's decision to move into the higher-frequency-range products. On the positive side, the 5-GHz range of 802.11a encounters less signal interference from devices such as microwave ovens and thus should be capable of higher transmission speeds.

The final disadvantage of wireless systems in general is related to security. Security on a wireless LAN was originally provided by Wired Equivalent Privacy (WEP). Although WEP is based on fairly modern encryption techniques, many critics have felt that it is not powerful enough to stop wiretapping and that it has several flaws. In any case, despite the existence of security protocols such as WEP, reports have shown that roughly half of the wireless LANs in existence are not even WEP-enabled, thus leaving the network transmissions wide open to interception. (Newer security protocols have been developed to replace WEP and will be discussed in more detail in Chapter Twelve.) Although concerns about transmission distance and security remain, it appears that wireless local area networks are here to stay and will see only increases in speeds, availability, and applications.

Despite the disadvantage of wireless security, wireless LANs continue to grow in popularity.

Wireless CSMA/CA

The contention-based medium access control protocol that supports wireless local area networks has two interesting differences from the older CSMA/CD protocol found on hub-based wired LANs. First, there is no collision detection. In other words, the transmitter does not listen during its transmission to hear if there was a collision with another signal somewhere on the network. Three reasons for this involve the cost of producing a wireless transmitter that can transmit and listen at the same time, there is no wire on which to listen to an increase in voltage (the collision of two signals), and that if very often two workstations are so far apart that they cannot hear each other's transmission signal, then they will not hear a collision. Instead, the algorithm of the protocol supporting wireless LANs limits when a workstation can transmit, in an attempt to reduce the number of collisions. The type of algorithm that tries to avoid collisions is called carrier sense multiple access with collision avoidance (CSMA/CA). How does the algorithm limit when a workstation can transmit? Part of that answer is tied to the second interesting difference—priority levels. In an attempt to provide a certain level of priority to the order of transmission, the CSMA/CA algorithm has been modified to function according to the following rule: If a user device wishes to transmit and the medium is idle, the device is not allowed to transmit immediately. Instead, the device is made to wait for a small period of time called the interframe space (IFS). If the medium is still idle after this interframe space, the device is then allowed to transmit. How does the interframe space provide a priority system? There are up to three different interframe space times. The first IFS time—short IFS—is used by devices that require an immediate response, such as an acknowledgment, a clear to send, or a response to a poll. The second IFS time—midlength IFS—is used by the access device when it is issuing polls to the user devices. The third IFS time—long IFS—is used as a minimum delay for ordinary user devices when they are contending for access to the network. Thus, before a standard user device can transmit, it must wait and give higher priority devices a chance to transmit first.

If the medium is initially busy, the device simply continues to listen to the medium. When the medium becomes idle, the user device delays for the interframe space. If the medium is still idle after the interframe space, the user device selects a random backoff factor. When the backoff counter reaches zero, the device transmits the packet. This procedure helps prevent a number of users from hearing an idle medium, transmitting at the same moment, and causing a collision.

Details ▶

CSMA/CD Persistence Algorithms

Suppose a workstation wishes to transmit data and listens to the medium. What happens if the medium is busy? The workstation does not transmit, but waits. How long does the workstation wait? What degree of listening persistence does the workstation exhibit? Three different persistence algorithms have been created: non-persistent, 1-persistent, and p-persistent.

With the non-persistent algorithm, if the workstation finds the medium is busy, it waits for a random amount of time (t), then listens again. What if the medium became free immediately after the workstation had listened and found it busy? That would be too bad; if the workstation had been more persistent, it would have learned sooner that the medium had become free, and it would not have wasted time waiting.

The 1-persistent algorithm takes this condition into account: with it, the workstation listens continuously until the medium is free, then transmits immediately. What happens if two workstations following the 1-persistent algorithm both are listening and waiting? They will probably try to transmit at the same moment and cause a collision.

With the p-persistent algorithm, if the medium is busy, the workstation continues listening. When the medium becomes idle, the workstation transmits with probability p or delays the standard random amount of time with probability $1 - p$. The p-persistent algorithm is a compromise between the non-persistent and 1-persistent algorithms. The workstation continuously listens until the medium becomes free, but then does not transmit immediately. It transmits only with probability p. If $p = 0.1$, then nine times out of ten the workstation waits, and one time out of ten it transmits immediately. The addition of probability to the algorithm makes collisions less likely to occur. In other words, the odds are now more likely that if two or more workstations are waiting for a free medium, they both will not begin transmitting at the exact moment they hear that the medium has become idle. The selection of the value of p is often determined by the number of workstations on the network. The larger the number, the smaller the p value, which should decrease the probability of two workstations transmitting at the same moment.

IEEE 802

When ISO created the OSI model in the 1970s, local area networks were just beginning to appear. In order to better support the unique nature of local area networks and to create a set of industry-wide standards, the IEEE produced a series of protocols under the name 802 (some of which you have already encountered in your reading). One of the first things the IEEE 802 protocols did was to split the data link layer into two sublayers: the medium access control sublayer and the logical link control sublayer (Figure 7-21). The medium access control (MAC) sublayer works more closely with the physical layer and contains a header, computer (physical) addresses, error-detection codes, and control information. Because of this closeness with the physical layer, there is not a strictly defined division between the MAC sublayer and the physical layer. The logical link control (LLC) sublayer is primarily responsible for logical addressing and providing error control and flow control information.

Figure 7-21
Modification of OSI model to split data link layer into two sublayers

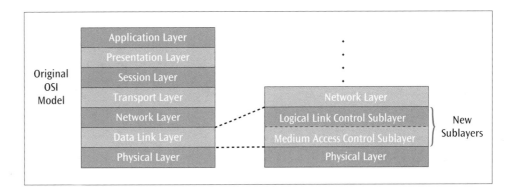

The medium access control sublayer defines the layout or format of the data frame, simply called the frame. There are a number of different frame formats, depending on the type of local area network. For example, CSMA/CD LANs have one frame format, while the now-defunct token ring LANs had another format. Within this frame format are the fields for error detection, workstation addressing, and various types of control information. Thus, the MAC sublayer is a very important layer when it comes to describing a local area network. Let us examine the most common MAC sublayer frame format: IEEE 802.3 format for CSMA/CD networks.

IEEE 802.3 frame format

The IEEE 802.3 standard for CSMA/CD uses the frame format shown in Figure 7-22. The preamble and start of frame byte fields combine to form an 8-byte flag that the receiver locks onto for proper synchronization. The destination address and source address are the 6-byte addresses of the receiving computer and sending computer. More precisely, each network interface card in the world has a unique 6-byte (48-bit) address. When CSMA/CD sends data to a particular computer, it creates a frame with the appropriate NIC address of the intended computer. The data length is simply the length in bytes of the data field, which is the following entry. The PAD field adds characters to the frame (pads the frame). The minimum size frame that any station can transmit is 64 bytes long. Frames shorter than 64 bytes are considered runs, or frame fragments, that resulted from a collision, and these are automatically discarded. Thus, if a workstation attempts to transmit a frame in which the data field is very short, PAD characters are added to ensure that the overall frame length equals at least 64 bytes. Finally, the checksum field is a 4-byte cyclic redundancy checksum.

Figure 7-22
Frame format for IEEE 802.3 CSMA/CD

Preamble	Start of Frame Byte	Destination Address	Source Address	Data Length	Data	PAD	Checksum
7 bytes of 10101010	10101011	6 bytes	6 bytes	2 bytes	0–1500 bytes	0–46 bytes	4 bytes

Our discussion of local area network technology started with an examination of the main types of network topologies: bus, star-wired bus, and wireless. Then the two major categories of medium access control protocols that operate on these different topologies were introduced: contention-based (CSMA/CD) and collision avoidance (CSMA/CA). Finally, we turned our attention to the actual products or local area network systems found in a typical computer environment. Let us conclude this chapter with some examples of actual local area network installations.

LANs In Action: A Small Office Solution ▶

Hannah is the computer specialist for a small business on the west side of Chicago. Her company currently has approximately 35 to 40 workers, each with his or her own workstation. The employees use the computers mainly for word processing, spreadsheets, and occasional database work. The owner of the company would like to update computer services by offering the following applications to employees:

- ▶ Internal e-mail
- ▶ Shared access to high-quality, black-and-white and color laser printers
- ▶ Access to a centralized database

After hearing these requests from the owner, Hannah concludes that the best way to offer these services is to install a local area network. Because all the workstations are only one to two years old, they do not need to be replaced. The workstations, however, will need the appropriate network interface cards (NICs) if they did not come new with them, as well as wiring, servers, and software to connect them to a local area network. The problem for Hannah is deciding which network to install. After reading the literature and talking to some fellow computer specialists, Hannah creates the following list of possible local area network systems:

1. 100-Mbps CSMA/CD (Ethernet)
2. 1000-Mbps CSMA/CD (Ethernet)
3. Wireless CSMA/CA (Ethernet)

The 100-Mbps Ethernet seems like a good candidate. It is the most popular local area network system and, therefore, has some of the lowest prices available. Additionally, obtaining technical support should not be any problem, and neither should finding matching hardware and software. An Ethernet network can easily support 40 users, but collisions might be a problem if hubs are used. If the users put a high demand on the network, they may notice some performance problems and complain. This would be a good time to consider switched Ethernet to eliminate collisions and increase throughput. Even though this first candidate looks like a good choice, Hannah does not want to eliminate other potentially good solutions, so she continues evaluating her list.

Might the 1000-Mbps Ethernet be a better solution? The cost of 1000-Mbps Ethernet compared to 100-Mbps Ethernet will be noticeably higher, especially when dealing with 1000-Mbps switches. Plus, there is the issue that a standard workstation does not experience a significant increase in performance when it is connected to 1000-Mbps Ethernet. Given the size of Hannah's company and its limited budget and relatively uncomplicated network needs, 100-Mbps Ethernet might be a more economically suitable solution.

Finally, Hannah considers wireless Ethernet. The wireless NICs and access points are a little more expensive than the wired components, but their costs should be offset by the elimination of the need to install wires. On the other hand, installing wires in this scenario should not be an extremely costly endeavor, because all the offices are on the same floor and in adjacent rooms (Figure 7-23). The walls and ceiling do not impose any unusual requirements, which also makes a wired solution relatively simple. Nonetheless, there is something attractive about not being tied to physical wire. Unfortunately, a wireless data transfer rate of roughly 25 Mbps (approximately one-half the theoretical rate of 54 Mbps) could be a problem. This data rate is roughly one-quarter that of wired 100 Mbps networks. Hannah decides to go with 100-Mbps wired CSMA/CD using hubs. She could have chosen switches over hubs but was not sure if the demand of her users would warrant the extra cost of switches. Prices of 100-Mbps equipment are quite reasonable, and these systems are fairly simple to install and maintain. Because Hannah, like you, is learning about local area networks, she figures she had better keep things as straightforward as possible.

Figure 7-23
Office layout for
Hannah's company

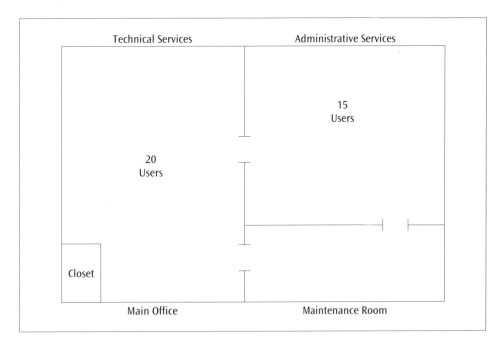

The next thing Hannah has to decide is how to configure the system. She knows that the NICs in each workstation plug into a hub and that a common hub supports 24 workstations. Because Technical Services has 20 employees and workstations and approximately 15 more employees and workstations are assigned to Administrative Services right next door, Hannah decides to connect the 20 Technical Services computers to one hub and the 15 Administrative Services computers to a second hub. The two hubs are then interconnected.

A rule known as the 5-4-3 rule helps guide the design of a shared-access Ethernet. More precisely, the **5-4-3 rule** states that between any two nodes on a network, there can be a maximum of only five segments (the sections between repeaters), connected through four repeaters, and only three of the five segments may contain user connections. A segment that does not contain any user connections is considered a link segment (unpopulated). Hannah's network has three segments: the Technical Services segment, the Administrative Services segment, and the unpopulated segment between the two hubs. Therefore, Hannah's network complies with the 5-4-3 rule.

The office space, which now has the configuration shown in Figure 7-24, dictates some of Hannah's next decisions. Category 6 unshielded twisted pair should work fine as the wiring between workstations and hubs. This is because no distance here needs to be longer than 100 meters (328 feet), and there is no equipment generating large amounts of electromagnetic noise. To interconnect the two hubs, Hannah decides to install fiber-optic cable, because the distance between hubs may be close to the 100-meter twisted pair limit.

Figure 7-24
Wiring diagram of Hannah's office space showing the placement of hubs and servers

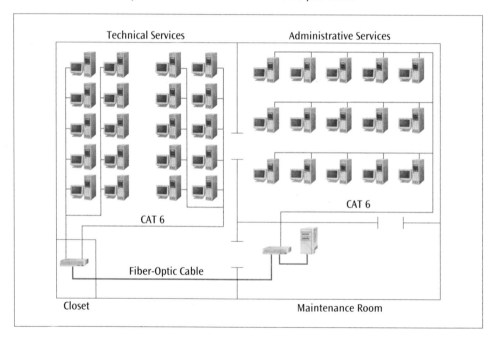

Hannah decides to place one hub in a closet just off the main room, where the 20 Technical Services employees work. This closet can be locked for security and has ample ventilation to keep the equipment cool. She decides to place the second hub and the network server in a small maintenance room adjacent to Administrative Services. The maintenance room can also be locked and has ample ventilation to keep the electronic equipment and room cool. All of the ceilings are false, so the wiring can run up the walls and through conduits between workstations and hubs.

Hannah is off to a good start with setting up a new LAN in her company's small office. Now that the network is operating as expected, Hannah is called into the owner's office. The owner of the company is pleased with the results thus far and wants to move into the next phase: giving all employees access to the Internet. This will allow them to download Web pages and send e-mail to anyone inside or outside the company. Hannah's next step is to install a router and get everyone connected to the Internet. She orders a router, then contacts an Internet service provider to acquire Internet access and approximately 40 Internet addresses. Hannah also calls the telephone company to install a high-speed line, such as a digital T-1 service, to support the connection between the router and the Internet service provider. These additions are shown in Figure 7-25.

Figure 7-25
The modified network with a router and high-speed phone line

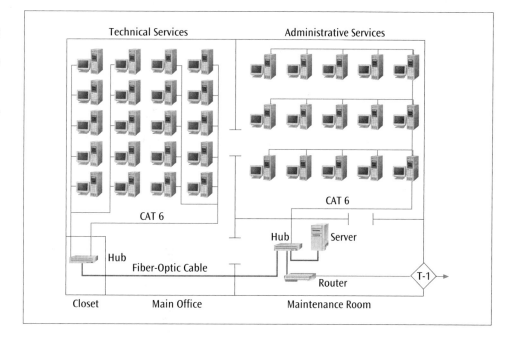

When the router arrives, Hannah installs and programs it. She is surprised at how involved routers are. She quickly learns that most routers operate in unprivileged mode and privileged mode. In unprivileged mode, a user can be shown only some of the router settings. In privileged mode, a user can be shown all router settings and configure the router. (Because Hannah is the lone network administrator, she selected privileged mode.) Configuring a router includes naming the router, designating a DNS server, setting the password for access to the privileged mode, assigning an IP address to each port, and telling the router to perform static routing or dynamic routing. After the router is operating correctly and Hannah makes the necessary changes to each employee's workstation, the employees are informed that they now have access to the Internet from their office workstations.

After all the systems are up and running and employees have access to the Internet (including the World Wide Web), the company's local area network and Internet usage starts growing, as Hannah feared, at a phenomenal rate. As the employees begin to learn the possibilities of sharing local files, databases, and printers, as well as accessing the vast amount of material available on the Internet, an increasing demand is placed on the local area network and server. Network response time grows slower and slower, and employees begin to complain. It is now time to examine the network and perform an upgrade.

After a number of hours of observation and talking to employees, Hannah determines that roughly half the employees use the business database quite heavily, and the other half use the Internet heavily. Almost everyone does word processing and some work with spreadsheets, but this work does not significantly contribute to the heavy network load. E-mail usage, however, is heavy and demands more and more storage space on the server.

To accommodate the heavy e-mail traffic, Hannah decides first that she needs a new server to handle just e-mail. If she can keep the database traffic on the database server and the e-mail traffic on the e-mail server, traffic flow should improve. Furthermore, if she goes to a full-duplex system instead of the half-duplex system she originally installed, the system will benefit from additional improvements because there will no longer be any collisions. Fortunately, during the initial installation, Hannah had the technicians install the standard four pairs of Category 6 twisted pair. As a consequence, she can now move up to full-duplex without having to change the original wiring. To create the full-duplex system, the two hubs are replaced with switches, the switches are programmed to support full-duplex, and all NICs are checked to make sure they support full-duplex connections. Figure 7-26 shows the upgraded network.

Figure 7-26
The upgraded network with an additional server and switches in place of hubs

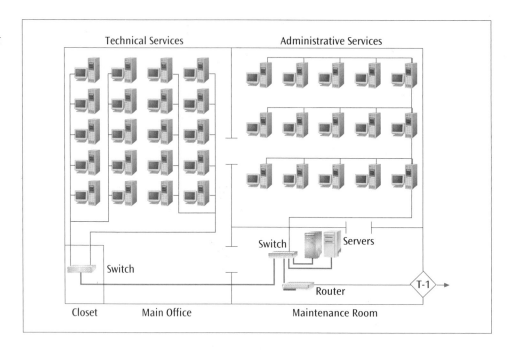

The upgrades Hannah makes to the network work very well. Network congestion is reduced significantly. Both e-mail users and database users notice a reasonable reduction in wait time. But because computer usage and corresponding network usage almost always increase over time, Hannah's solutions are more than likely temporary until the next problem arises. Nonetheless, Hannah's fellow employees are content for the present.

◆ ◆

LANs In Action: A Home Office Solution ▶

Sam had a nice computer system set up at home, but was finding it difficult to share with his wife and children. Because he was not willing to give up time on the computer, his only other option was to purchase a second computer. Having purchased a second computer, Sam realizes that he does not want to purchase another printer and most certainly does not want to install another broadband Internet access line. Yet both computers have a need for printing and access to the Internet. The solution for Sam is what is called a **Small Office/Home Office (SOHO) local area network**. The SOHO LAN is one the fasting growing segments of the networking market and is geared toward the small office or home office that has between 2 and 50 users. Often you can purchase a single package containing the NICs, cabling, switch/router device, and software necessary to establish a small local area network.

To install a SOHO LAN solution, Sam first has to make a number of decisions. Many computers now come from the manufacturer with NICs installed. If Sam's did not, he would have to decide whether he wants a system with NICs that plug into ISA (Industry Standard Architecture) or PCI (Peripheral Component Interconnect) slots in his computers. So he opens the cases on both computers to see what slots are available. The ISA bus was an early technology that allowed all the components within an IBM personal computer to talk to one another. The PCI bus is a newer technology that allows faster bus transmission speeds. Most IBM-compatible microcomputers offer one or two card slots of both ISA and PCI compatibility. Because Sam also wants to share an Internet connection, he has to make sure he orders a system that has a combination switch and router, because it is the router that allows you to connect to the Internet via some form of broadband access. Once he has made all these decisions, Sam orders the SOHO LAN and waits for its arrival.

The SOHO LAN finally arrives and Sam tears into the package, eager to get started. Installation of the NICs is straightforward, as is installing the cabling and the switch/router. Installing the software is a little more challenging, but not terribly difficult. In a relatively short time, Sam has created a system that allows both of his workstations to access the Internet, a high-quality printer, and a common disk storage area (Figure 7-27).

Figure 7-27
Sam's SOHO local area network solution for his home computer system

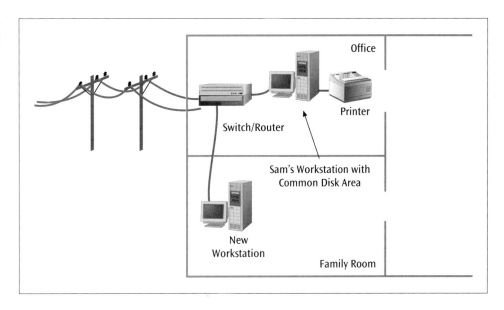

SUMMARY

> A local area network is a communications network that interconnects a variety of data communications devices within a small area and transfers data at high transfer rates with very low error rates. The primary functions of a LAN are to enable the sharing of data, software, and peripherals and to provide common services such as file serving, print serving, support for electronic mail, and process control and monitoring in office, academic, and manufacturing environments.

> Local area networks have numerous advantages, including resource sharing, separate component and network evolution, high data transfer rates, and low error rates.

Local area networks also have numerous disadvantages, including relatively high costs, a high degree of maintenance, and the constant need for upgrades.

► A local area network can be configured as a star-wired bus or a wireless network.

► The star-wired bus has replaced the baseband bus and broadband bus LANs.

► The wireless topology allows a highly flexible placement of workstations and requires no wiring to transmit and receive data.

► For a workstation to place data onto a local area network, the network must have a medium access control protocol. The two basic forms of medium access control protocols are contention-based, such as CSMA/CD, which is found on star-wired bus local area networks, and CSMA/CA, which is found on wireless local area networks.

► CSMA/CD works on a first-come, first-served basis, supports half-duplex and full-duplex connections, and is clearly the most popular access protocol, but it suffers from collisions of data frames during high-usage periods when hubs are employed.

► To standardize the medium access control protocols, IEEE created the 802 series of network standards.

► The most popular types of local area network systems are Ethernet (CSMA/CD) and wireless Ethernet. Ethernet LANs have the most variations of products and continue to dominate the local area network market.

► A hub is a device that interconnects multiple workstations within a local area network and, like most interconnection devices, can be either managed or unmanaged.

► Although many hubs are being replaced with switches, a switch can provide a significant decrease in interconnection traffic and increase the throughput of interconnected networks, while still using conventional cabling and adapters. A switch replaces a hub and isolates the traffic flow between segments of the network by examining the address of the transmitted frame and directing the frame to the appropriate port.

► A switch is transparent in that it builds its own forwarding tables by observing the flow of traffic on the networks, a process referred to as backward learning.

► A switch that employs a cut-through architecture is the opposite of a store-and-forward device, in that the data frame is leaving the switch almost as soon as it begins to enter the switch.

► Switches can create shared segments in which all workstations hear all the traffic or dedicated segments in which other workstations do not hear the local traffic. Switches can also be made to operate in full-duplex mode and can be used to create virtual LANs.

► Routers interconnect local area networks with wide area networks. A router's most common function is to route data packets between two networks, one of which uses the addresses in the medium access control sublayer, while the other uses the addresses in a layer other than the medium access control sublayer. Routers provide all users on a local area network with access to outside networks. Routers perform more slowly than switches and require more processing, because they have to dig deeper into the data frame for control information.

KEY TERMS

1-persistent algorithm
5-4-3 rule
access point
backplane
backward learning
Basic Service Set (BSS)
bidirectional
bus/tree local area network
carrier sense multiple access
　　with collision avoidance
　　(CSMA/CA)
carrier sense multiple access
　　with collision detection
　　(CSMA/CD)
collision
collision window
contention-based protocol
cut-through architecture
dedicated segment network
deterministic protocol
Ethernet
　　10Base5
　　10Base2
　　1Base5
　　10BaseT
　　10Broad36
　　10 Gbps Ethernet
　　100BaseTX
　　100BaseT4
　　100BaseFX
　　1000BaseSX

1000BaseLX
1000BaseCX
1000BaseT
10GBase-fiber
10GBase-T
10GBase-CX
Extended Service Set (ESS)
Fast Ethernet
file server
filter
firewall
full-duplex switch
gateway
Gigabit Ethernet
HiperLAN/2
hot swappable
hub
IEEE 802 suite of protocols
IEEE 802.11a
IEEE 802.11b
IEEE 802.11g
IEEE 802.11n
interframe space (IFS)
internetworking
local area network (LAN)
logical design
logical link control (LLC)
　　sublayer
managed hub
medium access control
　　protocol

medium access control (MAC)
　　sublayer
multiple input multiple output
　　(MIMO)
network interface card (NIC)
nondeterministic protocol
non-persistent algorithm
passive device
physical design
p-persistent algorithm
power over Ethernet (PoE)
print server
round-robin protocol
runts
shared network
shared segment network
Small Office/Home Office
　　(SOHO) local area network
star-wired bus local area
　　network
store-and-forward device
switch
tap
transparent
trees
unmanaged hub
virtual LAN (VLAN)
Wired Equivalent Privacy
　　(WEP)
wireless LAN

REVIEW QUESTIONS

1. What is the definition of a local area network?
2. List the primary activities and application areas of local area networks.
3. List the advantages and disadvantages of local area networks.
4. What are the basic layouts of local area networks? List two advantages that each layout has over the others.
5. What is meant by a "passive device"?
6. What is meant by a "bidirectional signal"?
7. What are the primary differences between baseband technology and broadband technology?
8. What purpose does a hub serve?
9. What is the difference between a physical design and a logical design?
10. What is a medium access control protocol?
11. What are the basic operating principles behind CSMA/CD? CSMA/CA?
12. What is meant by a "nondeterministic" protocol?
13. What does the term "100BaseT" stand for?

14. What is the difference between Fast Ethernet and regular Ethernet?

15. What are the latest 10-Gbps Ethernet standards?

16. What is the primary advantage of power over Ethernet? The primary disadvantage?

17. How does a transparent switch work?

18. What is backward learning?

19. How does a switch encapsulate a message for transmission? (For a refresher on encapsulation, see Chapter One.)

20. When referring to a hub or switch, what is a port?

21. What are the basic functions of a switch?

22. How does a switch differ from a hub?

23. What is cut-through architecture?

24. How is a full-duplex switch different from a half-duplex switch?

25. How is an ad-hoc wireless LAN different from the others?

EXERCISES

1. What properties set a local area network apart from other forms of networks?

2. Describe an example of a broadband bus system.

3. Of all the local area networks introduced in this chapter, is any system capable of supporting a full-duplex connection? Which one(s)?

4. State the primary advantage of a deterministic local area network protocol over a nondeterministic local area network protocol. Give a real-life example of this advantage.

5. What are the primary functions of the medium access control layer?

6. What are the primary functions of the logical link control layer?

7. Is a hub a passive device? Explain.

8. Which of the Ethernet standards (10 Mbps, 100 Mbps, 1000 Mbps, 10 Gbps) allow for twisted pair media? What are the corresponding IEEE standard names?

9. If a network were described as 1000BaseT, list everything you would know about that network.

10. In the IEEE 802.3 frame format, what is the PAD field used for? What is the minimum packet size?

11. List one advantage of IEEE 802.11a over IEEE 802.11g. List one advantage of IEEE 802.11g over IEEE 802.11a.

12. Suppose workstation A wants to send the message "HELLO" to workstation B. Both workstations are on an IEEE 802.3 local area network. Workstation A has the binary address 1, and workstation B has the binary address 10. Show the resulting MAC sublayer frame (in binary) that is transmitted. Do not calculate a CRC; just make upone.

13. What is the difference between the physical representation of a star-wired bus LAN and its logical representation?

14. Which of the wireless LAN protocols can support data rates as high as 54 Mbps? Theoretical or actual?

15. Your company wants to create a wireless network for the entire office building. The building is 10 stories high, and the company wants to incorporate IEEE 802.11a by placing one access point on the tenth floor. Will this layout work? Explain.

16. Give two examples of how a company might use an ad hoc wireless layout.

17. In wireless CSMA/CA, if a user device is trying to transmit standard data, and an access point device is trying to issue a poll at the same time, which device will transmit first, and why?

18. Explain the difference between 1000BaseSX and 1000BaseLX.

19. Describe an application that would operate more effectively using a wireless local area network (ad hoc design).

20. What are the advantages that Ethernet (CSMA/CD) has over other local area network forms?

21. What are the disadvantages unique to Ethernet?

22. Modify Hannah's network solution in the In Action section so that it violates the 5-4-3 rule.

23. Explain whether or not each of the following is a reason to segment a LAN into smaller segments:
    a. Large number of network collisions
    b. Administrators decide it is time to add an additional Web server
    c. Users complain of very slow response time
    d. A new network operating system is installed

24. How is a hub similar to a switch? How are they different?

25. Are hubs and switches interchangeable? Explain.

26. a. The local area network shown in Figure 7-28 has two hubs (X and Y) interconnecting the workstations and servers. Which workstations and servers will receive a copy of a packet if the following workstations/servers transmit a message:
    ▸ Workstation 1 sends a message to Workstation 3.
    ▸ Workstation 2 sends a message to Server 1.
    ▸ Server 1 sends a message to Workstation 3.
    b. Replace Hub Y with a switch. Now which workstations and servers will receive a copy of a packet if the following workstations/servers transmit a message:
    ▸ Workstation 1 sends a message to Workstation 3.
    ▸ Workstation 2 sends a message to Server 1.
    ▸ Server 1 sends a message to Workstation 3.

Figure 7-28
Network example to accompany Exercise 9

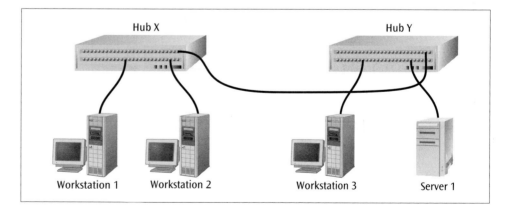

Hub X   Hub Y

Workstation 1   Workstation 2   Workstation 3   Server 1

27. A transparent switch is inserted between two local area networks, ABC and XYZ. Network ABC has Workstations 1, 2, and 3, and Network XYZ has Workstations 4, 5, and 6. Both forwarding tables start off empty. Show the contents of the two forwarding tables in the switch as the following packets are transmitted:
    a. Workstation 2 sends a packet to Workstation 3.
    b. Workstation 2 sends a packet to Workstation 5.

c. Workstation 1 sends a packet to Workstation 2.
d. Workstation 2 sends a packet to Workstation 3.
e. Workstation 2 sends a packet to Workstation 6.
f. Workstation 6 sends a packet to Workstation 3.
g. Workstation 5 sends a packet to Workstation 4.
h. Workstation 2 sends a packet to Workstation 1.
i. Workstation 1 sends a packet to Workstation 3.
j. Workstation 1 sends a packet to Workstation 5.
k. Workstation 5 sends a packet to Workstation 4.
l. Workstation 4 sends a packet to Workstation 5.

28. Give an example of a situation in which a virtual LAN might be a useful tool in a business environment. What about in an educational environment?

29. What does it mean when a switch or device is cut-through? What is the main disadvantage of a cut-through switch? Is there a way to solve the disadvantage of a cut-through switch without losing its advantages? Defend your answer.

30. Give a common business example that mimics the differences between a shared segment network and a dedicated segment network.

31. Your company's switch between its two networks has just died. You have a router lying on your desk that is not currently being used. Will the router work in place of the broken switch? Explain.

32. A CSMA/CD network is connected to the Internet via a router. A user on the CSMA/CD network sends an e-mail to a user on the Internet. Show how the e-mail message is encapsulated as it leaves the CSMA/CD network, enters the router, and then leaves the router.

THINKING OUTSIDE THE BOX

1 A retail department store is approximately square, 35 meters (100 feet) on each side. Each wall has two entrances equally spaced apart. Located at each entrance is a point-of-sale cash register. Suggest a local area network solution that interconnects all eight cash registers. Draw a diagram showing the room, the location of all cash registers, the wiring, and the server. What type of wiring would you suggest?

2 You work for a small advertising company with approximately 200 employees. Scattered around the company are a number of separate computer workstations that perform word processing, graphics design, spreadsheet operations, and market analysis. Your boss has asked you to consider installing some form of local area network to support computer operations. Create a list of possible computer applications that could operate over a local area network and would support employee and daily business functions. What type of local area network might you suggest? What would be the topology? The medium access control method? What kind of support equipment (hubs, servers) might you need? Where would that support equipment be located?

3 You have three computers at home, and you want to network all three together. Two computers are on the main floor of the house, but the third is upstairs in a bedroom. List as many ways as possible to interconnect the three computers so that they could operate on one local area network.

4 An office complex is four stories high. Each floor is roughly 75 meters (yards) by 75 meters. The company wants to create a wireless LAN for the entire complex.

Which wireless LAN technology would you recommend? Where would you place the access points? Where would you place the wired backbone?

5 A large company has many different departments housed in a large office complex in the city. Each department has its own database of local information. The company wants to let all the employees from around the company access any of these databases. What type of network would you recommend for such a situation?

6 You are working for a very small company with only a dozen employees. Five computer workstations need to be connected to the Internet. List the options that are available. What are the advantages and disadvantages of each option? What is your recommendation?

7 After the latest round of corporate reorganizations, you now have six internal divisions where you used to have seven. The technology department tries to organize the network layout by keeping users in each division on their own segment of the network, but it is getting tired of rearranging the wires in the closets to accommodate these changes. What might you suggest that the technology department use instead to support the six internal divisions?

8 You are working for a company that is composed of three departments: general support, marketing, and sales. General support occupies the first floor, while marketing and sales are on the second floor. General support has 28 workstations, marketing has 10 workstations, and sales has 30 workstations. Some applications require that data be transferred between departments, but generally each department has its own applications. Everyone needs access to the Internet, the company internal Internet (intranet), and e-mail. Design a local area network solution for this company. Show the locations of all workstations and interconnecting devices, such as hubs, switches, and routers, if any are used.

Show also the connection to the outside phone service that provides Internet access. What type of local area network wiring would you recommend? What type of local area network topology and protocol would you recommend? Draw a floor plan for each floor. If possible, show both floor plans on one page.

Use the following assumptions:

1. Available hubs and switches have a maximum of 24 ports.

2. Some software applications and large data sets reside on departmental servers.

3. To support e-mail, a server is needed.

4. To support the company intranet, a server is needed.

5. The company does not have an unlimited budget but is willing to invest in quality technology.

HANDS-ON PROJECTS

1. Find the IEEE (or other) Web site and report on the latest advances on the 802 standards. Do any additional standards exist for >10-Gbps Ethernet or wireless LANs? Any new proposals for systems not mentioned in this chapter? Explain what you find.

2. Besides the CSMA/CD protocol, there is a CSMA protocol. In what kind of systems is the CSMA protocol used? Why is CSMA preferred over CSMA/CD in these systems?

3. Have any other local area networks completely passed into oblivion? If so, what caused their demise? Was it technological, financial, or political?

4. Aside from the transmission speed, is Gigabit Ethernet the same as 10-Mbps Ethernet? Explain your response.

5. Are there plans for a 100-Gbps Ethernet?

6. Using the Internet, investigate Ethereal (Wireshark). Download a copy and use it to observe some traffic on an Ethernet connection.

7. Using the Internet or other external sources, collect literature on an Ethernet switch. What are the specifications (such as port speeds, number of ports, cable types supported, backplane speed, etc.) of the switch?

8. Create a map of your company or school's local area network. Are there any hubs? Where are the hubs located? Are there any switches? If so, where are the switches located? Are there any routers? Show where these are located.

9. Does anyone sell a bridge anymore? If so, what kind of networks is the bridge most commonly being used to interconnect these days?

8

Local Area Networks: Software and Support Systems

◆ ◆

IN 1998, THE U.S. Department of Justice and Attorneys General from 18 states filed an antitrust suit against the Microsoft Corporation. At issue was the required inclusion of the Microsoft Web browser, Internet Explorer, in the Windows operating system. A second concern was that Microsoft's business practices limited innovation. The claim was that the company's practice of bundling suites of applications had led to the decline of diversity, thus concentrating the market share in Microsoft's suite.

After a lengthy and oftentimes nasty court battle, the U.S. District Court for the District of Columbia found Microsoft guilty on April 3, 2000, and ordered the company to split into two companies: an operating system company and an applications company. At the time, this was likened to the breakup of AT&T in 1984, when AT&T had to sell off its local telephone companies but was allowed to keep its long-distance telephone service and its research center (Bell Labs). In the summer of 2001, however, a federal appeals court threw out the District Court judge's order to break up Microsoft. Nevertheless, the federal ruling did uphold the core of the government's case that Microsoft had violated the Sherman Antitrust Act. The appeals court left open the possibility of a breakup, but the judges narrowed the antitrust allegations against Microsoft—increasing the chances for an out-of-court settlement.

But this was not the end of the story. On March 24, 2004, the European Commission ordered Microsoft to pay a $613 million fine, to release a version of its Windows operating system without the Windows Media Player software, and to reveal details of Window's software codes and thus make it easier for other software companies to produce compatible software. Microsoft vowed to appeal this ruling but lost its appeal on September 17, 2007, and agreed to pay. Unfortunately, that was not the end for Microsoft. The European Union fined Microsoft again on February 27, 2008, stating Microsoft still had not resolved the software problem. This time, it might cost them $1.44 billion.

What is an operating system? What is a network operating system? Are they the same thing?

What choices are available when selecting a network operating system?

Is an operating system so important that it could cause one of the largest U.S. companies to be broken into pieces?

Source: Standards Engineering Society, "Microsoft Anti-Trust Litigation—The Case for Standards," downloaded from www.csrstds.com/WSD2000.html on November 7, 2001.

"Microsoft Breakup Order Reversed," downloaded from www.washtech.com on November 8, 2001.

"Microsoft Hit by Record EU Fine," BBC News UK Division, March 24, 2004, downloaded from news.bbc.co.uk on June 30, 2005. "European Union Microsoft Competition Case," downloaded from Wikipedia on October 13, 2009.

Objectives ▶

After reading this chapter, you should be able to:

▶ Identify the main functions of network operating systems

▶ Identify the basic features of past and present network operating systems including Novell NetWare/OES, Windows 2008, Unix, Linux, and Mac OS X Server

▶ Compare and contrast the Novell NetWare/OES, Windows 2008, Unix, Linux, and Mac OS X Server network operating systems

▶ Recognize the importance of the network server and the different types of network servers available

▶ Identify the different levels of RAID

▶ Identify common examples of network utility software and Internet software

▶ Enumerate the various components of software licenses

▶ Identify the different types of support devices commonly found on local area networks

Introduction ▶

Chapter Seven began the discussion of local area networks by introducing the basic topologies supporting most local area networks and the medium access control techniques that allow a workstation to transmit data onto the network. The hub, switch, and router are the three basic tools that connect devices within a local area network and connect a local area network with other networks. This chapter will conclude the discussion of local area networks by introducing the software that operates and supports the local area network. We will concentrate on two basic areas: network operating systems and the application software that supports the network.

Although there are many more types of network software—such as diagnostic and maintenance tools, utilities, and programming development environments—network operating systems and network support software are two of the most important. Network operating systems are essential if the network is going to allow multiple users to share resources. The network operating system provides users with password protection on their accounts and network administrators with services that help them control access to network resources as well as use and administer the network. Local area network operating systems, much like the hardware that supports them, continue to evolve into more powerful and elaborate tools every day. The more popular network operating systems include Microsoft Windows Server 2008, Unix, Linux, and Mac OS X Server. This chapter will outline each of these operating systems' basic features and capabilities and compare their advantages and disadvantages.

After the discussion of network operating systems, we will spend a few moments discussing the various hardware and software components that support the network operating system. The first such component will be the network server, which is the primary device that stores and executes the network operating system.

Like network operating systems, network support software is an essential tool. The more common support applications include antivirus software, anti-spam software, anti-spyware software, backup software, crash protection software, network-monitoring software, remote access software, security assessment software, and uninstall software. We will also examine some Internet-support software, such as Web browsers and Web server software. We will follow up this discussion with an introduction to the different software licensing agreements that apply to network software.

Finally, local area networks require many different types of support devices. As you saw in Chapter Seven, hubs, switches, and routers are important support devices for the segmentation and interconnection of local area networks. Other support devices that will be considered in this chapter include uninterruptible power supplies, tape drives, printers, media converters, network storage devices, and workstations. Each of these devices will be introduced, along with examples that show how the device can be used.

Network Operating Systems

What is a network operating system (NOS)? What functions does it perform? How is a network operating system different from an individual PC's operating system (OS)? The best way to begin understanding a network operating system is to learn the basic functions of an OS. Once you understand the basic functions of an operating system, you can then begin to understand the additional functions of a *network* operating system.

An operating system is the program initially loaded into computer memory when the computer is turned on; it manages all the other programs (applications) and resources (such as disk drives, memory, and peripheral devices) in a computer. Even after an application starts and is being executed, the application makes use of the operating system by making service requests through a defined application program interface (API). In addition, users can interact directly with the operating system through an interface such as a graphical user interface, command language, or shell.

An operating system can perform a number of services, most of which are crucial to the proper operation of a modern computer system. One of the most important services is determining which applications run in what order and how much time should be allowed for each application before giving another application a turn. In a multitasking operating system, multiple programs can be running at the same time. In this case, the operating system schedules each task and allocates a small amount of time for the execution of that task. In reality, a multitasking operating system runs only one program at a time, but it jumps from one program to the next so quickly that it appears as if multiple programs are being executed at the same time.

An equally important and difficult task for the operating system is handling the very complex operations of input and output to and from attached hardware devices, such as hard disks, printers, and modems. Hard disk drives are so complex and elaborate that it would be cruel and unusual punishment to require a user to specify the precise location of each record of data as it is written to or read from a hard disk. Just as complex as controlling the storage space on a hard disk drive is controlling the storage space in the computer's main memory. Because users are quite likely to run multiple applications at the same time, the operating system must be capable of allocating the limited amount of main memory in a way that provides each application with a sufficient amount of memory to operate.

A modern operating system must also provide various levels of operating system security, including directory and file security, memory security, and resource security. As applications and their users become more sophisticated, the capability of an operating system to provide protection from unscrupulous users is becoming more crucial.

Finally, a function that may not be as obviously essential as multitasking, memory and storage management, or security—but is in fact equally important—is communicating about the status of operations. An operating system sends messages to applications, an interactive user, or a system operator about the status of the current operation. It also sends messages about any errors that may have occurred.

A number of popular operating systems exist on different types of computer systems. For example, popular operating systems for microcomputers include Mac OS X, Unix, Linux, and the various forms of Windows XP/Vista/7. Popular operating systems for larger computers (minicomputers and mainframes) include IBM's OS/390 and AS/400, DEC's VMS (OpenVMS), and again Unix and Linux.

A **network operating system (NOS)** is a large, complex program that can manage the common resources on most local area networks, in addition to performing the standard operating system services mentioned previously. Table 8-1 summarizes the functions of a network operating system. The resources that a network operating system must manage typically include one or more network servers, multiple network printers, one or more physical networks, and a potentially large number of users who are directly, and sometimes remotely, connected to the network. Among these resources, the network server is critical. The server, as you will learn shortly, is usually a high-powered workstation that maintains a large file system of data sets, user profiles, and access information about all the network peripherals.

Table 8-1
Summary of network operating system functions

Network Operating System Functions

Manage one or more network servers
    Maintain a file system of data sets, applications, user profiles, network peripherals
    Coordinate all resources and services available
    Process requests from users
    Prompt users for network login, validate accounts, apply restrictions, perform
        accounting functions

Manage one or more network printers
Manage the interconnection between local area networks
Manage locally connected users
Manage remotely connected users
Support system security
Support client/server functions
Support Web page development and Web server operations

A network operating system also performs a wide variety of network support functions, including the coordination of all resources and services on the network and the processing of requests from "client," or user, workstations. It also supports user interaction by prompting a user for network login, validating user accounts, restricting users from accessing resources for which they have not been granted access, and performing user accounting functions.

Another important network operating system function is the capability of the operating system to interact with the Internet. More and more users working on a local area network expect to find a seamless connection between their workstations and the Internet. If these users have to create Internet Web pages, they would like to have the tools available to create and maintain those pages. The network operating system then should provide a seamless interconnection to the Web-based software that stores Web pages and makes them accessible to internal corporate users and external users of the Internet. Thus, a network operating system should be capable of supporting the applications and tools necessary to support Internet operations.

Interestingly, many current desktop operating systems now incorporate many of the features described previously, thereby enabling individual users to create network operating systems for their homes and small businesses. Thus, the line between a desktop operating system (such as Windows 7) and a network operating system (like Windows 2008) has blurred in recent years. Even so, there are still, as we will see shortly, significant differences between the two types of systems.

Network Operating Systems Past and Present

As you have just learned, a network operating system supports many functions that we, as users, normally take for granted. Understanding these basic functions will allow you to examine actual network operating systems more closely, compare these products by how well each supports the basic functions, and evaluate their strengths and weaknesses. Having examined the basic functions, let us turn our attention to several of the more popular network operating systems on the market and see how each supports these basic functions. But before we do that, let's start with an important network operating system pioneer no longer on the market—NetWare.

Novell NetWare

Novell was founded in 1983 and was one of the first developers of network operating systems. Over the years, the company has produced a number of influential products, including NetWare Directory Services (NDS), which is an intelligent system that authenticates users and includes a distributed database of information about every application, user, server, and resource on a network. NetWare 6 was the last version of Novell's network operating system before the product was renamed Open Enterprise Server (more on that shortly). At one point, more than 70 percent of the LAN market used Novell's network operating system. Despite NetWare's departure, the product introduced many features and concepts that have been incorporated into other network operating systems. Thus, it is worthwhile to spend a few moments examining the NetWare legacy.

An interesting feature shared by all versions of NetWare is that the user interface is virtually invisible to the user. When a workstation is part of a Novell network operating system, the workstation must still run a workstation-based operating system such as Windows XP or Mac OS X. The user performs Windows-type operations such as opening folders and double-clicking application icons without realizing that the application being used or the file being requested is physically located somewhere else on the network—on a network server rather than on the user's own workstation. The network administrator instructs Novell to hide such details from the user and perform all the network functions, such as client/server requests and network printing operations, with little or no knowledge on the part of the user.

For example, suppose you are using a word processor and wish to print a document. To do this, you click the printer icon in the word-processing program. If you have (or a network administrator has) instructed your workstation to redirect all printer output to a network printer, your print job will not print on the printer in your office (assuming you have one) but will be redirected to another printer. Once this **redirection** has been set, all future print requests will automatically be forwarded to the network printer in an operation that is transparent to both you and the word-processing application. The Novell network operating systems were the first to perform this redirection on a local area network, and now all other network operating systems incorporate this feature.

The first version of NetWare to gain widespread popularity was Version 3, which was released in 1989 and is still occasionally found in use today. One of Version 3's main features is the bindery. The **bindery** is a structure (similar to a database) that contains the usernames and passwords of network users and groups of users authorized to log in to that server; it consists of three linked, nonidentical files that are encrypted for security reasons. A bindery contains the data that pertains only to the server it resides in. It also holds information about other services provided by the server to the client, such as printer, modem, and bridge/router access information. Each server on a multiserver network requires and maintains its own bindery, and users are required to log in to a particular server. A user requiring access to two servers would need an account on each server and would have to log in to each server separately.

NetWare Version 4 represented a significant change from NetWare 3. Version 4 introduced NetWare Directory Services (called Novell Directory Services in later versions), which replaced the bindery. NetWare Directory Services (NDS) is a database that maintains information on and access to every resource on the network, including users, groups of users, printers, data sets, and servers. NDS is based on a well-known standard for directory services called X.500. In the NDS, all network resources are considered to be objects, regardless of their actual physical location. NDS is global to the network and is replicated on multiple servers to protect it from failure at a single point. Because NDS is global, users can log in from any location on the network and access any resources for which they have been granted permission. Every user who is allowed to log in to the network is entered into the NDS by the network administrator. Likewise, all network support devices such as printers are also entered into the NDS. For example, if user X wants to use printer Y, the network administrator has to assign the appropriate permissions to allow user X to access printer Y. These permissions are also entered into the NDS. Every time user X logs on and tries to send a print job to printer Y, the NDS is referenced for the necessary permissions.

The basic idea underlying NDS (and all other tree-based directories, such as Microsoft's Active Directory) is that the network administrator must create a hierarchical tree that represents the layout of the organization. This hierarchical structure actually resembles an inverted tree, with the root at the top and the users and network resources—the leaves—at the bottom. This tree could correspond to the physical layout of the organization: for example, workstations 1 through 20 will be on the third floor of an office building, workstations 21 through 40 on the second floor, and workstations 41 through 60 on the first floor. A more powerful and flexible hierarchical tree can be created based on a logical layout. For example, a logical layout could describe the organization in terms of its departmental structure: the engineering department (which could be scattered over floors 1 through 3), the sales department (which could be situated on floors 2, 6, and 7), and marketing (which might be physically located in two different buildings).

Creating a good and appropriate tree design is not a trivial task; however, the following four basic steps will give the network administrator a good start. First, gather the appropriate corporate documents so that you know about the available hardware and software and the employee departments and divisions. As part of this first step, you should obtain an organizational chart.

Second, design the top section of the tree before you design any lower sections. To design the top section of the tree, you give the top level of the tree the name of the organization object (in most cases, the name of the company) and then create the first layer of organizational units, or container objects. An organizational unit (OU) is an object that is further composed of additional objects (examples of which include servers, printers, users, or groups of users). For example, a division in a company that is composed of multiple departments would be considered an organizational unit. A department is also an organizational unit because it too is further composed of objects, such as employees. If your network is very small, there may not be a need for any organizational units.

Third, you design the bottom section of the tree, which includes the remaining hierarchy of organizational units and leaf objects. Leaf objects are not composed of any objects and are usually entities such as the users, peripherals, servers, printers, queues, and other network resources.

The fourth and final step in creating a tree is to design the security access rights—determine who has rights to the appropriate objects. For example, if you create a new user and place that user in a particular location on the tree, what rights will that user have? What printers and directories will that user be able to access?

Throughout the entire tree-creation process, it is important to review your draft design for accuracy, flexibility, and completeness. Because an NDS tree is going to be used by every server on the network and is probably going to be in place for a number of years, the creation of a well-designed tree is extremely important. A tree that is designed improperly will lead to difficulties in the future when you try to add new users, user groups, and resources to the network. A poorly designed tree may also create a sluggish system with poor response times.

Figure 8-1 shows an example of an appropriate tree design. Note that the design of a tree should ideally be similar to a pyramid. Fewer container objects should be at the top of the tree than at the bottom. Wide, flat trees are usually not a good design, as this layout causes too many inner-container communications, and users may have trouble finding the appropriate resources. Likewise, trees that are too narrow and tall may also be poorly designed, because they incorporate too few containers, which can lead to future problems when a network administrator wants to add new users or resources.

Figure 8-1
Possible design for a network operating system tree

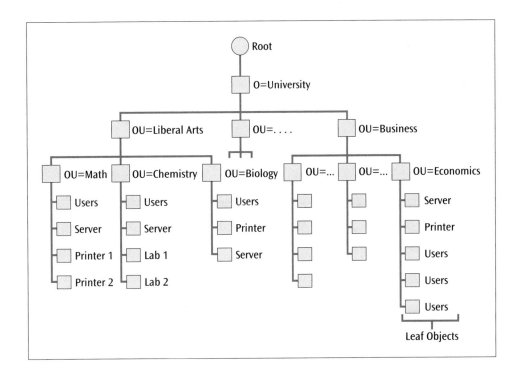

NetWare 5, the next major version of the NetWare network operating system, retained the NDS distributed database of network resources, which had grown in popularity and flexibility as NetWare products continued to be used as network operating systems. The administrative functions of creating and supporting the NDS tree and utilities still had a graphical user interface in Version 5, as they had in Version 4.

The next major version of NetWare was Version 6. Once again, the new NetWare retained many of the popular features found in earlier versions, such as the NDS and great administrative tools. In this latest version, Novell has added a couple of very impressive features. The first feature gives any authorized client anywhere on the Internet the ability to print and use storage services from a NetWare 6 server without having previously loaded a single byte of Novell's Client32 software onto his or her machine. On most network operating systems, before a client can access any network resources, the client workstation must be loaded with a substantial set of software. With NetWare 6, all a user needs on his or her client workstation is a Web browser. The Web browser will download and install the necessary client software.

The final modification of NetWare's Version 6 (winter of 2008) was NetWare 6.5. This version combined the previous features of NetWare with the Linux operating system. The software was designed to simplify the movement of data and resources from older versions of NetWare to a stable system that is capable of supporting large numbers of users. Finally, NetWare morphed into the product called Open Enterprise Server (OES) in 2003. OES is basically a set of the more popular applications from NetWare: the NDS structure, now called eDirectory, their popular print software called iPrint, and a set of core operating system services called NetWare Core Protocol. OES is offered on both Linux and NetWare platforms. Even though the company was thinking of dropping the NetWare platform, dedicated users of NetWare keep asking the company to maintain the status quo.

Details ▷

Additional Suggestions for Designing a Network Tree

When designing a tree directory structure such as Microsoft's Active Directory or NetWare's NDS, some network designers feel that the top of the tree should be based upon the corporate wide area network structure. For example, if a company has offices in San Francisco, Dallas, and New York, then the top three OUs should be containers representing those three cities (see Figure 8-2). Or, if the company is smaller and is located in only one city but occupies multiple buildings, each building should be one container object. If, however, the company is smaller still and occupies only a single building, then there is no need to define the tree based on the corporate wide area network. Instead, you can design the upper portion of the tree based upon a division/department/workgroup design.

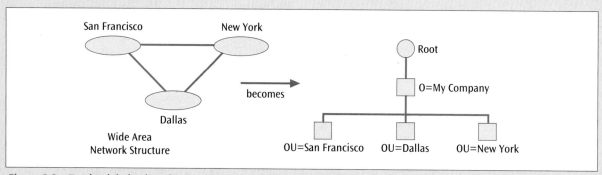

Figure 8-2   *Top-level design based upon the corporate wide area network structure*

Although a pioneer in the design of network operating systems, NetWare has, in recent years, practically disappeared from the network operating system market. In its place are a number of other network operating systems, one of which is Microsoft's.

Microsoft Windows NT and Windows Server 2000, 2003, and 2008

To compete in the local area network operating system market, Microsoft announced a new operating system at the October 1991 Comdex meeting. The new operating system's name was Windows NT Version 3. Microsoft had been working on this operating system for roughly three years, and the new product, despite having a name similar to the desktop operating system Windows, represented a significant departure. Since its introduction, Windows NT has gone through a number of versions and has become very serious competition for Novell's NetWare operating system.

Like NetWare, Windows NT and its later incarnations, Windows Server 2000/2003/2008 (hereafter called Windows Server) are network operating systems designed to run over a network of microcomputer workstations and provide file sharing and peripheral sharing. Also like the most recent versions of NetWare, Windows Server was designed to offer the necessary administrative tools to support multiple users, multiple servers, and a wide range of network peripheral devices. In addition, Windows Server supports many of the applications that allow users to create, access, and display Web pages, as well as the software that allows a network server to act as a Web server. One feature that sets the Microsoft operating systems apart from NetWare is that the Windows operating system works seamlessly with the hugely popular Microsoft applications tools.

If you did design the top layer of the tree using the corporate wide area network structure, you would then design the next portion of the tree according to the corporate local area network structure at each wide area network site. Designers often consult corporate organizational charts and LAN maps to create these lower tree portions. For example, Dallas' container object from Figure 8-2 is expanded to include Marketing, Engineering, Sales, and Human Resources (see Figure 8-3).

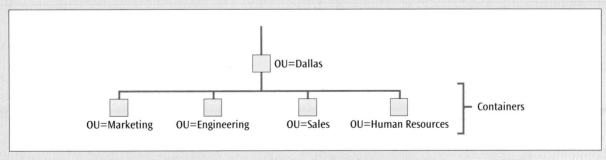

Figure 8-3 *Lower-level tree containers under the upper-level Dallas container*

Windows NT Version 4

First released in August 1996, Windows NT Version 4 contained a number of significant features that quickly made Windows NT a serious competitor in the network operating system market. One of the most noticeable features of NT Version 4 was the choice of user interface. To be precise, the Microsoft Windows 95 operating system's user interface for single-user personal computers was incorporated into Windows NT, making the server interface easier to use and consistent with other Microsoft products. Windows NT Version 4 also offered the Microsoft Management Console and administrative wizards. This new software allowed administrators to create task-based consoles that could be delegated to the appropriate administrator. These tools were supplemented by the new network monitor and diagnostic tools, which allowed the administrator to follow network traffic, perform traffic analysis, and make network changes and repairs. Despite being offered the new management console, many network administrators complained that the administration of an NT system was still too scattered over many management tools, and thus difficult to perform.

Nevertheless, Windows NT Version 4 quickly became a successful network operating system. It did a very good job of supporting most, if not all, of the network operating system functions summarized in Table 8-1. Whether the applications were on the user workstations (the front end) or were database systems located on the server (the back end), Windows NT provided a smooth marriage of network operating system with Microsoft applications.

One noticeable difference between Windows NT Version 4 and NetWare was NT's directory structure. While NetWare made extensive use of container objects and leaf objects, NT had only the domain. The NT domain was a container object that contained users, servers, printers, and other network resources. There was no way of taking the users within a domain and creating a series of sub-domains or subcontainers. It was possible to create a network with multiple domains, but these domains were *not* hierarchical, and in many cases they increased the level of administration (Figure 8-4).

In addition to the relative inflexibility of the domain models, the instability of the operating system was another area that many network professionals felt was a weakness of Windows NT Version 4. This instability was partially due to the high number of bugs in the NT Version 4 software. But many of these bugs were the kind that are common in new software, and in all fairness, the early versions of NetWare were not as stable as NetWare Version 3. Having been available for a longer period of time, NetWare had had more opportunities to be debugged of problems. In light of these issues, Microsoft worked hard at eliminating most of its system's instability problems, as well as other problems inherent with such a large piece of software.

Figure 8-4
Two Windows NT domain models

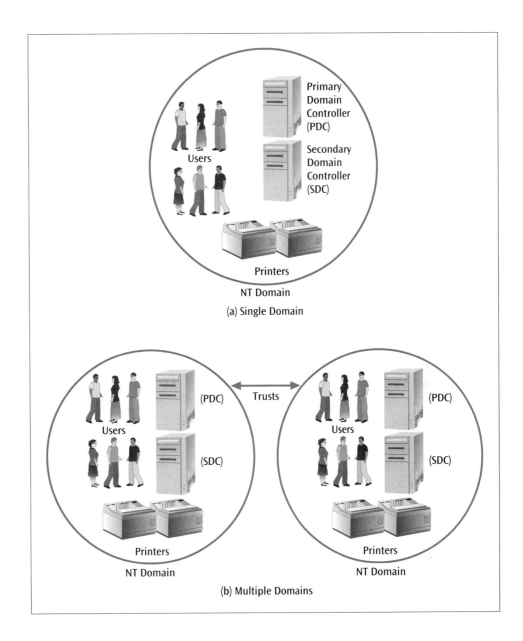

(a) Single Domain

Trusts

(b) Multiple Domains

Windows Server 2000

Windows Server 2000, released in the year 2000, was the next generation of the Windows NT operating system and represented a significant improvement over the earlier versions of NT. Windows Server 2000 incorporated Microsoft's answer to the highly popular NetWare NDS directory system: Active Directory. **Active Directory** stores information about all the objects and resources in a network and makes this information available to users, network administrators, and application programs. Like NetWare's NDS, Active Directory creates a hierarchical structure of resources. Microsoft felt it was in its best interest to create a directory service based on existing standards rather than to create a new proprietary directory service. Thus, you may hear network specialists talk about how Active Directory is built around the Internet's Domain Name System (DNS), which is introduced in Chapter Ten, and a second standard—Lightweight Directory Access Protocol (LDAP).

To construct an Active Directory hierarchy, you create a tree design similar to the tree design of NDS. Objects, such as users, groups of users, computers, applications, and network devices, are the leaf items within the tree. Leaf items are grouped in organizational units similar to the container objects in NetWare's NDS. One or more organizational units can be grouped together into a domain. As in Windows NT, the main object in Windows Server 2000 is the domain. Unlike NT domains, however, Windows Server 2000 domains can be hierarchically organized as a tree, and a collection of trees is then a forest. For example, two of the departments within an organization may be marketing and engineering (Figure 8-5). Both of these departments are organizational objects. Within both marketing and engineering are users and network servers. Grouping objects within the directory tree allows administrators to manage objects and resources on a macro level rather than on a one-by-one basis. With a few clicks of a mouse, a network administrator can allow all the users within engineering to access a new engineering-related software application.

Many Windows NT network administrators were pleased with the advances in Windows Server 2000, and thus migrated their systems to the newer operating system. A number of NetWare administrators also moved over to Windows Server 2000, because it now incorporated a tree-based directory structure. But a number of administrators—much like the users who have remained loyal to NetWare Version 3 and still use it instead of Version 6—did not see the need to upgrade to the newer system and remained users of the older Windows NT. Like its predecessor Windows NT Version 4, Windows Server 2000 performs the necessary functions of a local area network operating system as summarized in Table 8-1 and has the backing of the largest software company in the world.

Figure 8-5

Examples of a tree design of Windows 2000 Active Directory

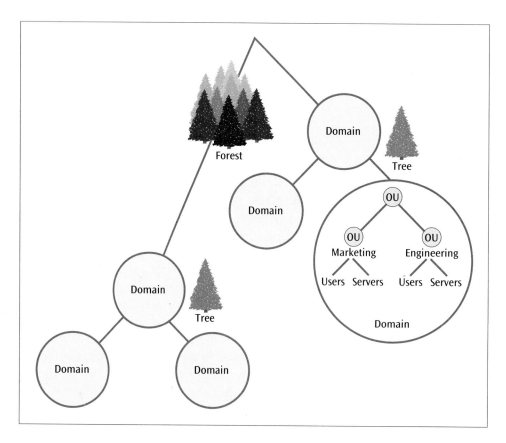

Windows Server 2003

The next version of the Windows network operating system was Windows Server 2003. Although it was not a major reorganization of the operating system—as Windows Server 2000 was, when compared to Windows NT—Windows Server 2003 offered many new features that network administrators found useful. Some of these features were:

- ▶ Updates to the Active Directory, including a new management tool that unifies all the tasks related to group policy
- ▶ Capability to interconnect (cluster) up to eight Windows 2003 servers for better user and application support
- ▶ New and improved file and print support services, including increased reliability, wider range of supported printers, and remote document sharing
- ▶ Support for Internet Protocol version 6 (IPv6)
- ▶ Better security features for files, networks, servers, the operating system, and Web-based transactions

It also appeared that Microsoft adopted a new policy that Windows Server 2003 would be the core operating system for future generations of Microsoft products. Thus, if a company decides to install Windows Server 2003, it will be able to incorporate future versions of Microsoft's products (such as its Office applications suite, Web-based servers, and real-time communications systems, to name a few) more easily.

Windows Server 2008

The most recent addition to the Windows Server family is Windows Server 2008. While it maintains many of the same features and characteristics of the 2003 server, it does have a number of new features. Some of these features include an expanded Active Directory (which includes certificate, identity, and rights management services), a new server core (which allows Windows Server 2008 to act as a number of different types of servers, including Windows virtual server), a self-healing file server that can fix corrupted files and/or folders, improvements in processing speed, and advancements in network security. Clearly, Microsoft is staying competitive in the network operating system market and will more than likely continue to offer products that are attractive to network administrators.

UNIX

Unix is a popular operating system that can be found on single-user workstations but primarily functions well on mainframe computers and network servers. It is most often found with a text-based interface, although graphical user interfaces are available. Unix was initially designed at Bell Labs and was first implemented on a PDP-7 minicomputer in 1970. It is a relatively streamlined system, which explains why it operates quickly. Shortly after its introduction, the Unix software was rewritten in the popular C programming language. Owing to characteristics of the C programming language and Unix's design, the internal code of Unix is relatively easy to modify. For this reason, and because early versions of the operating system were given away for free, Unix became extremely popular in academic institutions.

Because Unix is one of the older operating systems and has grown in processing power over many years, it is quite stable. Another consequence of its age is that Unix runs on the widest variety of hardware of any network operating system. It handles network operations well, and a broad range of applications have been written to run in Unix. Today, many experts consider Unix to be one of the best operating systems for supporting large applications such as multiuser database systems and Web-based servers.

One of the biggest drawbacks of Unix is the user interface. Many users today expect a graphical user interface and find Unix's text-based, command-line interface old-fashioned and cumbersome. A second disadvantage is that, although Unix adequately supports a directory system that can maintain data sets, applications, and user profiles, its directory system is not as powerful as Microsoft's Active Directory or NetWare's NDS.

As a network operating system, Unix has many strengths and supports all of the features of a network operating system that are summarized in Table 8-1. Furthermore, Unix has a stable security system that is also based on many years of experience. Many applications have been created over the years for Unix that support client/server functions as well as Web page development and Web server operations. Because it has a devout following that will ensure its evolution in the face of competition from newcomers, Unix will probably be available for many years.

Linux

Since its early days, Unix has evolved into many versions and executes on a number of computer platforms. One version that generated a lot of interest in the 1990s was Linux. Linux, while based on the Unix concept, is a complete rewrite of the Unix kernel and borrows additional functions from the well-established Free Software Foundation's GNU toolset and from the even larger free software community. Linux shares many of the same advantages and disadvantages of Unix and performs similarly to Unix as a network operating system. A number of features, however, set Linux apart from Unix as well as from all other network operating systems. First and foremost is the cost of Linux—if downloaded from the Internet, the software is and always has been free.

These days, however, many companies are choosing to purchase Linux from a vendor that specializes in providing the latest version of Linux along with all the supporting applications and utilities, such as Web page support tools, a graphical user interface, and the latest versions of peripheral drivers. Even when Linux is purchased from a vendor, the cost of it is insignificant compared to the cost of purchasing either Unix or Windows Server 2008. Whereas Unix and Windows Server 2008 can cost thousands to tens of thousands of dollars, depending upon the number of user workstations, commercial versions of Linux can cost as little as a few hundred dollars.

A second advantage of Linux over other network operating systems is that when you purchase or download Linux, you can receive the original source code along with the compiled code for free. Having the original source code gives organizations a great amount of control over the software. In the hands of an experienced programmer, the Linux source code can be modified in almost limitless ways to provide a customized operating system. The ability to customize the software, however, is a double-edged sword. In inexperienced hands, customized source code can be a source of constant trouble.

A final advantage of Linux is the size of the code. Since its debut, Linux has been able to operate on a system as small as an Intel 386 processor with only 4 MB of main memory. Although Linux will certainly run on larger systems, many administrators use a Linux system on an old personal computer to support simpler network functions such as providing network security (a firewall). These systems apparently can run unattended with little or no support.

Linux shares many other advantages with Unix, such as fast execution (due to its relatively small size and efficient code), network support functions, and an option that gives the software the feel of a product with a graphical user interface.

Although supporters of other network operating systems in the past claimed that Linux was not yet ready for "prime time," more and more companies are running Linux on at least one server in order to support a particular application. Currently, Linux software is commonly used to support e-mail servers, Web servers, FTP servers, file/print servers, firewalls, proxy servers, and Domain Name System (DNS) servers. Regarding interfaces, Linux operating systems can support USB, SCSI, RAID, and various forms of video interfaces; they are also capable of operating in plug-and-play mode (where the device driver is automatically loaded when the device is attached).

Another competitive advantage that Linux has over other network operating systems is that it is part of a growing family of open source software that is rather highly regarded within the business and educational industries. Linux can be coupled with popular software applications such as the Apache Web server, MySQL database server, AXIS, Jboss, Jetty, Saxon, and Tomcat to create a relatively complete network system that can serve the needs of many types of businesses.

Linux does have three potentially important negative features. First, because the free versions of Linux are not sold through large companies like Microsoft, many network professionals fear that technical support for Linux systems would be inadequate. Even if a business were to purchase Linux from a well-known vendor, it may still experience problems procuring technical support because the other applications that run under Linux are often also open source. Second, although more and more software companies are announcing that they are producing or will produce a version of their software for Linux, applications that run on a Linux system are not as plentiful as applications that run on Windows network operating systems. Third, a substantial level of expertise is required to install a Linux system. Despite these drawbacks, Linux continues to evolve into a system capable of supporting larger and larger systems and to gain market share. Inforworld magazine stated that in 2008, Linux servers were found in 29 percent of the worldwide server shipments. As this evolution continues, more network management personnel will look at Linux as a viable large-scale network operating system.

Novell Linux

As the NetWare market share was eroding down to single digits, Novell embarked on a bold new campaign in 2003—moving into the open-source world and becoming a vendor of Linux. After purchasing two Linux software companies in 2003 and 2004, Novell began to offer four versions of Linux. The first two products are desktop versions of the Linux operating system—one for the office environment (Novell Linux Desktop) and one for the home user (Suse Linux Professional). Novell's third version is the Suse Linux Enterprise server, which is a scalable, open-source network operating system for business systems. Finally, the fourth version of Linux is the Novell Open Enterprise Server, which contains Novell's famous NetWare products in Linux form. Many observers in the industry are curious about whether Novell's reputation and the past quality of NetWare products can propel Linux into becoming a serious contender in both the network operating system and the desktop operating system markets. Time will tell.

Mac OS X Server

During a time when Microsoft products dominate the business systems market, Apple Computer continues to create a unique space for itself in the K-12 educational market. To support its line of Macintosh computers (as well as non-Macintosh computers), Apple has created the Mac OS X Server (essentially the tenth version of the Macintosh operating system). This server is capable of supporting both Macintosh and Windows workgroups and is based on software created by the open-source community. In other words, Mac OS X Server is based on the Unix concept, and thus shares some characteristics with both the Unix and Linux operating systems, such as fast, efficient, and stable code. The operating system also incorporates a number of well-known open-source software applications, such as the Apache Web server, Kerberos security, SpamAssassin, OpenLDAP directory services, and the Samba file and print server.

Because many people used to and continue to believe that Apple products can interface only with other Apple products, the designers of Mac OS X Server have gone out of their way to make sure their network operating system is compatible with a wide variety of devices. For example, Mac OS X's file and printer sharing software provides secure access to Macintosh, Windows, and Linux client workstations. In addition, the directory services software, which is called Open Directory, can interface with Microsoft's Active Directory. Despite these efforts, the Mac OS X Server has a long battle ahead if it wants to catch up with the already entrenched Microsoft Windows Server family.

Summary of network operating systems

To conclude this section, let us summarize the network operating systems and compare them using the following criteria: range of compatible hardware, performance, corporate acceptance, installed base, directory services power, stability, software cost, TCP/IP support, and greatest strength. Table 8-2 compares some of the features of the network operating systems discussed.

For the widest range of compatible hardware, Unix is hard to beat. This operating system runs on a large number of processor types, both large and small. After Unix would be Linux, followed by Windows, NetWare (Open Enterprise Server), and Mac OS X Server. With unique pieces of hardware, you may have to look into whether these systems are compatible, but in general it is safe to assume that device drivers have been written for a large majority of peripherals.

Performance-wise, all the network operating systems presented are very good. One network operating system may outperform the others in a particular area, however. For example, Window's operating systems interface well with other popular Microsoft applications such as Microsoft Word, Excel, and PowerPoint. NetWare was very fast at serving files to requesting workstations. Unix, Linux, and Mac OS X Server are fast operating systems in general, partly due to their streamlined codes.

Except for Mac OS X Server, all the network operating systems discussed have very large installed user bases and can be found in a wide variety of businesses, both large and small.

With respect to directory services power, NetWare's NDS is considered by most experts to be the best, with Microsoft's Active Directory following close behind.

All the network operating systems in Table 8-2 are quite stable. The one exception was Windows NT, and this was due, in part, to the fact that NT was one of the newest players in the market when it was released and thus still experiencing a shakedown. Windows Server 2008, however, is much more stable than NT.

It is a little difficult to compare all five operating systems according to cost, because there are different ways to determine the cost of a network operating system. For one, the cost of a network operating system can vary according to how many "seats" are required. For example, if you want to use the operating system for 100 users, you might be expected to pay for a 100-seat license. Windows, NetWare, and Unix are often sold with per-seat licenses. But there are alternative pricing structures. Mac OS X Server, for instance, currently can be purchased in either a 10-seat license or an unlimited-seat license. Linux, as we have seen, can be much less costly than most operating systems—it may even be free. To mitigate the attractiveness of this, the other operating system companies often argue that even though Linux might be free initially, the cost of operating it and maintaining it is not free. In other words, the total cost of operation (TCO) is not zero dollars. Linux supporters agree that their system's TCO is not zero, but still maintain that it is far less than the TCO of Windows and NetWare products.

All these network operating systems support TCP/IP protocols. Most operating systems support TCP/IP directly, but NT supports its own proprietary network protocol (NetBEUI) and layers TCP/IP on top of it.

Finally, each network operating system has a particular strength. NetWare is an excellent file server. NetWare is also hard to beat for directory services power, and thus holds a slight advantage over Windows. Unfortunately, however, NetWare is nearing extinction. Windows Server is an excellent application server, providing client/server capabilities to a wide range of applications. Unix and Linux offer both power and flexibility by providing a wide range of services to many different types of network applications. Mac OS X Server is a powerful, stable system in the Apple environment and is slowly but steadily gaining acceptance in the PC market.

Table 8-2

Comparison of network operating system products

Criteria	NetWare/OES	Windows Server	Unix	Linux	Mac OS X Server
Range of compatible hardware	Moderately wide	Moderately wide	Very wide	Wide	Moderately wide
Performance	High	High	High	High	High
Corporate acceptance	Fading away	Wide	Very wide	Wide	Modest
Installed base	Not too many left	Millions	Millions	Millions	Thousands
Directory services power	Very high	High	High	High	High
Stability	High	High	High	High	High
Software cost	Moderate to high	Moderate to high	Moderate to high	Low to moderate	Moderate to high
TCP/IP support	Yes	Yes	Yes	Yes	Yes
Strength	File server, NDS	Application server, Microsoft name	Speed, flexibility, stability	Cost, speed, flexibility, stability	Speed, stability

Now that we have examined the most popular local area network operating systems available today, let us take a closer look at the device that holds a major portion of the network operating system—the network server.

Network Servers

A network operating system needs a host machine from which to operate. Although a part of a network operating system resides in each client computer, the bulk of the operating system operates in a network server. More precisely, a network server is the computer that stores software resources such as the network operating system, computer applications, programs, data sets, and databases, and either allows or denies workstations connected to the network access to these resources. Let us examine the different forms of network servers, along with some of the typical hardware and software features that they employ.

Network servers range in size, from small microcomputers to mainframe computers. Typically, the network server is a powerful microcomputer workstation with redundant components. *Redundant* is a key word here—the network server has redundant disk drives, redundant power supplies, and even redundant cooling fans. The network server workstation typically houses hundreds of megabytes of random access memory, one or more large-storage hard disk drives (each drive providing hundreds of gigabytes of storage), and at least one high-speed microprocessor. The hard disk drives typically are hot swappable and have an interface, such as the Small Computer System Interface. The Small Computer System Interface (SCSI), which was introduced in Chapter Four, is a specially designed interface that allows for a high-speed transfer of data between the disk drive and the computer. As you may recall from Chapter Seven, a device that is hot swappable is capable of being removed from the computer while the power is still on, which makes performing maintenance and repairs simpler. In addition to the disk drives, the power supplies and cooling fans on most modern servers are also hot swappable.

To protect the server from catastrophic disk failure, the disk drives on most network servers support one of the redundant array of independent disks (RAID) techniques. Redundant array of independent disks (RAID) is a collection of techniques for interfacing multiple hard disk drives to a computer. With the exception of the first RAID technique, RAID-0, RAID is used primarily for storing data redundantly on multiple hard disk drives. Some of the more common RAID techniques include:

- ► RAID-0, in which the data is broken into pieces, and each piece is stored on different disk drives. This technique is known as striping. There is no redundancy of data in this technique, so if one disk drive fails, some of the data is lost. The advantage of this technique, however, is the speed at which data can be read or written across multiple disks at the same time.

- ► RAID-1, in which the data is stored on at least two disk drives, in duplicate, to provide a level of redundancy (or fault tolerance), should one disk become corrupted. This technique is also known as disk mirroring.

- ► RAID-3, in which the data is redundantly stored across multiple disk drives (striping), and error-checking information concerning the stored data is kept on a separate disk. This error-checking information can be used to detect errors and possibly reconstruct the data should some of it become corrupted.

- ► RAID-5, in which data is broken into pieces (stripes) and stored across three or more disks. Parity information (error-checking code) is stored along with the striped data, not on a separate disk. RAID-5 is the most popular of the RAID techniques.

Several more RAID techniques exist, but they are essentially variations on the above four techniques.

Along with the RAID techniques, most servers also feature a powerful set of management software. This software allows a network administrator to monitor the status of the server, remotely deploy network applications, update the necessary drivers, install and fine-tune the network operating system, and configure the RAID storage system.

With respect to the hardware side of a server, many of the higher-power network servers support either multiple processors and/or multi-core processors. Because the processor is the driving engine within the computer, a server or workstation with multiple processors or multiple cores can simultaneously execute multiple requests from other workstations. A network server that boasts symmetric multiprocessing can evenly distribute the requests it receives over multiple processors, thus effectively utilizing the full potential of the multiprocessing system.

A recent technological advance for network servers and servers in general is server virtualization. With server virtualization, it is possible to make one computer (or server) act as if it were multiple computers (or servers). The advantage to this is that each server thinks it is the only server running on that computer. It has access to all the necessary resources it needs, and if the software experiences difficulty, the difficulties will not effect the other virtual servers running on that machine. The disadvantages of server virtualization include additional software complexity and an added level of human management.

Another member of the server family (that is seeing a reduction in popularity due to server virtualization) is the server appliance. A server appliance is a single unit or box that supports many networking functions such as Internet sharing, intranet Web serving, firewall security, FTP services, file and print serving, e-mail service, and virtual private network configurations. A small company with up to 50 users may find this type of server the perfect tool. Because so many functions are located in a single unit, the individual components are not designed for heavy use, as might happen in a company with more than 50 users. Also, you should be aware that there are a number of devices with similar names that are usually designed for single applications. For example, a cache server appliance is designed to provide a mirror or backup service for other servers. A Web server appliance is designed primarily to function as a Web server. A storage server appliance is designed solely for providing disk storage for a network.

Another type of server that appeared at the turn of the century is the server blade. A server blade is a server that has no cabinet or box, but resides on a single printed circuit card. A company that desires to install many servers in a relatively small location can install multiple server blades in a card frame. Although individual server blades may not be as powerful as a stand-alone server, they are quite compact, and when combined into dozens or hundreds of units, they can make very powerful servers. Plus, if one server blade fails, unplugging the server blade and replacing it with a working unit is fairly simple.

A number of miscellaneous support devices are associated with a network server. A network server requires at least one connection to the network in the form of a network interface card. To safeguard the contents of the disk drives, a tape backup system is usually employed to back up hard disk contents on a regular and automatic basis. In addition, some form of battery backup system (uninterruptible power supply) is used to maintain power to the computer for various periods of time, should electrical power be lost. All of these support devices will be examined in more detail later in this chapter.

Client/server networks vs. peer-to-peer networks

Most local area networks are composed of user workstations that communicate to and rely upon one or more servers. This type of LAN is called a client/server network. The client, or user workstation, requests something such as a database record from a server. The server accepts the request, retrieves the data, and returns a response. Although a majority of LANs are client/servers, another form of LAN is growing in popularity: the peer-to-peer local area network. A peer-to-peer network may not have any servers but allows communications between workstations, as if the workstations were all equals. Quite often, however, peer-to-peer networks do possess servers, although the servers may function in a different role than that of a server in a client/server network.

When peer-to-peer networks first appeared roughly 30 years ago, they were server-less networks. All workstations communicated with each other. Because there was no dedicated server, each workstation essentially acted as a server. This type of peer-to-peer setup could support only networks with just a handful of workstations. Plus, without the security of a shared file system, businesses did not trust the peer-to-peer approach. Thus, this form of network practically disappeared. Today's peer-to-peer networks are trying to dispel these old shortcomings. Instead of defining a specific type of architecture or topology, they are focused on applications and their users.

The kinds of applications that operate on modern peer-to-peer networks include collaboration tools, content-management products, distributed file sharing, and distributed processing (in which unused cycle time is "borrowed" from peer processors). To understand how peer-to-peer networks work, consider, for example, a company with multiple offices around the country or world. Each office has a unique collection of files or databases. The company would like to give any employee at any office access to any set of files. Rather than creating one central repository for the information in all the offices (as would be true for a client/server setup), the company could develop a peer-to-peer network and give each office access to the file system of all other offices. Note that in this scenario each office would still maintain its own file or database server. One of the most famous examples of peer-to-peer applications was Napster, in which users could download music from other Napster users on the Web.

Peer-to-peer networks still face several obstacles, however. Security, as mentioned previously, is probably one of the biggest challenges. The other obstacles include performance, management, interoperability, and a serious lack of standardization. Without the development of an accepted set of standards for supporting peer-to-peer systems, it will be very difficult for companies to mix software applications from multiple vendors.

Now that we have discussed the various network operating systems and the network servers that run those systems, let us turn our attention to the many other pieces of network software that support a local area network. In particular, we will examine software utilities and software for accessing the Internet.

Network Support Software

Even though the network operating system is clearly the most important piece of software on a local area network, the operating system cannot work alone. Two kinds of local area network software that work with and support the network operating system are utilities and Internet software. Let us examine both of these in turn.

Utilities

Utilities are software programs that operate in the background and support one or more functions to keep the network running at optimal performance. To support a local area network and its operating system, a wide variety of utility programs are available. Sometimes these utility programs come bundled with an operating system, but many times they are separate and have to be purchased individually. As with a network operating system, it is important, when purchasing utility software (as well as any software product), to pay attention to the licensing agreements. Many utility programs are licensed for a single machine. If 200 workstations are on your local area network, you may have to purchase 200 licenses before your usage is within the letter of the law.

Some of the more common groups of network utility software are:

▸ antivirus software

▸ anti-spam software

▸ anti-spyware software

▸ backup software

▸ crash protection software

▸ network-monitoring software

▸ remote access software

▸ security assessment software

▸ uninstall software

Antivirus software is designed to detect and remove viruses that have infected your memory, disks, or operating system. Because new viruses appear all the time, it is important to continuously update antivirus software with the latest version. Many times, owners of purchased antivirus software can download these updates from the software company's Web site for no additional charge. The only time it is necessary to purchase a new version of the antivirus software is when the software has undergone a major revision. Some antivirus manufacturers offer corporate deals in which a flat annual fee is paid, and the company can disperse and update as many copies of the software as desired.

Spam, or unsolicited commercial bulk e-mail, has become a major nuisance to corporate users as well as individuals. Some professionals estimate that tens of billions of spam messages are sent each day. This large volume of spam wastes the time of workers who have to delete these messages, consumes billions of bytes of storage in e-mail servers, and congests the networks that transfer the data. Anti-spam software is used to block this unwanted e-mail and is available at many levels:

▸ Desktop software, which blocks the spam at the user's workstation or home computer

▸ Corporate software, which blocks the spam at the corporate e-mail server or at the Internet service provider level

▸ Services that block spam before it even encounters the corporate server or Internet service provider

Recently, business and home computer users have been victims of another type of intrusion from unscrupulous outsiders—spyware. Spyware is software that a user unknowingly downloads from the Internet; and when this software is executed on the user's machine, it begins spying on the user. Spyware software may take the form of a remote control program operated by a hacker, or it may be a program launched by a retailer hoping to gather your shopping habits for itself or to share with other retailers. In the latter case, the program may simply be a harmless type of market research tool. Even so, most users feel that spyware is intrusive and should be blocked or eliminated. Antispyware software can locate and clean the spyware programs found in a computer's memory and hard disk drive. Home computer users as well as network administrators should install antispyware software and either execute it on a regular basis or program the operating system to perform it on a timed schedule. You should be careful, however, when searching the Internet for free spyware programs as many of them (but not all) are spyware programs themselves.

Backup software allows network administrators to back up data files currently stored on the network server's hard disk drive. Typically, this backup is written to a tape system, but there are other possibilities, such as performing a backup to a remote site via the Internet. Most backup systems can be completely automated so that the software executes at times when few, if any, users are using the system—for example, the early morning hours. Some backup software also includes recovery capabilities in the event of a system failure.

The primary goal of crash protection software is to perform crash stalling, or to try to keep the operating system running long enough to perform a graceful exit. Crash protection software is also known as rollback software. To take advantage of this utility, a company must have the software installed on both the user's workstation and the network server. When an application running on a user workstation is about to crash, that application transmits a signal to the operating system stating that it is experiencing trouble and needs to be shut down. A crash protector intercepts this message and attempts to fix the problem while keeping the application running. Even if the severity of the problem is beyond repair, the crash protector will often create a pause that allows users to save important data and exit the program safely. Because a crash protector runs constantly in the background, it does create a performance slowdown. Currently, most crash protectors degrade overall system performance anywhere from 2 to 8 percent. Although a loss of 8 percent in processing time may not seem like much, if you are a serious computer user (the kind for whom this software is designed), you are likely to notice the loss in computing speed.

Network-monitoring software incorporates a fairly large number of network support tools. For example, there is software that can monitor network servers and report on CPU utilization, network activities, and server requests. Devices called sniffers can be used on both wired networks and wireless networks. Sniffers can "listen" to traffic on a network and determine if invalid messages are being transmitted, report network problems such as malfunctioning NICs, and detect traffic congestion problems. Wireless sniffers can perform similar operations and can also detect how far wireless signals reach. Thus, if you are having wireless communication problems, a wireless sniffer can tell you if the signal is too weak at a particular location. Conversely, wireless sniffers can also tell you if your wireless signals go too far, such as outside your building or down the street, where they might be exploited by unauthorized users.

Remote access software allows a user to access all of the possible functions of a personal computer workstation from a mobile or remote location. The two most common users of remote access software are nomadic users, who must access software and data on their work computers while traveling or working at home, and support personnel, who need to enter a user's computer system to troubleshoot problems and suggest or make repairs. Many types of remote access software also create a virtual private network (VPN) between the remote user and the work computer. In Chapter Ten, you will learn how VPNs use tunneling protocols and encryption software to create secure connections between the remote computer and the work computer.

Security assessment software is designed to scan an IP address or range of IP addresses for any type of security weakness. Weaknesses may include ports that are open, improperly designed shares and trusts, suspicious processes running in the background, and known vulnerabilities in the operating system that have not yet been corrected. You will learn more about network security in Chapter Twelve.

Software applications continue to grow in size and in the number of support files necessary to execute the program. If someone decides to remove an application, it can be virtually impossible to locate all the files associated with the application. Uninstall software works with the user to locate and remove applications that are no longer desired. As the uninstall software locates each associated file, it prompts the user to decide whether the file should be removed. Because some files may be shared by multiple applications, the user may not want to perform the remove operation. Most uninstall programs keep a backup of removed files to correct problems if a mistake is made and the wrong file is removed. Uninstall programs can also locate orphan files that no longer belong to any application and prompt the user for removal. Finally, many uninstall programs can locate duplicate files and remove the duplicates.

Performing essentially the opposite function of uninstall software is automatic distribution of software. A number of years ago, if a network administrator wanted to load a copy of a software program onto every machine on the network, the administrator would have to visit each machine physically and install the software. Now, software exists that allows the network administrator to load the software remotely onto any and all machines on a network.

Internet software

One of the important segments of the software market is Internet software, the tool set to support Internet-related services. These services and applications include Web browsers, Web server software, and Web page publishing software, among other applications.

Browsers allow users to download and view World Wide Web pages. Browsers also allow users to access internal intranet pages and corporate extranet pages. Most users are familiar with Microsoft's Internet Explorer. Although a few other browsers—such as Firefox, Opera, Chrome, and Lynx—exist, Internet Explorer controls a majority of the market. A number of tools are available to support browsers, such as spelling checkers, pop-up blockers, and download managers that can handle Web page downloads smoothly and efficiently no matter what the connection speed.

Web server software is the application or set of programs that stores Web pages and allows browsers from anywhere in the world to access those Web pages. When a user running a browser enters or clicks a Web page address (its Uniform Resource Locator, or URL), the browser transmits a page request to the Web server at the address cited. The Web server receives the request, retrieves the appropriate Web page, and transmits the requested Web page back across the Internet to the browser. Web server software is capable of supporting secure connections. A secure connection allows the system to transfer sensitive data such as credit card information with confidence that the integrity of the data will not be violated.

When users wish to create one or more Web pages that will be stored on a Web server, Web page publishing software helps them prepare the necessary files. Most Web page publishing programs allow users to insert static images, animated images, various forms of script, and Java-based code into HTML files.

Now that we are familiar with the various types of local area network software, we need to examine an important legal issue involving software.

Software Licensing Agreements

The licensing agreement that accompanies a software product is a legal contract and describes a number of conditions that must be upheld for proper use of the software package. Most licensing agreements specify conditions in the following areas:

▶ Software installation and use—Specifies the number of computers on which a user may legally install and use this software

▶ Network installation—Indicates if the package may be installed on a computer network, and if so, whether additional licenses are necessary for each machine on the network

▶ Backup copy—Informs the user if making a backup copy is acceptable

▶ Decompilation—Specifies that a user may not decompile, disassemble, or reverse engineer the software code in an attempt to retrieve its high-level language to make modifications to the code

▶ Rental statement—States that a user is not allowed to rent or lease the software to a third party

▶ Upgrades—If the software program is an upgrade of a previous version, informs the user that the previous version needs to be purchased before the upgrade can be installed

▶ Copyright—Informs the user that all documentation, images, and other materials included in the package are copyrighted and under the protection of copyright laws

▶ Maintenance—Informs the user whether product support is included in the purchase of the software package

One of the most important issues that affects most users is software installation and use. When a software package is sold, it is usually intended for a particular type of installation, referred to as the user license. Software companies establish user licenses so that an individual in a company does not purchase one copy of a program and install it on, for example, 200 machines, thus cheating the software company out of 199 purchase fees, or royalties. These royalties pay for the cost of creating the software and help support the future cost of supplying upgrades and maintenance.

Several forms of user licenses exist: single-user licenses, interactive user licenses, system-based licenses, site licenses, and corporate licenses. Under the terms of one of the most common user licenses, a single-user-single-station license, the software package may be installed on a single machine, and then only a single user at one time may be using that machine. Some software packages actually count how many times the software has been installed and allow only a single installation. To move the software to another machine requires running the software's uninstall utility. A single-user-multiple-station license is designed for the user who might have a desktop machine at work and a laptop machine for remote sites, or another desktop machine at home. The user is on his or her honor to allow only one copy of the software to be in use at one time. For example, if the user is at work and operating a particular word-processing program with the single-user-multiple-station license, no one should be running the same program on the user's laptop at the same time.

An interactive user license, operating system user license, and controlled number of concurrent users license all refer to essentially the same situation. When a software package is installed on a multiuser system, such as a network server on a local area network, it is possible for multiple users to execute multiple copies of the single program. Many multiuser software packages maintain a counter for every person currently executing the program. When the maximum number of concurrent users is reached, no further users may access the program. When a software counter is not used, the network administrator must estimate how many concurrent users of this software package are possible at any one moment and take the necessary steps to purchase the appropriate number of concurrent user licenses.

System-based licenses, cluster-wide licenses, and network server licenses are similar to the interactive user licenses. However, with a system-based or network server license, there is rarely a software counter controlling the current number of users.

A site license allows a software package to be installed on any and all workstations and servers at a given site. No software installation counter is used. The network administrator must guarantee, however, that a copy of the software will not leave the site.

A corporate license allows a software package to be installed anywhere within a corporation, even if installation involves multiple sites. Once again, the network administrator must ensure that a copy of the software does not leave a corporate installation and, for example, go home with a user.

Every software company may create its own brand of user licenses and may name them something unique. It is the responsibility of the person installing the software—either the user, system administrator, or network administrator—to be aware of the details of the user license and follow them carefully.

What happens if someone does not behave in accordance with the user license and installs a software package in an environment for which it was not intended? In these situations, some packages simply will not work. Other packages will not work correctly if the installation does not follow the license agreement. For example, some software packages maintain a counter and will not allow more copies than were agreed upon during the purchase. Some software packages, such as database systems, may be installable on multiple machines or may give the appearance of allowing multiple users access, but if they were not designed for multiple-user access, they may not work correctly. It is, in fact, important during multiuser database access that the database system lock out any other users while one user is currently accessing data records. If the system is intentionally not designed for concurrent multiuser access, violations of the licensing agreement may lead to unknown results.

Installing the software on more machines than have been agreed to in the user license is illegal. If you install more copies than the number for which you have licenses and the software manufacturer discovers this fact, you and your company may face legal consequences. This is a risk, therefore, that is definitely not worth taking. Be sure any software that you are considering installing has the proper user license before you install it.

Up to this point, the chapter has introduced the software of local area networks—network operating systems, utilities, and other tools—and software licenses. The hardware side of local area networks contains many different types of devices that also warrant further investigation.

LAN Support Devices

A local area network is composed of many elements, both software and hardware components. In Chapter Seven, hub, switches, and routers—the hardware that interconnects devices on the network—were discussed in detail. Let us look at a number of other LAN support devices—including uninterruptible power supplies, tape drives, network attached storage, printers, media converters, and workstations—that you may encounter on a typical local area network.

An **uninterruptible power supply (UPS)** is a valuable device in the event of a power failure. A UPS is a battery backup device that can maintain power to one or more pieces of equipment for short periods of time (usually less than one hour). In a corporate environment, UPS devices are often employed to back up local area network servers and mainframe computers. Most companies use UPS systems to maintain power long enough for the server or mainframe to be properly shut down. Some people even use UPS devices in their homes to protect themselves from data loss in case the power is interrupted while they are working on a document. In general, however, most home users do not use UPS devices. Instead, many protect their computers by using simple electrical surge protectors that isolate their computer equipment from electrical surges.

Tape drives are excellent backup devices. It is quite common to install a tape drive into a network server so that modified files can be backed up one or more times a day. Many companies back up their files nightly, when all the employees are at home and network use is at a minimum. As new technologies such as Blu-Ray DVDs become more powerful, the tape drive backup system may be replaced. Currently, however, tape systems still hold more data than DVDs.

Network attached storage (NAS) is a computer-based device that provides a large amount of storage to users on a network. A stripped-down operating system running on the NAS is designed for the sole purpose of retrieving data for users. Typically a NAS device does not have a keyboard or monitor. Access to its control operations are usually through an assigned Web address.

NAS should not be confused with SAN (**storage area network**). A SAN is a simpler storage device and uses network protocols such as iSCSI and Fibre Channel to store and retrieve data. Another way to look at the difference between NAS and SAN is that NAS is powerful enough to use its own file system protocols, whereas SAN is simpler (just a storage device) that relies on network-level protocols for creating the file system.

Printers tend to evolve at a rate faster than that of other network devices. A number of years ago, a wide selection of dot-matrix printers, daisy-wheel printers, chain printers, ink-jet printers, line printers, and laser printers was available. Today, the printer market has essentially shrunk to just two basic formats: ink-jet and laser. Both formats are capable of printing black and white and color with very high resolution.

Media converters are handy devices if it is ever necessary to connect one type of medium with another. For example, if different portions of one network were installed at different times using different media, it may be more reasonable to use a media converter to connect the two types of media than to remove all the older media and replace them with the newer media. There are many different types of media converters, just as there are many different combinations of media. For example, there are media converters specifically for connecting coaxial cable to twisted pair, and twisted pair to fiber-optic cable.

Like computer peripherals, the computer workstation itself has evolved over the years. Workstations now support hundreds of megabytes (even gigabytes) of random access memory and have hundreds of gigabytes of hard disk storage. Some applications, however, do not require a workstation with a disk drive. A thin client workstation is a computer with no disk drives of any sort. Any software that operates on the thin client is downloaded from the network server to the thin client. The idea behind the thin client is to minimize workstation maintenance and reduce hardware costs. Security is also improved in a thin client system, because it is not possible to insert a CD-ROM or DVD for uploading or downloading software or data. The idea of saving hardware money by using a thin client instead of a fully configured workstation lost some momentum, however, when manufacturers of fully configured workstations dropped their prices to under $1000 for entry-level machines. Nonetheless, thin clients still address important security considerations, and their maintenance costs are lower because of the lack of disk drives.

Having examined the important software and hardware components of local area networks, we are now ready to look at a business example.

LAN Software In Action: A Small Company Makes a Choice

As we have seen in the previous chapter, Hannah is the network administrator for a small company. Her responsibility was to create a fairly simple local area network that provided inter-office e-mail, access to a local database, and print serving. She next updated the network to include access to the Internet, full-duplex connections, and switches in place of hubs. Along with deciding the layout and placement of workstations, cables, and interconnecting devices, Hannah now has to decide which network operating system (NOS) to install on all her company's workstations and servers. As Hannah understands it, from reading the literature, she has the following five choices:

1. Windows Server 2008
2. Unix
3. Linux
4. Mac OS X Server
5. Open Enterprise Server (NetWare)

To help make the decision, Hannah asks herself the following questions:

▸ What are the primary uses (applications) of the current system? Would the primary uses of the system change if a particular NOS were installed?

▸ How would the choice of a particular NOS affect maintenance and support?

▸ Are finances an issue in the selection of a NOS?

▸ Does her company's existing system have any unusual hardware or software that might influence the NOS choice?

▸ Will the network be located in a single location or in multiple locations?

▸ Are there any political pressures to select a particular NOS?

Let us examine each of these questions to see how they affect the decision that Hannah has to make with regard to selecting a network operating system for her company.

Primary uses of current system

As we have seen earlier, the company that Hannah works for is using its workstations and network for spreadsheets, word processing, local and remote e-mail, and Web page access. All of these applications are fairly common and should pose no unusual network requirements. In addition, the company does not have any legacy or home-grown applications that might work well with one type of operating system but not with another.

The only application currently used by the employees that may be of concern is the database system. Database software packages can cause difficulties, depending upon where the actual data store is kept. If the company wishes to maintain local data storage on each individual workstation, then the choice of package should not pose any special requirements in terms of the network operating system, because a number of database packages will support this setup. If the company wishes to maintain a central data storage, however, fewer database package choices are available. Care should be taken to make sure the central storage database package chosen functions with a particular operating system. For example, Microsoft's database storage server SQL Server best operates on Microsoft networks and may not interoperate with other networks such as OES/NetWare.

Network maintenance and support

Hannah is concerned about the level of maintenance that will be necessary to support the corporate local area network. Because her company is small, with fewer than 40 workstations, Hannah will probably be doing most of the network administration herself. Thus, she wants (naturally) to select a network operating system that will require the least amount of work. NetWare has some of the best administrative tools. With a single NetWare administrative program, a network administrator can manage nearly 95 percent of all network operations. Windows, unfortunately, requires the services of many administrative programs.

Hannah may want to receive training in order to support her NOS choice. One can become a certified network administrator in both Windows and OES/NetWare. The most common certification for Windows is the MCSE (Microsoft Certified Systems Engineer), and for Linux, it is the LPI (Linux Professional Institute). Alternately, it should be rather straightforward to hire an individual with either or both of these certifications.

One scenario that Hannah is trying to avoid is having multiple operating systems. It is possible, and fairly common, for networks to support more than one type of network operating system. For example, one could create a Microsoft database server and install it on a Microsoft NOS, and also install a Mac OS X Server to handle network files and print requests. This solution, however, would require Hannah to learn two network operating systems and perform two sets of administration—not a position she wants to be in.

Cost of the NOS

When considering the cost of a network operating system, Hannah will need to consider the administrative costs of running a system in addition to the initial outlay of funds needed to purchase the software. Also, she will need to factor in any significant downtime costs in case the NOS she chooses turns out not to be as stable as was expected. When a new system is installed, significant training costs are also incurred. Many people consider Windows Server easier to set up than either OES/NetWare or Linux because of its many software helper programs, called wizards. Mac OS X Server is also easy to configure. But once the network is up and running, administration is a different concern. Some consider Windows more difficult to manage for day-to-day operations, while OES/NetWare was considered one of the easiest. Creating the directory structure, whether it is Active Directory, NetWare's NDS, or Apple's Open Directory requires a high degree of planning. Linux and Unix are more difficult to set up, but once set up properly, are extremely reliable operating systems. Regardless of the particular system, networks require administration, maintenance, and training.

When Hannah examines the initial costs of a network operating system, Linux is the clear winner (anywhere from zero dollars to a couple of hundred). Mac OS X Server is also quite attractive with its current pricing formula. OES/NetWare, Windows, and Unix are more expensive and roughly equivalent to each other in overall purchase cost. But Hannah has, from her research, learned that there are other costs that are not always readily apparent that she should be aware of. For example, although a single-user-single-station license for Windows is less expensive than a single-user-single-station license for OES/NetWare, using a Windows NOS also requires the purchase of a Windows server license. No such cost is associated with a NetWare server. Likewise, Linux, Unix, and Mac OS X Server do not have such separate client and server licenses.

Any unique hardware choices affecting NOS decision

None of the network operating systems Hannah is considering requires any unique hardware. Conversely, the hardware that Hannah has chosen to support the network will not impose any restrictions on the choice of an NOS. One interesting feature of Linux is that it can, depending upon the operation it is to perform, run in a relatively small memory space and on fairly old microprocessor technology (such as an Intel 386 microprocessor). The other network operating systems require substantially more memory and hard disk space, as well as modern processor technology.

Single location or multiple locations

At the present time, all of Hannah's company's resources are in one location. But if the company wants to add remote or mobile users, OES/NetWare, with its simpler client software requirements, is a very attractive option. OES/NetWare's and Mac OS X Server's administrative functions make dealing with both local and remote users easy. Although the other operating systems can also handle both types of users, their administration functions are more challenging.

Political pressures affecting decision

No apparent political pressures are forcing Hannah to select one particular NOS over another. The company administration has no allegiance to a particular system and has little or no previous experience with a particular NOS. Likewise, most users are not strongly in favor of one system over another. The only potential problem is that most users have computers at home and are already familiar with a Windows-like environment. Thus, they may have difficulty adapting to a non-Windows environment such as Linux. Also, many people (wrongly) think that the Mac OS X Server works only with other Apple devices, and thus may be resistant to the idea of using it for a business-type application.

Final decision

When you are considering one network operating system over another, it is most important that the decision be well thought out. All factors must be weighed, and hype and advertising should not be part of the consideration. Fortunately, all the network operating systems Hannah has considered are high-quality products. Also, all five systems are popular and powerful, and they each have strong followings. One concern associated with NetWare/OES is how much longer the system will be around. But given the strength of the field, it is not too likely that Hannah can make a poor choice, as long as she considers the factors described previously and the needs of her company.

Wireless Networking In Action: Creating a Wireless LAN for Home

Chris has three computers and a high-speed Internet connection at home. He would like to connect all three computers to the Internet, but he does not want to install any wires because the computers are located on different floors of his house. So Chris has decided to install a wireless local area network. Let us follow Chris as he makes the many decisions necessary to install a wireless network.

Chris first needs to determine which wireless LAN technology he is going to purchase. Recall that four approved technologies exist for wireless LANs: IEEE 802.11b, 802.11a, 802.11g, and 802.11n. IEEE 802.11b was the first standard approved (in 1999) for use with wireless LANs. It transmits signals at 2.4 GHz for roughly 50 meters (150 feet) between access point and wireless device. Will the transmission distance of 50 meters be enough? If Chris places the high-speed Internet connection and access point on the main floor of his house, will it reach upstairs and downstairs? Chris has done some rough measuring and believes the signals will reach all floors. If the distances were too great and Chris still wanted to use his wireless from *every* location in his house, he would have had to extend the high-speed Internet cable to a second point in the house and install another wireless access point.

IEEE 802.11b theoretically transfers data at 11 Mbps. Its actual data transfer speed, however, is approximately 5 Mbps. Will this be fast enough for Chris' needs? For now, Chris plans on using the wireless LAN to transfer Web pages and small files between computers. Five million bits per second should be sufficient, but Chris decides to consider the faster technologies, just in case.

IEEE 802.11a was the second standard approved in 2002. It transfers data at a theoretical speed of 54 Mbps (with an actual speed of roughly 18 Mbps), using the 5-GHz frequencies. Because of the higher frequencies, 802.11a can transmit for only about half the distance of 802.11b. To complicate matters, 802.11a and 802.11b are not compatible. You can purchase hardware that transmits both 802.11a and 802.11b signals, but the cost (with double the circuitry and double the antennas) is higher.

The third standard approved was IEEE 802.11g, in 2003. This standard transmits at the same frequencies as 802.11b (2.4 GHz) but has a theoretical transfer speed of 54 Mbps (with an actual speed of approximately 18 Mbps). It is compatible with 802.11b and has the same transmission distances, roughly 50 meters. Finally, 802.11n was approved in 2009 and can transfer data in the 100s of Mbps. Even though 802.11n is more expensive (being the newest), Chris decides to go with this standard. He likes the higher speeds, which may be important in the near future (especially if he ever decides to transmit wireless audio and video from the Internet to his home entertainment center). So the next step for Chris is to purchase 802.11n-enabled network interface cards (NICs) for his computers. Before Chris purchases anything, however, he checks to see if any of his computers came with wireless NICs installed. Although many if not most of the computers (in particular, laptops) sold these days come with wireless interface cards, Chris discovers that each of his computers will need a card.

Now Chris has to buy a wireless access point, but he has read that there are also devices called wireless routers and gateways. What is the difference? If you already have a wired network at home (a modem to connect to the high-speed Internet and a router to interconnect the various workstations), then all you need is an access point. Plug in the access point, connect it to your existing router, install the security software, and you should be all set. If you have the high-speed Internet connection but nothing else, you can purchase a wireless router. A wireless router acts as both a router and a wireless access point. If you don't even have a high-speed Internet connection yet, you might consider purchasing a gateway. A gateway is often a combination of a high-speed modem, router, and wireless access point—all three devices rolled into one. A potential problem with buying your own high-speed modem (as part of a gateway) is compatibility with the high-speed Internet service. Make sure you check with your Internet service provider before you purchase your own modem.

Chris already has a high-speed Internet service, so he decides to purchase a wireless router, making sure the router has security options such as WEP or WPA (see Chapter Twelve). When installing the router, he immediately changes the default network name and administrator's password to minimize the chance of someone hacking into his wireless network. One additional security option Chris looked for was Stateful Packet Inspection (SPI). A router running SPI makes sure that every incoming packet corresponds to an outgoing request that Chris has made, and not someone next door "stealing" Chris' free radio transmissions.

What about the operating system on his workstations? Does Chris need a network operating system such as Windows Server or OES/NetWare? No, Chris does not need anything as powerful as a network operating system, even though he is creating a network of wireless devices. Many recent desktop operating systems such as Windows XP/Vista/7 have a provision for supporting wireless workstations. Chris will probably have to use the Control Panel in the XP operating system to select Internet Protocol TCP/IP for the connection, and tell the XP operating system how the IP address for this device will be obtained. Instructions like these are commonly included with the wireless router and network interface cards and are not difficult to follow.

Does Chris need any additional network support software? He will definitely need some form of antivirus software to protect his machines from viruses, and it may not be a bad idea to install anti-spam and anti-spyware software at the same time. Security assessment software is always useful, especially with wireless devices and Internet access. Chris also plans on using the firewall software that came with the router and setting the security options to protect his computers and his data.

Chris should be ready to use his new wireless network. All wireless NICs have been installed, as well as the wireless router and security software. Now Chris can surf the Internet from any room in his house, maybe even from the backyard, next to the pool.

◆ ◆

SUMMARY

▶ A network operating system has several additional functions not normally found in an operating system. For example, it can:
  ▶ manage one or more network servers
  ▶ maintain a file system of data sets, applications, user profiles, and network peripherals
  ▶ coordinate all resources and services available

> ▸ process requests from users
> ▸ prompt users for network login, validate accounts, apply restrictions, perform accounting functions
> ▸ manage one or more network printers
> ▸ manage the interconnection between local area networks
> ▸ manage locally connected users; manage remotely connected users
> ▸ support system security; support client/server functions
> ▸ support Web page development and Web server operations

▸ Novell NetWare (now called Open Enterprise Server) is a network operating system with a powerful directory service (NDS) and is very good at performing file and print serving.

▸ Windows NT was another popular network operating system, was very good at supporting client/server applications, and was based on the domain; Windows Server 2000/2003/2008 represents a significant advancement over NT and includes the powerful directory service, Active Directory.

▸ Unix is an older operating system that is stable, fast, and capable of running on a variety of platforms.

▸ Linux is a derivative of Unix and shares Unix's features of stability and speed as well as low cost and the capability to run on a variety of platforms. Novell now offers commercial versions of Linux.

▸ Mac OS X Server is another derivative of Unix and, like Linux, shares Unix's features of stability and speed. It supports both Macintosh and PC workgroups.

▸ The network server is the computer that stores software resources such as the network operating system, computer applications, programs, data sets, and databases, and either allows or denies workstations connected to the network access to these resources.

▸ Many network servers can perform one or more levels of RAID. RAID is designed to provide redundant backup of data onto multiple hard disk drives.

▸ Many types of software programs support a local area network. These include utility programs and Internet software tools.

▸ Software licensing agreements are an important part of local area network software installation. The most common forms are: single-user-single-station, single-user-multiple-station, interactive user, system-based, site, and corporate licenses.

▸ Many types of hardware devices are necessary to support a local area network, including hubs, switches, and routers; uninterruptible power supplies and surge protectors; tape drives; network attached storage; printers and print servers; media converters; workstations; and network servers.

KEY TERMS

Active Directory
anti-spam software
anti-spyware software
antivirus software
application program interface (API)
backup software
bindery
corporate license
crash protection software

◆ disk mirroring
◆ domain
◆ interactive user license
◆ Internet software
◆ leaf object
◆ licensing agreement
◆ media converters
◆ multitasking operating system
◆ NetWare Directory Services (NDS)

◆ network attached storage
◆ network-monitoring software
◆ network operating system (NOS)
◆ network server
◆ network server license
◆ organizational unit (OU)
◆ operating system
◆ peer-to-peer network
◆ redirection

redundant array of independent
    disks (RAID)
remote access software
security assessment software
server appliance
server blade
server virtualization

- single-user-multiple-station license
- single-user-single-station license
- site license
- sniffers
- spam
- spyware
- storage area network

- striping
- thin client
- uninstall software
- uninterruptible power supply (UPS)
- utilities
- Web server software

REVIEW QUESTIONS

1. List the six basic functions of an operating system.
2. What distinguishes a multitasking operating system from a non-multitasking operating system?
3. List the primary differences between a network operating system and an operating system.
4. What is an application program interface?
5. What are the basic functions of a network operating system?
6. What is meant by disk mirroring?
7. What is an organizational unit?
8. What is the function of OES/NetWare's NDS?
9. What are the primary features of OES/NetWare?
10. What are the main advantages of Windows Server 2000 over NT Version 4?
11. What are the differences between Windows Server 2008 and Windows Server 2000?
12. What is the function of Windows' Active Directory?
13. What are the strengths of Unix?
14. List the reasons for Linux's popularity.
15. What are the disadvantages of Linux?
16. What are the advantages and disadvantages of Mac OS X Server?
17. What is the importance of a network server?
18. What different kinds of network servers are available?
19. What are the different levels of RAID, and what does each do?
20. What are the issues often stated in a software license agreement?
21. List the types of software license agreements.
22. What are the nine most common groups of network utility software?
23. What are the primary functions of an Internet Web page server?
24. What are the different types of hardware support devices for local area networks?
25. What is the difference between NAS and SAN?

EXERCISES

1. In a client/server system, a client transmits a request to a server, the server performs a processing operation, and the server returns a result. List all the possible problems that can occur with transmission in this scenario.
2. What are the primary advantages of NetWare Version 4 over NetWare Version 3? Are there any disadvantages?

3. You want to create a local area network that protects the contents of the network server's hard disks from disk crashes. List all the different techniques for providing this protection that have been presented thus far.

4. What is the main structure (container) used when designing a Windows NT network?

5. Windows Server 2000/2003/2008 uses Active Directory for its directory service, and NetWare/OES uses NDS. How are the two directory services alike? How are they different?

6. When using either OES/NetWare's NDS or Window's Active Directory, an administrator can control resources on a macro level. Explain what controlling resources on a macro level means, and show an example.

7. You are working for a company that has three divisions: marketing, research, and sales. Each division has many employees, each with his or her own workstation. Each division also has its own network server and a number of high-quality printers. Draw the Windows NT domain diagram, Windows Active Directory diagram, and OES/NetWare NDS diagram that support this company's network structure.

8. In what ways are Unix and Linux similar? In what ways are they different?

9. Why is the stability of Linux so high when compared to other network operating systems?

10. The problem with antivirus software is that new viruses are being created every day. If you have antivirus software installed on your machine or on your network, how do you keep your software up to date?

11. How does a single-user-single-station software license differ from a single-user-multiple-station license? Does one have an advantage over the other in the business world? In the home computer world?

12. What kind of software applications might a company consider as likely candidates for a site license?

13. Consider the following software licensing scenario: Office Suite 1 costs $229 per single-user-single-station license, while Office Suite 2 costs $299 per interactive user license. You have 200 users on your network, and you estimate that at any one time only 60 percent of your users will be using a suite application. Determine the best license solution. At what level of interactive use will the cost of the interactive user license break even with that of the single-user-single-station licenses?

14. Is a thin client computer more advantageous in a corporate setting or in a user's home? Explain.

THINKING OUTSIDE THE BOX

1 A small business with 100 computer workstations is installing a new local area network. All of the users perform the usual operations of e-mail, word processing, Internet browsing, and some preparation of spreadsheets. Approximately one-quarter of the employees perform a large number of client/server requests into a database system. Which local area network operating system would you recommend? State your reasoning.

2 Create an NDS or Active Directory tree structure for one of the following environments:

a. A segment of the company at which you are employed

b. The whole company at which you are employed

c. A segment of the school at which you are enrolled

    d. The whole school at which you are enrolled

    e. A hypothetical corporation that has multiple departments, servers, peripherals, and users

3 Using the same environment chosen in the previous problem, create a Windows NT solution for that environment.

4 You are creating a network at home that consists of multiple computers, a high-quality printer, and a router with access to a high-speed Internet connection. Surely you do not need Windows Server 2008 or OES/NetWare, but is there a particular desktop operating system that can be installed on each computer that will optimize the operations of your home network?

HANDS-ON PROJECTS

1. Who makes thin client machines? What are their specifications, features, and prices?

2. Are there any network operating systems other than those listed in this chapter? Are they holding their own, or are they slipping into oblivion?

3. A local area network operating system that may still hold a very small share of the market is OS/2. What is the current state of this system? Is it still being produced? Is it still being supported? Can you find an estimate of how many networks are using OS/2?

4. BSD is another free, popular operating system available as a download from the Internet. What are its advantages and disadvantages? Is it a network operating system or simply a single-machine operating system?

5. What is the difference between a copyright and a patent when applied to computer software? How long are copyrights valid? How long are patents valid?

6. In this chapter, you learned about four RAID levels: RAID-0, RAID-1, RAID-3, and RAID-5. Find out and list all the levels (or versions) of RAID that exist. What is the primary function of each level?

9

Introduction to Metropolitan Area Networks and Wide Area Networks

◆◆

VINTON CERF, THE CO-CREATOR of TCP (a major protocol used on the Internet), has turned his attention to the creation of a network that can span a very wide area—the solar system. Cerf joined a team from NASA to create a wireless network that would allow users on Earth, satellites, space probes, and eventually astronauts to talk with one another. This network is called the Interplanetary Internet (IPN).

Cerf says: "I realized it had taken 20 years for the Internet to take off: from 1973 to 1993. So I wondered what I should be doing to prepare for our needs in the future. An interplanetary backbone was the answer."

Currently, the two rovers on Mars have their own IPN address, a .mars domain name, and a newly created protocol that is similar to TCP/IP but streamlined to interoperate with the peculiarities of transmitting data through outer space. Some of the peculiarities that IPN will have to overcome are high levels of noise, weak signals, small power supplies, and long propagation delays.

For his next project, Cerf is hoping to put a telephone satellite in orbit by the end of the decade that will connect the two planets. Will NASA be able to pull this off? Cerf is optimistic. He feels that once space travel becomes a commercial endeavor (as the Internet eventually became), IPN and interplanetary communications will take off. Like a rocket.

The Internet and the Interplanetary Internet are wide area networks. What is a wide area network, and how does it differ from a local area network?

Do all wide area networks share the same features?

Source: Jeffrey Davis, "Vint Cerf is taking the Web into outer space—reserve your .mars address now," **Wired Magazine**, January 2000, downloaded from *www.wired.com/wired/archive/8.01/solar.html.*

Michael Singer, "Vint Cerf, 'A Father of the Internet,'" June 18, 2004, downloaded from *www.internetnews.com/infra/article.php/3370411.*

Objectives ▶

After reading this chapter, you should be able to:

▶ Distinguish local area networks, metropolitan area networks, and wide area networks from each other

▶ Identify the characteristics of metropolitan area networks, and explain how they compare and contrast with wide area and local area networks

▶ Describe how circuit-switched, datagram packet-switched, and virtual circuit packet-switched networks work and how they compose network cloud

▶ Identify the differences between a connection-oriented network and a connectionless network, and give an example of each

▶ Describe the differences between centralized routing and distributed routing, and cite the advantages and disadvantages of each

▶ Describe the differences between static routing and adaptive routing, and cite the advantages and disadvantages of each

▶ Document the main characteristics of flooding, and use hop count and hop limit in a simple example

▶ Discuss the basic concepts of network congestion, including quality of service

A local area network, as you may recall, is typically confined to a single building or set of buildings that are in close proximity (as in a campus). What happens when a network expands into a metropolitan area, across a state, or across the entire country? A network that expands into a metropolitan area and exhibits high data rates, high reliability, and low data loss is called a metropolitan area network (MAN). In this chapter, we will examine metropolitan area networks and see how they compare and contrast to other forms of networks.

What happens when a network is larger than a metropolitan area? A network that expands beyond a metropolitan area is a wide area network. Wide area networks share a few characteristics with local area networks: they interconnect computers, use some form of medium for the interconnection, and support network applications. There are some differences, however, between wide area networks and local area networks. For example, wide area networks include both data networks (such as the Internet) and voice networks (such as telephone systems), whereas local area networks in most cases include only data networks—but this, as we shall see in Chapters Ten and Eleven, is slowly changing with the advent of VoIP (voice over IP). Wide area networks can interconnect thousands of workstations (devices), tens of thousands, or more, in such a way that any one workstation can transfer data to any other workstation. As the name implies, wide area networks can cover large geographic distances, including the entire Earth. In fact, as you learned in the opening vignette to this chapter, there are even plans under way to network the planet Mars as more technology is dropped on the planet and the need for returning signals to Earth becomes greater. Thus, wide area networks may indeed some day cover the entire solar system!

Because of the differences between local area networks and both metropolitan area and wide area networks, the latter two network forms merit their own discussion. Let us begin with a discussion of metropolitan area networks, and then introduce the basic terminology of wide area networks. We will then examine the differences between circuit-switched and packet-switched wide area networks. A brief introduction to routing is also in order, because wide area networks use routing extensively to transfer data. Subsequent chapters of this text will deal with unique types of wide area networks, in particular, the Internet.

Metropolitan Area Network Basics

Many of the same technologies and communications protocols found in local area networks (and wide area networks) are used to create metropolitan area networks. Yet MANs are often unique with respect to topology and operating characteristics. MANs can be used to support high-speed disaster recovery systems and real-time transaction backup systems. They can also provide interconnections between corporate data centers and Internet service providers, and support high-speed connections among government, business, medical, and educational facilities. MANs are almost exclusively fiber-optic networks, and thus capable of supporting data rates into the tens of millions and hundreds of millions of bits per second. For the same reason, they are advertised as networks with very low error rates and extremely high throughput. Although these characteristics are not that different from those of many local area networks, a few characteristics distinguish MANs from LANs. The first characteristic is that MANs cover much greater distances than LANs do. As the name implies, metropolitan area networks are quite capable of supporting entire metropolitan areas, such as New York, Chicago, and Los Angeles. Local area networks rarely extend beyond the walls of a single building, and thus are smaller than MANs.

A second characteristic that distinguishes MANs from LANs (but not necessarily from WANs) is that most MANs can recover very quickly from a link or switch/router failure. MANs are designed to have highly redundant circuits so that in the event of a component failure, the network can quickly reroute traffic away from the failed component. This ability to reroute in the event of a failure is called failover, and the speed at which a failover is performed is the failover time. While not all MANs have low failover times, achieving them is certainly the goal of any company that offers a MAN service.

A third characteristic that distinguishes many MANs from both LANs and WANs is that some MAN topologies are based on a ring. The ring MAN is unique in that, unlike the now pretty-much-deceased token ring LAN, it is a ring both logically *and* physically. Thus, not only is data passed around in a ring fashion, but also the network routers and switches are interconnected in a ring fashion (Figure 9-1).

Figure 9-1
A physical ring used to support a metropolitan area network

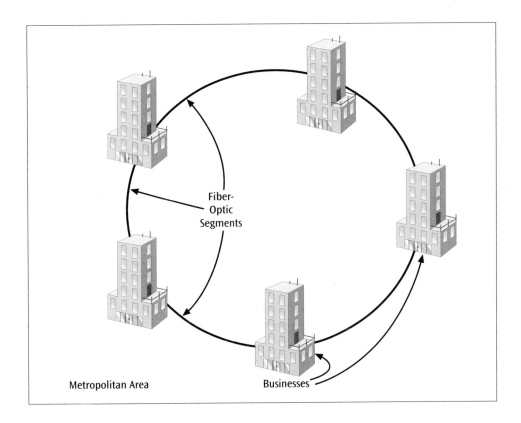

Last, a feature that is beginning to appear in MANs but that neither LANs nor WANs currently have is the ability of a user to dynamically allocate more bandwidth on demand. Let us say you are working for a business and have a MAN connection between your corporate office and an Internet service provider. You anticipate, perhaps because you are about to unveil a new customer service offering, that the demand on the MAN link will grow substantially within the next few days. Thus, you place a telephone call—or, in some cases, access a Web page—and request that the bandwidth of your MAN connection be increased by a particular amount. The MAN service provider receives your request, immediately increases your bandwidth, and bills you accordingly. Perhaps at some point in the future, all networks—local, metropolitan, and wide area—will have this powerful feature. For now, it is offered only by certain MAN service providers.

Now that we have examined some of the basic characteristics that distinguish MANs from LANs and WANs, let us turn our attention to the technologies that support MANs.

SONET vs. Ethernet

Almost all MANs are based on one of two basic forms of supporting technology: SONET or Ethernet. SONET, as we saw in Chapter Five, is a synchronous time division multiplexing technique that is capable of sending data at hundreds of millions of bits per second. The network topology is a ring, but this ring is actually composed of multiple rings that enable the network to provide backup in the event of a segment failure (Figure 9-2). This is one of the characteristics of SONET rings that allows them to have a very low failover time. At the present time, many MANs are supported by SONET ring technology.

Figure 9-2
SONET systems are comprised of multiple rings

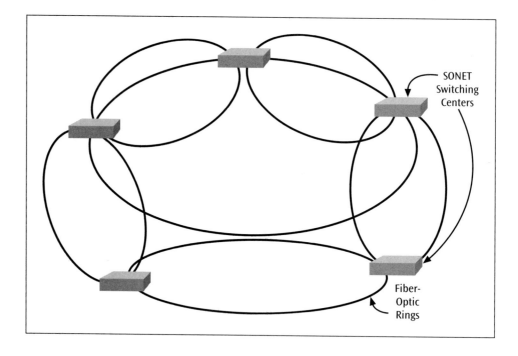

Unfortunately, SONET has a number of disadvantages. It is a complex, fairly expensive technology that cannot be provisioned dynamically. Furthermore, SONET was designed to support multiple streams of voice channels (such as multiple T-1s, which transmit at 1.544 Mbps) and thus does not scale nicely into the 1-Mbps, 10-Mbps, 100-Mbps, and 1000-Mbps chunks that are typically used with data transmissions. These shortcomings have, interestingly, revived interest in a technology that is older than SONET but a newcomer in the field of metropolitan area networks: Ethernet.

Ethernet MANs are less expensive than SONET systems, well understood, easily scalable from 10 Mbps to 100 Mbps to 1000 Mbps to 10 Gbps, and the best technology for carrying IP traffic (the type of traffic that runs over the Internet). One disadvantage of Ethernet in the MAN is higher failover time. Ethernet MANs do not recover as quickly as SONET rings and can potentially leave customers without service for seconds. Although SONET rings typically have a failover time of 50 milliseconds, Ethernet failover times can be higher. Nonetheless, Ethernet MANs have a number of attractive characteristics and are growing in popularity. Figure 9-3 shows a typical layout of an Ethernet MAN topology. Note how the network is a mesh design with redundant paths between endpoints.

Figure 9-3
The Ethernet MAN topology

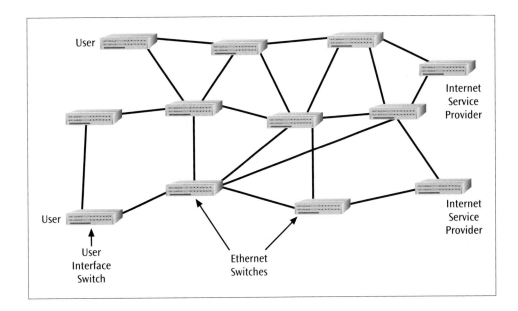

Ethernet MANs have given rise to a newer service whose popularity has grown in the last few years: Metro Ethernet. Metro Ethernet is a data transfer service that can connect your business to another business (or businesses) using a standard Ethernet connection. With Metro Ethernet, you may connect your company directly to another company using a point-to-point connection, or, for example, to two other companies using two point-to-point connections, as is shown in Figure 9-4 (a). Alternatively, you may connect your company to multiple companies as though they were all part of a large local area network, as shown in Figure 9-4 (b). The former connection is the same as having a private connection between two points. A common example of this type of Metro Ethernet connection is found when a company is connected to an Internet service provider. All the traffic on this connection is between only two locations. The latter connection is an example of a multipoint-to-multipoint connection. Here, any company can talk to one or more (or all) connected companies. Thus, a company needs to send out only one packet to ensure that multiple companies receive this data.

Figure 9-4 (a)
*Two point-to-point
connections*

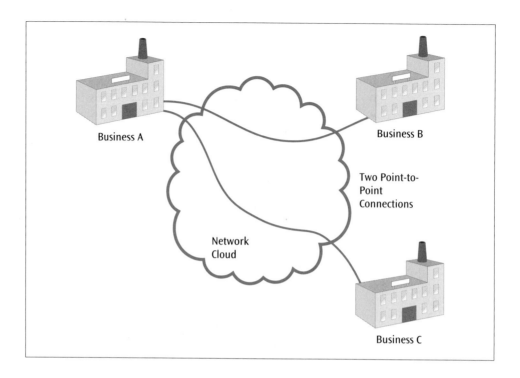

Figure 9-4 (b)
*Multipoint-to-
multipoint connections*

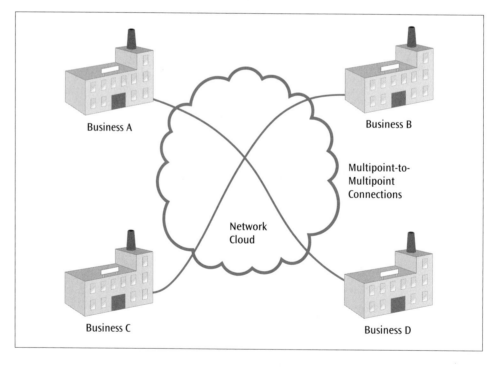

An interesting aspect of Metro Ethernet is that the users of the service can seamlessly connect their company Ethernet local area networks to the Metro Ethernet service. Because all the involved networks are Ethernet, there is no need for time-consuming and sometimes clumsy conversions from one format to another. Thus, a corporate LAN running at 100 Mbps can nicely connect to a Metro Ethernet service also running at 100 Mbps. The same cannot be said if one is connecting a 100 Mbps LAN to SONET, with speeds that typically start at 51.48 Mbps.

A company that has a connection with Metro Ethernet can create a bandwidth profile for that connection. This bandwidth profile describes various characteristics about the connection, such as basic data transfer rates, basic burst rates (a burst, as you may recall, is a large surge of data that is transmitted for a short period of time), excess data transfer rates, and excess burst rates. Local area network Ethernet, as we saw in Chapter Seven, does not allow users to set their own data transfer rates—they must simply accept the one rate corresponding to the particular brand of Ethernet (such as 100 Mbps Ethernet) they have chosen. This profile feature of Metro Ethernet is an interesting and powerful option for companies that want to tailor a connection to a particular application. For example, if a company is launching a new Web application and expects a wide range of user response, the company's network personnel can set a basic data transfer rate for the average response anticipated, and then set an excess burst rate for the peak periods.

Now that we have examined MAN basics, let us look at the basics of wide area networks.

Wide Area Network Basics

A wide area network (WAN) is a collection of computers and computer-related equipment interconnected to perform a given function or functions and typically using local and long-distance telecommunications systems. The types of computers used within a wide area network range from microcomputers to mainframes. The telecommunications lines can be as simple as a standard telephone line or as advanced as a satellite system. Wide area networks are typically used to transfer bulk data between two endpoints and provide users with electronic mail services, access to database systems, and access to the Internet. Wide area networks can also assist with specialized operations in many fields, such as manufacturing, medicine, navigation, education, entertainment, and telecommunications.

As you may recall, a local area network works as a bus-based network in that clusters of workstations are connected to a central point (hub or switch) through which workstations can transmit messages to one another. Because there are so many workstations in a wide area network and they are spread over large (possibly very large) distances, this type of interconnection is not feasible. Likewise, a network in which each workstation is connected to every other network workstation is also impractical, as there would be so many connections into each workstation that the technology would be totally unmanageable. Instead, a wide area network connects its workstations through the use of a mesh design and requires routing to transfer data across the network. A network that is connected in a mesh is one in which neighbors are connected only to neighbors (Figure 9-5). Thus, to be transmitted across a mesh network, the data has to be passed along a route from workstation to workstation.

Figure 9-5
*A simple mesh
network*

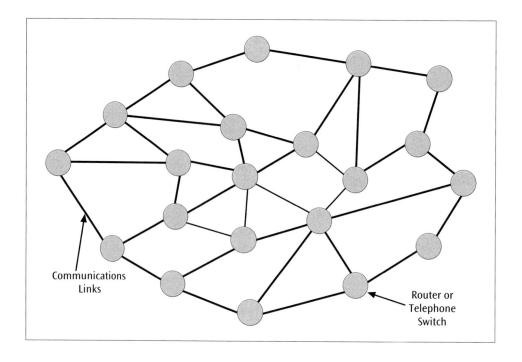

All wide area networks are collections of at least two basic types of equipment: a station and a node. A station is a device that a user interacts with to access a network, and it contains the software application that allows someone to use the network for a particular purpose. Very often, the station is a microcomputer or workstation, but it can also be a terminal, a cell phone, or a mainframe computer.

A node is a device that allows one or more stations to access the physical network and is a transfer point for passing information through the network. When data or information travels through a network, the information is transferred from node to node through the network. When the information arrives at the proper destination, the destination node delivers it to the destination station. In a network that is designed to transfer computer data, such as the Internet, the node is usually a very fast and powerful router with multiple ports. Incoming data arrives on one port and, depending on the route the data must take to reach its destination, is retransmitted out a second port. In a network that is designed to transfer voice signals, the node is usually a powerful telephone switch that performs functions similar to those of a router.

The support structure of a wide area network is the sub-network, or network cloud. A sub-network, or network cloud is a collection of nodes and interconnecting telecommunications links, as shown in Figure 9-6 (notice how the sub-network is drawn resembling a cloud). The type and number of interconnections between nodes and the way network data is passed from node to node are the responsibility of the network cloud. It helps to think of the wide area network and the network cloud as being two almost separate entities. The wide area network is the entire system: the nodes, the stations, the communications lines, the software, and perhaps even the users. The network cloud is the underlying physical interconnection of nodes and communications lines that transfer the data from one location to another. All these components work in concert to create the network. A user sitting at a workstation and running a network application passes his or her data to the network through a station, which passes the data to the cloud. The network cloud is responsible for getting the data to the proper destination node, which then delivers it to the appropriate destination station. Clearly, a network would not exist without a network cloud, but it should not matter to the network what the inside of the network cloud looks like. The network cloud is simply the vehicle for getting the data from sender to destination.

Figure 9-6
Network cloud, nodes, and two end stations

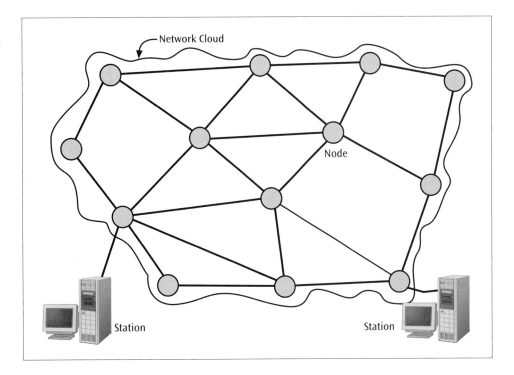

The topics introduced in the rest of this chapter—network clouds, routing, and congestion—are handled by the network layer of a network architecture model, regardless of whether the model is OSI or TCP/IP. Let us first examine the topic of network clouds in detail, and then explore routing and congestion.

Types of network clouds

A wide area network's cloud may be categorized by the way it transfers information from one end of the cloud to the other. The three basic types of clouds are circuit-switched, packet-switched, and broadcast. Let's examine each of these in turn.

Circuit-Switched Network

A circuit-switched network is a network cloud in which a dedicated circuit is established between sender and receiver, and all data passes over this circuit. One of the best examples of a circuit-switched network is the dial-up telephone system. When someone places a call on a dial-up telephone network, a circuit, or path, is established between the person placing the call and the recipient of the call. This physical circuit is unique, or dedicated, to this one call and exists for the duration of the call. The information (the telephone conversation) follows this dedicated path from node to node within the network, as shown in Figure 9-7. A wide area network in which information follows a dedicated path from node to node within the network is a circuit-switched wide area network.

Figure 9-7

*Two people carrying
on a telephone
conversation using
a circuit-switched
network*

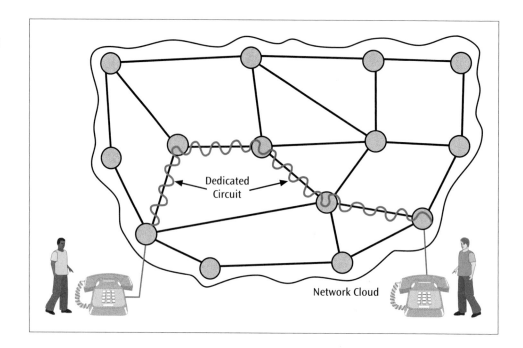

When a telephone call is placed over a circuit-switched network, the network needs time to establish the circuit and to tear down the circuit. But once the circuit is established, all subsequent data travels quickly from node to node. A circuit-switched network has two key disadvantages. First, each circuit is dedicated to only one connection. Second, when the circuit is used to transfer data (as opposed to voice), it is probably not being utilized fully, because computer data transfer is often sporadic.

Despite the widespread use of circuit-switched networks throughout the twentieth century, many of the circuit-switched networks operated by telephone companies have been or are being replaced with a more efficient network design—the packet-switched network. But the transition from circuit-switched to packet-switched network with regards to the public telephone networks is nowhere complete, thus the need for us to continue studying both forms, at least for a little while longer.

Packet-Switched Network

The packet-switched network is found in networks designed to transfer computer data (such as the Internet and in many modern telephone networks). In a **packet-switched network**, all data messages are transmitted using fixed-sized packages, called packets, and no unique, dedicated physical path is established to transmit the data packets across the sub-network. (To distinguish between a piece of data processed at the data link layer and a piece of data processed at the network layer, the term "frame" is used at the data link layer, and the term "packet" is used at the network layer.) If the message to be transferred is large, it is broken into multiple packets. When the multiple packets arrive at the destination, they are reassembled into the original message.

The two types of packet-switched networks are the datagram and the virtual circuit. In a **datagram** packet-switched network, each data packet can follow its own, possibly unique, course through the cloud. As each packet arrives at a node, a decision is made as to which path the packet will follow next. This dynamic decision making allows for great flexibility should the network experience congestion or failure. For example, if a group of data packets is currently being routed through Dallas on its way to Phoenix, and Dallas experiences router problems, the network can reroute the packets through Denver instead. The problem with datagram networks is that when a large group of packets is addressed to the same destination, the network's nodes have to examine each packet individually and determine each packet's next path. This can lead to inefficiency, or simply to wasted time.

To solve this problem, the virtual circuit packet-switched network was created. In a virtual circuit packet-switched network, all packets that belong to a logical connection can follow the same path through the network. For example, one station may want to transfer a large amount of data, such as the entire electronic contents of a book, across the network to another station. To accomplish this, a virtual circuit breaks the large amount of data into n packets and determines an optimal temporary path through the network. Each router along the path is then informed that it will be participating in a particular virtual circuit. When the data arrives with the address of that particular virtual circuit, the router simply sends the data out the router connection that is associated with that virtual circuit. When the data transfer is complete, the temporary path is dissolved (that is, each router tosses that virtual circuit information). This type of packet-switched network is called a virtual circuit because the path followed by the packets acts like a circuit but is not an actual, physical circuit like a telephone circuit.

Although this type of network sounds similar to a circuit-switched network, in that all data packets follow a fixed path, it is substantially different. The path in a virtual circuit packet-switched network exists only in the software and only when the network creates the necessary routing tables at the appropriate nodes. These routing tables are similar to the port tables used by switches to determine a path through a local area network. Another difference between a virtual circuit and a circuit-switched network is that the path in a virtual circuit network may be shared by other traffic. To appreciate the significance of this, recall that when you place a call on a circuit-switched network, you have a *dedicated* circuit from one end of the connection to the other. Thus, if you could see data as it travels through the wire in your circuit, you would see that the only data in this circuit is your data. In contrast, each user in a virtual circuit network is sharing one or more circuits with other users. If you could see into a wire in this circuit, you might see data from several users, or data from one user, then another, traveling across the network. Thus, the various wires that constitute a circuit in a virtual circuit network are carrying data streams from multiple users, and it is the software that keeps each data stream separate from the others.

To summarize, packet-switched networks break computer messages and telephone voice transmissions into fixed-sized packets and are thus designed to support computer data and voice transmissions. In a datagram packet-switched network, each packet is an entity by itself. When a large message has been broken into n packets, each packet in a datagram network enters a node, where a unique routing decision is made. Because there is no fixed circuit to follow, the network spends no time in creating a circuit. Time is spent, however, in determining each packet's route. Nevertheless, datagram networks are quite flexible, as they can react quickly to network changes. The other type of packet-switched network, the virtual circuit packet-switched network, is a marriage of the packet-switched network and the circuit-switched network. When a large message in this network is broken into multiple packets, all these packets follow the same path through the network. This path is determined before the first packet is transmitted—an activity that requires circuit setup time—and a virtual circuit, with its internal routing tables, helps route the packets from source to destination.

The telephone company Sprint announced in 2003 that it began replacing its circuit-switched networks with packet-switched networks. The use of packet-switched networks in the transmission of telephone conversations involves digitizing voice data and converting it into packets. Once voice data is in packet form, it can be transferred over a packet-switched network along with data and video. This type of conversion will lead to more economical communications because all the forms of data being transmitted will be sharing one form of network. One of the biggest obstacles to using packet-switched networks to transfer voice (that is, to make a phone call) is that some packet-switched networks such as the Internet are not always the best for delivering the voice packets in real-time on a consistent basis, and therefore the voice conversation can break up. If you are a communications company hoping to offer a new or alternative service, this can result in unhappy customers and, consequently, a loss of business.

Broadcast Network

As in the case of a node in the broadcast design of most local area networks, when a node on a wide area network **broadcast network** transmits its data, the data is received by all the other nodes. This form of wide area network is, at the moment, relatively rare. Some systems, however, are in existence. These systems use radio frequencies to broadcast data to all workstations and typically operate as radio broadcast networks in rural areas or in areas where there are many islands surrounded by large bodies of water. Some of the new wireless Internet access services such as Wi-Max (Chapter Three) are also based on a broadcast network, but they are more often considered to be metropolitan area networks than wide area networks. Because broadcast networks are not as common as circuit-switched and packet-switched networks, the remaining discussions in this chapter will not include them.

In summary, the physical design of a wide area network, or its network cloud, has three basic forms: circuit-switched, packet-switched (datagram and virtual circuit), and broadcast. The characteristics of these forms are summarized in Table 9-1. Note how circuit-switched and virtual circuit networks require path setup time and cannot dynamically reroute packets should a network problem occur. It is also worth noting from Table 9-1 that the circuit-switched network was designed primarily for voice signals and is the only network that offers a dedicated path.

Table 9-1
Summary of network cloud characteristics

Characteristic	Circuit-Switched	Datagram Packet-Switched	Virtual Circuit Packet-Switched	Broadcast
Path setup time?	Yes	No	Yes	No
Routing decision for each packet?	No	Yes	No	Typically no routing
Dedicated path?	Yes	No	No	No
Can dynamically reroute if problems occur?	No	Yes	No	Typically no routing
Connection dedicated to your transfer only?	Yes	No	No	No

Having examined the physical design of wide area networks, let us now turn our attention to their logical design.

Connection-oriented vs. connectionless network applications

The network cloud of a wide area network is the physical infrastructure and thus consists of nodes (routers or telephone switches) and various types of interconnecting media. What about the logical entity that operates over this physical infrastructure? This logical entity often takes the form of a software application. For example, if you are using an e-mail application to send a message to a friend across the country, the e-mail application is the logical entity that uses the network's physical infrastructure or cloud to deliver the message. Many different types of applications are found on wide area networks, including e-mail, Web browsing, and other commercial applications. Let us categorize all the network applications, or logical entities, into two basic categories: connection-oriented applications and connectionless applications.

A connection-oriented network application, such as the one that performs a file transfer using FTP, provides some guarantee that information traveling through the network will not be lost and that the information packets will be delivered to the intended receiver in the same order in which they were transmitted. Thus, a connection-oriented network application provides what is called a reliable service. To provide a reliable service, the network requires that a logical connection be established between the two endpoints. If necessary, connection negotiation is performed to help establish this connection. For example, consider the following scenario: A bank wants to transfer a large sum of money electronically to a second bank. The first bank creates a connection with the second bank. As part of establishing this connection, the two banks agree to transfer the funds using data encryption. After the first bank sends the transfer request, the second bank checks the request for accuracy and returns an acknowledgment to the first bank. The first bank will wait until the acknowledgment arrives before doing anything else. All packets transferred during this period are part of this connection and are acknowledged for accuracy. If this is the only electronic transfer, the first bank will say goodbye, and the second bank will acknowledge the goodbye. Note that the *type* of sub-network used in this transfer process is not an immediately relevant issue. The network could have been circuit-switched or packet-switched. All the application required was that a reliable connection be used to transfer the funds. You can perform online banking requests (that is, use a connection-oriented network application) over a local area network (a packet-switched network) at work or school, just as you can perform them over a dial-up telephone connection (a circuit-switched network) from home.

A connectionless network application does not require a logical connection to be made before the transfer of data. Thus, a connectionless application does not guarantee the delivery of any information or data. Data may be lost, delayed, or even duplicated. No connection establishing or terminating procedures are followed, because there is no need to create a connection. Each packet is sent as a single entity and not as part of a network connection.

A common example of a connectionless network application is DNS, a program that converts a URL, such as *www.cs.depaul.edu*, into an IP address. When you request a Web page using its URL, no connection is created between you and the DNS system. You simply click the browser button, and DNS converts the URL of the request into an IP address and sends the request along. Connectionless applications do not negotiate a connection, and the transfer of data is rarely, if ever, acknowledged. Additionally, if you send a second URL request, it has no relationship (network-wise) to the first. As in the case of a connection-oriented network application, the underlying network cloud of a connectionless application is, again, not really an issue. It can be either a circuit-switched network or a packet-switched network. You can send a URL request from work or school over a local area network (packet-switched network), just the same as you can from home over a dial-up connection (circuit-switched network).

Another good example that illustrates the difference between connection-oriented and connectionless networks is the relationship between the telephone system and the postal service. When you call someone on the telephone, that person will answer the telephone if he or she is available. Once the telephone has been answered, a connection is established. The conversation follows, and when one person or the other has finished with the conversation, some sort of ending statement is issued, both parties hang up, and the connection is terminated (Figure 9-8). Thus, when you use the telephone system, you are using a connection-oriented network that provides a reliable service.

Figure 9-8
Connection-oriented telephone call

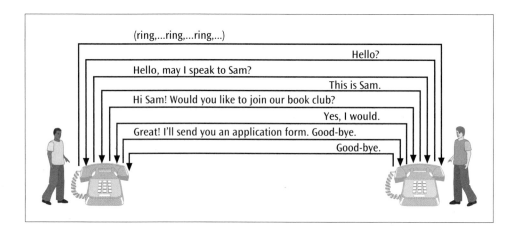

In contrast, if you send a standard letter to someone through the U.S. Postal Service, it will most likely be delivered, but there is no guarantee. There is also no guarantee of when it will arrive. If the letter is lost, you will not know until some time passes and you begin to think, "I haven't heard from Kathleen yet; did she receive my letter?" The postal service, then, is similar to a connectionless network that provides (no offense intended) an unreliable service. The unreliable service offered by the connectionless postal "network" requires that you take additional actions if you want to ensure that the letter is delivered as intended. Furthermore, a postal patron never knows for sure which route a letter will take through the postal network. Under some circumstances, mail is transferred by truck; under other circumstances, it is transferred by airplane (Figure 9-9).

Figure 9-9
Connectionless postal network

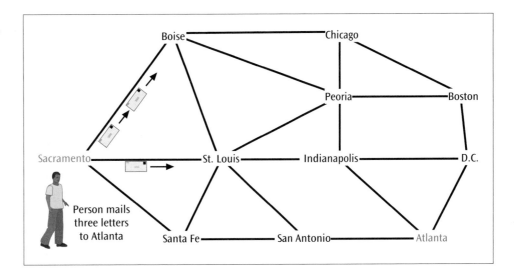

Routing

Recall that a wide area network's underlying network cloud consists of multiple nodes, each with multiple possible connections to other nodes within the network. Each node is a router that accepts an input packet, examines the destination address of the packet, and forwards the packet onto a particular communications line. In the case of multiple-linked nodes, there may be one or more paths into a node as well as one or more paths out of a node. If most of the nodes in the network have multiple inputs and outputs, numerous routes from a source node to a destination node may exist. Examining Figure 9-10, you can see a number of routes are possible between Node A and Node G: A-B-G, A-D-G, A-B-E-G, and A-B-E-D-G, to list a few.

Figure 9-10
A seven-node network showing multiple routes between nodes

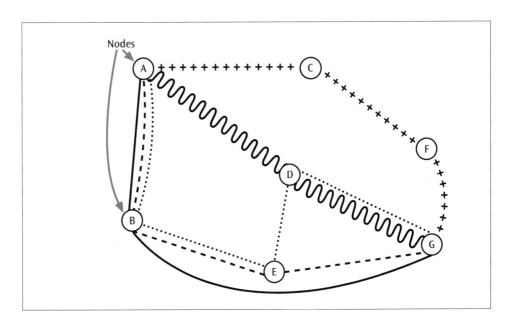

How is routing through a wide area network accomplished? Consider the Internet as an example: It is a massive collection of networks, routers, and communications lines (various types of telephone lines). When a data packet enters a router, the router examines the IP address encapsulated in the network layer of the packet and determines where the packet should go next. When there are multiple routes through a network such as the Internet, how is any one particular route selected? Although routing on the Internet is fairly complex, it is possible to examine the basic routing techniques that all types of wide area networks employ. But keep in mind that a wide area network does not use only one form of routing. The routing algorithms used within the Internet, for example, are actually combinations of several types of basic routing techniques.

To begin to understand the often complex issue of routing, it is helpful to think of the network cloud as a graph consisting of nodes (computers, routers, or telephone switches) and edges (the communications links between the nodes), as shown in Figure 9-11. In this network graph, the edge between each pair of nodes can be assigned a weight or associated cost, as has been done in Figure 9-11, to form a structure called a weighted network graph.

You can assign many meanings to the weights in a weighted network graph. For example, a weight can correspond to the dollar cost of using the communications link between two nodes. A weight could also represent the time-delay cost associated with transmitting data on that link between the source and destination nodes. Another factor that is commonly represented as a weight is the size of the queue that has backed up while waiting for a packet to be transmitted onto a link. Each of these weights may be useful for determining a route through a network.

Figure 9-11
A simple example of a network graph

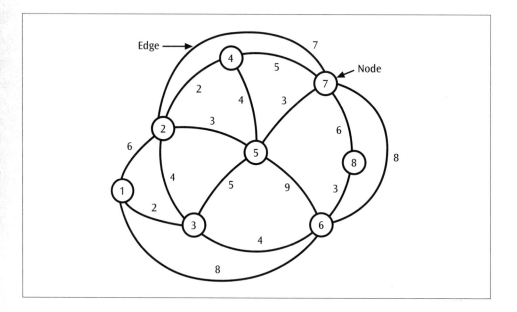

Once you consider the network cloud as a graph and assign weights to the paths between nodes, you can develop an algorithm for traversing the network. There are, in fact, many algorithms for selecting a route through a network. Often, algorithms strive for an optimal route through a network, but there are different ways to define "optimal." For example, one algorithm might define the optimal route as one that generates the least dollar cost. Another algorithm might consider the path with the least time delay to be the optimal route. A third algorithm might define the optimal route as the one having the smallest queue lengths at the nodes along the path.

Some algorithms use criteria other than optimality. For example, they might try to balance the network load over a number of different paths. Another kind of algorithm might favor one type of traffic over another, for example, real-time traffic over non-real-time traffic. A third type may try to remain robust, responding to changing network demands as nodes and communications links fail or become congested. Yet a fourth kind may try to remain static and not switch between possible paths. As you can see, routing is a complicated topic. To get a feel for routing in wide area networks, let us examine several of the most commonly used routing algorithms (Dijkstra's least-cost algorithm and flooding) and several techniques for managing routing information (centralized vs. distributed routing; adaptive vs. fixed routing). Most wide area networks use a combination of these routing techniques to achieve a routing algorithm that is fair, efficient, and robust, but at the same time stable.

Dijkstra's least-cost algorithm

One possible method for selecting a route through a network is to choose a route that minimizes the sum of the costs of all the communications paths along that route. A classic algorithm that calculates a least-cost path through a network is **Dijkstra's least-cost algorithm**. This algorithm is executed by each node, and the results are stored at the node and sometimes shared with other nodes. Because this calculation is time-consuming, it is done only on a periodic basis or when something in the network changes—for example, when there is a connection or node failure.

Let us look at an example. Figure 9-12 depicts the same network cloud shown in Figure 9-10, but has modified it to include an arbitrary set of associated costs on each link. Path A-B-G has a cost of 9 (2 + 7), A-D-G has a cost of 10 (5 + 5), and path A-B-E-G has a cost of 8 (2 + 4 + 2). To ensure that you find the least-cost route, you need to use a procedure that calculates the cost of every possible route, starting from a given node. Although the human eye can quickly pick out a path through a network graph, there are a number of drawbacks to "eyeballing" the data to find a solution:

- ▶ You can easily miss one or more paths.
- ▶ You may not find the least-cost path.
- ▶ Wide area networks are never as simple as the network in Figure 9-12; thus, eyeballing the data could never be sustained as a reliable, long-term procedure.

Most wide area networks use some form of Dijkstra's algorithm to determine a least-cost route through a network, whether that cost is a measure of time or money. To learn more about how this algorithm works, see the article titled "Understanding Dijkstra's Algorithm" on the student online companion to this text or at the author's Web site: *facweb.cs.depaul.edu/cwhite*.

Figure 9-12
*Network with costs
associated with
each link*

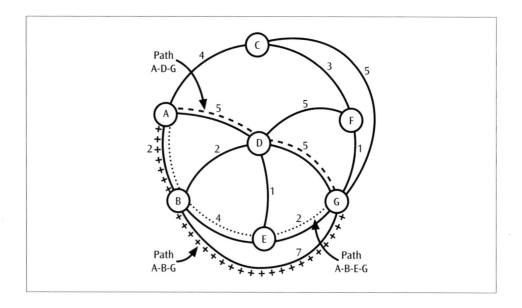

Flooding

Compared to Dijkstra's least-cost routing algorithm, the flooding technique seems simple. Flooding states that each node takes the incoming packet and retransmits it onto every outgoing link. For example, assume a packet originates at Node A, as shown in Figure 9-13. Node A simply transmits a copy of the packet on every one of its outgoing links. Thus, a copy of the packet (the first copy, or copy 1) is sent to Nodes B, C, and D.

Figure 9-13
*Network with flooding,
starting from Node A*

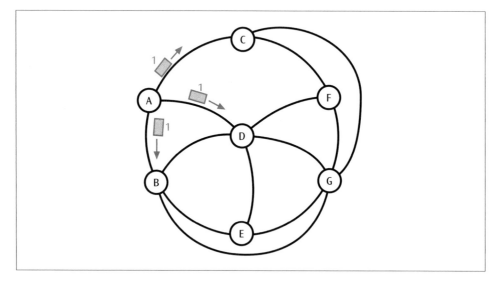

When the packet arrives at Node B, B simply transmits a copy of the packet to each of its outgoing nodes (D, E, and G). Likewise, Node C will transmit a copy of the packet to each of its outgoing nodes (F and G). Node D will also transmit a copy of the packet to each of its outgoing nodes (B, E, F, and G). Figure 9-14 shows the second copies of the packets leaving Nodes B, C, and D. It should not take you long to realize that the network will very quickly be flooded with copies of the original data packet.

Figure 9-14
Flooding has continued to Nodes B, C, and D

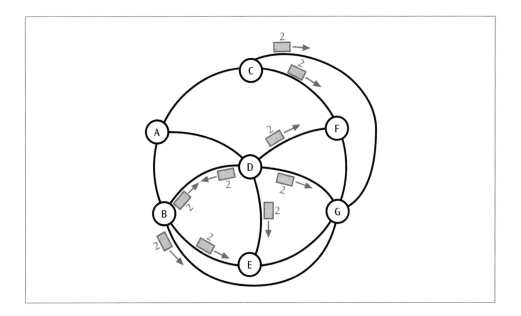

To prevent the quantity of copied packets from becoming overwhelming, two common-sense rules can be established. First, a node need not send a copy of the packet back to the link from which the packet just arrived. Thus, when Node A sends a copy of the packet to Node C, C does not need to send a copy immediately back to A.

Second, a network limit, called the hop limit, can be placed on how many times any packet is copied. Each time a packet is copied, a counter associated with the packet increases by one. This counter is called the hop count. When the hop count equals the network hop limit, this particular packet will not be copied anymore. For example, suppose the network has a hop limit of 3. When Node A first sends copies to B, C, and D, each of the three copies has a hop count of 1. When the packet arrives at Node C, copies with hop counts of 2 will be sent to F and G (Figure 9-14). When the copy arrives at Node F, two copies with hop counts of 3 will be transmitted to D and G, and the packets that arrive at D and G with hop counts of 3 will go no farther.

Although flooding may seem like a strange way to route a packet through a network, the procedure does have its merits. If a copy of a packet *must* be sent to a particular node, flooding *will* get it there, assuming, of course, there is at least one active link to the receiving node, and the network hop limit is not set at too small a value. Flooding is also advantageous when a copy of a packet needs to get to all nodes—for example, when emergency information or network initialization information is sent. The major disadvantage of flooding is the large number of copied packets distributed throughout the network.

Centralized vs. distributed routing

Centralized and distributed routing are not so much algorithms for routing data packets through a network as they are techniques for providing routing information. Centralized routing involves storing all the routing information at one central location (Table 9-2). Whenever any router in a network needs routing information, this central location is queried and the routing results are returned. For example, if a packet arrives at Node A and is destined for Node G, Table 9-2 conveys that it should be sent to Node B next. These days, centralized routing is rarely if ever used in wide area networks. Instead, most wide area networks now employ distributed routing. Distributed routing is a technique that uses a routing algorithm, such as a least-cost algorithm, to generate routing information and dictates that this information be stored (in the form of routing tables) at distributed locations—typically, routers—within the network. When a data packet enters the network at node *x*, that node consults its own routing table to determine the next node that should receive the packet. In this scheme, each node needs routing information only for its own locale. For example, the routing table for Node C would be Table 9-3. From this table, you can see that if a data packet arrives at Node C and is destined for Node G, the packet should be sent to Node F next. (Once the packet arrives at Node F, Node F's routing table will have to be examined for the next hop in the path.)

Table 9-2
Routing table kept at a centralized network site

		Destination Node						
		A	**B**	**C**	**D**	**E**	**F**	**G**
	A	-	B	C	B	B	C	B
	B	A	-	A	D	D	D	D
	C	A	A	-	A	F	F	F
Origination Node	**D**	B	B	F	-	E	E	E
	E	D	D	G	D	-	G	G
	F	C	G	C	G	G	-	G
	G	E	E	F	E	E	F	-

Details

RIP

Let us consider the following example of how Routing Information Protocol (RIP) works. Suppose that Router A has connections to four networks (123, 234, 345, and 789) and has the following current routing table:

Network	Hop Cost	Next Router
123	8	B
234	5	C
345	6	C
789	10	D

Table 9-4 *Current routing table for Router A*

Now suppose Router D sends out the following routing information (note that Router D did not send Next Router information, because each router will determine that information for itself):

Network	Hop Cost
123	4
345	5
567	7
789	10

Table 9-5 *Routing information sent by Router D*

Table 9-3
Local routing table for Node C

		Destination Node						
		A	B	C	D	E	F	G
Origination Node	C	A	A	-	A	F	F	F

One of the primary advantages of distributed routing is the fact that no single node (or central router) is responsible for maintaining all routing information. This situation confers a number of benefits. First, if any node crashes, it will probably not disable the entire network. Second, a node will not need to send a request to a central router, because each node has its own table. By (1) eliminating the request packets that, in centralized routing, are transmitted to the node holding the routing table and (2) eliminating the resulting packets leaving the routing table node, distributed routing does away with much of the data traffic on a network that a centralized routing scheme can generate.

One disadvantage of distributed routing is related to the problems that arise if the routing tables need to be updated. When all the routing information is in one place (that is, in a single table), it is simple to make updates. When routing information is scattered throughout a network, getting the appropriate routing information to each node is a complex problem. Another consequence of storing routing information at multiple locations is that at any given point in time there may be one or more routing tables that contain old or incorrect information.

Router A will look at each entry in Router D's table and make the following decisions:

1. Router D says Network 123 is 4 hops away (from Router D). Because Router D is 1 hop away from Router A, Network 123 is actually 5 hops away from Router A. That is better than the current entry of 8 hops in Router A's table, so Router A will update the entry for Network 123.
2. Router D says Network 345 is 6 hops away (5 hops from Router D plus the 1 hop between Router A and Router D). That is currently the same hop count as shown in Router A's table for Network 345, so Router A will not update its table.
3. Router D says Network 567 is 8 hops away (7 hops from Router D plus the 1 hop between Router A and Router D). Because Router A has no information about Network 567, Router A will add this entry to its table. And because the information is coming from Router D, Router A's Next Router entry for network 567 is set to D.
4. Router D says Network 789 is 11 hops away (10 hops from Router D plus the 1 hop between Router A and

Router D), which is worse than the value in Router A's table. Nothing is changed.

Router A's updated routing table will thus look like the following:

Network	Hop Cost	Next Router
123	5	D
234	5	C
345	6	C
567	8	D
789	10	D

Table 9-6 *Updated routing table for Router A*

Router A now has an updated routing table (based upon the information sent from Router D). It has found a shorter path to Network 123 through Router D and has learned that Network 567 is 8 hops away from itself.

Adaptive vs. fixed routing

Centralized and distributed routing are methods for sending routing information. They are typically used in conjunction with some form of least-cost routing algorithm. Regardless of whether routing information is centralized or distributed, when networks change, routing information needs to change too.

When routing tables adapt to network changes, the routing system is called adaptive. Adaptive routing is a dynamic technique in which routing tables react to network fluctuations, such as congestion and node/link failure. When a problem occurs in a network with adaptive routing, the appropriate information is transmitted to the routing tables, and new routes that avoid the problem areas are created. Adaptive routing raises some questions and issues: How often should information be shared, and how often should routing tables be updated? How much additional traffic is generated by messages transmitting routing information?

Unfortunately, adaptive routing can add to network congestion. Each time the network experiences a change in congestion, information about this change is transmitted to one or more nodes. The transmission of this information adds to the congestion, possibly making it worse. In addition, if the network reacts too quickly to a congestion problem and reroutes all traffic onto a different path, it can create congestion problems in a different area. The network might then detect the congestion in this different area and possibly reroute all traffic back toward the first problem area. This back and forth rerouting produces a yo-yo effect, affecting network stability and decreasing efficiency.

The opposite of adaptive routing is fixed routing. With fixed routing, routing tables are created once, typically when the network is installed, and then never updated again. While this method is simple and eliminates the need for routers to talk to one another (thus avoiding additional traffic), it can also yield networks with out-of-date information and thus inefficient or slow routing. It is debatable if fixed routing exists anymore on wide area networks (or for that matter, on any other type of network).

Details ▶

OSPF

In order for OSPF to handle routing in an efficient manner, a system (a network or collection of networks) is divided into areas. An area may be a collection of networks, routers, and hosts. All routers inside an area flood all the other routers in the same area with their information. Special routers called area border routers interconnect two or more areas. One area is selected as the backbone, and all other areas must connect to this backbone area. The routers inside the backbone are called backbone routers.

Unlike RIP, in which the cost associated with moving through a network is simply the number of router hops, OSPF can use any metric assigned by a network administrator (although using a timestamp as a metric is common). Also, each router floods other routers in its area with its information only when something in the network changes. Thus, compared to RIP, which transmits every 30 seconds, OSPF transmits fewer packets of routing information.

Link state advertisements are the packets of routing information that are passed between routers. More precisely, the routing information is similar to a snapshot of the current state of all the router links (to other networks) that are connected to that router. According to OSPF, there are four types of router links. The first type is a point-to-point link, which is a single connection between two routers. The second, a transient link, is a network with multiple routers attached to it. The third type of link is a stub link, in which there is only one router on the network. The fourth type of link is a virtual link. A virtual link is one that is created by an administrator when a link between two routers fails. More than likely, the virtual link passes through multiple routers. Once a router collects all the link state advertisements, it uses Dijkstra's least-cost algorithm to compute the optimal route to all other networks. Thus, OSPF is a dynamic protocol that can adjust node routing tables as new information becomes available.

Routing examples

The Internet is covered in detail in Chapter Ten, but let us take a few minutes to examine two of the routing algorithms that have been used within the Internet over the years. By examining these algorithms, you will see how a real-life routing protocol is actually composed of many of the algorithms and techniques introduced in the preceding sections.

The first routing algorithm used within the Internet (when it was still called ARPANET) was called a distance vector routing algorithm. The distance vector routing algorithm was an *adaptive* algorithm in which each node maintained a routing table called a vector. Because each node maintained its own routing table, the routing algorithm was also a *distributed* algorithm. Every 30 seconds, each node exchanged its vector with its neighbors. When all its neighbors' vectors came in, a node would update its own vector with the *least-cost* values of all the neighbors. This adaptive, distributed algorithm had the formal name Routing Information Protocol (RIP). This protocol had two unfortunate side effects. The first was that good news (routing information that indicated a shorter path) moved relatively slowly through the network, one router at a time. The second side effect was that bad news, such as a router or link failure, very often moved even more slowly through the network.

The next routing protocol that was implemented on the Internet (in approximately 1979) was called a link state routing algorithm. Link state routing essentially involves four steps. The first step is to measure the delay or cost to each neighboring router. For example, each router can send out a special echo packet that gets bounced back almost immediately. If a timestamp were placed on the packet as it left and again as it returned, the router would know the transfer time to and from a neighboring router. The second step is to construct a link state packet containing all this timing information. The third step is to distribute the link state packets via *flooding*. In addition to using flooding, the link state routing algorithm is a *distributed algorithm*. The fourth and final step is to compute new routes based on the updated information. Once a router collects a full set of link state packets from its neighbors, it creates its routing table, usually using Dijkstra's least-cost algorithm. Open Shortest Path First (OSPF) protocol is a link state algorithm and is still used today by many Internet routers. To learn more about this algorithm, check out the Details box titled "OSPF".

Now that we know how data is routed in a network, we need to look at a common side effect of routing too much data at a time—congestion.

Network Congestion

When a network or a part of a network becomes so saturated with data packets that packet transfer is noticeably impeded, network congestion occurs. Congestion may be a result of a short-term problem, such as a temporary link or node failure, or it may be a result of a longer-term problem, such as inadequate planning for future traffic needs or poorly created routing tables and routing algorithms. As with so many things in life, the network is only as strong as its weakest link. If network designers could properly plan for the future, network congestion might exist only in rare instances. But the computer industry, like most other industries, is filled with examples of failures to plan adequately for the future. Computer networks are going to experience congestion, and no amount of planning can avoid this situation. Thus, it is important to consider effective congestion-avoidance and congestion-handling techniques.

The problems associated with network congestion

Networks experience congestion for many reasons. A network failure—either failure on a communications link between nodes, or the failure of the node itself—may lead to network congestion. If the network cannot quickly detect the point of failure and dynamically route around this point, it may experience a wide range of congestion problems, from a small slowdown on an individual link to total network collapse. Even if the network were to begin the rerouting process, it might still experience congestion because one less network path would be available. But communications link and node failures are not the only causes of congestion. Insufficient buffer space at a node in a network can also cause network congestion. It is not uncommon to have hundreds or even millions of packets arriving at a network node each second. If the node cannot process the packets quickly enough, incoming packets will begin to accumulate in a buffer space. When packets sit in a buffer for an appreciable amount of time, network throughput begins to suffer. If adaptive routing is employed, this congestion can be recognized, and updated routing tables can be sent to the appropriate nodes (or to a central routing facility). But changing routing tables to deflect congestion might provide only a temporary fix, if any fix at all. What is needed is a more permanent solution. Two possible more permanent solutions would be increasing the speed of the node processor responsible for processing the incoming data packets and increasing the amount of buffer space in the node. Unfortunately, both of these solutions may take a large amount of time and money to implement. Perhaps less costly alternatives are possible.

What happens if the buffer space is completely full and a node cannot accept any additional packets? In many systems, packets that arrive after the buffer space is full are discarded. Although this is not a very elegant solution, it momentarily solves the problem of too many packets. Unfortunately, this is like bad medicine—it treats the symptoms, but not the disease. What is needed is a solution that reacts quickly to network congestion and addresses the real problem—too many packets. Let us examine several possible solutions that have been suggested or tried in the past several years.

Possible solutions to congestion

Solutions to network congestion generally fall into two different categories. The first category contains solutions that are implemented after congestion has occurred. The second category of solutions contains techniques that attempt to avoid congestion before it happens. Let us assume that congestion has already occurred and examine the first category of solutions. Most, if not all, of the solutions in the first category involve telling the transmitting station to slow down or stop its transmission of packets into the network. For example, if an application on the network suddenly realizes that its packets are being discarded, the application may inform the transmitting station to slow down until further notice. Because the application is simply observing its own throughput and not relying on any special types of signals coming from the network, this is called implicit congestion control.

Often, however, the network itself sends one or more signals to a transmitting station, informing the station to slow down or stop its insertion of packets into the network. When the network signals the transmitting station to slow down, this is called explicit congestion control. For an example, consider frame relay (discussed in more detail in Chapter Eleven), which is designed to use two explicit congestion control techniques: forward explicit congestion notification and backward explicit congestion notification. With forward explicit congestion notification (FECN), when a frame relay router experiences congestion, it sends a congestion signal (inside the data frames) forward to the destination station, which in turn tells the originating station to slow down the transfer of data. With backward explicit congestion notification (BECN), the frame relay router experiencing congestion sends a signal back to the originating station, which then slows down its transmission.

Other congestion control methods are based on simpler techniques such as the flow control methods introduced in Chapter Six. Flow control at the data link layer allows two adjacent nodes to control the amount of traffic passing between them. When the buffer space of a node becomes full, the receiving node informs the sending node to slow or stop transmission until further notice. We will see in Chapter Eleven how the Internet handles end-to-end flow control using the TCP software. As in node-to-node flow control at the data link layer, the receiver in end-to-end flow control returns a value to the sender informing the sender to either maintain the current speed, slow down, or stop the transmission of packets.

Now let us consider the second category of congestion control techniques—those that try to prevent congestion before it even happens. The old saying "An ounce of prevention is worth a pound of cure" certainly applies to computer networks. Avoiding congestion not only leads to fewer lost and delayed packets but also to many happier customers. One possible solution to controlling the flow of packets between two nodes is buffer preallocation. In buffer preallocation, before one node sends a series of n packets to another node, the sending node inquires in advance whether the receiving node has enough buffer space for the n packets. If the receiving node has enough buffer space, it sets aside the n buffers and informs the sending node to begin transmission. Although this scheme generally works, it does introduce extra message passing, additional delays, and possible wasted buffer space if all n packets are not sent. But the alternative—discarding packets due to insufficient buffer space—is worse.

More recent network technologies such as Asynchronous Transfer Mode (ATM), which is discussed in more detail in Chapter Eleven, approach network congestion in a very serious fashion. Because ATM networks transfer data at very high speeds, congestion can occur rather quickly and be devastating. Thus, it is extremely important to keep congestion from occurring in the first place.

ATM uses a congestion avoidance technique that appears to work quite well. This technique—connection admission control—avoids congestion by requiring users to negotiate with the network regarding how much traffic they will be sending, or what resources the network must provide to satisfy the user's needs before the user sends any data. If the network cannot satisfy the user's demands, the user connection is denied. In negotiating the viability of a connection, users and networks must resolve questions such as the following:

  ► What is the average (or constant) bit rate at which a user will transmit?
  ► What is an average peak bit rate at which a user might transmit?
  ► At what rate might a network start discarding packets in the event of congestion?
  ► What is the average bit rate that the network can provide?
  ► What is the average peak bit rate that the network can provide?

Many networks relate these issues to quality of service (QoS), a concept in which a network user and the network agree on a particular level of service (acceptable guidelines for the proper transfer of data). For example, a user who requires a very fast, real-time connection to support live action video will negotiate with the network for a particular quality of service. If the network can provide this level of quality, a contract is agreed on, the user is charged accordingly, and a connection is established. If a second user requires a slower connection for e-mail, a different level of quality is agreed on, and this connection is established. Very often these agreements between service provider and a service user are formalized in a service level agreement, a legally binding, written document, that can include service parameters offered in the service, various types of service/support options, incentives if the service levels are exceeded, and penalties if service levels are not met.

Unfortunately, only one network technology successfully supports connection admission control and quality of service, and that is Asynchronous Transfer Mode. Most ATM systems can provide a range of services from high-speed constant bit rate down to slower-speed bit rate on demand. CSMA/CD, the most popular LAN protocol, does not provide different levels of service, nor does the Internet, with its TCP and IP protocols. (As we will see in the next chapter, however, the newer Internet protocol, IPv6, does include some form of labeling to support customer-specified connections.) We hope that in the future we will have new or updated protocols that contain some form of quality of service.

Now that we have covered all the necessary technical details of WANs, let us look at a business example.

WANs In Action: Making Internet Connections ▶

Accessing the World Wide Web via the Internet is so commonplace these days, it is easy to forget that not too long ago it was considered to be a remarkable technological achievement. Many people have a personal computer at home and connect to the Internet via an Internet service provider (ISP) and either a standard telephone line, a cable modem, or a DSL service. If you are a student enrolled in a course at a college, you are likely to have Internet access through the college and its campus computer laboratories. Likewise, many business employees have Internet access over a corporate local area network, which allows them to download Web pages. Although the end result is the same—someone gets to view Web pages—the underlying types of network clouds and connections can vary, depending on whether you are browsing the Web from home, from school, or from work. Let us examine the different types of network clouds that support applications commonly found on the Internet.

A home-to-Internet connection

Let us consider the first scenario, in which you connect to the Internet from home. Although a majority of the users that connect to the Internet from home now use either DSL or a cable modem, there are still users who connect to the Internet using a dial-up modem and a standard telephone service. To initiate dial-up service, the communications software on your personal computer dials a (one hopes) local telephone number, and the local telephone system creates a dedicated circuit-switched connection between your modem and your ISP (Figure 9-15).

Figure 9-15
*User at home using
a dial-up telephone
line (circuit-switched
network) to run a
connection-oriented
application
(Web browser)*

When a user running a browser clicks a link, HTTP (Hypertext Transfer Protocol) creates one or more connections between the user's browser and the server that holds the requested Web page. Using these connections, the server transmits various pieces of the requested Web page, such as text, graphics, links, JavaScript, or Java applications. When the last piece of the requested Web page is transmitted, the last connection is dissolved. This is basically how a connection-oriented application runs over a circuit-switched network.

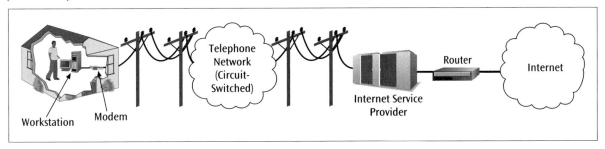

Suppose that while browsing the Web, the user enters a URL into her browser. The DNS system has to convert the URL into an IP address. The DNS system, as you may recall, is connectionless. DNS transmits the URL request without first informing the recipient that a message is coming, nor does the DNS system wait for a reply (an acknowledgement of the request). When the URL request travels over the telephone line between a house and an Internet service provider, you have a connectionless application running over a circuit-switched network. When the URL request is passed from the Internet service provider onto the packet-switched networks of the Internet, you have a connectionless application running over a packet-switched network.

A work-to-Internet connection

Now let us assume you are at work and using the company computer system. (This scenario would be very similar to one in which you were at school using the campus Internet connection.) More than likely your workstation is connected to a local area network, which connects to a router, which then connects to an ISP over a leased high-speed telephone line, such as a T-1 (Figure 9-16).

Figure 9-16
*User at work using a
local area network to
access the Internet*

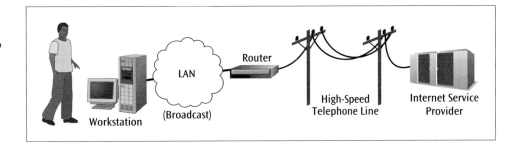

For simplicity, let us suppose your company's local area network is hub based. Recall that a hub-based local area network is a broadcast network. It does not use circuit switching or packet switching. Instead, the data (the Web page request or URL request) is packaged into a frame and broadcast over a medium to a router. (Note, however, that if the local area network were switch based, it would act more like a packet-switched wide area network.) The router recognizes the destination address and determines that the frame has to leave the company network and traverse the high-speed telephone line to the Internet service provider. In this case, the high-speed telephone line happens to be a T-1 line. The data traverses the T-1 and arrives at the Internet service provider. A router within the Internet service provider looks at the data packet and determines that it has to be sent over a high-speed line to another location somewhere in the country. We do not know what kind of high-speed line or service will carry our data, so for now let us simply say that it is some type of packet-switched network. Thus, whether the user is browsing the Web (a connection-oriented network application) or requesting a URL be converted to an IP address (a connectionless network application), the application is traveling first over a broadcast network, next over a T-1 connection, and then over a packet-switched network on its way through the Internet.

These examples show how both connection-oriented network applications and connectionless network applications can operate over packet-switched networks, broadcast networks, and circuit-switched networks.

◆ ◆

SUMMARY

- A network that expands into a metropolitan area and exhibits high data rates, high reliability, and low data loss is called a metropolitan area network (MAN). Many of the same technologies and communications protocols found in local area networks (and wide area networks) are used to create metropolitan area networks.

- Metropolitan area networks are based upon either SONET or Ethernet backbones. SONET backbones consist of fiber-optic rings, while Ethernet backbones are mesh networks.

- A Metro Ethernet service provides an Ethernet interface to a business and can transfer data at high rates over metropolitan areas.

- Wide area networks cover larger geographic areas than both local area networks and metropolitan area networks, and they are based on potentially different physical sub-networks, or network clouds: circuit-switched, packet-switched, and broadcast.

- A circuit-switched network creates a dedicated circuit between sender and receiver, and all data passes over this circuit.

- A packet-switched network transmits fixed-sized packages of data called packets. Packet-switched networks fall into two subcategories: datagram networks and virtual circuit networks. The datagram packet-switched network transmits each packet independently of every other packet. Each packet is considered a single entity and is not part of a larger grouping of packets. The virtual circuit packet-switched network creates a virtual circuit using routing tables and transmits all packets belonging to a particular connection over this virtual circuit.

- A broadcast network transmits its data to all workstations at the same time. Broadcast networks are more often used in local area networks than in wide area networks.

- The network application that runs over a network can be either connection oriented or connectionless. A connection-oriented network application provides some guarantee that the information traveling through the network will not be lost and that the information packets will be delivered to the intended receiver in the same order in which they were transmitted. To provide this service, a logical connection is established before any data transfer takes place. A connectionless network application does not require a logical connection to be made before data is transferred.

▶ Selecting the optimal route for the transfer of a data packet through a network is a common service of many networks. This optimal route is obtained by combining two or more of the many routing algorithms and techniques available today.

▶ One possible way to select an optimal route through a network is to choose a path whose total path costs have the smallest value. This technique is based on Dijkstra's least-cost algorithm and is a common method for determining the optimal route.

▶ Flooding is a routing technique that requires each node to take the incoming packet and retransmit it onto every outgoing link. If a copy of a packet must be sent to a particular node, flooding will get it there, but it will also create a very large number of copied packets that are distributed throughout the network.

▶ Centralized routing is a technique for providing routing information that dictates that the routing information, which is generated by a method such as the least-cost algorithm, be stored at a central location within the network.

▶ Another technique for providing routing information, distributed routing allows each node to maintain its own routing table.

▶ Adaptive routing allows a network to establish routing tables that can change frequently, as network conditions change.

▶ RIP and OSPF are two important routing protocols found within the Internet.

▶ When a network or a part of a network becomes so saturated with data packets that packet transfer is noticeably impeded, network congestion has occurred. Congestion may be the result of too much traffic, network node failure, network link failure, or insufficient nodal buffer space. Remedies for network congestion include implicit congestion control, explicit congestion control, flow control, preallocation of nodal buffers, and connection admission control parameters.

▶ Quality of service (QoS) parameters can be used by network users and the service providers to establish acceptable guidelines for the proper transfer of data. These guidelines can cover transmission speed, level of errors, and overall network throughput.

KEY TERMS

adaptive routing
backward explicit congestion notification (BECN)
bandwidth profile
broadcast network
buffer preallocation
centralized routing
circuit-switched network
connection-oriented network application
connection admission control
connectionless network application
datagram
Dijkstra's least-cost algorithm

distributed routing
explicit congestion control
failover
failover time
flooding
forward explicit congestion notification (FECN)
hop count
hop limit
implicit congestion control
Metro Ethernet
metropolitan area network (MAN)
network cloud
network congestion

node
Open Shortest Path First (OSPF) protocol
packet-switched network
reliable service
Routing Information Protocol (RIP)
service level agreement
station
sub-network
virtual circuit
wide area network (WAN)
weighted network graph

REVIEW QUESTIONS

1. What are the main differences between a local area network and a wide area network?

2. How does a metropolitan area network differ from a wide area network? How are they similar?

3. What is meant by failover time?

4. What are the two types of technology that support metropolitan area networks?

5. What is Metro Ethernet, and how does it relate to metropolitan area networks?

6. What is a sub-network, or network cloud, and how does it differ from a wide area network?

7. What is the difference between a station and a node?

8. What are the main characteristics of a circuit-switched network? What are its advantages and disadvantages?

9. What are the main characteristics of a datagram packet-switched network? What are its advantages and disadvantages?

10. What are the main characteristics of a virtual circuit packet-switched network? What are its advantages and disadvantages?

11. How does a connectionless network application differ from a connection-oriented network application?

12. Is a connectionless network application reliable or unreliable? Explain.

13. What are the various combinations of circuit-switched and packet-switched sub-networks and connection-oriented and connectionless network applications?

14. How does a weighted network graph differ from a network graph?

15. For a weighted network graph, how many different definitions of weight can you list?

16. What are the basic goals of Dijkstra's least-cost algorithm?

17. How can flooding be used to transmit a data packet from one end of the network to another?

18. How are the hop count and the hop limit used to control flooding?

19. What are the main advantages and disadvantages of:

    a. centralized routing

    b. distributed routing

    c. adaptive routing

20. What are the differences between RIP and OSPF?

21. What can cause network congestion?

22. How can network congestion be avoided?

23. What does quality of service have to do with network congestion?

EXERCISES

1. State three advantages of a SONET-based metropolitan area network over an Ethernet-based metropolitan area network.

2. State three advantages of an Ethernet-based metropolitan area network over a SONET-based metropolitan area network.

3. Which type of network application requires more elaborate software: connection oriented or connectionless? Explain.

4. Create an analogy similar to the "making a telephone call vs. sending a letter" analogy that demonstrates the differences between connection-oriented and connectionless network applications.

5. Explain the difference between a network node and a network station.

6. Does a datagram network require any setup time before a packet is transmitted? Explain your answer.

7. Does a virtual circuit network require any setup time before a packet is transmitted? If so, when, and how often?

8. List the steps involved in creating, using, and terminating a virtual circuit.

9. You are downloading a file over the Internet. Is the download a connectionless application or a connection-oriented application? Explain.

10. Using the concept of flooding and the graph shown in Figure 9-17, how many packets will be created if a packet originates at Node A and there is a network hop limit of three?

11. How do you determine the hop limit in flooding?

Figure 9-17
Sample weighted network graph for Exercise 10

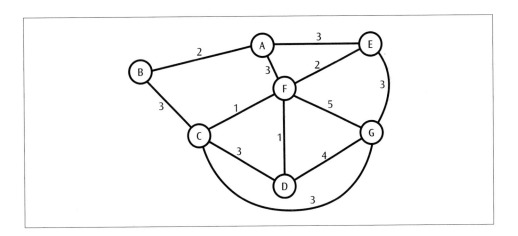

12. Explain how RIP routing forwards bad news slowly.

13. What can be done to protect a centralized routing network if the computer hosting the central routing table fails?

14. What happens in a virtual circuit packet-switched network if a node or communications link along the virtual path fails?

15. Does an Ethernet network support quality of service? Explain.

16. Frame relay supports two kinds of congestion avoidance procedures. Does the destination station learn of the congestion in both procedures?

17. Assume that two stations on a typical ATM network that is transmitting data at a speed of 150 Mbps are at opposite ends of the United States (the propagation time to transmit between the two stations is approximately 46×10^{-3} seconds). Congestion begins to occur at one station. If the congested station wants to send a signal back to the other station to slow down or stop, how many bits will be dumped before the other station slows down or stops its transmission?

THINKING OUTSIDE THE BOX

1 Your company has a number of interconnected local area networks within a single building. Also in the building is a router that connects the LANs to an Internet service provider using a couple of T-1 telephone lines. Your company is considering replacing the T-1s with a connection to a metropolitan area network. Show the interconnection of the LANs, the MAN, and the Internet, assuming the MAN is SONET based. Would the interconnection be different if the MAN were Ethernet based? Explain.

2 A large wide area network covers the United States and has multiple nodes in every state. The node in Denver crashes completely. How do the other nodes in the country find out about the crash if RIP is used? If OSPF is used?

3 Your company is creating a new network application that allows employees to view their pay stubs electronically via the Internet. Should this application be connectionless or connection oriented? Defend your answer, and draw a series of sample screens that a user accessing his or her electronic pay stub might view.

4 One form of congestion avoidance is the permit system, in which a node has to have a permit before it can transmit. Suppose a wide area network is using the permit system to control congestion. What happens if, for some unknown reason, all the permits disappear? How can this event be detected? How can this event be repaired?

5 When we talk about routing in wide area networks, we often introduce the concept of how does a network decide which path particular data traffic will follow? We often mention the words "fairness" and "balancing" when trying to decide network routing and loading conditions. But what if the network was not fair? What if it favored one particular type of traffic, or traffic from a particular type of user? Would that still be fair? Does this have anything to do with "network neutrality"? Explain.

HANDS-ON PROJECTS

1. Given a network with *n* nodes, create a formula that calculates the number of messages transmitted between nodes if one node contains a centralized routing table and each of the other nodes refers to this routing table once every second.

2. Does the Internet use flooding in some situations? Prove your answer.

3. Asynchronous Transfer Mode is popular for its quality of service capabilities. How does it support quality of service?

4. Write a computer program, using the language of your choice, that inputs the data for a weighted network graph and computes the least-cost paths using Dijkstra's algorithm. The details of Dijkstra's algorithm can be found in the student online companion for this text.

5. Using the Dijkstra's algorithm procedure described in the student online companion that accompanies this text and the weighted network graph shown in Figure 9-17, find the least-cost route from Node A to all other nodes.

6. What is/are the current routing algorithm(s) used by the Internet? Are any of the basic algorithms and techniques introduced in this chapter used? Which one(s)?

10

The Internet

ACCORDING TO Ipsos Reid, a Canadian polling company, Canadian teenagers spend more time surfing online than watching television. The study states that Internet use has recently increased by about 46 percent to 12.7 hours a week, up from 8.7 hours in 2002. "Young Canadians Spend More Time on Internet Than Watching TV," August 16, 2005, downloaded from *www.prorev.com*.

Amazon.com, the Seattle-based Internet retailer, reported that its net sales for the first quarter of 2005 were $1.9 billion. Compared to $1.53 billion for the same quarter in 2004, this represents an increase of 24 percent. "Amazon.com announces first quarter financial results," *Internet Ad Sales*, May 2, 2005, downloaded from *www.internetadsales.com*.

If the online auction house eBay were a country, it would be one of the top 20 largest nations in the world, with its 157 million registered users. At any given time, eBay has approximately 55 million items for sale. "UnWired Buyer Allows eBay Users to Bid via Any Mobile Phone," September 19, 2005, downloaded from *www.prnewswire.com*.

With so much money at stake, you have to ask, "If something goes wrong with the Internet, who or what is in charge of fixing it?" The answer, surprisingly, is nobody and nothing. The Internet is managed by a loose confederation of businesses and nonprofit organizations. The players are scattered around the globe and motivated to achieve consensus out of enlightened self-interest. Interestingly, this structure makes the Internet the only nonsovereign global association.

What is this thing called the Internet?

What services are offered on the Internet?

What basic mechanisms keep the Internet working?

Source: Patsuris, P., "Who Is Running This Joint," *Forbes Digital Tool*, *Forbes.com*, November 1998.

Objectives ▶

After reading this chapter, you should be able to:

▶ Discuss the responsibilities of the Internet Protocol (IP) and how IP can be used to create a connection between networks

▶ Discuss the responsibilities of the Transmission Control Protocol (TCP) and how it can be used to create a reliable, end-to-end network connection

▶ Identify the relationships between TCP/IP and the protocols ICMP, UDP, ARP, DHCP, NAT, and tunneling protocols

▶ Cite the basic features of HTML, dynamic HTML, and XML, and describe how these three languages differ from one another

▶ Describe the responsibility of the Domain Name System and how it converts a URL into a dotted decimal IP address

▶ Describe the major Internet applications and services

▶ Discuss the business advantages of the World Wide Web

▶ Recognize that the Internet is constantly evolving and that IPv6 and Internet2 demonstrate that evolution

Introduction ▶

During the late 1960s, a branch of the U.S. government titled the Advanced Research Projects Agency (ARPA) created one of the country's first wide area packet-switched networks, the ARPANET. Select research universities, military bases, and government labs were allowed access to the ARPANET for services such as electronic mail, file transfers, and remote logins.

In 1983, the Department of Defense broke the ARPANET into two similar networks: the original ARPANET and MILNET, for military use only. Although the MILNET remained essentially the same over time, the ARPANET was eventually phased out and replaced with newer technology. During this period, the National Science Foundation funded the creation of a new high-speed, cross-country network backbone called the NSFnet. The backbone is the main telecommunications line through the network, connecting the major router sites across the country. It was to this backbone that smaller regional or mid-level (statewide) networks connected. A set of access or "campus" networks then connected to these mid-level networks (Figure 10-1). Eventually this collection of networks became known as the Internet.

Figure 10-1
Old NSFnet backbone and connecting mid-level and campus networks

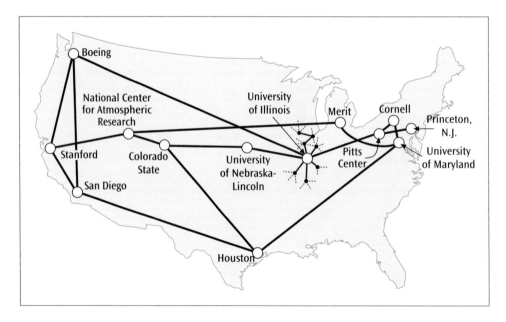

During the early 1990s, the government essentially withdrew all direct support for the Internet and turned it over to private industries and universities. Thus, there is no longer a single backbone but multiple backbones supported by different businesses and organizations, all of which are in competition with one another. Current estimates suggest that there are more than 450 million hosts (computer sites that store and deliver Web pages) connected to the Internet and that more than 1 billion people around the world access the Internet on a regular basis. Although the number of users cannot be verified, one thing is certain: Use of the Internet has grown at a phenomenal rate. Its early creators did not—could not—envision the Internet we have today.

Supporting the Internet are a host of protocols. Two of the most common protocols are the Internet Protocol (IP) and the Transmission Control Protocol (TCP). These protocols are themselves supported by a host of secondary protocols, which include Internet Control Message Protocol (ICMP), User Datagram Protocol (UDP), Address Recognition Protocol (ARP), Dynamic Host Configuration Protocol (DHCP), and Network Address Translation (NAT).

The application of the Internet that has gained the broadest appeal is the World Wide Web. By running a Web browser and clicking links, users can examine text- and image-based Web pages on virtually any topic from virtually any location in the world. To transfer Web pages, the Internet uses the Hypertext Transfer Protocol (HTTP), and to create Web pages, it uses a combination of markup languages named HTML, dynamic HTML, and XML. Uniform Resource Locators (URLs) are used to select unique documents from anywhere in the world. The Domain Name System converts the URLs to a form recognized by routers—IP addresses. The impact of the World Wide Web and other Internet functions on the business community has been monumental. By the late 1990s, an entirely new field of business had come into existence: e-commerce. Now, you can buy practically anything on the Internet. Entertainment, medicine, groceries, real estate, mortgages, and automobiles are just a few of the products available online through e-commerce. Along with e-commerce, businesses have embraced two other forms of the World Wide Web—intranets and extranets.

Many people think the Internet is merely the application that allows a person to browse Web pages and click links, but the Internet actually comprises a wide range of services. One of the first, and still one of the most popular, services offered on the Internet, is electronic mail, or e-mail. Other services made possible by the Internet include File Transfer Protocol, remote login, Internet telephony, listservs, Usenet, and streaming audio and video.

In order to support the many different types of Internet services and to support the layers of the TCP/IP protocol suite, numerous protocols have been created. Let us begin our study of the fascinating world of the Internet by examining several of the more important Internet protocols.

Internet Protocols

From simple e-mail to the complexities of the Web, many services are available on the Internet. What enables these varied services to work? What enables the Internet itself to work? The answer to these questions is Internet protocols. Although the operation of the Internet depends on many protocols, several stand out as the most commonly used: Internet Protocol (IP), Transmission Control Protocol (TCP), Internet Control Message Protocol (ICMP), User Datagram Protocol (UDP), Address Resolution Protocol (ARP), Dynamic Host Configuration Protocol (DHCP), and Network Address Translation (NAT).

Recall the layers of the TCP/IP protocol suite that were introduced in Chapter One—this hierarchy was originally established by the Department of Defense (Figure 10-2). Note that the layers are designed so that an Internet user who needs to use an application (such as e-mail, file transfer, or remote login, all of which are discussed later in the chapter) requires the use of the application layer, which in turn requires the use of the transport layer, which requires the use of the network layer, which requires the use of the network access layer. In other words, the layers build on one another in this hierarchy, and the user is not exposed to the details of those layers with which the user does not have direct contact.

Figure 10-2
*Hierarchy of layers
as created by the
Department of Defense*

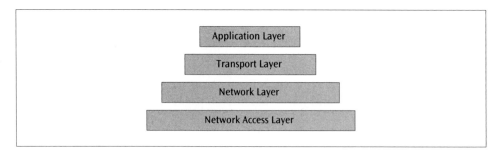

The protocol that resides at the network layer in the TCP/IP protocol suite is called Internet Protocol (IP). IP's primary function is to perform the routing necessary to move data packets across the Internet. IP is a connectionless protocol that does not concern itself with keeping track of lost, duplicated, or delayed packets, or packets delivered out of order. Furthermore, the sender and receiver of these packets may not be informed that these problems have occurred. Thus, IP is also referred to as an unreliable service. If an application requires a reliable service, then the application needs to include a reliable transport service "above" the connectionless, unreliable packet delivery service. The reliable transport service is provided by software called Transmission Control Protocol (TCP), which turns an unreliable network into a reliable network, free from lost and duplicate packets. This combined service is known as TCP/IP.

The use of TCP/IP proved so effective at connecting networks on the Internet that eventually many universities and corporations began to use TCP/IP for the interconnection of their internal networks. In fact, the combination of TCP operating at the transport layer and IP operating at the network layer is found on almost all networks supporting the global Internet. Because TCP/IP is essentially two protocols, we will examine these protocols independently. We will also, in the upcoming sections, examine a number of protocols that support TCP, IP, and the Internet.

The Internet Protocol (IP)

The Internet Protocol (IP) provides a connectionless data transfer service over heterogeneous networks by passing and routing IP datagrams. IP datagram is essentially another name for a data packet. To be passed and routed on the Internet, all IP datagrams or packets that are passed down from the transport layer to the network layer (the connectionless packet delivery layer) are encapsulated with an IP header (see the upper-left corner of Figure 10-3) that contains the information necessary to transmit the packet from one network to another. The format of this header will be explained in the next few paragraphs.

Consider once again the example of a workstation performing a network operation such as sending an e-mail message to a distant workstation, a process that is depicted in Figure 10-3. Suppose both workstations are on local area networks and that the two local area networks are connected via a wide area network. As the local workstation sends the e-mail packet down through the layers of the first internal network, the IP header is encapsulated over the transport layer packet, creating the IP datagram. This encapsulation process is similar to the examples of encapsulation presented in Chapters One and Seven. The appropriate MAC layer headers are encapsulated over the IP datagram creating a frame, and this frame is sent through LAN 1 to the first router. Because the router interfaces LAN 1 to a wide area network, the MAC layer information is stripped off, leaving the IP datagram. At this time, the router may use any or all of the IP information to perform the necessary internetworking functions. The necessary wide area network level information is applied, and the packet is sent over the WAN to Router 2. When the packet arrives at the second router, the wide area network information is stripped off, once again leaving the IP datagram. The appropriate MAC layer information is then applied for transfer of the frame over LAN 2, and the frame is transmitted. Upon arrival at the remote workstation, all header information is removed, leaving the original data.

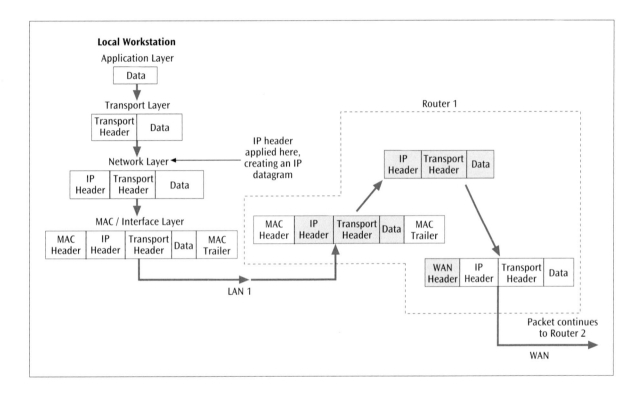

Figure 10-3
Progression of a packet from one network to another

When a router has the IP datagram, it may make several decisions affecting the datagram's future. In particular, the router must perform the following functions:

- ▹ Make routing decisions based on the address portion of the IP datagram
- ▹ Fragment the datagram into smaller datagrams if the next network to be traversed has a smaller maximum packet size than the current size of the packet
- ▹ Decide that the current datagram has been hopping around the network for too long and delete the datagram

To perform these functions, the router needs address information, datagram size, and the time the datagram was created. This information is found in the IP header, which was the information applied to the transport packet at the network layer. Each of these router functions, along with the IP header fields that support these functions, will be examined in more detail in the upcoming section.

IP Datagram Format

Figure 10-3 shows that the network layer of the communications software added an IP header to the transport layer packet—thereby creating an IP datagram—before passing the packet on to the next layer of software. The information included in this IP header and the way the header is packaged allow the local area networks and wide area network to share data and create internetwork connections.

Exactly what is in this IP header that allows this internetworking to happen? Figure 10-4 shows the individual fields of the IP header in more detail. Even though all 14 fields are important, let us examine the field that specifies the version of the Internet Protocol and those fields that affect three of the primary functions of IP: fragmentation, datagram discard, and addressing. By examining these fields, you will begin to discover how IP works and why it is capable of interconnecting so many different types of networks.

Figure 10-4
Format of the IP datagram

Version	Hlen	Service Type	Total Length	Identification	Flags	
4 bits	4 bits	8 bits	16 bits	16 bits	3 bits	

Fragment Offset	Time to Live	Protocol	Header Checksum	Source IP Address	
13 bit	8 bits	8 bits	16 bits	32 bits	

	Destination IP Address		IP Options	
	32 bits		Variable Length	

Padding	Data	
Optional	Variable Length	

The first field of interest is the Version field. The Version field contains the version number of IP being used, just in case a new version becomes available. Currently, many networks involved in the Internet use IP version 4. IP version 6, which was created during the late 1990s and is discussed in detail in a later section of this chapter, should eventually replace version 4. The Version field is important because it tells the router how to interpret the IP datagram.

The next three fields of interest—Identification, Flags, and Fragment Offset—are used to fragment a datagram into smaller parts. Why would we want to fragment a datagram? When the Internet Protocol was created, the maximum packet size of some older networks that were still in existence was small. This maximum size was limited by network hardware, software, or other factors. Because IP was designed to work over practically any type of network, it had to be able to transfer datagrams of varying sizes. Rather than limit IP to the smallest maximum packet size in existence (who even knows what that is?), the Internet Protocol allows a router to break or fragment a large datagram into smaller fragments so it will fit onto the next network. Fortunately, hardly any modern networks have a maximum packet size that is small enough to be a concern. Thus, someday fragmenting a datagram into smaller packets will no longer be an important issue. In fact, as you will learn shortly, IP version 6 does not even have a field in the header to perform fragmentation. (If you want to see how these fields actually perform the fragmentation, see the Details section titled "IP Fragmentation.")

The next field we will examine is the Time to Live field, which enables the network to discard a datagram that has been traveling the Internet for too long. The

Time to Live field indicates how long a particular datagram is allowed to live—bounce from router to router—within the system. When a packet is first created, this field is set to its maximum value: 255. Each router along the route from source to destination decreases the number in the Time to Live field by 1. When the value of the Time to Live field reaches zero, a router deletes the datagram. The Time to Live field is analogous to the hop count and hop limit introduced in Chapter Nine. The next field—the Header Checksum—performs an arithmetic checksum on the header only portion of the packet.

The final two fields of interest, Source IP Address and Destination IP Address, contain, respectively, the 32-bit IP initial source and final destination addresses of the datagram. A 32-bit address uniquely defines a connection to the Internet—usually a workstation or device, although one workstation or device may support multiple Internet connections. As an IP datagram moves through the Internet, the Destination IP Address field is examined by a router. The router, using some routing algorithm, forwards the datagram onto the next appropriate communications link. The details of IP addresses will be covered in the section titled "Locating a document on the Internet."

The Internet Protocol is definitely one of the most important communications protocols. Because of its simple design, it is relatively easy to implement in a wide variety of devices. And because of its power, it is capable of interconnecting networks of virtually any type. Even though the Internet Protocol is powerful, its primary objective is getting data through one or more networks. It is not responsible for creating an error-free, end-to-end connection. To accomplish this, the Internet Protocol relies on the Transmission Control Protocol, or TCP.

Details ▶

IP Fragmentation

For an example of IP fragmentation, consider the IP datagram shown in Figure 10-5(a). The data portion of the datagram is 960 bytes in length. This datagram needs to traverse a particular network that limits the data portion of a datagram to 400 bytes. Thus, there is a need to divide the original datagram into three fragments of 400, 400, and 160 bytes. The three fragments created have almost identical IP headers.

The Fragment Offset field contains the offset, which is the count, in units of 8 bytes, from the beginning of the original datagram data field to this particular fragment. Figure 10-5(b) shows these offsets. The first fragment, which is not offset from the beginning, has an offset value of 0. Because the first fragment is 400 bytes long, the second fragment must start 400 bytes into the original datagram. We are counting in units of 8 bytes, so 400 divided by 8 equals an offset of 50. The first two fragments total 800 bytes, so the third and final fragment starts 800 bytes into the original datagram. Dividing 800 bytes by an 8-byte unit equals an offset of 100.

The next field that assists in fragmentation is the Flags field. Within the Flags field is a More bit, which indicates if there are more fragments of the original datagram to follow. Figure 10-5(b) also shows the contents of the More bit. A More

bit of 1 indicates, as you can see in the first two packets, that more fragments are to come; a More bit of 0 signals the last fragment of a packet. The final field that assists in fragmentation is the Identification field. The Identification field contains an ID unique to these three fragments that is used to reassemble the three fragments back into the original datagram.

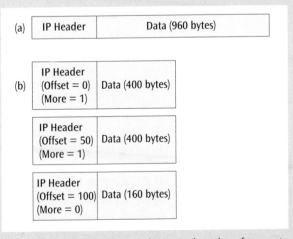

Figure 10-5 *Division of an IP datagram into three fragments*

The Transmission Control Protocol (TCP)

Perhaps one of the most common examples of a transport layer protocol is the other half of the popular TCP/IP protocol. The primary function of Transmission Control Protocol (TCP) is to turn an unreliable network (such as the one created by IP) into a reliable network that is free from lost and duplicate packets. Thus, TCP essentially fills in some holes created by IP. But how can a transport layer protocol make an unreliable network reliable? To make a network more reliable, TCP performs the following six functions:

> ▸ Create a connection—The TCP header includes a port address that indicates a particular application on a machine. Used in conjunction, the port address and the IP address identify a particular application of a particular machine. When TCP creates a connection between a sender and a receiver, the two ends of the connection use a port number to identify the particular application's connection. This port number is found within the TCP datagram and is passed back and forth between sender and receiver.

> ▸ Release a connection—The TCP software can also dissolve a connection after all the data has been sent and received.

> ▸ Implement flow control—To make sure the sending station does not overwhelm the receiving station with too much data, the TCP header includes a field, called the Window value, that allows the receiver to tell the sender to slow down. This Window value is similar in operation to the sliding window used at the data link layer. The difference between the two window operations is that the data link layer's sliding window operates between two nodes or between a workstation and a node, while the TCP window operates between the two endpoints (sender and receiver) of a network connection.

> ▸ Establish multiplexing—Because the TCP header includes a port number instead of an IP address, it is possible to multiplex multiple connections over a single IP connection. This multiplexing can be done by creating a different connection that has a port number different from a previous connection.

> ▸ Perform error recovery—TCP numbers each byte for transmission with a sequence number. As the packets of bytes arrive at the destination site, the receiving TCP software checks these sequence numbers for continuity. If there is a loss of continuity, the receiving TCP software uses an acknowledgment number to inform the sending TCP software of a possible error condition.

> ▸ Establish priority—If the sender has to transmit data of a higher priority, such as an error condition, TCP can set a value in a field (the Urgent Pointer) that indicates that all or a portion of the enclosed data is of an urgent nature.

To perform these six functions, TCP places a header at the front of every data packet that travels from sender to receiver, or from one end of the connection to the other. As we did with the IP header, let us examine the more important fields in the TCP header.

TCP Datagram Format

A user is sitting at a workstation and running a network application—for example, an e-mail program. When the user wants to send an e-mail message, the e-mail program takes the e-mail message and passes it to the transport layer of the software.

If the e-mail is heading out onto the Internet, the transport layer adds a TCP header to the front of the e-mail message. The information in this header is used by the TCP layer at the receiving workstation to perform one or more of the six transport functions. The TCP header contains the fields shown in Figure 10-6. Let us examine only those fields that assist TCP in performing the six functions listed earlier.

Figure 10-6
The fields of the TCP header

Source Port	Destination Port	Sequence Number	
16 bits	16 bits	32 bits	
Acknowledgment Number	Hlen	Reserved	Flags	Window
32 bits	4 bits	6 bits	6 bits	16 bits
Checksum	Urgent Pointer	Options	Padding
16 bits	16 bits	Variable Length	Optional
Data			
Variable Length			

The first two TCP header fields, Source Port and Destination Port, contain the addresses of the application programs at the two ends of the transport connection. These port addresses are used in creating and terminating connections. The port number can also be used to multiplex multiple transport connections over a single IP connection.

It is important to note the difference between an IP address and a port number. The IP address identifies a device connected to the Internet, while the port number identifies an application on that device. Working together, the two create what is called a **socket**—a precise identification of a particular application on a particular device. What if your company has one server that handles both e-mail and FTP connections? The server would have one IP address but two different port numbers—one for the e-mail application and one for the FTP application. Now let us add the fact that this server is more than likely on a local area network, and thus has a network interface card with a unique 48-bit NIC address. Now we have three addresses. The NIC address is used only on the local area network to find a particular device. The IP address is used to move the data packet through the Internet. The port number is used to identify the particular application on a device.

The Sequence Number field contains a 32-bit value that counts bytes and indicates a packet's data position within the connection. For example, if you are in the middle of a long connection in which thousands of bytes are being transferred, the Sequence Number tells you the exact position of this packet within that sequence. This field can be used to reassemble the pieces at the receiving workstation and determine if any packets of data are missing.

The Window field contains a sliding window value that provides flow control between the two endpoints. If one end of the connection wants the other end of the connection to stop sending data, the Window field can be set to zero. The Checksum field is the next field and provides for an arithmetic checksum of the header and the data field that follows the header. The Urgent Pointer is used to inform the receiving workstation that this packet of data contains urgent data.

Like its counterpart IP, TCP is a fairly streamlined protocol. Its primary goal is to create an error-free, end-to-end connection across one or more networks. TCP and IP do have their shortcomings, however, and these have typically led to problems such as security weaknesses, quality of service issues, and congestion control. Thus, both of these protocols will need to continue to evolve. But even so, their basic components should remain in existence for a long time.

Internet Control Message Protocol (ICMP)

As an IP datagram moves through a network, a number of things can go wrong. As a datagram nears its intended destination, a router may determine that the destination host is unreachable (the IP address is wrong, or the host does not exist), the destination port is unknown (there is no application that matches the TCP port number), or the destination network is unknown (again, the IP address is wrong). If a datagram has been on the network too long and its Time to Live value expires, the datagram will be discarded. Also, there could be something wrong with the entire IP header of the datagram. In each of these cases, it would be nice if a router or some other device would send an error message back to the source workstation, informing the user or the application software of a problem. The Internet Protocol was not designed to return error messages, so something else is going to have to perform these operations.

Details ▶

Multiprotocol Label Switching

Throughout the textbook so far, encapsulation has been a key concept. Recall that encapsulation is the layering of one layer's header over the existing data packet. As we have just seen, in the TCP/IP protocol suite the TCP header is layered on top of (placed in front of) the packet that comes from the application layer. The information in the TCP header is used to perform transport layer connection functions at the endpoints of the connection. Then the IP header is layered on top of the TCP packet, and the information in the IP header is used to perform network routing and Time to Live operations. Normally, the next step would be to place the network access layer information on top of the IP packet. But wide area networks do not use network access layer information to route the data packets. Thus, routers must "dig into" a data packet and extract the IP information and use that for routing. This is typically not a fast process, at least not when compared to the forwarding operation that local area network switches can perform. If wide area networks could perform switching

operations quickly, the Internet (as well as other wide area networks) would operate much more efficiently.

In order to provide this level of switching at the network level, network designers have created Multiprotocol Label Switching. **Multiprotocol Label Switching (MPLS)** is a technique that enables a router to switch data from one path onto another path. To make this possible, one or more labels (headers with MPLS information) are encapsulated onto the front of an IP packet (Figure 10-7). Each label contains four fields:

▶ a 20-bit Label Value field (Label)

▶ a 3-bit Experimental field (reserved for future use) (Exp)

▶ a 1-bit Bottom of Stack Flag field (S=0)

▶ an 8-bit Time to Live field (TTL)

The 20-bit label value essentially tells the MPLS-enabled router which connection this packet belongs to and thus how

The **Internet Control Message Protocol (ICMP)**, which is used by routers and nodes, performs this error reporting for the Internet Protocol. All ICMP messages contain at least three fields: a type, a code, and the first eight bytes of the IP datagram that caused the ICMP message to be generated. A type is simply a number from 0 to n that uniquely identifies the kind of ICMP message, such as invalid port number or invalid IP address. A code is a value that provides further information about the message type. Together, ICMP and IP provide a relatively stable network operation that can report some of the basic forms of network errors.

User Datagram Protocol (UDP)

TCP is the protocol used by most networks and network applications to create an error-free, end-to-end network connection. TCP is connection-oriented in that a connection via a port number must be established before any data can be transferred between sender and receiver. What if you do not want to establish a connection with the receiver but simply want to send a packet of data? In this case, User Datagram Protocol is the protocol to use. **User Datagram Protocol (UDP)** is a no-frills transport protocol that does not establish connections, does not attempt to keep data packets in sequence, and does not watch for datagrams that have existed for too long. Its header contains only four fields—Source Port, Destination Port, Length, and Checksum—and it is used by a small number of network services, such as DNS, that do not need to establish a connection before sending data.

to forward the packet. The 1-bit bottom of stack flag indicates if there are multiple labels on this packet (0 if this is the last label and 1 if otherwise). The 8-bit Time to Live field works similarly to the Time to Live field in the IP header. When a packet is entered into the network, an MPLS-enabled router determines the connection the packet should be in and inserts one or more labels in the newly created MPLS header. The packet is now called a *labeled packet*. Subsequent MPLS-enabled routers look at the information in the MPLS header and forward the labeled packet in the appropriate direction. For each labeled packet that enters, MPLS-enabled routers can have built-in routing tables that tell the router what kind of operation to perform based upon the packet's topmost MPLS label. This operation is performed quickly, often in hardware. When the labeled packet arrives at the destination router, the MPLS-enabled router removes the MPLS header, leaving the original IP packet.

MPLS is gaining popularity in large networks and is an Internet Engineering Task Force standard. As we will see in Chapter Eleven, it may someday replace or at least aid in the replacement of other wide area network protocols.

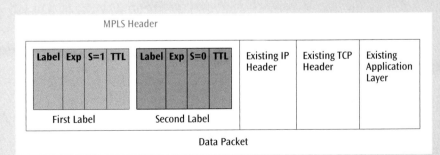

Figure 10-7 *An MPLS header and its four fields*

Address Resolution Protocol (ARP)

The Address Resolution Protocol is another small but important protocol that is used to support TCP/IP networks. **Address Resolution Protocol (ARP)** takes an IP address in an IP datagram and translates it into the appropriate medium access control layer address for delivery on a local area network. As mentioned earlier, every workstation that has a connection to the Internet is assigned an IP address. This IP address is what a packet uses to find its way to its intended destination. There is one problem, however, when a workstation is on, for example, an Ethernet or CSMA/CD local area network. Recall that the piece of data that traverses a CSMA/CD LAN is called a frame. This frame consists of a number of fields of information, none of which is an IP address. If the frame is supposed to go to a particular workstation with a unique IP address, but the frame does not contain an IP address, how does the frame know where to go? ARP provides the answer to this question. After an IP datagram enters a CSMA/CD LAN through a router and before its IP header is stripped off to leave only the CSMA/CD frame, ARP broadcasts a message on the LAN asking which workstation belongs to this IP address. The workstation that recognizes its IP address sends a message back saying, "Yes, that is my IP address, and here is my 48-bit CSMA/CD (NIC) address. Please forward that IP packet to me via my address." (The 48-bit NIC address is stored in a buffer just in case it will be needed again in the near future.)

Dynamic Host Configuration Protocol (DHCP)

When a company installs a number of computer workstations and intends to give them all Internet access, it must assign each of them an IP address. This IP address, as we have learned, allows a workstation to send and receive information over the Internet. Two basic methods are used to assign an IP address to a workstation: static assignment and dynamic assignment. With static assignment, somebody sits down at each machine and, using the network operating system, installs an IP address. The person installing the address must then record that IP address on paper somewhere so that the IP address is not accidentally assigned to another machine.

What happens if a workstation with an IP address that was statically assigned is then removed from service? Someone has to make sure the IP address assigned to that machine is removed for use in another machine. What if a mistake is made, and the same IP address is assigned to two machines? In this case, there would be an IP address conflict, and a network administrator would have to locate the two machines with the same IP address and rectify the situation. Another issue to consider is related to IP address procurement. If a company has 1000 workstations and each workstation has access to the Internet, then the company has to acquire (typically lease) 1000 IP addresses. This is not cost-effective, particularly if only one-half of the users are on the Internet (using their IP address) at a time. Static assignment of IP addresses can lead to a waste of resources. Dynamic assignment of IP addresses solves these three problems.

The most popular protocol that handles dynamic assignment is **Dynamic Host Configuration Protocol (DHCP)**. When a workstation running the DHCP client software needs to connect to the Internet, the protocol issues an IP request, which prompts the DHCP server to look in a static table of IP addresses. If this particular workstation has an entry, then that IP address is assigned to that workstation. But if there is no entry in the static table, the DHCP server selects an IP address from an available pool of addresses and assigns it to the workstation. The IP address assignment is temporary, with the default time limit being one hour. DHCP clients may negotiate for a renewal of the assignment if the workstation is still accessing the Internet when the temporary assignment is nearing expiration. Thus, with DHCP, the three problems introduced with static assignment are solved. No individual has to assign IP addresses to workstations, two workstations never get assigned the same IP address, and if only 200 workstations out of 1000 are ever using the Internet at the same time, the company can probably get by with acquiring only 200 IP addresses.

Users who dial in to the Internet from home often use DHCP without knowing it. Rather than assigning an IP address to every potential dial-in user, an Internet service provider's DHCP server assigns an IP address during the session creation period. The computer uses this temporary IP address until the user logs off, at which time the address is placed back into the pool, ready for the next dial-in user.

Network Address Translation (NAT)

Another protocol that is used to assign IP addresses is Network Address Translation (NAT). More precisely, NAT lets a router represent an entire local area network to the Internet as a single IP address. When a user workstation on a company local area network sends a packet out to the Internet, NAT replaces the IP address of the user workstation with a corporate global IP address. In fact, all packets that leave the corporate network contain this global IP address. Thus, the only IP address that anyone sees outside of the corporate network is the one global IP address. If all packets from all workstations leave the corporate network with the same IP address, how do the responses that come back from the Internet get directed to the proper machine? The NAT software maintains a cache listing of all IP packets that were sent out and who sent each packet. When a response comes back, NAT checks the cache to see who originally sent the request. When NAT finds the match, it removes the global IP address, reinserts the user workstation's IP address, and places the packet on the corporate network.

What happens if a packet arrives at the corporate NAT software and there is no cache entry with a matching outgoing request? In this case, the packet is destroyed. Apparently, someone has sent a packet to the corporate network that was not requested by a corporate workstation. There is one exception to this rule: If the company is supporting a server, such as a Web server, a packet may originate from a user somewhere out on the Internet. When the Web page request packet arrives at the NAT software, the IP address where the packet originated will not match any IP addresses in the local cache. Before the NAT software destroys the packet, it examines the message's destination port number. If the packet is a request for a Web page from the corporate Web server, the NAT software lets the packet in.

An interesting feature of using NAT is that, because the outside world never sees any of the IP addresses used within the corporate network, a level of security has been added. Additionally, the company does not need to use purchased IP addresses on the corporate network. To support this feature, a number of IP addresses have been designated as "phony" IP addresses. When a workstation with a phony IP address issues an Internet request, the NAT software replaces the phony IP address with the corporate global IP address. The use of NAT and phony IP addresses is another way to save money on leasing IP addresses.

Home and small business local area networks often use NAT to conserve IP addresses. By keeping track of each request in its internal cache, NAT allows multiple workstations to access the Internet with only one IP address. If the computers are also using DHCP, a home user can dial in to the Internet, have an IP address dynamically assigned, and use that IP address for the NAT operation. NAT is so useful that many routers now incorporate it as a standard security (firewall) feature.

Tunneling protocols and virtual private networks (VPNs)

One of the more serious problems with the Internet is its lack of security. Whenever a transmission is performed, it is susceptible to interception. Retailers have solved part of the problem by using encryption techniques to secure transactions dealing with credit card numbers and other private information. Businesses that want their employees to access the corporate computing system from a remote site have found a similar solution—virtual private networks. A **virtual private network (VPN)** is a data network connection that makes use of the public telecommunications infrastructure but maintains privacy through the use of a tunneling protocol and security procedures. A **tunneling protocol**, such as the Point-to-Point Tunneling Protocol (PPTP), is the command set that allows an organization to create secure connections using public resources such as the Internet. Proposed by CISCO Systems, PPTP is a standard sponsored by Microsoft and other companies. It is an extension of the Internet's **Point-to-Point Protocol (PPP)**, which is used for communication between two computers using a serial connection. The most common example of a serial connection is a dial-up modem connection between a user's workstation and an Internet service provider. (PPP is usually the standard preferred over an earlier protocol, Serial Line Internet Protocol, or SLIP.)

Employees who are located outside a company building can use PPTP to create a tunnel through the Internet into the corporate computing resources. Because this connection runs over the Internet and may be used to transmit confidential corporate data, it must be secure. The security of the connection is often supported by IPsec. **IPsec**, an abbreviation for IP Security, is a set of protocols developed by the Internet Engineering Task Force to support the secure exchange of data packets at the IP layer. In order for IPsec to work, both sender and receiver must exchange public encryption keys (explained in detail in Chapter Twelve).

Besides being a secure connection, a tunnel is also a relatively inexpensive connection, because it uses the Internet as its primary form of communication. Alternatives to creating a tunnel are acquiring a dial-up telephone line or leasing a telephone line—both of which can be more expensive.

Now that we have an understanding of the protocols that enable the Internet to function, let us look closer at the World Wide Web, the area of the Internet most familiar to all of us.

The World Wide Web

The **World Wide Web (WWW)** is a vast collection of electronic documents that are located on many different Web servers and contain text and images that can be accessed by simply clicking links within a browser's Web page. Using a Web browser, you can download and view Web pages on a personal computer. Of all the Internet services, the World Wide Web is probably the one that has had the most profound impact on business. Internet retail sales and service support have exploded with the use of personal computers and Web browsers. Virtually any and every imaginable product and service is now sold on the Web. In a single day, you can purchase clothing and groceries, buy an airline ticket, select and purchase an automobile, plan a funeral, submit an auction bid on a toy you had as a child, find a new job, rent a videotape, and order pizza for dinner—all online. To do this, all you need is a personal computer with a connection to the Internet and a Web browser. The Web pages you download can consist of text, graphics, links to other Web pages, sometimes music and video, and even executable programs.

Web pages are created using Hypertext Markup Language (HTML), which can be generated manually with a text-based editor such as Notepad, or through using a Web page authoring tool. A Web page authoring tool is similar to a word processor, except that instead of creating text documents, you create HTML-based Web pages. The Web page authoring tool has a graphical user interface that allows you to enter text and insert graphics and other Web page elements and also arrange these elements on the page using drag-and-drop techniques. You will learn more about using a markup language to create a Web page later in this section.

Once you have created your Web page, you store it on a computer that contains Web server software and has a connection to the Internet. The Web server software accepts Hypertext Transfer Protocol (HTTP) requests from Web browsers connected to the Internet, retrieves a requested Web page from storage, and returns that Web page to the requesting computer via the Internet. The Hypertext Transfer Protocol, or HTTP, is an application layer protocol. As was explained in Chapter One, when a user sitting at a workstation running a browser clicks a link on a Web page, the browser software sees this click and creates an application layer HTTP command to retrieve a Web page. This HTTP command passes through the transport layer, network layer, and network access layer before it is placed on the network medium. When the Web page request is received at the Web site server, the Web page is retrieved and returned across the Internet to the user's browser, where it is displayed on the monitor.

Now that we have an understanding of how documents are passed across the Internet, the question remains: How are Internet documents addressed and found? To answer this question, we need to examine URLs, the Domain Name System, and IP addresses.

Details ▶

HTTP

The entire purpose of HTTP is to send and receive Web pages. To accomplish this operation, HTTP can perform a number of different commands called *methods*. A few of the more common methods are the following:

- ▶ GET—Retrieves a particular Web page, which is identified by a URL
- ▶ HEAD—Uses a given URL to retrieve only the HTTP headers (not the document body) of the Web page
- ▶ PUT—Sends data from a user's browser to a remote Web site (this method is used, for example, to send a buyer's credit card number to a Web merchant during a purchasing transaction)

- ▶ DELETE—Requests that a server delete the information corresponding to a given URL

As an example of how these methods are used, suppose you click a link during a Web browser session. The browser software creates a GET command and sends that command to the Web server. The Web server then returns the Web page to the browser. Now suppose you happen to be purchasing a product via the Web, and you are filling out a form. When you click the Send button, HTTP creates a PUT command and uses it to send the data to a Web server.

Locating a document on the Internet

When a user is running a browser on a workstation and clicks on a link, the browser attempts to locate the object of the link and bring it across the Internet to the user's workstation. This object can be a document, a Web page, an image, an FTP file, or any of a number of different types of data objects. How does the Internet locate each object? Stated simply, every object on the Internet has a unique English-based address called its Uniform Resource Locator (URL). The Internet, however, does not recognize URLs directly. In order for the Internet to find a document or object, part of the object's URL has to be translated into the IP address that identifies the Web server where the document or object is stored. This translation from URL to IP address is performed by the Domain Name System (DNS).

The IP address itself is not as simple as it appears. The assignment of IP addresses is complex and requires an understanding of the different classes of addresses and of a concept called subnet masking. Let us examine each of these concepts—URLs, DNS, IP addresses, and subnet masking—further to gain a better understanding of how the Internet finds one document from among billions of documents.

Uniform Resource Locator (URL)

In order for users to be able to find something on the Internet, every object located on the Internet must have a unique "address." This address, or Uniform Resource Locator (URL), uniquely identifies files, Web pages, images, or any other types of electronic documents that reside on the Internet. When you are using a Web browser and click a link to a Web page, you are actually sending a command out to the Internet to fetch that particular Web page from a specific location that is based on the Web page's URL.

All Uniform Resource Locators consist of four parts, as shown in Figure 10-8. The first part, indicated by a 1 in the figure, is the service type. The service type identifies the protocol that is used to transport the requested document. For example, if you request a Web page, then Hypertext Transfer Protocol (http://) is the service type used to retrieve the Web page, as shown in Figure 10-8(a). If you are requesting an FTP document, then the FTP protocol (ftp://) is used, as in Figure 10-8(b). Other service types include telnet:// to perform a remote login, news:// to access a Usenet group, and mailto:// to send an electronic mail message.

The second part of the URL, indicated by a 2 in Figure 10-8, is the domain name. This portion of the URL specifies a particular server at a particular site that contains the requested item. In the example in Figure 10-8, *cs.depaul.edu* is one of the servers supporting the computer science program at DePaul University. Starting on the right, *edu* is the top-level domain and indicates that the Web site is an educational site. Other top-level domains are com (commercial), gov (government), mil (military), org (nonprofit organization), net (network-based), biz, name, info, pro, museum, aero, and coop. Each country also has its own top-level domain name. For example, Canada is ca and the United Kingdom is uk. The domain name at the next level—called the mid-level domain name—is usually the name of the organization (often a company or school) or host, such as *depaul*. Any other lower-level domains are further subdivisions of the host and are usually created by the host. For example, suppose a company named FiberLock applies to the agency handling domain name registration to ask for the mid-level domain name *fiberlock*. Because FiberLock is a commercial business, its top-level domain name will be

.com. If no one else is using *fiberlock*, the company will be granted *fiberlock.com* as its domain name. The company may then add more domain levels, such as www or email, to create entities such as *www.fiberlock.com* or *email.fiberlock.com*, which could correspond to the company's Web page server and e-mail server, respectively.

The third part of a URL, labeled 3 in Figure 10-8, is the directory or subdirectory information. For example, the URL *http://cs.depaul.edu/public/utilities/ada* specifies that the requested item is located in the subdirectory *ada*, under the subdirectory *utilities*, under the subdirectory *public*.

The final part of the URL, specified by the number 4 in Figure 10-8, is the filename of the requested object. In this case, it is a document titled *example.htm*. If no filename were specified in the URL, then a default file, such as default.htm or index.html, would be retrieved.

Figure 10-8
The parts of a Uniform Resource Locator (URL) for HTTP (a) and FTP (b)

http://cs.depaul.edu/public/utilities/ada/example.htm

   1   2   3   4

(a)

ftp://gatekeeper.dec.com/pub/games/starwars.exe

   1   2   3   4

(b)

Domain Name System (DNS)

When referencing an Internet site, we often refer to its domain name. Computers, however, do not use domain names. They use 32-bit binary addresses called IP addresses. For example, a valid IP address has the following form:

1000 0000   1001 1100   0000 1110   0000 0111

To make IP addresses a little easier for human beings to understand, these 32-bit binary addresses are represented by dotted decimal notation. This dotted decimal notation is created by converting each 8-bit string in the 32-bit IP address into its decimal equivalent. Thus, the IP address above becomes 128.156.14.7, as shown here:

1000 0000   1001 1100   0000 1110   0000 0111
   128   156   14   7

But even the decimal equivalent to the IP address is not convenient for us humans. Because computers use 32-bit binary addresses and almost all human beings use the domain name form, the Internet converts the binary forms into English-based domain names, and vice versa. To do this, it uses the Domain Name System (DNS), which is a large, distributed database of Internet addresses and domain names. This distributed database consists of a network of local DNS servers, mid-level DNS servers, and higher-level DNS servers. To keep the system manageable, the DNS database is distributed according to the top-level domains: edu, gov, com, mil, and so on.

Converting a domain name into a binary IP address can be simple or complicated. The level of complexity depends on whether or not a local network server on the originating local area network recognizes the domain name. If a network server cannot resolve an address locally, it will call upon a higher authority. A local DNS server will send a DNS message to the next higher DNS server until the address is found, or it is determined that the address does not exist. If the address does not exist, an appropriate message is returned.

Consider, first, what happens when a local server recognizes a domain name. When the new domain name appears, an application program, such as a Web browser, calls a library procedure named the resolver. The resolver sends a DNS message to a local DNS server, which looks up the name and returns the IP address to the resolver. The resolver then returns the IP address to the application program.

But what happens if the local DNS server does not have the requested information? It may query other local DNS servers, if there are any. The information concerning the existence of other local servers and remote server locations would be kept in a file on the local computer network. If the answer again is no, the local DNS server tries the next level up—perhaps a mid-level server. If the mid-level server does not recognize the domain name, or there is no mid-level server, the top-level name server for the domain is queried. If the top-level name server does not recognize the domain name, it will either return a "URL Not Found" message or go down a level and query a local DNS server. To understand this better, let us look at an example.

Consider a scenario in which a user at *cs.waynestate.edu* in Wayne State University wants to retrieve a Web page from *www.trinity.edu* at Trinity University. The message originates from *cs.waynestate.edu* and goes to the *waynestate.edu* name server. The *waynestate.edu* name server does not recognize *www.trinity.edu*, because the domain name *www.trinity.edu* is not in a list of recently referenced Web sites. Therefore, the *waynestate.edu* name server sends a DNS request to *edu-server.net*. Although *edu-server.net* does not recognize *www.trinity.edu*, it does recognize *trinity.edu*, so it sends a query to *trinity.edu*. The *trinity.edu* name server recognizes its own *www.trinity.edu* and returns the result to *edu-server.net*, which returns the result to the *waynestate.edu* name server, which returns the result to the user's computer, which then inserts the result—that is, the appropriate 32-bit binary IP address—into the browser request.

IP Addresses

Now let us talk a little more about the IP address. When IP and IP addresses were created in the 1960s, an IP address belonged to a particular class. This type of addressing was called classful addressing, and it was based, as we shall see shortly, on five different classes. In approximately 1996, a new form of addressing became available: classless addressing. Although classless addressing is more common today, it is still important to understand how both of these systems work. To start things off, we will examine classful addressing.

As you have already learned, IP addresses are currently 32 bits long. In classful addressing, the addresses are not, however, simple 32-bit integers. Instead, they can consist of three specific pieces of information. The size and value of these three pieces of information depend on the basic form of the address. There are five basic forms of an IP address: Class A, B, C, D, and E (Table 10-1).

Table 10-1
Five basic forms of a 32-bit IP address

Address Type	Beginning Bit Pattern	Network Address (net ID)	Host Address (host ID)
Class A	0	128 addresses (7 bits)	16,777,216 addresses (24 bits)
Class B	10	16,384 addresses (14 bits)	65,536 addresses (16 bits)
Class C	110	2,097,152 addresses (21 bits)	256 addresses (8 bits)
Class D	1110	Multicast address	
Class E	1111	Reserved addresses	

As mentioned earlier, each IP address can consist of three parts:

- A 1-, 2-, 3-, or 4-bit identifier field (also known as a beginning bit pattern)
- A net ID, which indicates a particular network
- A host ID, which indicates a particular host, or computer, on that network

As Table 10-1 shows, given a beginning bit pattern of 0, there are 128 Class A addresses, or networks, in existence. Each Class A address can have 16,777,216 hosts, or computers. Clearly, 128 is not very many networks; in fact, all the 128 Class A addresses were assigned a long time ago. Another impractical feature of the Class A address type is the allocation of 16,777,216 computers per network. Attaching 16,777,216 computers to one network is beyond imagination. As we have seen, many local area networks rarely have more than a few hundred computers attached to them. Unfortunately, for this reason, many Class A addresses go unused. Class B addresses allow for 16,384 net IDs, or networks, each supporting 65,536 host IDs—meaning that each of the 16,384 networks can have 65,536 host computers attached to it. Class C addresses allow for 2,097,152 net IDs, or networks, and 256 host IDs. In the case of the Class C address type, the number of host computers allowed is too small to accommodate any networks but the smallest.

Class D addresses are available for networks that allow multicasting of messages. IP multicasting is the capability of a network server to transmit a data stream to more than one host at a time. Consider a scenario in which a company wants to download a streaming video of a training exercise to 20 users sitting at separate workstations. If a server has to transmit 20 individual copies to the 20 workstations (unicast), a very high bandwidth signal will be necessary. If the server could instead multicast one copy of the video stream to the 20 workstations, a much smaller bandwidth would be necessary. To multicast, the server could insert a Class D address into the IP datagram, and each of the 20 workstations could tell its IP software to accept any datagrams with this Class D address.

Although IP multicasting has some very promising advantages, it suffers from a lack of security. Because it is relatively easy for a workstation to tell its IP software to accept a particular Class D address, any workstation—even one that does not have permission—could potentially receive the multicast. Nonetheless, new applications that address the shortcomings of IP multicasting have been announced, opening the door for future possibilities.

When a company applied for a set of Internet addresses that used classful addressing, a number of options were possible. First, the company could apply for a Class B address. With a Class B address, it could allocate 65,536 workstations on its Class B network. If this number of workstations were too large, the company could consider a

Class C address. A Class C address allows for 256 computers on a single network. But if the company had 400 to 500 users, a Class C address would be too small. Therefore, the company could apply for two Class C addresses.

Suppose that a company did apply for and receive a large number of Class B addresses. How could it use these efficiently? In this case, the company can break its IP addresses into subnets using subnet masking, a technique that allows for easier network management. The basic idea behind **subnet masking** is to take the host ID portion of an IP address and further divide it into a subnet ID and a host ID. Using this technique, an ISP and a company can take a large number of host IDs and break them into subnets. Each subnet can then support a smaller number of hosts.

With classless addressing, companies (users) do not apply for a particular class of addresses. Instead, a company will get its IP addresses from an Internet service provider (ISP). Most ISPs have already applied for a large number of IP addresses and are willing to lease those addresses to companies. Returning to our previous example, instead of applying for two Class C addresses, a company with 400 to 500 users could contact an ISP and request to lease 512 IP addresses (the number of addresses requested has to be a power of 2). These addresses, at least from the company's perspective, are not identified by any class—they are simply a contiguous block of IP addresses. As you might have guessed, classless addressing has led to a much more efficient allocation of the IP address space. A company can now lease nearer the exact number of addresses it requires, without wasting unused addresses.

When classless addressing is combined with DHCP and NAT, a company can efficiently use a smaller number of IP addresses. This is due to the fact that all three concepts more effectively make use of the IP address space. Even though a 32-bit IP address supports more than 4 billion addresses, classful addressing was so wasteful that there was serious concern among industry experts that the supply of IP addresses would soon be depleted. The advent of classless addressing (along with the DHCP and NAT protocols) has considerably helped slow the IP address drain. Despite this, there is a push to incorporate a new version of the Internet Protocol, IPv6, which has a much larger address space. We will investigate this new protocol later in the chapter.

Creating Web pages

To transmit a Web page, a Web browser, Web server, and the Internet use the Hypertext Transfer Protocol. HTTP is not, however, used to display the Web page once it reaches the intended destination. To control how a Web page is displayed, another specification is used—Hypertext Markup Language (HTML). Although HTML was the original and is still the most commonly used method for controlling the display of Web pages, two additional forms of HTML have emerged that offer more power and flexibility in Web-page creation—dynamic HTML and eXtensible Markup Language. Let us examine each of these three specifications and discuss their importance to businesses and consumers.

Before we begin, however, the question of why you should know the basic methods of constructing a Web page is important and should be addressed. The Web page is the fundamental element of the World Wide Web. If you understand what is involved in creating a Web page, you can communicate better with individuals who create Web pages, and you can create Web pages yourself if you are ever called upon to do so. A simple Web page can, for instance, be an effective tool in demonstrating an idea to fellow employees. And in some small businesses, you may be the employee who not only designs the product and creates the marketing campaign, but also designs and creates the supporting Web pages.

Markup Languages

In order to create and display Web pages, some type of markup language is necessary. While there are many types of markup languages, we will briefly introduce three common types here: Hypertext Markup Language (HTML), dynamic Hypertext Markup Language (dynamic HTML), and eXtensible Markup Language (XML). HTML, D-HTML, and XML (as well as a fourth language mentioned at the end of this section, XHTML) are members of a family of markup languages called Standard Generalized Markup Language (SGML). Despite the name, SGML itself is not a markup language, but a description of how to create a markup language. To put it another way, SGML is a metalanguage.

Hypertext Markup Language (HTML) is a set of codes inserted into a document that is intended for display on a Web browser. The codes, or markup symbols, instruct the browser how to display a Web page's text, images, and other elements. The individual markup codes are often referred to as tags and are surrounded by brackets (< >). Most HTML tags consist of an opening tag, followed by one or more attributes, and a closing tag. Closing tags are preceded by a forward slash (/). Attributes are parameters that specify various qualities that an HTML tag can take on. For example, a common attribute is HREF, which specifies the URL of a file in an anchor tag (<A>).

Figure 10-9 shows an example of a small HTML file on the left, and a list of descriptive comments on the right. Each line in the file includes an HTML tag, which tells the browser how to display text or images. Line 1's opening <HTML> tag begins every HTML document. HTML documents are broken into HEAD and BODY sections. Line 2 starts the HEAD section. Line 3 generates the title that appears on the title bar of the browser; the title statement always appears within the HEAD section. Line 4 denotes the end of the HEAD section. Line 5 denotes the beginning of the BODY section. Line 6 is a text line with a Break tag, <BR>, at the end. A Break tag inserts a line break, and any text following it starts on the next line. Line 7 begins a new paragraph on a new line. Lines 8 and 9 are Headings, which display text in larger size or bolder typeface than the normal typeface. Headings come in different sizes, with sizes ranging from H1 (the largest) to H6 (the smallest). Line 10, <HR>, generates a Horizontal Rule, which is a graphic dividing line that reaches across the page. Lines 11 and 12 are examples of statements for bolding and italicizing text, respectively.

Line 13 is a statement that places an image at this point in the Web page. The line consists of two parts: the image tag <IMG> and the SRC attribute. The SRC attribute specifies the location (or source) of the image file to be displayed. In this example, the image is in a directory called *images* and has the filename *banner.gif*. All images in HTML documents must be either .gif or .jpg files.

Line 14 is an example of a link or hyperlink. A hyperlink consists of the anchor tag (<A>. . .</A>), the HREF attribute that specifies the URL of the file you want to link to, and text that will be highlighted. In this case, the text "DePaul CS Page" will be highlighted (the default highlight color is blue) and underlined. When a user browsing the Web moves his or her mouse pointer over this highlighted text, the mouse pointer turns into a *clickable hand*. If the user then clicks the highlighted text, the browser tries to load the Web page that is identified by the URL specified in the HREF attribute, which is *http://www.cs.depaul.edu*, in this case.

Figure 10-9
Example of an HTML file

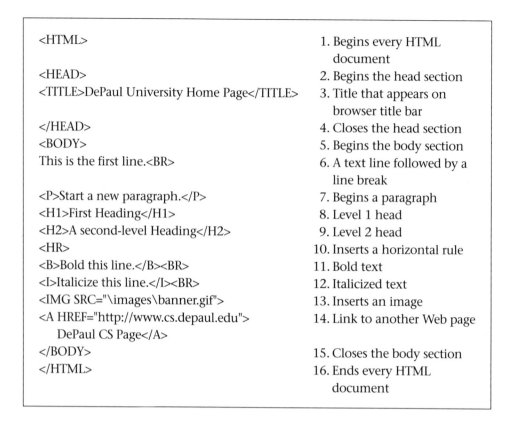

`<HTML>`	1. Begins every HTML document
`<HEAD>`	2. Begins the head section
`<TITLE>DePaul University Home Page</TITLE>`	3. Title that appears on browser title bar
`</HEAD>`	4. Closes the head section
`<BODY>`	5. Begins the body section
`This is the first line. `	6. A text line followed by a line break
`<P>Start a new paragraph.</P>`	7. Begins a paragraph
`<H1>First Heading</H1>`	8. Level 1 head
`<H2>A second-level Heading</H2>`	9. Level 2 head
`<HR>`	10. Inserts a horizontal rule
`<B>Bold this line.</B> `	11. Bold text
`<I>Italicize this line.</I> `	12. Italicized text
`<IMG SRC="\images\banner.gif">`	13. Inserts an image
`<A HREF="http://www.cs.depaul.edu">` `DePaul CS Page</A>`	14. Link to another Web page
`</BODY>`	15. Closes the body section
`</HTML>`	16. Ends every HTML document

Lines 15 and 16 denote the end of the BODY section and the end of the HTML document, respectively. Figure 10-10 shows the Web page as it would appear in a browser.

Figure 10-10
The Web page generated by the example in Figure 10-9 as it would appear in a browser

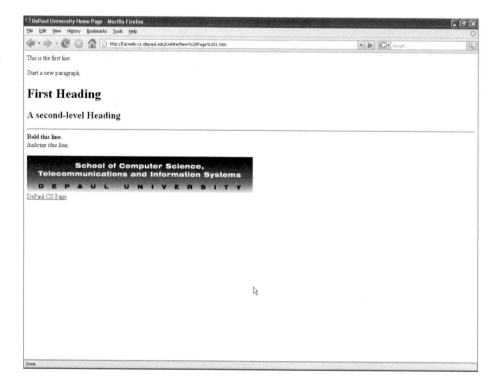

Although HTML is relatively simple to use, it suffers from a number of short-comings. One of the most serious shortcomings is HTML's inability to allow a user to place text or an image at a precise position on the Web page. With HTML, it is also difficult for a user to specify and switch between sets of font styles and colors easily and quickly. To better support these types of functionality, dynamic HTML was created. Rather than functioning as a single specification, dynamic HTML (DHTML) is a collection of newer markup tags and techniques that can be used to create more flexible and more powerful Web pages. HTML pages are simple, static text documents that browsers read, interpret, and display on the screen. In contrast, dynamic HTML pages have additional functionality that allows them to be, among other things, interactive. Stated simply, dynamic HTML can grab any element on a Web page and change its appearance, content, or location on the page dynamically. Features of dynamic HTML include:

- active pop-ups (when a user moves his or her mouse pointer over an area on a page, additional text can be made to appear on the page)
- the live positioning of elements or layers (in other words, the placement of an object on a Web page can be specified by using x,y coordinates)
- cascading style sheets (CSS), which allow a Web page author to incorporate multiple styles (fonts, styles, colors, and so on) in an individual HTML page

Another member of this family of markup languages is eXtensible Markup Language (XML). eXtensible Markup Language (XML) is a subset of SGML and is a specification for how to create a document—the specification covers both the definition of the document and the contents of the document. That is, whereas HTML determines only how the content of a document is to be displayed by a Web browser, XML also defines the content of the document. When an XML document is passed between two entities, the document contains the data and a detailed description of the data. This dual construction eliminates the need for sending additional earlier documents that describe the format of the data.

The syntax of XML is fairly similar to that of HTML; however, there are a number of very important differences. First, XML is extensible, which means a user can define his or her own tags. You can create tags that define entire data structures. For example, the entries in an auto parts catalog might require tags such as <PARTNAME>, <MAKE>, <MODEL>, <YEAR>, <DESCRIPTION>, and <PARTCOST>. Second, XML is much less forgiving than HTML. XML documents have many more precise rules for the creation of tags and the elements within a document. For example, all tags must be properly nested, all attribute values must have quotation marks around them, and all tags with empty content must end in "…/>". Unlike an HTML document, a document created in XML will not be displayed if the coding contains a mistake. A mistake in HTML coding is often ignored by the browser, and the rest of the document is displayed.

For a business, the biggest advantage XML offers is its capability of performing data interchange, the transfer of data records between two companies. Because different companies, and sometimes even different parts of the same organization, rarely create a standard on a single set of tools, it takes a significant amount of intercommunication for two groups to reach a point where they can send data records back and forth. XML simplifies the sending of structured data across the Web so that nothing gets lost in translation. For example, Company A can receive XML-tagged data from Company B, and Company B can receive XML-tagged data from Company A. Neither company has to know how the other company's data is organized, because the XML tags define the data. If another supplier or company wants to be included in this data transfer, the new company does not have to write any computer code to exchange data with its system. Instead, the new company simply has to observe the data definition rules spelled out in the XML tags.

One additional markup language that should be mentioned is eXtensible Hypertext Markup Language (XHTML). XHTML combines HTML, dynamic HTML, and XML into one standard and should eventually replace HTML.

If the future growth of the World Wide Web continues at even a fraction of its current growth rate, the Web is going to remain a vital element in the daily lives of businesses and individuals. To tap that potential, an individual creating a Web page must be familiar with HTML, DHTML, and XML. As these markup languages continue to evolve, Web pages will continue to grow in quality and power.

Now that we have learned how Web pages are created, let us look at some of the other services provided by the Internet.

Internet Services

When the Internet came into existence as the ARPANET, most people used it for e-mail, file transfers, and remote logins. In addition to studying those services, let us examine several of the more popular services that the Internet provides today.

Electronic mail (e-mail)

Electronic mail, or e-mail, is the computerized version of writing a letter and mailing it at the local post office. Many people are so committed to using e-mail that if it were taken away tomorrow, some serious social and economic repercussions would be felt throughout the United States and the rest of the world.

Many commercial e-mail programs are in existence, as well as a number of free ones that can be downloaded from the Internet. Although each e-mail program has its own unique feel and options, most offer the following services:

- ▶ Creating an e-mail message
- ▶ Sending an e-mail message to one recipient, multiple recipients, or a mailing list
- ▶ Receiving, storing, replying to, and forwarding e-mail messages
- ▶ Attaching a file, such as a word-processing document, a spreadsheet, an image, or a program to an outgoing e-mail message

Most e-mail systems consist of two parts: (1) the user agent, which is the portion of the e-mail program that allows a user to create, edit, store, and forward e-mail messages and (2) the message transfer agent, which is the portion of the e-mail program that prepares and transfers the e-mail message. Each transmitted e-mail also consists of two basic components: an envelope, which contains information describing the e-mail message, and the message, which is the contents of the envelope.

Most messages consist of plain text and are written in simple ASCII characters. What if you want to send (or attach) a nontext-based item, such as a spreadsheet, a database, or an image? The e-mail program then creates a Multipurpose Internet Mail Extensions (MIME) document and attaches it to the e-mail message.

Once the e-mail and optional attachment have been created, it is time to transmit the message. The Simple Mail Transfer Protocol (SMTP) is an Internet protocol for sending and receiving e-mail and is used to perform the transfer. To send the e-mail message, the source computer establishes a TCP connection to port number 25 (typically) on the destination computer. The destination computer has an e-mail daemon—a program that is always running in the background and waiting to perform its function—that supports the SMTP protocol. The e-mail daemon watches port 25, accepts incoming connections, and copies messages to the appropriate mailbox. The European and Canadian equivalent to SMTP is the X.400 protocol.

How many times do you receive an e-mail message even though your machine is not turned on? When you turn your computer on and run your e-mail program, a message informs you that you have *n* new e-mail messages waiting. What software performs this operation? Post Office Protocol version 3 (POP3) is the software that allows the user to save e-mail messages in a server mailbox and download them when desired from the server. POP3 is useful if you do not have a permanent connection to a network and must dial in using a temporary Internet connection. POP3 will hold your e-mail messages until the next time you dial in and access your mailbox. Thus, POP3 software is commonly found on mobile laptop computers or home computers without permanent network connections, but also on computer systems that have permanent connections.

An alternative to POP3 is the more sophisticated Internet Message Access Protocol (IMAP). IMAP (the latest version is IMAP4) is a client/server protocol in which e-mail is received and held for you at your Internet server. You can view just the heading of the e-mail or view the sender of the message and then decide if you want to download the mail. You can also create and manipulate folders or mailboxes on the server, delete old e-mail messages, or search for certain parts of an e-mail message.

Many e-mail packages allow a user to encrypt an e-mail message for secure transfer over an internal local area network or over a wide area network such as the Internet. When an e-mail message is encrypted, it is virtually impossible to intercept and decode the message without the proper encryption algorithm and key. Important e-mail messages can also have a digital signature applied so that, in the future, the owner of an e-mail message can prove that the e-mail message belongs to no one but him or her. Encryption techniques will be discussed in more detail in Chapter Twelve.

File Transfer Protocol (FTP)

The File Transfer Protocol, or FTP, was one of the first services offered on the Internet. Its primary functions are to allow a user to download a file from a remote site to the user's computer and to upload a file from the user's computer to a remote site. These files could contain data, such as numbers, text, or images, or executable computer programs. Although the World Wide Web has become the major vehicle for retrieving text- and image-based documents, many organizations still find it useful to create an FTP repository of data and program files. Using a Web browser or specialized FTP software, you can easily access an FTP site. If you desire privacy and wish to restrict access to an FTP site, the site can be designed to require a user ID and password for entry.

To access an FTP site via a Web browser and download a file, you need at least three pieces of information. First, you must know the name of the FTP site. Second, you must know the name of the directory or subdirectory in which to look for the file. Third, you must know the name of the file that you want to download. Thus, downloading FTP files is not a "browsing" activity. You must have a good idea of what you are looking for and where it is located.

As an example, let us say you are reading an article in a computer magazine, and the article describes a free utility program that organizes your time, keeps you on schedule, and helps you make new friends. You must have this program, but it is on some FTP site in the middle of nowhere. No problem. All you need is the utility program's URL and a Web browser. As explained in an earlier section, every document on the Internet has a unique URL. For our example, pretend the URL of this program is:

ftp://ftp.goodstuff.com/public/util/

and the name of the file is *perfect.exe*. At the Web browser's location or address prompt, you type this URL. When you arrive at the *ftp.goodstuff.com* FTP site and are in the public/util subdirectory, you should see the file *perfect.exe* listed. Clicking the filename and specifying the receiving location of the downloaded material starts the downloading process. The FTP software sends the appropriate FTP request command to the FTP site, which opens a file transfer connection between your workstation and the remote site. The file is broken into packets, and the packets are transferred one by one over the Internet to your workstation. When all the packets for the file have arrived, the connection is dissolved. Actually, two connections were created. The first connection transferred the control information, which set up the file transfer, and the second connection transferred the actual data.

Details ▶

How the MIME System Works

Five different MIME type headers provide information about a file that has been attached to an e-mail message: MIME version, content description, content ID, content type, and content transfer encoding. The MIME version states the version number of the MIME protocol being used. The content description is the actual document being attached. The content ID identifies the MIME attachment and its corresponding e-mail message. The content-type header describes whether the attachment is a text, image, audio, video, application, message, or multipart file.

The content-transfer-encoding header describes how the non-ASCII material in the attachment is encoded, so the non-ASCII characters do not create havoc when transmitted. An e-mail program can use one of five different schemes when encoding non-ASCII material for transmission:

▶ Plain 7-bit ASCII, the more common form of ASCII.

▶ 8-bit ASCII, which is not supported by many systems.

▶ Binary encoding, which is machine code. Binary encoding is dangerous to use, because a particular binary value may signal a special control character, such as carriage return, which might cause the receiving device to behave erratically.

▶ Base64 encoding, which is the best method. To perform Base64 encoding, the computer breaks 24 bits of binary data into four 6-bit groups, then encodes each 6-bit group into a particular ASCII character (based on the standard ASCII character chart).

▶ Quoted printable encoding, which uses 7-bit ASCII. In quoted printable encoding, any character with a decimal ASCII value greater than 127 is encoded with an equal sign (=) followed by the character's value as two hexadecimal digits.

Remote login (Telnet)

Remote login, or Telnet, is a terminal emulation program for TCP/IP networks, such as the Internet, that allows users to log in to a remote computer. The Telnet program runs on your computer and connects your workstation to a remote server on the Internet. Once you are connected to the remote server or host, you can enter commands through the Telnet program, and those commands will be executed as if you were entering them directly at the terminal of the remote computer.

There are three reasons for using a Telnet program. First, Telnet allows a user to log in to a personal account and execute programs no matter where he or she is. For example, you might have a computer account at two different companies or two different schools. Although you may be physically located at one site, you can use Telnet to log in to the other computer site. Through this login, you can check your e-mail or run an application. Second, remote login allows a user to access a public service on a remote computer site. For example, the Colorado Alliance of Research Libraries (CARL) provides a large suite of databases, which include bibliographic indexing and abstracting services as well as full-text files. You can gain access to CARL through a Telnet connection or by using a Web browser. Third, Telnet enables a network administrator to control a network server and communicate with other servers on the network. This control can be performed from a remote distance, such as another city or the network administrator's home.

Voice over IP

One of the newer services that is attracting the interest of companies and home users alike is the sending of voice signals over an IP-based network such as the Internet. The practice of making telephone calls over the Internet has had a number of different names, including packet voice, voice over packet, voice over the Internet, Internet telephony, and Voice over IP (VoIP). But it appears the industry has settled on the term "Voice over IP," in reference to the Internet Protocol (IP), which controls the transfer of data over the Internet. Whatever its title, Voice over IP has emerged as one of the hottest Internet services and has certainly drawn the attention of many companies.

Because this technology is relatively new, the implementation of Voice over IP varies greatly, depending on your level of involvement. Due to the allure of a potentially large market, many companies are offering complete packages that can be installed on a local area network system. These packages involve large amounts of equipment, such as VoIP servers, high-speed switches, special IP-enabled telephones, and routers that can direct telephone calls. At the other end of the spectrum, some telecommunications companies offer a service only for individual/small business users and home users. These services use traditional telephone lines, or cable television services, and telephones but require special adapters (that convert traditional telephone signals to IP packets and back) to be inserted between the line and the phone.

When a reliable and high-quality telephone system is already available, why explore new technology that provides the same service? One of the earlier advantages of Voice over IP was simply related to the fact that long-distance calls, especially overseas calls, cost money, while sending data—or voice—over the Internet is essentially free. As it turns out, this advantage has become less important for many corporate users. One reason for this is that long-distance telephone rates have dropped significantly over the years; another is that the call quality of long-distance

VoIP is often worse than conventional long-distance. But many companies are now finding other, more important advantages in being able to treat voice data like other forms of data. For one, if both voice and data can travel over the same network, companies would realize savings in both equipment and infrastructure. This would contribute to yet another significant advantage: the need for separate telephone management personnel, and local area network management personnel would be reduced or even eliminated. Thus, a company's management can be simplified. As we will see shortly, running voice and data over the same system opens the door for interesting and powerful applications.

Voice over IP has a number of disadvantages as well. The statement that sending data over the Internet is essentially free is misleading. Nothing, of course, is free. All Internet users must pay an Internet service provider for access, the interconnecting phone line, and any necessary hardware and software. Also, even more additional hardware and software is necessary to handle the transmission of voice packets over a corporate data network. Nonetheless, if you already have high-speed Internet access, adding VoIP may be a reasonable way to obtain both local and long-distance telephone service.

A second, and more important disadvantage, is that transmitting voice over a corporate network can be demanding on the network's resources. If the current corporate network system is straining to deliver data, adding voice to this system can cause severe service problems. These service problems can be compounded because voice systems require networks that can pass the voice data through in a relatively small amount of time. A network that delays voice data by more than 20 milliseconds from end to end will introduce a noticeable echo into the transmission. If the delay becomes longer than 250 milliseconds (that is only a quarter of a second), the system will be basically unusable.

Further compounding this delay problem is the fact that the Internet is essentially a large datagram packet-switched wide area network. Recall from Chapter Nine that in a datagram packet-switched network, each packet of data is routed individually. This individual routing can introduce a routing delay for each packet, even with the use of newer protocols such as MPLS. (For more information on this protocol, see the Details section titled "Multiprotocol Label Switching.") In contrast, a circuit-switched network (for example, the telephone system) first establishes a route (a circuit) and then transmits all following packets down that established route. Because the route is fixed and dedicated, each packet does not experience a routing delay. To maintain a telephone conversation over the Internet, the voice packets must arrive at their destination in a continuous stream. If they do not, the voice conversation breaks up and is of poor quality. At the present moment, Voice over IP sometimes works well and sometimes does not.

Although VoIP may someday be one of the most common ways for establishing telephone connections, it is still a growing technology. Thus, serious analysis would be required before a company decided to invest time and resources in Voice over IP.

For companies considering Voice over IP, one important and interesting fact to keep in mind is that the technology does not need to involve using the Internet. A company can use IP for transmission of data *within* its own network, but use traditional telephone lines outside the company network. Many people are now beginning to call such systems private VoIP. Because these systems do not use the Internet but instead remain internal, packet delays are minimal, and this makes

Voice over IP attractive. New systems are beginning to appear that can support both telephone operations and computer data operations over the same set of wires. This area of study—computer telephony integration—is an exciting new marriage of computer systems and telephone systems.

In order to gain a little better understanding of VoIP, let us examine the various steps necessary to create a VoIP packet from someone's voice and introduce the more common protocols that are being used to support this new technology. The first step in converting voice to IP packets is to perform an analog-to-digital conversion using a codec. Once the voice is digitized, it is compressed into a much smaller package and then converted into datagram format. The voice datagram is then encapsulated with the appropriate UDP and IP headers and sent over the IP network.

A common device that performs the conversion from an analog telephone call (voice *and* signals) into the packetized IP data is the VoIP gateway. The **VoIP gateway** can perform the digitization, compression, and encapsulation required, and it controls the setup of VoIP calls between the calling device and the called device. Currently, two basic sets of protocols support all these steps. The first set of standards, H.323, is from ITU-T and was first issued in 1996. **H.323** is actually a set of protocols named packet-based multimedia protocols, and it was designed for a wide range of applications (audio and video). Interestingly, H.323 was not originally designed for TCP/IP networks but for X.25 and ATM networks.

The second protocol, Session Initiation Protocol, is the one many believe will become the primary VoIP standard, but apparently this has not happened yet. **Session Initiation Protocol (SIP)** was introduced in 1998 by the Internet Engineering Task Force specifically for supporting the transfer of voice over the Internet. SIP is essentially an application layer protocol that can create, modify, and terminate voice sessions between two or more parties. These voice sessions are not limited to simple telephone calls, but can also include conference calls and multimedia transmissions. SIP also calls upon other protocols to support VoIP. One of these protocols is ENUM. **ENUM** is a protocol that converts telephone numbers to fully qualified domain name addresses. For example, if you want to call the number 555-1212 in area code 312, ENUM will convert it to a domain name address of the form 2.1.2.1.5.5.5.2.1.3.1.e164.arpa (this is the fully qualified number 1(312)555-1212 reversed, with e164.arpa added to the end).

Finally, while not exactly the same as Voice over IP, **Voice over Wireless LAN (VoWLAN)** (and also known as Voice over Wi-Fi) is very similar. Instead of transmitting a voice over an wired IP network, Voice over WLAN allows a user to roam within the building or campus and, using a wireless handset, communicate with one or more people using the wireless LAN network.

There are a number of attractive reasons for using VoWLAN. First, while many people think of cell phones when they think of talking on a wireless handset, many cell phone signals don't travel too deeply into large buildings. Because many businesses already have or are planning wireless LAN networks, their signals are typically much more pervasive within large buildings. Another advantage of VoWLAN over cell phones is the cost—cell phone service is provided by a cell phone provider, whom you have to pay. Talking over the company's existing wireless LAN network is essentially "free".

Of course, VoWLAN is not without its drawbacks. One potential problem is the handoff—when the mobile user moves from one basic service set to another within the network. In order to provide a quality telephone call, the handoff has to be quick and seamless. Many local area networks can become quickly saturated with multiple voice users using the wireless signals. Nonetheless, this is yet one more technology that we should keep an eye on in the near future.

Listservs

A listserv is a popular software program used to create and manage Internet mailing lists. Listserv software maintains a table of e-mail addresses that reflects the current members of the listserv. When an individual sends an e-mail to the listserv address, the listserv sends a copy of this e-mail message to every e-mail address stored in the listserv table. Thus, every member of the listserv receives the e-mail message. Other names for listserv or other types of listserv software include *mailserv*, *majordomo*, and *almanac*.

To subscribe to a listserv, you send a specially formatted message to a special listserv address. This address is different from the address for sending an e-mail to all the listserv members. For example, you could subscribe to the University of Southern Michigan at Copper Harbor's listserv about erasers by sending the message SUBSCRIBE ERASERS to *listserv@copper.usm.edu*. To send an e-mail message to all the subscribers, you would send an e-mail to *erasers@copper.usm.edu*.

Listservs can be useful business tools if you are looking for information on a particular topic, or if you wish to be part of an ongoing discussion on a particular topic. By subscribing to a listserv on your topic of interest, you can receive continuous e-mail from other members around the world who are interested in the same topic. Businesses can also use listservs to communicate with their customers about products and services. To find out what listservs are available, visit the Web site *www.lsoft.com/lists/listref.html*.

Streaming audio and video

Streaming audio and video involves the continuous download of a compressed audio or video file, which can then be heard or viewed on the user's workstation. Typical examples of streaming audio are popular and classical music, live radio broadcasts, and historical or archived lectures, music, and radio broadcasts. Typical examples of streaming video include prerecorded television shows and other video productions, lectures, and live video productions. Businesses can use streaming audio and video to provide training videos, product samples, and live feeds from corporate offices, to name a few examples.

To transmit and receive streaming audio and video, the network server requires the space necessary to store the data and the software to deliver the stream, and the user's browser requires a streaming product such as RealPlayer from RealNetworks to accept and display the stream. All audio and video files must be compressed, because an uncompressed data stream would occupy too much bandwidth and would not travel in real time (recall the compression techniques from Chapter Five).

Real-Time Protocol (RTP) and Real-Time Streaming Protocol (RTSP) are two common application layer protocols that servers and the Internet use to deliver streaming audio and video data to a user's browser. Both RTP and RTSP are public-domain protocols and are available in a number of software products that support streaming data.

Instant Messages, Tweets, and Blogs

Instant messaging (IM) allows a user to see if people are currently logged in on the network and, if they are, to send them short messages in real time. Many users, especially those in the corporate environment, are turning away from e-mail and using instant messaging as a means of communicating. The advantages of instant messaging include real-time conversations, server storage savings (because you are not storing and forwarding instant messages, as you would e-mails), and the capability to carry on a "silent" conversation between multiple parties. Service providers such as AOL, Microsoft's MSN, and Yahoo!, as well as a number of other software companies, incorporate instant messaging into their products.

Several of the issues involving the creation and support of instant messaging systems include:

- The proper identification of users
- How to tell if the user you are trying to contact is present
- Whether it is possible to limit who can see your presence and when
- Whether you can just send short one-sentence messages or something more involved
- Whether the system can support a conversational mode between two users
- Whether the system can support a chat-room mode involving multiple users

One of the more recent services offered on the Internet is Twitter. Twitter is a service that allows individuals to send short messages (maximum 140 characters) to multiple users. It is essentially a cross between instant messaging and blogging (blogging is when an individual posts an ongoing commentary—a blog—for one or more readers). The messages sent in Twitter are termed tweets. The creator of the tweets can allow only select users to read or can allow the general public to read the tweets.

Now that we have a technical understanding of the Internet, let us look at how the Internet can have an impact on business.

The Internet and Business

Throughout this chapter, numerous references have been made to how the Internet affects a business. The term that has come to represent a business's commercial dealings over the Internet is e-commerce. E-commerce can also be defined as the buying and selling of goods and services via the Internet, and in particular via the World Wide Web. To understand the important issues and trends associated with this intersection between technology and business, let us subdivide e-commerce into the following four areas:

- E-retailing—E-retailing is the electronic selling and buying of merchandise using the Web. Virtually every kind of product imaginable is available for purchase over the Web. Sophisticated Web merchandisers can track their customers' purchasing habits, provide online ordering, allow credit card purchases, and offer a wide selection of products and prices that might not normally be offered in a brick-and-mortar store.
- Electronic data interchange (EDI)—Electronic data interchange (EDI) is the electronic commercial transaction between two or more companies.

For example, a company wishing to purchase a large number of cell phones may send an electronic request to a number of cell phone manufacturers. The manufacturers may bid electronically, and the company may accept a bid and place an order electronically. Bank funds will also be transferred electronically between companies and their banks.

▸ Micro-marketing—**Micro-marketing** is the gathering and use of the browsing habits of potential and current customers, which is important data for many companies. When a company knows and understands its customers' habits, it can target particular products to particular individuals.

▸ Internet security—The security systems that support all Internet transactions are also considered an important part of e-commerce.

Many analysts predict that, despite the so-called dot-com bust of the early twenty-first century, e-commerce is not going to go away—that, in fact, it will continue to experience tremendous growth in the coming years. There is little doubt that e-commerce has the potential to be an incredible industry that will produce a wealth of jobs.

Cookies and state information

One feature of the Web that many businesses use and that has received a good deal of negative publicity is the cookie. A **cookie** is data created by a Web server that is stored on the hard drive of a user's workstation. This data, called state information, provides a way for the Web site that stored the cookie to track a user's Web-browsing patterns and preferences. A Web server can use state information to predict future needs, and state information can help a Web site resume an activity that was started at an earlier time. For example, if you are browsing a Web site that sells a product, the Web site may store a cookie on your machine that describes what products you were viewing. When you visit the site at a later date, the Web site may extract the state information from the cookie and ask if you wish to look again at that particular product or at products like it. Other common applications that store cookies on user machines include search engines, the shopping carts used by Web retailers, and secure e-commerce applications.

The information on previous viewing habits that is stored in a cookie can also be used by other Web sites to provide customized content. One recent use of cookies involves the storage of information that helps a Web site decide which advertising banner to display on the user's screen. With this technique, a Web site can tailor advertisements to the user's profile of past Web activity.

Although Web site owners defend the use of cookies as being helpful to Web consumers, many users feel that the storing of information by a Web site on their computers is an invasion of privacy. Users concerned about this privacy issue can instruct their browsers to inform them whenever a Web site is storing a cookie and provide the option of disabling the storage of the cookie. Users can also view their cookies (which may be unreadable) and delete their cookie files. The Firefox browser stores cookies in a file called Cookies.txt; Internet Explorer keeps each cookie as a separate file in a special cookie subdirectory.

Intranets and extranets

One of the more powerful advantages of the Internet is that any person anywhere in the world can access the information you post, be it a Web page or an FTP file. For many people, this accessibility is exactly what is desired. A company, however, may not want the entire world to view one or more of its Web pages. For example, a company may want to allow its employees easy access to a database, but to disallow access by anyone outside the company. Is it possible to offer an Internet-like service inside a company, to which only the company's employees have access? The answer is yes, and the service is called an intranet. An intranet is a TCP/IP network inside a company that allows employees to access the company's information resources through an Internet-like interface. Using a Web browser on a workstation, an employee can perform browsing operations, but the applications that can be accessed through the browser are available only to employees within the company. For example, a corporate intranet can be used to provide employee e-mail and access to groupware software, a corporate client/server database, corporate human resources applications, or custom applications and legacy applications.

Intranets use essentially the same hardware and software that is used by other network applications such as Internet Web browsing, which simplifies installation, maintenance, and hardware and software costs. If a company desires, it can allow an intranet and its resources to be accessed from outside the corporate walls. In these circumstances, a user ID and password are often required to gain access. Companies can use an intranet to establish access to internal databases such as human resource information, payroll data, and personnel records; to interface into corporate applications; to perform full-text searches on company documents; and to allow employees to access and download training materials and manuals, and perform daily business operations, such as filling out travel expense reports. Chances are you probably registered for classes using your college's intranet.

When an intranet is extended outside the corporate walls to include suppliers, customers, or other external agents, it becomes an extranet. Because an extranet allows external agents to have access to corporate computing resources, a much higher level of security is usually established. Essentially, an extranet is an interconnection of corporate intranets. The interconnection is usually performed over the Internet, using virtual private networks. Recall that virtual private networks use tunneling protocols to create connections that cannot be intercepted by outsiders.

Needless to say, business has embraced the Internet in its many forms. Many companies have profited from introducing the Internet into their business operations, and an equal number have suffered. The rise and fall of fortunes based upon the Internet and its associated technologies are probably too difficult for anyone to predict. Although we cannot predict the future of the business side of the Internet, we can take a look at the future of the technology of the Internet.

The Future of the Internet

The Internet is not a static entity. It continues to grow by adding new networks and new users every day. People are constantly working on updating and revising the Internet's myriad components. The driving force behind all these changes, as well as all Internet protocols, is the self-regulating government of the Internet.

Based on a committee structure, this government consists of many committees and groups, including:

> ► The Internet Society (ISOC)—A volunteer organization that decides the future direction of the Internet
>
> ► The Internet Architecture Board (IAB)—A group of invited volunteers that approves standards
>
> ► The Internet Engineering Task Force (IETF)—A volunteer group that discusses operational and technical problems
>
> ► The Internet Research Task Force (IRTF)—The group that coordinates research activities
>
> ► The World Wide Web Consortium (W3C)—A Web industry consortium that develops common protocols and works to ensure interoperability among those protocols
>
> ► Internet Corporation for Assigned Names and Numbers (ICANN)—An international organization of members who control the names and numbers associated with web addresses
>
> ► Many more steering groups, research groups, and working groups

The best place to look for recently completed changes and changes-in-progress is the Internet Engineering Task Force's Working Groups Web page. IETF's Web site is *www.ietf.org*; it contains all the available current information about the Internet and its protocols.

One of the biggest changes to affect the Internet will be the ongoing adoption of a new version of the Internet Protocol, version IPv6. Currently most of the Internet is using IPv4, which was the version presented earlier in the chapter. (In case you are wondering, IPv5 was not a version open to the public but one used for testing new concepts.) While IPv6 has been out for a number of years now, only a few companies and schools have begun to implement it. Even the United States government has been slow to adopt it, with some divisions stating implementation sometime during 2008. Let us take a closer look at the details of version 6 and see how they compare to the current version, IPv4.

IPv6

When the Internet Protocol was created in the 1960s, the computing climate was not the same as it is today. There was nowhere near the number of users currently using the Internet, and the telecommunications lines used to support the high-speed networks were not as fast and as error-free as they are today. Also, the applications transmitted over the Internet involved smaller data packets, and there was not such a demand to transmit them in real time. As these demands on the Internet began to grow, the designers decided it was time to create a more modern Internet Protocol that took advantage of the current technology. Thus, IPv6 was created.

Several notable differences exist between IPv6 and the current IPv4. The first concerns addressing. As you have learned, IPv4 uses 32-bit addresses. With the explosive growth of the Internet, concern arose that the current system will run out of IP addresses in the near future. Consequently, IPv6 calls for addresses to be 128 bits long. With 128-bit addresses, the Internet will be able to assign one million IP addresses every picosecond (10^{-12} seconds) for the known age of the universe! Needless to say, with IPv6, we will never run out of addresses.

Significant changes were also made to the IP header between version 4 and IPv6. As you may recall from Figure 10-4, the IP header in version 4 contains 14 fields. In IPv6, the IP header contains eight fields, plus the payload (data) and optional extension headers, as shown in Figure 10-11. As with the IPv4 header, the first field is four bits long and represents the version number (6 for IPv6 and 4 for IPv4). The second field is the Priority field, which allows values 0 through 7 for transmissions that are capable of slowing down, and values 8 through 15 to indicate real-time data. The third field is Flow Label; this field allows a source and destination to set up a pseudoconnection having particular properties and requirements. This would allow certain applications and connections to more quickly route their packets through the Internet. The fourth field is Payload Length, which identifies how many bytes follow the 40-byte header. The fifth field is called the Next Header and tells which, if any, of six extension headers follow. If no extension headers follow Next Header, then Next Header simply tells which transport layer header (TCP, UDP, and so on) follows. The sixth field is Hop Limit, which says how long this particular datagram is allowed to live—hop from network to network—within the system; it provides the same information as the Time to Live field in IPv4. The last two fields are the Source Address and the Destination Address.

Figure 10-11
The fields in the IPv6 header

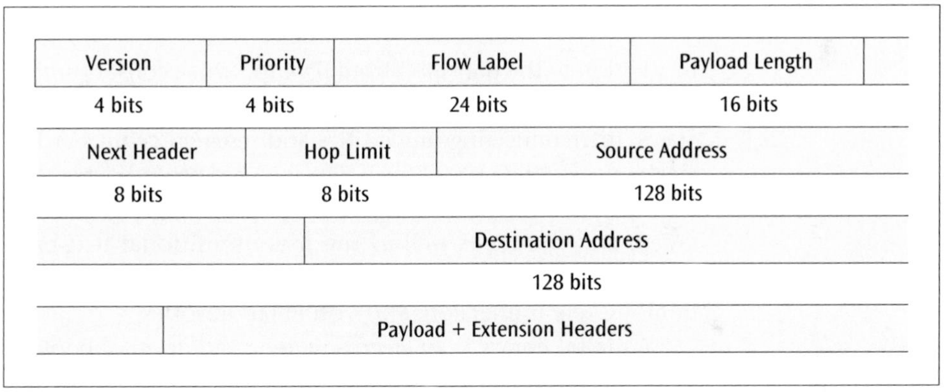

Version	Priority	Flow Label	Payload Length	
4 bits	4 bits	24 bits	16 bits	
Next Header	Hop Limit	Source Address		
8 bits	8 bits	128 bits		
	Destination Address			
	128 bits			
	Payload + Extension Headers			

Which fields in IPv4 are no longer present in the IPv6 header? IPv6 does not have a Header Length field. Because the header in the new version is a fixed length, a length field is unnecessary. Another field not included in IPv6 is the field that allows fragmentation. The creators of IPv6 felt that most of the networks in operation today can handle packets of very large sizes, and that it would no longer be necessary to break (or fragment) a packet into smaller pieces. Thus, even though IPv6 can perform fragmentation, this function is only available as an option in an extension header. Yet another field whose absence from IPv6 is surprising is the Checksum field. Doing away with Checksum fields appears to be a trend in most modern networks, as both network quality improves and higher-level applications assume the role of error detector.

Although modifications to the IP header represent profound changes in the IP protocol, there are even more significant differences between IPv4 and IPv6. Namely, IPv6 has:

▸ Better support for options using the extension headers

▸ Better security, with two extension headers devoted entirely to security

▸ More choices in type of service

This last improvement to the IP protocol relates to quality of service (QoS), which is an important part of modern networks. It is a very useful tool if a user can specify a particular level of service, and the network can support that level. IPv6 should deliver much better quality of service than IPv4.

Unfortunately, despite these improvements, the jury is still out on when IPv6 will replace IPv4. Although the U.S. Office of Management and Budget announced in 2005 that it intends to require support for the protocol across government agencies by 2008, corporate America has been very slow in adopting the newer protocol. There are a number of reasons for this. Many users have complained that the new features offered by IPv6 are not terribly different from the features that have been added to IPv4 over the past few years. In addition, although the 128-bit address space offers an unimaginable number of addresses, it consumes a whopping 16 bytes of packet space. If a device needs to send a small packet, the address space alone will account for a huge portion of the packet. Finally, many network managers like using the NAT and DHCP protocols of IPv4. These protocols enable them to use the company's IP address space conservatively, while keeping the internal IP address structure secret from the outside world. It appears that IPv6 is yet another technology that we will have to keep our eyes on in the future.

Internet2

In addition to the transition from IPv4 to IPv6, a consortium of universities, businesses, and the government created a very high-speed network that will cover the United States, interconnecting universities and research centers at transmission rates up to a gigabit per second (1000 Mbps). The new high-speed network is called Internet2.

Internet2's creators aimed to provide high-speed access to digital images, video, and music, as well as the more traditional text-based items. In particular, Internet2 has targeted a number of primary application areas, including digital libraries, tele-immersion, and virtual laboratories.

A digital library is an electronic representation of books, periodicals, papers, art, video, and music. A patron accessing a digital library can quickly retrieve any document, using a powerful query language. Even art, video, and music can be retrieved by specifying one or more keywords that describe the contents of the work.

Tele-immersion enables users at geographically diverse locations to collaborate in real time within a shared, simulated environment. The technology has high-power interactive audio and video capabilities that enable it to make users feel as though they are all in the same room. Some of these high-power tools include three-dimensional environment scanning and projective, tracking, and display technologies.

With Internet2's virtual laboratories, it is possible to create realistic lab surroundings without the expense of brick-and-mortar facilities. Through virtual laboratories, students and researchers are able to conduct experiments that cannot be conducted in the real world, such as atomic explosions. Plus, if virtual laboratories are combined with the application area of tele-immersion, students and researchers from around the world would be able to collaborate on one or more projects.

In terms of the technology, Internet2 is not simply a physical network, as the current Internet is, and it will not replace the Internet. Instead, Internet2 is result of a partnership of universities (more than 200 participating schools), industry (dozens of leading companies), and the government; these organizations share the goal of developing new technologies and capabilities that will eventually be accessed by a wide number of users. Both businesses and the academic community will benefit from the creation and application of these new technologies and capabilities.

Now, let us look at an example that applies what we have learned about the technological and business-related aspects of the Internet.

The Internet In Action: A Company Creates a VPN ▶

CompuCom, a fictitious workstation reseller, was looking for a way to make its salespeople more productive and thus save money. One possible option was to allow the salespeople to work from home. Using a home computer, each employee could access the corporate network to send and retrieve messages as well as access the company database. The management at CompuCom learned that there were a number of ways the company could arrange for employees to access the corporate network. For one, employees could use a dial-up modem to connect with the company system. The problem with this solution was that dial-up modems are slow, and the company would need one modem on site for each employee dialing in (Figure 10-12). Another option was to set up each employee to use DSL or a cable modem, both of which are much faster than dial-up modems. The employee's DSL or cable modem would connect to an Internet service provider, through the Internet, and into the corporate network. The biggest problem here was security. Unless special measures were taken, any data transferred over the Internet would be not secure. CompuCom wanted to take advantage of the Internet, because it was so widespread and economically priced, but the company also needed a way to create a secure connection.

Figure 10-12
CompuCom employees dialing directly in to the corporate computing center

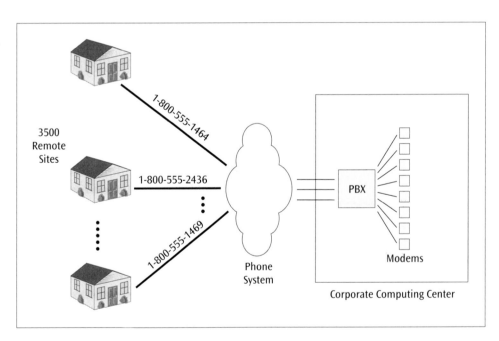

Ultimately, CompuCom decided to create a virtual private network. A virtual private network (VPN) is a data network connection that makes use of the public telecommunications infrastructure but maintains privacy through the use of a tunneling protocol and security procedures. A tunneling protocol, such as the Point-to-Point Tunneling Protocol (PPTP), is the command set that allows an organization to create secure connections using public resources such as the Internet.

Note that this communications link is not a completely private connection, as a leased telephone line would be. To create a network with a leased line, CompuCom would need the telephone company to install a special circuit between two locations that no one else would be able to use. In contrast, a VPN over the Internet uses the public circuits that thousands of users share every second. With special routing information and high-security measures, however, a VPN can be created that only you can access and understand.

By using the Internet as the transfer medium, a VPN offers a number of advantages. Because the Internet is practically everywhere, a VPN can be created between any two points. Because the Internet is relatively economical, the VPN is also relatively economical. Once a connection to the Internet has been achieved (via an Internet service provider), data transferred over the Internet—and over the VPN—is free. The Internet—and thus the VPN—maintains a relatively stable environment with reasonable throughput.

With CompuCom's VPN, an employee at home connects to his or her local Internet service provider through a DSL or cable modem connection. Once the user is connected to the Internet, tunneling software creates a secure connection across the Internet to the corporate computing center (Figure 10-13).

Figure 10-13
CompuCom's employees using a tunnel across the Internet into the corporate computing center

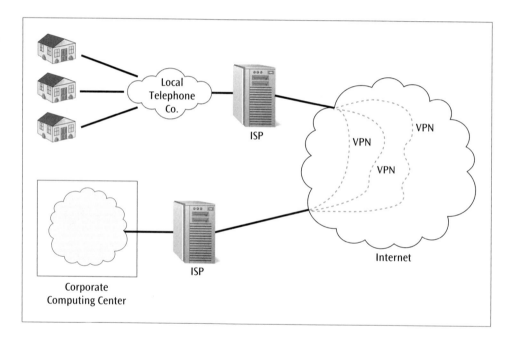

A VPN is one example of a creative use for the extremely popular and powerful Internet. As businesses continue learning to use the Internet, they will be able to branch out into new territories and improve on existing communications strategies.

✦ ✦

SUMMARY

▶ To support the Internet, many protocols, such as IP, TCP, ICMP, UDP, ARP, DHCP, and NAT, are necessary.

　▶ The Internet Protocol (IP) provides a connectionless transfer of data over a wide variety of network types.

- ▶ Transmission Control Protocol (TCP) resides at the transport layer of a communications model and provides an error-free, end-to-end connection.
- ▶ Internet Control Message Protocol (ICMP) performs error reporting for the Internet Protocol.
- ▶ User Datagram Protocol (UDP) provides a connectionless transport layer protocol, in place of TCP.
- ▶ Address Resolution Protocol (ARP) translates an IP address into a CSMA/CD MAC address for data delivery on a local area network.
- ▶ Dynamic Host Configuration Protocol (DHCP) allows a network to dynamically assign IP addresses to workstations as they are needed.
- ▶ Network Address Translation (NAT) allows a network to replace local IP addresses with one, global-type IP address.
- ▶ Tunneling protocols allow a company to create virtual private network connections into a corporate computing system.
- ▶ The World Wide Web is a vast collection of electronic documents containing text and images that can be accessed by simply clicking a link within a browser's Web page. The browser uses HTTP to transmit and receive Web pages and HTML to display those Web pages. HTML and dynamic HTML are markup languages used to define how the contents of a Web page will be displayed by a Web browser. XML is a set of rules for defining your own markup language. By creating your own markup language, you can create a document that contains both the data and a definition of that data.
- ▶ To locate a document on the Internet, you usually refer to its Uniform Resource Locator (URL). A URL uniquely identifies each document on every server on the Internet. One component of the URL is the site address of the requested document. Each site address consists of a network ID and a host or device ID. The site address is converted to a 32-bit IP address by the Domain Name System (DNS).
- ▶ The Internet consists of many commonly used network applications.
  - ▶ Electronic mail, or e-mail, is a standard requirement for most business operations and can transfer standard text messages and include MIME-encoded attachments. Protocols such as Simple Mail Transfer Protocol (SMTP), Post Office Protocol version 3 (POP3), and Internet Message Access Protocol (IMAP) support the operations of e-mail.
  - ▶ File Transfer Protocol (FTP) is useful for uploading or downloading files across the Internet.
  - ▶ Remote login using Telnet allows an individual to log in to a remote computer site and perform operations as if the user were physically located at the remote site.
  - ▶ VoIP (Internet telephony) offers an inexpensive alternative to long-distance calling, but with questionable quality. The Internet was not designed to transfer real-time data, which is a capability that is necessary to support interactive voice. Nevertheless, many businesses are embracing VoIP internally as a way to deliver combined voice and data applications.
  - ▶ A listserv is a popular software program used to create and manage Internet mailing lists.
  - ▶ Streaming audio and video are the continuous downloading of a compressed audio or video file, which is then heard or displayed on the user's workstation. Streaming audio and video require support protocols such as Real-Time Protocol (RTP) and Real-Time Streaming Protocol (RTSP).
  - ▶ Instant messaging is growing in popularity as a way to maintain real-time communications between multiple users.
  - ▶ Twitter and blogs are two more Internet applications that allow individuals to post information for one or more users.

▶ E-commerce, a rapidly growing area of the Internet, is the buying and selling of goods and services electronically. Many companies are investing heavily in e-commerce in hopes that it will increase their market share and decrease their costs.

▶ Cookies store state information on a user's hard drive and provide a way for Web sites to track a user's Web-browsing patterns and preferences.

▶ An intranet is an in-house Internet with Web-like services that are available only to a company's employees or to customers and suppliers through an extranet.

▶ The Internet continues to evolve, with a new Internet Protocol version 6 as well as a completely new, higher-speed Internet2.

KEY TERMS

Address Resolution Protocol (ARP)
ARPANET
blog
cascading style sheets (CSS)
cookie
datagram
domain name
Domain Name System (DNS)
Dynamic Host Configuration Protocol (DHCP)
dynamic HTML (DHTML)
e-commerce
electronic data interchange (EDI)
electronic mail (e-mail)
ENUM
e-retailing
eXtensible Hypertext Markup Language (XHTML)
eXtensible Markup Language (XML)
extranet
File Transfer Protocol (FTP)
H.323
Hypertext Markup Language (HTML)

Hypertext Transfer Protocol (HTTP)
instant messaging (IM)
Internet2
Internet Control Message Protocol (ICMP)
Internet Message Access Protocol (IMAP)
Internet Protocol (IP)
intranet
IP multicasting
IPsec
IPv6
listserv
micro-marketing
MILNET
Multiprotocol Label Switching (MPLS)
Multipurpose Internet Mail Extensions (MIME)
Network Address Translation (NAT)
Point-to-Point Protocol (PPP)
Post Office Protocol version 3 (POP3)
private VoIP
Real-Time Protocol (RTP)

Real-Time Streaming Protocol (RTSP)
remote login
Session Initiation Protocol (SIP)
Simple Mail Transfer Protocol (SMTP)
socket
streaming audio and video
subnet masking
Telnet
Transmission Control Protocol (TCP)
tunneling protocol
tweet
twitter
Uniform Resource Locator (URL)
User Datagram Protocol (UDP)
virtual private network (VPN)
Voice over IP (VoIP)
VoIP gateway
Voice over wireless LAN (VoWLAN)
World Wide Web (WWW)

REVIEW QUESTIONS

1. What was the precursor to the present day Internet?
2. List the main responsibilities of the Internet Protocol.
3. List the main responsibilities of the Transmission Control Protocol.
4. Explain the relationship of the port number to an IP address.
5. Is UDP the same thing as TCP? Explain.
6. What is the purpose of MPLS?
7. What is the relationship between IP and ICMP?
8. Is DHCP the same thing as NAT? Explain.
9. What is the relationship between the Internet and the World Wide Web?

10. What is HTTP used for?

11. How is a Web markup language different from a programming language?

12. What features were added to HTML to produce dynamic HTML?

13. What can a cascading style sheet add to a Web page?

14. How does XML differ from HTML and dynamic HTML?

15. What is the purpose of the Uniform Resource Locator (URL)?

16. List the four basic parts of the URL.

17. How does the Domain Name System translate a URL into a 32-bit binary address?

18. What are the different classes of IP addresses?

19. What is the File Transfer Protocol (FTP) used for?

20. What is the main function of Telnet?

21. Is VoIP more reliable within a business or over the Internet? Explain.

22. What other options are available for transmitting voice via Internet technology besides voice over IP?

23. List the basic features of a common electronic mail system.

24. What are the duties of SMTP, POP3, and IMAP, when referring to e-mail systems?

25. What is a listserv used for?

26. What tools are necessary to support streaming audio or video?

27. How does instant messaging differ from e-mail?

28. What are the two competing protocols used to support instant messaging?

29. What is the relationship between electronic data interchange and e-commerce?

30. What is a cookie? Who makes cookies, and where are they stored?

31. How do intranets and extranets compare with and differ from the Internet?

32. How can a business use a virtual private network and tunneling to support an off-site connection?

33. How will IPv6 differ from the current version (4) of IP?

34. What are the main features of Internet2?

35. What is the difference between VoIP and VoWLAN?

EXERCISES

1. Does the U.S. government support the Internet? Explain.

2. Given an IP packet of 540 bytes and a maximum packet size of 200 bytes, what are the IP Fragment Offsets and More flags for the appropriate packet fragments?

3. The Hop Limit field in IPv6 is 8 bits long (the same size as the equivalent Time to Live field in IPv4). Given that this hop count is reduced each time an IP datagram enters a router, what are the implications of such a small field size?

4. Suppose there is a small commercial retail building in your town that has essentially one room. On one side of the room is a real estate agency, and on the other side of the room is a guy who sells tea. How does this situation serve as an analogy for the concept of IP addresses with TCP port numbers?

5. Somewhere in the middle of the United States are two Internet routers with routing tables that are all messed up. The two routers keep sending their packets back and forth to each other, nonstop. Will an error message ever be generated from this action? If so, who will generate the error message, and what might it look like?

6. Why is ARP necessary if every workstation connected to the Internet has a unique IP address?

7. If I dial in to the Internet from home, is it likely that my workstation is using DHCP? Explain.

8. If someone on the Internet sends you an e-mail, will NAT delete it before it gets delivered to you? Why, or why not?

9. If three users on one local area network all request a Web page at the same time, how does NAT know which results go to which workstation?

10. If your computer workstation has an IP address of 167.54.200.32, is it a Class A, B, C, D, or E IP address?

11. What are the advantages and disadvantages of using dynamic IP address assignments?

12. Locate and label the service, host name, directory, and filename in the following hypothetical URL: *http://www.reliable.com/listings/pages/web.htm.*

13. Using a Web browser, go to the site *www.gatekeeper.dec.com*, and list three of the sub-directories in the directory pub.

14. If you use your computer primarily to access a remote computer via Telnet, will you have a large long-distance telephone bill? Explain.

15. What is the difference between voice over the Internet and private VoIP?

16. Convert the telephone number 413 555-2068 into ENUM form.

17. With respect to e-mail, what is the relationship between SMTP and POP3? Can one operate without the other? Explain.

18. Your company has asked you to keep abreast of workplace ethics. Which Internet service(s) introduced in this chapter would best help you accomplish this task?

19. Are HTTP and HTML two protocols with the same function? Explain.

20. Given the following section of HTML code, what will be displayed on the screen when the HTML code is read by a browser?

    <HTML>

    <BODY>

    <H1> Chapter Ten </H1>

    <HR>

    <IMG SRC="amerflag.gif" ALIGN=RIGHT>

    <A HREF="http://www.everyone.edu"> Everyone College Home Page </A>

    </BODY>

    </HTML>

21. Which dynamic HTML concept can be used to print the phrase *Computer Network* multiple times on the screen, each one larger than the previous one, and each one slightly overlapping the previous one?

22. Is HTML a subset of XML, or is it the other way around? Explain.

23. Some of the new protocols, such as Internet Protocol version 6, are not including any kind of error-detection scheme on the data portion of the packet. What is the significance of this trend?

THINKING OUTSIDE THE BOX

1 You are working for a company that wants to begin electronic data interchange (EDI) with two other companies that supply parts. Your company produces mobile telephones, and the two other companies produce batteries and telephone keypads. When your company places an order with the battery company, it specifies battery size, battery type, battery power, and quantity. When your company places an order with the keypad company, it specifies keypad configuration, power consumption, keypad dimensions, and keypad color. How might XML be used to support EDI between these companies?

2 Two banks want to establish an electronic link between themselves, over which they can transmit money transfers. Can they use a virtual private network and a tunneling protocol, or is a better technique available? Defend your answer.

3 You are working for a company that sells high-end bicycles and has decided to create a series of Web pages to promote sales. What kind of services might you offer in your Web pages? Will your Web pages use HTML, dynamic HTML, or some other language? How might your company use cookies to its advantage?

4 You are thinking about creating a network solution for a small business of approximately 10 users. You give each user a workstation and access to the Internet. Which Internet protocols will your network and workstations need to support? Which optional IP protocols might you implement?

5 A company currently has a telephone system and a separate system of local area networks. It is thinking about converting its telephone system to VoIP and running both voice and data over the system of local area networks. What must the company consider before making this move?

HANDS-ON PROJECTS

1. Using your computer at home or at school, locate all the cookies stored on your hard disk drive. Can you tell by examining the cookies who put them there? Why, or why not?

2. On the Web, locate a free copy of a streaming audio or video player and download it to your machine. Then, locate a Web site that delivers a streaming audio or video program. How difficult was the download and installation of the player? How good is the quality of the audio or video stream?

3. Either by using a Web page editor or by writing HTML code manually, create a Web page that has an H1 heading, some text, one or more links, and one or more images.

4. Find an example of a company that performs electronic data interchange. What companies are involved with the interchange? What data is transferred during the interchange? What other details can you describe?

5. If you have access to the Internet at work, school, or home, does your workstation have a Class B IP address or a Class C IP address? Is it statically assigned or dynamically assigned? Did your IP address come from an Internet service provider or from somewhere else? Explain your reasoning.

6. Who are the "players" involved with the new Internet2? What does it take to become one of these players?

7. What are the other protocols that are used to support H.323? SIP?

11

Voice and Data Delivery Networks

◆◆◆

BACK IN THE OLD DAYS, most houses and apartments had one telephone. That telephone had a number, and if you never moved, you probably kept that telephone number for a long time. And because there were no cellular telephones, your home phone number was the only phone number you had. Today, people tend to move more often, and virtually everyone has a cellular phone. If you move from one city to another, chances are you will need a new home telephone number.

Changing telephone numbers has always been a hassle, until recently. The federal government has given cellular telephone companies the authorization to allow subscribers to keep their cellular telephone number if they change services, and there is much speculation that the non-cellular telephone companies may eventually follow suit. If and when this happens, it might be possible

for someone to have one telephone number for an entire lifetime, no matter where he or she lives.

Imagine that one day you are at the hospital for the birth of your child. While completing the usual hospital paperwork, you come across three forms that stand out from the rest. The first is a form for your newborn to receive a Social Security number. The second is for the baby to receive a telephone number. The third is for your child to receive a URL for his or her future Internet needs.

Does this seem far-fetched? Could the hospital assign just one number for all services?

How important is the common telephone system?

Are there other telephone services that may be of use to us?

Objectives ▶

After reading this chapter, you should be able to:

▶ Identify the basic elements of a telephone system and discuss the limitations of telephone signals

▶ Describe the composition of the telephone industry before and after the 1984 Modified Final Judgment and explain the differences

▶ Describe the difference between a local exchange carrier and an interexchange carrier and list the services each offers

▶ Differentiate between the roles of the local telephone company before and after the Telecommunications Act of 1996

▶ Describe the basic characteristics of a 56k modem

▶ List the types of leased lines that are available and their basic characteristics

▶ Identify the main characteristics of digital subscriber line, and recognize the difference between a symmetric system and an asymmetric system

▶ Identify the main characteristics of a cable modem

▶ List the basic characteristics of frame relay, such as permanent virtual circuits and committed information rate

▶ Identify the main characteristics of Asynchronous Transfer Mode, including the roles of the virtual path connection and the virtual channel connection, the importance of the classes of service available, and ATM's advantages and disadvantages

▶ List the basic characteristics of MPLS and VPN tunnels and how their growth in the industry is having an effect on frame relay and ATM.

▶ Describe the concept of convergence, and identify several examples of it in the networking industry

In Chapter Nine, we were introduced to wide area networks. Then, in Chapter Ten we discussed one of the most famous wide area networks—the Internet. While the Internet can transfer conventional data and voice data, it is not the only wide area network that can do this. In fact, many other types of networks have been delivering data and voice over local and wide distances for many years. The most obvious example of such a network is the telephone system. Although it originally was designed to transmit voice signals, the telephone network now transfers more data than voice. In addition to the telephone system, a number of other important networks can deliver both data and voice over both local and wide areas. Unlike the telephone network, however, these networks were originally designed for data. More precisely, these networks provide some form of data delivery service. A customer contacts a service provider and requests a connection between two geographic locations. Once the connection is in place, the customer can transfer data at contracted speeds between the two endpoints. As we will see shortly, some of these data delivery networks work over local distances only, while others work over local and long distances. But before we examine these newer data delivery services, let us examine the telephone system.

The Basic Telephone System

The basic telephone system, or plain old telephone service (POTS), has been in existence since the early 1900s. During most of those years, POTS was an analog system capable of supporting a voice conversation. It was not until the 1970s that POTS began carrying computer data signals as well as voice signals. The amount of data transmitted on POTS eventually grew so large that near the end of the twentieth century the system carried more data than voice. Even though it has experienced a number of technological changes in its long life, such as the increasing use of digital signals in place of analog signals, POTS is still basically a voice-carrying medium. Because of the requirements associated with carrying voice, POTS has limitations. Let us start our discussion of telephone networks by examining the composition of the basic telephone system, followed by how the telephone system was affected by two major events in telephone history—the breakup of AT&T in 1984 and the Telecommunications Act of 1996.

Telephone lines and trunks

Amid all the legal and industry-based changes that have occurred in the field of telecommunications over the years, one thing that has remained relatively constant is the basic telephone line. The local loop is the telephone line that leaves your house or business and consists of either four or eight wires (Figure 11-1). Although only two wires are required to complete a telephone circuit, more and more residential customers are demanding that multiple lines and services be connected to their premises. These extra lines and services require additional pairs of wires. To avoid having to send out a technician to install each additional set of lines, a telephone company often anticipates future user demand and installs extra wires during the initial installation. On the other end of the local loop is the local telephone company's central office. The central office (CO) contains the equipment that generates a dial tone, interprets the telephone number dialed, checks for special services, and connects the incoming call to the next point.

Figure 11-1
The local loop as it connects your house to the telephone company's central office

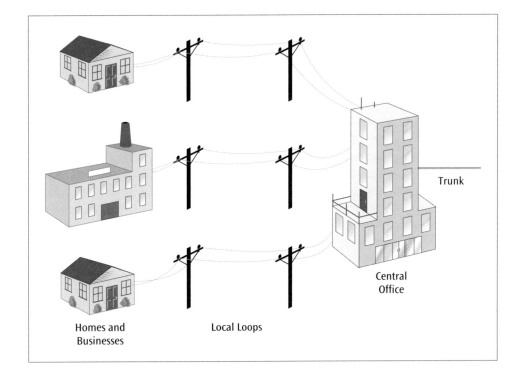

Trunk

Central Office

Homes and Businesses   Local Loops

From the central office, the signal can go to one of three places:

▶ Over another local loop to a house in your neighborhood, if the call is local

▶ To another central office in a neighboring community, if the call is not in the same neighborhood but is not long distance

▶ To a long-distance provider, if the telephone call is a long-distance call

How does the telephone network know when a telephone call is long distance? North America is divided into local access transport areas. A local access transport area (LATA) is a geographic area such as a large metropolitan area or part of a large state. As long as a telephone call remains within a LATA (an intraLATA call), the telephone call is local and is handled by a local telephone company. If the telephone call passes out of one LATA and into another (an interLATA call), the telephone call is long distance and must be handled by a long-distance telephone company. The difference between local calls and long-distance calls is beginning to disappear, however. Many, if not most cellular telephone companies now offer "one-rate" plans that do not distinguish between local and long-distance calls. We are now seeing this type of service appear on noncellular telephones (wired lines) as well. If this continues, someday the function of the LATA will be superfluous.

Returning to the basics, the telephone network consists of two basic types of telephone lines: a subscriber loop, or standard telephone line, and a trunk. A subscriber loop, such as the wire that runs between a house and the central office, has a unique telephone number associated with it. A trunk does not have a telephone number associated with it, because the trunk can carry hundreds of voice and data channels. These hundreds of voice and data channels are typically carried between central offices and other telephone switching centers (the buildings where telephone connections are made between trunks). Most trunks transmit digital signals, and consequently require repeaters every few kilometers. In contrast, subscriber loops usually transmit analog signals, which require amplifiers every few kilometers.

Now we have looked at the basic telephone system. Before we introduce the numerous services that are offered by telephone companies, let us examine two recent major events in the history of telephone networks.

The telephone network before and after 1984

Prior to 1984, AT&T (American Telephone and Telegraph) owned all the long-distance telephone lines in the United States, a majority of local telephone systems, and Bell Laboratories. In the 1970s, the federal government took AT&T to court, citing antitrust violations. AT&T lost the case, and in 1984 the court's ultimate ruling, known as the Modified Final Judgment, required the divestiture, or breakup, of AT&T into separate companies. This breakup allowed AT&T to keep the long-distance lines and Bell Labs, but the company had to divest itself of all local telephone companies, and had to allow other long-distance telephone providers access to their offices.

At the time, AT&T consisted of 23 Bell Operating Companies (BOCs), which provided local telephone service across the country. There were also as many as 14,000 other local telephone companies not owned by AT&T, but these typically served very limited areas. As part of the divestiture, the 23 BOCs were separated from AT&T and, so that they would survive, were reorganized into 7 Regional Bell Operating Companies (RBOCs), as shown in Figure 11-2. These seven RBOCs then competed with the remaining 14,000-plus local telephone companies to provide local service to each home and business in the country. As a result of competition and the sheer power and size of the seven RBOCs, the number of local telephone companies has dropped dramatically over the years. Mergers have also taken a toll on the industry, and as of year-end 2009 only three RBOCs remain: AT&T (formerly Southwestern Bell, BellSouth, and Ameritech), Qwest (formerly US West), and Verizon (formerly Bell Atlantic).

Figure 11-2

The original 23 Bell companies and the newer 7 regional telephone companies

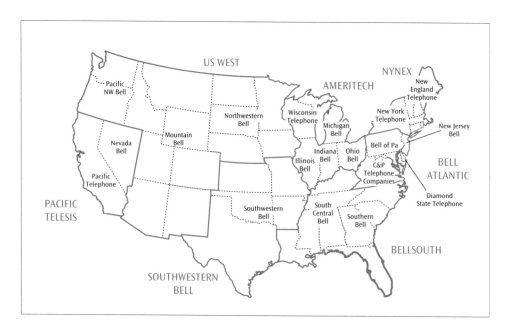

A number of other outcomes of the 1984 divestiture completely changed the landscape of the U.S. telephone system. For one, the United States was divided into the local access transport areas (LATAs) we just learned about. These LATAs determined when a telephone call was local or long distance. The breakup also allowed long-distance telephone companies other than Bell Telephone, such as MCI and Sprint, to offer competing long-distance services. To support this competition among long-distance providers, the local phone companies were required to give all long-distance telephone companies equal access to the telephone lines running into each home and business. Also, people could purchase their own telephones. Before this time, an individual could only lease a telephone from the telephone company. Finally, the local telephone companies became known as local exchange carriers (LECs), and the long-distance telephone companies became known as interexchange carriers (IECs or IXCs).

A local exchange carrier offers a number of services, including Centrex, private and tie lines, and many other telecommunications services such as call waiting and conference calls. A Centrex (central office exchange service) is a service from local telephone companies in which up-to-date telephone facilities at the telephone company's central (local) office are offered to business users, so that they do not need to purchase their own facilities. Businesses are spared the expense of having to keep up with fast-moving technology changes, because the telephone company is providing the hardware and the services, and the business is simply paying a monthly fee. If a business does not want to lease a Centrex service from a local telephone company, it can instead purchase its own Private Branch Exchange, the main competition to Centrex. A Private Branch Exchange (PBX) is a large, computerized, self-contained telephone system that sits in a telephone room on a company's premises. A PBX handles all in-house calls and places calls onto outside telephone lines. A PBX can also offer many telephone services such as voice mail, call forwarding, and dialing plans that use the least expensive local and long-distance telephone circuits.

Private lines and tie lines are leased telephone lines that require no dialing. They are permanent direct connections between two specified points. Consider a company that has two offices in the same city that are always transferring data back and forth. To connect these offices, the company could use a dial-up telephone line with two modems, but because many telephone companies charge for all calls made, a dial-up line would be very expensive. A leased line might offer a less expensive alternative, and it will always be connected—which means an employee in one office never has to dial a telephone number to contact an employee in the other office.

An interexchange carrier, or long-distance telephone company, can also offer a large number of services, including credit card and calling card dialing; 700, 800, 888, and 900 access; international access; and operator and directory assistance.

Telephone networks after 1996

A second major event in the recent history of the telecommunications industry occurred in 1996, with the passing of the Telecommunications Act of 1996. Passed by Congress and administered by the Federal Communications Commission, or FCC, the Telecommunications Act of 1996 paved the way for *anybody* to offer a local telephone service to homes and businesses. New providers of local telephone services were called competitive local exchange carriers (CLECs), and they could include interexchange carriers (long-distance telephone companies), cable television operators, small companies with virtually no equipment (called non-facilities-based CLECs), and even the electric power company. The reasoning behind the development of CLECs was fairly straightforward and meant to better accommodate the way phone services were already being delivered. For example, in some areas of the country, local cable television companies already offered local telephone service. Because a majority of homes and businesses in the United States are already wired for cable television, allowing cable companies to offer local telephone service over those same cable TV lines seemed like a reasonable idea. Termed *cable-telephone* or *cable telephony*, telephone services over cable television lines are now offered by most of the major cable television companies—including Cox Communications and Comcast, as well as many smaller cable companies—in most major cities across the country. In fact, large cable companies in the U.S. are achieving penetration rates of as high as 10 to 25 percent in their particular markets. Of new subscribers, 95 percent are dropping their old local telephone service in favor of the new cable service, and some reports indicate that the number of cable telephony subscribers in 2008 exceeded 34 million, with 48 million predicted by the end of 2010. The medium that delivers cable telephony is a hybrid of fiber and coaxial cable called Hybrid Fiber Coax, or HFC. This technology can also deliver cable modem service into homes and businesses.

Unfortunately, there is a problem with allowing all these new local telephone providers into a market: the telephone lines. It is prohibitively expensive for a new telephone provider to install new telephone lines into each home and business. Cable TV companies have the advantage of using the cable lines that already run into their subscribers' houses, and power companies can use the cables that deliver electrical service, but other companies trying to break into the local telephone market do not have this advantage. To compensate for this high cost of installation, the Telecommunications Act of 1996 mandated that the existing local telephone companies, now called incumbent local exchange carriers (ILECs), must give CLECs access to their telephone lines. Furthermore, the ILECs must give competitors access to telephone numbers, operator services, and directory listings; access to poles, ducts, and rights-of-way; and physical co-location of equipment within ILEC buildings—and they must give these services at wholesale prices.

Needless to say, the ILECs were not thrilled with this arrangement and have been resisting implementing it for some time. Over the last few years, the FCC has made some changes to the Telecommunications Act of 1996 that is altering this requirement, forcing the CLECs to pay more or even full price for access to the local lines. The loss of wholesale prices and the continued resistance of the ILECs has caused most CLECs that rely on the ILEC phone lines and equipment to abandon their services. Already, many CLECs that offered DSL service have given up and closed shop, giving the ILECs and the cable telephone companies a continued larger share of the DSL market. In conclusion, we see that the Modified Final Judgment of 1985 opened up the long-distance telephone markets to competition, while the Telecommunications Act of 1996 was meant to open up the local-distance telephone markets to competition.

Limitations of telephone signals

Telephone systems were originally designed to transmit the human voice. As we have already seen in Chapter Two, the average human being has a frequency range of roughly 300 to 3400 Hz (a bandwidth of 3100 Hz). Thus, the telephone network was engineered to transmit signals of approximately 3100 Hz. In practice, the telephone system actually allocates 4000 Hz to a channel and uses filters to remove frequencies that fall above and below each 4000-Hz channel. The channels and their frequencies are depicted in Figure 11-3.

Figure 11-3
The various telephone channels and their assignment of frequencies

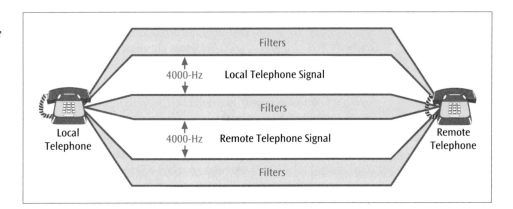

A 4000-Hz channel for voice is significantly smaller than the 20,000-Hz channel required for CD-quality music. Thus, to make the telephone system capable of transmitting CD-quality music, each of its channels would need to be modified to support 20,000-Hz signals. All this talk about frequencies and voice transmission leads to two important points:

▶ The more information you wish to send over a medium, the higher the frequency of the signal you need to represent that information. (See Chapter Two for a detailed discussion of this point.)

▶ If you want to send data over a telephone line, it has to be able to travel in one or more 4000-Hz channels.

When you put these two statements together, a painful fact about using voice communication lines for data transmission emerges: any data transmission that is performed over a standard telephone line must fit within one or more fairly narrow bands of 4000 Hz, which means the data transmission rate will also be limited. Recall Shannon's formula from Chapter Two, which shows that the frequency, noise level, and power level of an analog signal determine the maximum rate of data transmission. With various analog modulation techniques, it has been possible to push data transmission speeds on telephone lines up to 33,600 bps. Newer digital transmission techniques have pushed data transmission speeds up to 56,000 bps over a single 4000 Hz channel, but this speed can be achieved only under ideal conditions. Let us see what those conditions are.

The 56k Dial-Up Modem

When the 33,600-bps modem became available in approximately 1995, many industry experts believed that this was the fastest speed a modem would ever achieve using the standard telephone lines that connect our homes and businesses to the telephone network. This belief was based on two facts: the first was that the telephone connection into a home or business is an analog-modulated connection (using a technique such as phase shift keying), and the second was that the telephone signal is transmitted with a certain signal power level and a given amount of background noise. (To understand how these facts contributed to creating a limit on the data transfer speed that was possible, review the Details section in Chapter Two, "The Relationship Between Frequency and Bits Per Second.") Approximately two years after the 33,600-bps modem became available, the 56,000-bps, or 56k, modem was introduced. Did something in the telephone system change to allow the faster transmission speed, or were the industry experts wrong? The experts were correct—but one important fact had changed with the advent of 56,000-bps modems: digital signaling was introduced.

The 56k modems are a hybrid design, combining analog signaling and digital signaling. The upstream connection from modem to remote end still uses conventional analog signaling and modulation techniques, and thus is limited to a maximum transmission speed of 33,600 bps. The downstream link, however, is where the 56k modem shows an improvement. Instead of using analog signaling, the 56k modem employs digital signaling. The telephone system has actually been using digital signaling (for a reminder of how this works, see Chapter Two's section on pulse code modulation) over telephone lines for many years. In fact, the telephone system can send an 8-bit sample 8000 times per second, which corresponds to 64,000 bits per second, or 64 kbps. If the telephone company can transmit 64-kbps, does this mean that we users can receive a 64-kbps downstream signal? Unfortunately, the answer is "No." When the telephone company transmits a digital 64-kbps telephone signal, the signal is transmitted digitally from one switching center to another. But when a telephone signal is transmitted into our homes and small businesses, it must be adjusted so that it can traverse the local loop. The local loop, as shown in Figures 11-1 and 11-4, is the stretch of telephone wire that runs between a house (or small business) and the telephone company's central office. This local loop is analog and can support only analog signaling.

Figure 11-4
The analog local loop that runs from your house to the telephone company's central office

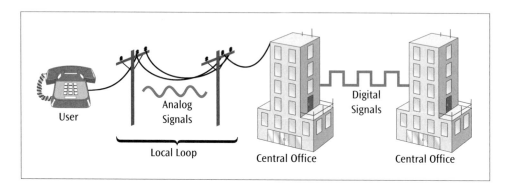

Before the telephone signal is transmitted over the local loop, the central office converts the digital signal to an analog signal. When the analog signal enters your house, your computer's 56k modem converts the analog signal back to digital data, because computers manipulate digital data (Figure 11-5). As you may recall from Chapter Two, when an analog signal is converted to digital data, quantizing noise is introduced. The presence of this noise is the reason it is not possible to transmit a 64-kbps data stream into the local loop. But a smaller data stream, of approximately 56 kbps, is possible.

Figure 11-5
The analog and digital forms of a telephone connection between a home and the central office

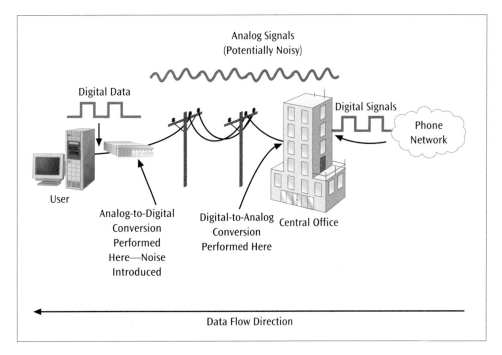

Analog Signals
(Potentially Noisy)

Digital Data

Digital Signals

Phone Network

User

Analog-to-Digital
Conversion
Performed
Here—Noise
Introduced

Digital-to-Analog
Conversion
Performed Here

Central Office

Data Flow Direction

As it turns out, receiving a 56,000-bps signal is not possible either. To ensure that the digital signals on one telephone line do not interfere with the digital signals on an adjacent telephone line, the FCC requires that the power level of the signal be lowered by a small amount. An effect of this lowering is that noise becomes an even bigger factor in the data transmission. Because noise is a bigger factor, the signal slows down more, to approximately 53,000 bps.

Unfortunately, the picture gets even worse from here. As we know, noise is everywhere. It should not be surprising, therefore, that the analog local loop is susceptible to noise. The more noise present, the more the transmission is slowed down. And the farther you are from the central office, the greater your chance for experiencing noise and an even slower transmission rate. Thus, if you are using a 56k modem, chances are your modem does not receive signals at a rate 56,000 bits per second or even 53,000 bps. If you happen to live near your central office or have very low noise on your telephone line, you might, at best, receive data in the upper 40,000-bps range.

The first standard to appear that supported 56,000 bps dial-up modems was the **V.90 standard**. Shortly after V.90 was introduced, ITU introduced the **V.92** modem standard. This standard is a slight improvement over the V.90 standard in two respects. First, the upstream link between the user and the telephone company is capable of supporting connections up to 48 kbps (as opposed to the 33,600 bps of the V.90 modem). This rate, of course, is possible only under ideal conditions of low noise on the telephone line. The second improvement is that a V.92 modem can place a data connection on hold should someone call the user's telephone number (call waiting). The user can then pick up the telephone and talk while the data connection is still active but on hold. When the user ends the standard telephone call, the data connection is resumed.

Despite the advantages offered by the V.90 and V.92 modems over the older 33,600-bps modems, many people found the 56-kbps speed too slow. In particular, many users find this speed noticeably slow when downloading large documents or documents with images. What we need are alternative ways to send data through the basic telephone system. In the following sections, we will examine several of these alternatives.

Digital Subscriber Line

Digital subscriber line (DSL) is a technology that allows *existing* twisted pair telephone lines to transmit multimedia materials and high-speed data. Transfer speeds can range from hundreds of thousands of bits per second up to several million bits per second. Although some of the larger local telephone companies were initially hesitant to offer DSL service, the newly emerging telephone companies (CLECs) were more than eager to do so. The hesitation of the larger telephone companies was partially based on the fact that they were already selling leased-line services such as T-1 (which will be covered later in this section) to many customers. To begin offering DSL, in order to hold on to those customers who might defect from T-1 to another provider's DSL, would have required these larger telephone companies to invest a large amount of money in DSL technology and training. For these companies, investing in DSL just to stay even did not make economic sense. Not long after the CLECs started attracting DSL customers, local telephone companies began changing their plans, and now most, if not all, telephone companies offer various levels of DSL service.

DSL basics

As with any technology that can deliver multimedia data, transmission speed is an important issue. DSL is capable of a wide range of speeds. The transfer speed of a particular line depends on one or more of the following factors: the carrier providing the service, the distance of your house or business from the central office of the local telephone company, and whether the DSL service is a symmetric connection or an asymmetric connection. The first of these factors, the carrier, determines the particular form of DSL technology and the supporting transmission formats, which are chosen by each carrier individually. The form of DSL and its underlying technology determine the speed of the service.

The effect of the second factor, distance, on the transfer speed of a line is relatively straightforward: the closer your house or business is to the central office, the faster the possible transmission speed. This dependency on distance is due to the fact that copper-based twisted pair is fairly susceptible to noise, which can significantly affect a DSL transmission signal. The longer the wire goes without a repeater, the more noise on the line, and the slower the transmission speed. The maximum distance a house or business can be from the central office without a repeater is approximately 5.5 km (3 miles).

The third factor affecting transfer speed is the type of connection: symmetric or asymmetric. A symmetric connection is one in which the transfer speeds in both directions are equal. An asymmetric connection has a faster downstream transmission speed than its upstream speed. For example, an asymmetric DSL service can provide download speeds into the hundreds of thousands of bits per second, while upload speeds may be less than 100,000 bps (100 kbps). An asymmetric service is useful for an Internet connection in which the bulk of the traffic (in the form of Web pages) comes down from the Internet to the workstation. Often the only data that goes up to the Internet is a request for another Web page, some relatively small e-mail messages, or an FTP request to download a file. Most residential DSL services are asymmetric.

Another characteristic of DSL is that it is an always-on connection. Users do not have to dial and make a connection. Thus, a home or business workstation can have a 24-hours-a-day connection to the Internet, or some other telecommunications service. Furthermore, the connection is charged at a flat, usually monthly, rate. The user does not pay a fee based on distance of connection and how long the connection is established.

Because DSL is an always-on connection, it uses a permanent circuit instead of a switched circuit. This permanent circuit is not dynamically configurable, but must be created by the DSL service provider. Because most current home and business users of DSL use the service to connect their computers to an Internet service provider, the absence of a dynamic connection is not a serious issue. But if the user wishes to move the DSL connection to another Internet service provider, a new DSL service may have to be contracted.

What does a business or home user need to establish a DSL connection? At the present time, four components are required. The local telephone company (LEC) must install a special router called a DSLAM (digital subscriber line access multiplexer) within the telephone company's central office. This device bypasses the central office switching equipment and creates and decodes the DSL signals that transfer on the telephone local loop (Figure 11-6). Next, the local telephone company may also install a DSL splitter on its premises, which combines or splits the DSL circuits (the upstream and downstream channels) with the standard telephone circuit of POTS. Some DSL systems transmit over the same telephone line that runs from a central office to a home or business. Because it is the same telephone line, DSL must share the line with a POTS signal.

If the DSL is of a particular form, such as DSL Lite (to be described shortly), then the DSL line does not carry a standard telephone circuit. Thus, there is no need to split the DSL signal from the telephone signal at either the sending or receiving end. When no splitter is used to separate the DSL signal from the POTS signal, then the service is called splitterless DSL.

On the user end, a DSL modem is required to convert the DSL signals into a form that the user workstation or network can understand. If the DSL circuit is also carrying a POTS telephone circuit, the user will also need a splitter to separate the regular telephone line from the DSL data line.

Finally, the DSLAM router at the telephone company's central office must be connected to an Internet service provider via a high-speed line. Because this high-speed line will be supporting the Internet service requests from multiple users, the line needs to be a very fast service, such as ATM. All these components are shown in Figure 11-6.

Figure 11-6

*The four necessary
components of a
DSL connection*

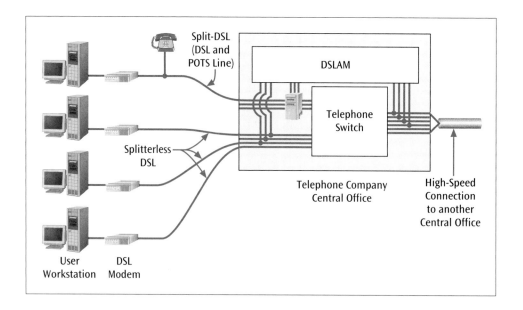

Let us consider a telephone company and the types of DSL service it provides. Most DSL providers offer a number of different types of service. The first service, designed for home users, typically provides a downstream speed of approximately 700 kbps and an upstream speed of 100 kbps for roughly $25.00 per month. For home users requiring more speed, a second service typically provides a downstream speed of 1.5 to 3.0 Mbps for approximately $50.00 per month. For businesses, which typically require more capacity, DSL services are available that can upload and download data at several million bits per second; these cost hundreds of dollars per month. The speeds and prices noted here are only a small sampling of those possible across the United States.

DSL formats

Digital subscriber line comes in a variety of formats. Although this variety provides carriers and users with a wide selection of technologies and speeds, it can also split users into numerous camps, making the process of settling on one format as a standard rather difficult. A number of these formats will, no doubt, fall by the wayside over time, but there is no knowing how long that will take. For now, an informed business network user should be aware of the various formats. (Please note that the data transmission rates provided will change often as the technology evolves.) Often collectively referred to as **xDSL**, six DSL formats are in use today:

 ▶ **Asymmetric digital subscriber line (ADSL)**—A popular format that transmits the downstream data at a faster rate than the upstream rate. Typical downstream data rates range from 600 kbps to 1500 kbps, while upstream data rates range from 300 kbps to 600 kbps.

 ▶ **DSL Lite**— slower format compared to ADSL; also known as Universal DSL, G.Lite, and splitterless DSL. Typical transmission speeds are in the 200 kbps range.

 ▶ **Very high data rate DSL2 (VDSL2)**—is a very fast format (roughly 100 Mbps downstream and upstream) over very short distances (less than 300 meters).

> ► Rate-adaptive DSL (RADSL)—RADSL is a format in which the transfer rate can vary, depending on noise levels within the telephone line's local loop.

DSL is not the only way that a home or small business user can contract a data delivery service to connect their computers to the Internet. Let us examine a second very popular technology—the cable modem.

Cable Modems

A **cable modem** is a high-speed communications service that allows high-speed access to wide area networks such as the Internet via a cable television connection. Technically speaking, a cable modem is a physical device that separates the computer data from the cable television video signal, but many people refer to the entire system (fiber-optic cables, neighborhood distribution nodes, coaxial cables, cable modem splitter, and interface card) as a cable modem service. This connection of components is shown in Figure 11-7.

Figure 11-7
Cable modem connecting a personal computer to the Internet via a cable television connection

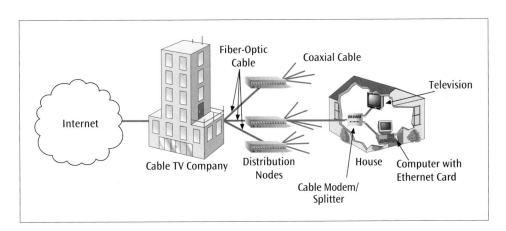

Most cable modems are external devices that connect to the personal computer through a common Ethernet network interface card, which is either provided by the cable company or purchased at most stores that sell computer equipment. As such, the connection is capable of transmitting megabits of data, with transmission rates ranging anywhere from 300 kbps to several mbps. These connections, similar to DSL, are typically asymmetric.

Cable modems provide high-speed connections to the Internet and the demand for them is growing rapidly. This growth has not come without some growing pains within cable companies. Like the telephone system, the cable system features a kind of local loop—that is, a cable that runs from the neighborhood distribution node into a home or business. To support the very high bandwidths required to provide high-speed Internet access to multiple users, cable companies have had to replace a large segment of their existing local loop cable, which is primarily coaxial, with fiber-optic cable. At some point, cable companies (and telephone companies too) may replace their copper wires with fiber-optic cable as well. In fact, a growing number of communities around the country are already beginning to reap the benefits of this newer technology.

A disadvantage of cable modems—quite possibly their only disadvantage—is related to the following trend: As traffic on Ethernet-based local area networks increases, there is a decrease in overall throughput—the ability to send or receive a complete message. Thus, as more customers within a local geographic area, such as a small number of neighborhood blocks, subscribe to cable modem service, traffic will increase to the point where throughput may suffer noticeably.

T-1 Leased Line Service

Since the 1960s and 1970s, businesses have needed high-speed connections between themselves and the telephone company, or between themselves and another corporate building. Dial-up telephone lines were simply not fast enough, and dialing the called party every time a connection was necessary was inconvenient. DSL and cable modems would have been good solutions in many cases, but they had not yet been invented. Instead, the telephone companies began offering leased line services. These leased lines came with various speeds of data transfer

Details ▶

Cable Modem Operation

Cable companies that offer a cable modem service use a standard called Data Over Cable Service Interface Specification (DOCSIS). DOCSIS (now DOCSIS version 3.0) was designed to include all the operational elements used in delivering a data service over a cable television system, including service provisioning, security, data interfaces, and radio frequency interfaces. The basic architecture of a cable modem system is shown in Figure 11-8.

The system consists of six major components. The first component, the Cable Modem Termination System (CMTS), is located at the main facility of the cable operator and translates the incoming Web page data packets into radio frequencies. The frequencies used for the transfer of the Web page data packets are unused frequencies in the 6-MHz range. The second component of the cable modem system is the combiner, which combines the frequencies of the Web page data packets with the frequencies of the common cable television channels. These signals are then sent over a fiber-optic line, which is the third component in the system, to a fiber distribution node, the fourth component in the system. The fiber distribution node resides in a user's neighborhood and can distribute the signals to 500–2000 homes. From the fiber distribution node, coaxial cables carry the signals into homes where a splitter, the fifth component, separates the Web page data packet frequencies from the other cable channel frequencies. The other cable channel frequencies go to the television set, and the Web page data packet frequencies that have been split off go to a cable modem. The cable modem (the sixth component of the system) converts the radio signals back to digital data packets for delivery to your computer.

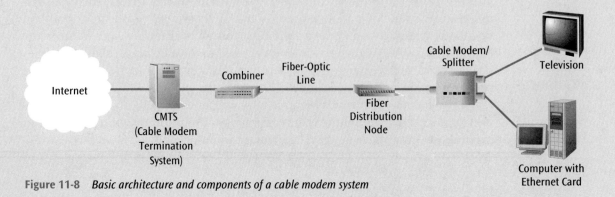

Figure 11-8 *Basic architecture and components of a cable modem system*

(from 56,000 bps up to 64 Mbps) and were fixed circuits—the circuit (or connection) was permanently installed between two locations. The company did not have to dial a telephone number because the circuit was always active. The telephone company charged the business an installation fee and then a monthly rental fee. Many companies used these leased lines to maintain a constant data connection between two locations, or they used them to connect a group of telephone lines to the telephone company's central office. Today, leased lines such as these are pretty much extinct.

One form of leased line service that was very popular and still exists today is the T-1. A T-1 service, as you may recall from earlier chapters, is an all-digital connection that can transfer either voice or data at speeds up to 1.544 Mbps (1,544,000 bits per second). Depending on the user's wishes, the T-1 line can support up to 24 individual telephone circuits, 24 individual data lines at 56,000 bits per second each, or various combinations of these options. If you do not need 1.544 Mbps to transmit your computer data, you can request a quarter-T-1 or a half-T-1, which, as the names imply, provide one-quarter and one-half of the 1.544 Mbps data rate, respectively. These quarter-T-1 and half-T-1 lines are called fractional T-1 services; they are not offered by all telephone companies. At the opposite end of the spectrum, customers who require transmission speeds higher than 1.544 Mbps can consider the T-3 service, which transmits data at 45 Mbps. Unfortunately, the difference in cost, like the difference in speeds, between T-1 and T-3 is dramatic.

Like all leased line services, a T-1 connection is a point-to-point service and is always active. IntraLATA T-1 lines typically cost approximately $350 to $400 per month, while interLATA T-1 lines can cost as much as $1200 per month plus $2.50 per mile for the connection.

Interestingly, the T-1 service, which was created in 1963, was designed to support only the telephone circuits that the telephone companies used to transmit data between switching centers. When businesses heard about the T-1 line, they began asking for it as a way to connect their offices to a telephone company switching center (central office). Demand for the T-1 line has grown steadily ever since. Recently, however, the T-1 line has started encountering stiff competition from new services that offer equal or faster transmission speeds with smaller price tags. We will examine some of these alternative services next and see how they compare to the T-1.

T-1 Alternatives

Consider the following scenario: A company wants a fast, reliable connection to another site. This site could be another building within the same company located at a remote site, or it could be a different company with which the first company wants to conduct business. The company wants this connection to transmit data, so the system it chooses should be designed for data and not voice. Sometimes the connection needs to be permanent, and sometimes there is a need for a temporary or switched connection. Finally, the connection needs to be private; the data is relatively sensitive and we do not want just anyone tapping into our connection. What should this company use? A dial-up telephone line? Definitely not. A dial-up telephone line was not designed for data, is not fast, and would be very expensive. What about a T-1 line? A T-1 would not be an acceptable solution either as it is also likely to be too expensive. DSL and cable modems would not work as they are typically designed to connect home users and businesses to the Internet. So what can we use to connect our two locations? Let's take a look at a few possible solutions to our problem, starting with one of the oldest, frame relay.

Frame Relay

Frame relay is a packet-switched network that was designed for transmitting data over fixed lines (not dial-up lines). If you or your company were interested in leasing a frame relay circuit, you would contact your local telephone company or frame relay service provider. Most long-distance telephone companies offer frame relay service over most of the country. The frame relay service can be either a local service or a long-distance service. Once the service is established, the customer needs only to transmit his or her data over a local link to a nearby frame relay station, shown in Figure 11-9 as a cloud. The frame relay network is then responsible for transmitting the user's data across the network and delivering it to the intended destination site.

Figure 11-9

Three businesses connected to the frame relay network via local connections

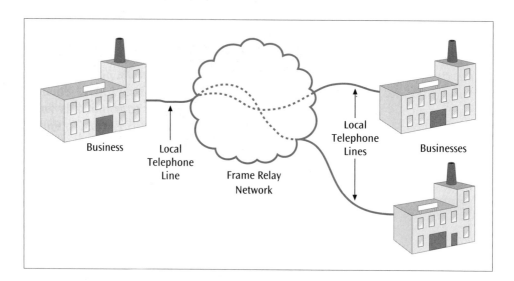

A frame relay service provides many attractive alternatives to leased lines. One of the first noticeable characteristics of a frame relay network is its very high transfer speeds. The data transfer speeds can be very fast, up to 45 Mbps. Along with the high data transfer rates, there is high throughput—the network as a whole is also very fast. Using fiber-optic cables, frame relay switches quickly transfer data so that it travels from one end of the network to the other in a relatively short period of time. Frame relay networks also provide a reasonable level of security, based on private network ownership along with fiber optic transmissions. (Note: it is recommended by many that all secure wide area network connections should use either encryption software or virtual private network software to provide a higher level of security.) In addition, frame relay connections can be both permanent and switchable. With permanent connections, the channel is always available. Switched connections, on the other hand, provide a dynamic ability to allocate bandwidth on demand.

Another advantage associated with frame relay (and other modern high-speed wide area networks) is that their error rates during transmission are low. In fact, the error rate of frame relay networks is so low that the network does not have any form of error control. If an error occurs, the frame relay network simply discards the frame. It is the responsibility of the application, and not the frame relay network, to perform error control.

Last, frame relay networks are reasonably priced. Fixed monthly pricing is based on three charges: a port charge for each access line that connects a business into the frame relay network; a charge for each permanent connection that runs through the frame relay network between two endpoints (companies using the service); and a charge for the access line, which is the high-speed telephone line that physically connects the business to the frame relay port. Because a port and access line are capable of supporting multiple permanent connections, a user may pay one port charge, one access line charge, and several permanent connection charges.

The permanent connection that is necessary to transfer data between two endpoints is called a permanent virtual circuit (PVC). (While the discussion is currently centered on frame relay, most of the following details also apply to other data delivery services.) When a company wants to establish a frame relay connection between two offices, the company will ask the frame relay provider to establish a PVC between the offices (Figure 11-10). For example, if a company has offices in Chicago and Orlando, the company can ask a frame relay provider to create a PVC between the Chicago and Orlando offices. Each office is connected to the frame relay network by a high-speed telephone access line through a port. This port and high-speed telephone line combine to create the physical connection. The PVC—the logical connection—runs from the Chicago office, through the frame relay network, to the Orlando office.

Figure 11-10
A frame relay connection between Chicago and Orlando, showing access lines, ports, and PVC

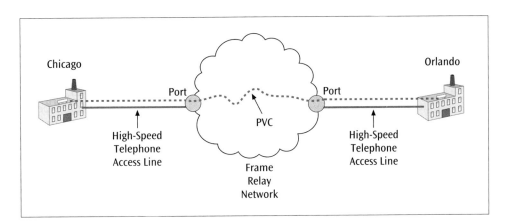

If the two offices in Chicago and Orlando have one application (such as sharing a database or transferring corporate records) running between them, then only one PVC is necessary. However, if two different kinds of applications are running between the two offices, it will probably be necessary to create two PVCs. Even though there are two PVCs, it may be possible to support both PVCs over the same access line and port. Supporting two PVCs over the same access line and port would be possible if the sum of the two transmission speeds of the PVCs was not greater than the capacity of the port and access line.

To illustrate this concept, let us assume that the Chicago–Orlando application required a 256-kbps connection. The company asks the frame relay provider for a 256-kbps PVC. Suppose the company had a 512-kbps port and a telephone line capable of supporting at least 512 kbps (such as a T-1). The frame relay provider can install a 256-kbps PVC between its Chicago and Orlando offices. Now suppose the two offices want to add a new application and therefore need an additional 256-kbps PVC. Because the sum of the two PVCs is not greater than 512 kbps, the company can ask its frame relay provider to add this additional 256-kbps PVC.

Frame relay is called a layer 2 protocol. Recalling the OSI model, the designation layer 2 protocol means that frame relay technology is only one part of a network application and resides at the data link layer. When you are creating a network solution for a particular application, the choice of frame relay is only one part of the solution. You still have to choose network and transport layer software, such as the software supporting the TCP and IP protocols, to run on top of the frame relay network, as well as the type of high-speed telephone line at the physical layer. The intended application—for example, a database retrieval system—then runs on top of TCP/IP.

Committed information rate (CIR) or service level agreements

When a customer establishes a virtual circuit with a frame relay carrier, both customer and carrier agree on a number of important parameters pertaining to the transfer of the customer data. With a frame relay service, this is called the committed information rate (CIR). In general, this is called a service level agreement (SLA). Most data delivery services require some form of service level agreement signed before any virtual circuits are established. For our frame relay example, if a customer seeks a CIR of 512 kbps, the carrier agrees to transfer the customer's data at this rate. The customer agrees that it will not exceed this rate. It is also agreed that if the customer does exceed its CIR, and the frame relay network becomes saturated, the carrier may drop any of the customer's frames that are in excess of the committed information rate. An additional part of the agreement is the burst rate, which allows the customer to exceed the committed information rate by a fixed amount for brief moments of time.

For each virtual circuit, the customer can specify a guaranteed throughput and a limited data delivery time. Having a guaranteed throughput means that the frame relay provider guarantees that a specified percent of all frames will be delivered, assuming that the data is not transferred faster than the committed information rate. Limited data delivery time means that all data frames will be delivered in an agreed-upon amount of time, if, once again, the customer does not exceed the CIR. The price of the virtual circuit will then be directly proportional to the CIR level. The higher the CIR level, the higher the price of the virtual circuit.

Consider a simple example. A business in New York has a high-speed (1.544 Mbps) access line into the frame relay network. It asks its carrier to create a virtual circuit from New York to Dallas with a CIR of 512 kbps. If the business stays within its CIR, the carrier guarantees 99.99 percent throughput and a network delay of no longer than 20 milliseconds. Now, what happens if the customer exceeds its CIR by transmitting data at 800 kbps? Most carriers will allow customers to exceed CIR up to a fixed burst rate for up to two seconds with no penalty. For this example, assume that the customer has a burst rate of 256 kbps. That means the customer can transmit at 512 kbps plus 256 kbps, or 768 kbps for two seconds. If the customer exceeds its CIR for longer than two seconds, the carrier sets a Discard Eligible (DE) bit in the frame header of the customer's frames. If network congestion occurs, DE-marked frames are *discarded* by network switches. It is then the responsibility of the customer's application to watch for discarded frames, because the frame relay network does not inform the user that frames are being discarded.

What is the trade-off between a low CIR and a high CIR? If you request a low CIR and your data rate continuously exceeds that CIR, some of your frames may be discarded. But if the network is not congested, you will not lose frames. If you pay for a high CIR, the higher data rate is guaranteed and you will not lose frames, regardless of whether the network is congested or not. The question comes down to how much of a gambler you are. You can save money, go with a lower CIR, and hope the network does not become congested. Or you can pay for the higher CIR and be fairly confident that no frames will be discarded. If you are supporting an application that should not lose any frames under any circumstances, it might be worth paying for the higher CIR. Interestingly, however, many frame relay customers choose to pay the lowest cost by selecting a *Zero CIR* option that provides no delivery guarantees.

Now we will turn our attention to another powerful and quite interesting data delivery service—Asynchronous Transfer Mode.

Asynchronous Transfer Mode (ATM)

Asynchronous Transfer Mode (ATM), like frame relay, is a very high-speed, packet-switched service that is offered by a number of communications companies. A business that wants to send data between two points (either within a building or across the country) at very high transfer rates might consider using ATM. The transfer rates are as fast as 622 Mbps, with even faster speeds possible.

ATM has several unique features that set it apart from frame relay and other packet services. In ATM, all data is sent in small 53-byte packages called cells. The cell size is kept small so that when a cell reaches a node (a switch point) in an ATM network, the cell will quickly pass through the node and continue on its way to its destination. This fast switching capability gives ATM its very high transfer rates. Furthermore, ATM is a fully switched network. As the cells hop from node to node through the network, each one is processed and forwarded by high-speed switches (for a refresher on this process, review the discussion on switches in Chapter Seven).

ATM networks were designed to simultaneously support voice, video, and data. As you will learn shortly, ATM can handle a wide range of applications, including live video, music, and interactive voice. In addition, ATM can be used as the transfer technology for local area networks, metropolitan area networks, and wide area networks. Because ATM, like frame relay, is a layer 2 protocol, it can be supported by many different types of physical layer media, such as twisted pair and fiber-optic cable. Finally, ATM can support different classes of traffic to provide different levels of service (quality of service, or QoS). ATM traffic classes will be described in detail in the next section.

As in frame relay, before ATM can transfer any data, you must first create a logical connection called a virtual channel connection (VCC). Unlike frame relay's virtual connection, however, this VCC must be created over a virtual path connection. A virtual path connection (VPC) is a bundle of VCCs that have the same endpoints. VCCs are established for data transfer, but they are also established for control signaling, network management, and routing. A VCC can be established between two end users, in which case the VCC would transmit data and control signals. A VCC can also be established between an end user and a network entity, in which case the VCC would transmit user-to-network control signaling. When a VCC transmits user-to-network control signaling, the connection is called the user-network interface. Finally, a VCC can be established between two network entities, in which case the VCC would transmit network management and routing signals. When a VCC transmits network management and routing signals, it is called the network-network interface.

ATM classes of service

One of the features of ATM that sets it apart from most other data and voice transmission systems is that it provides users with the ability to specify a class of service. A class of service is a definition of a type of traffic and the underlying technology that will support that type of traffic. For example, the real-time interactive class of service defines live video traffic. A real-time interactive application is one of the most demanding with respect to data transmission rate and network throughput. Transmitting a video stream over a medium requires a significant amount of bandwidth, and providing a real-time service requires the network to be fast enough to send and receive video so that a live application can be supported.

Details ▶

ATM Cell Composition

ATM transmits all data and control signals using 53-byte cells (Figure 11-11). The cells are kept small, so that the redirection of cells onto particular routes (cell switching) can be performed by hardware, which is the fastest means of cell switching. The ATM header is always 5 bytes long, which leaves 48 bytes of the cell for data. Cell fields and their functions include:

▶ Generic flow control, which is applied only at the user-network interface. Generic flow control alleviates short-term overload conditions by telling the user to slow down on sending data into the ATM network. Generic flow control is not currently used by many ATM systems, if any.

▶ Virtual path identifier, which is an 8-bit field (allowing for 256 combinations) that identifies a particular VPC.

▶ Virtual channel identifier, which is a 16-bit value that identifies a unique VCC. When a company creates an ATM connection between two locations, a virtual path supports

one or more virtual channels. To be identified properly, the path and the channels have unique ID numbers.

▶ Payload type, which consists of three bits. These three bits identify the type of payload, as either user data or network maintenance, and report whether congestion has been encountered in the network.

▶ Cell loss priority, which is a single bit and is set by the user. A value of 0 means this cell is of high priority, so the ATM network should think twice about discarding it in the event of network congestion.

▶ Header error control, which consists of an 8-bit cyclic redundancy checksum ($x^8 + x^2 + x + 1$). It provides error checking for the first four bytes of the header, but not for the data. The cyclic checksum can also provide error correction by fixing single-bit errors in the header.

Generic Flow	Virtual Path Identifier	Virtual Channel Identifier	
4 bits	8 bits	16 bits	

Payload Type	Cell Loss Priority	Header Error Control	Data
3 bits	1 bit	8 bits	

Data (continued)
48 bytes

Figure 11-11 *The 53-byte ATM cell with its individual fields*

If a network does not provide varying class of service levels, then all applications share the network bandwidth equally, which may not provide a fair distribution of bandwidth. For an example, let us examine the Internet, which using IPv4 does not offer a class of service feature. Suppose you and a coworker are sitting at computer workstations in adjacent cubicles. You are sending e-mail, while your coworker is participating in a live videoconference with a third coworker in another state. Both the e-mail and videoconferencing applications are using the Internet for data transmission. Even though your e-mail application often transmits very little data and time is not at all important, your e-mail has exactly the same priority on the Internet as your coworker's application, which is trying to run a very time-intensive and data-consuming videoconference. Although your e-mail program may work fine via the Internet, your coworker's videoconferencing is undoubtedly performing poorly. If the Internet could support different classes of service, your e-mail program would request a low class of service, and your coworker's videoconference program would request a high class of service. With a high class of service, the videoconference program would perform much better.

With ATM, the customer specifies a desired class of service for every virtual channel that is set up. As you might expect, the desired class of service determines the price for using a particular virtual channel. ATM has defined four classes of service that are functionally equivalent to leased line service, frame relay service, and Internet service. These classes of service determine the type of traffic an ATM network can carry:

- ▶ **Constant bit rate (CBR)**—Similar to a current telephone system leased line, CBR is the most expensive class of service. CBR acts very much like a time division multiplexed telephone service, but it is not a dedicated circuit like a leased telephone line. CBR delivers a high-speed, continuous data stream that can be used with transmission-intensive applications.

- ▶ **Variable bit rate (VBR)**—Similar to frame relay service, rt-VBR is used for real-time applications, such as compressed interactive video, and nrt-VBR is used non-real-time applications, such as sending e-mail with large, multimedia attachments. VBR applications often send bursts of data, and the ATM network guarantees that the VBR traffic is delivered on time. CBR and VBR are both analogous to flying first class. First class is expensive, but the customers get top priority by being guaranteed a seat and by being seated first.

- ▶ **Available bit rate (ABR)**—ABR is also used for traffic that may experience bursts of data, called "bursty" traffic, and whose bandwidth range is roughly known, such as that of a corporate collection of leased lines. The low and high transfer ranges are established through negotiation, but getting bit space for sending data through the ATM network with this class of service is similar to flying standby on an airline. You may get a chance to transmit your data, or you may have to wait. But once your data is in the network, it will be delivered. ABR is good for traffic that does not have to arrive in a certain amount of time. ABR also provides feedback that indicates if a part of the ATM network is experiencing congestion.

▶ **Unspecified bit rate (UBR)**—UBR is also capable of sending traffic that may experience bursts of data, but there are no promises as to when the data may be sent, and if there are congestion problems, there is no congestion feedback (as is provided with ABR). Unlike data transmitted in the other classes of service, data transmitted over a UBR connection may not make it to the final destination. If the bandwidth necessary to transmit your data is required by data using a higher class of service, your data may be discarded partway through the network connection. Imagine you are flying standby but will be tossed off of the airplane if, during a connection, your seat is needed by someone else. The UBR service is usually the least expensive service.

By offering different classes of service, the ATM service provider can also offer tailor-made transmission services and charge the customer accordingly. Also, the customer can request a particular class of service that is appropriate for a particular application.

Advantages and disadvantages of ATM

Due to its range of features, such as high transfer speeds, various classes of service, and ability to operate over many types of media and network topologies, ATM has a number of significant advantages. ATM can support a wide range of applications with varying bandwidths, at a wide range of transmission speeds. Cell switching, which is performed by ATM's high-speed, hardware-based switches that route cells down the appropriate path, is so fast that it provides short delays and high bandwidths. ATM's different classes of service allow customers to choose service type and pricing individually for each data connection (VCC). Finally, ATM is extremely versatile. It can carry voice, packet data, and video over the same facilities.

As you might expect, ATM also has a number of disadvantages. It is often more expensive than other data transmission options. The cost of ATM equipment is high. This is because the cell-switching equipment, to be capable of delivering such a fast connection, is by necessity relatively complex. Due to the complexity of ATM, there is a high learning curve for setting up and managing the network. Lastly, compatible hardware and software may not be widely available.

ATM is now being heavily used by the large telecommunications carriers (AT&T, Sprint, and so on) to provide voice and Internet services. Because of its complexity and cost, smaller businesses have been reluctant to use ATM. Alternate technologies that can offer reasonably comparable speeds at much lower costs have, however, attracted much attention from smaller businesses. Metropolitan area networks, as we saw in an earlier chapter, are seeing a growth in the metro area data delivery market. But what about data delivery over longer distances? As it turns out, there is one network that is causing grief for frame relay and ATM providers: the Internet.

MPLS and VPNs

Up to this point in the discussion, much has been said about data delivery services such as frame relay and ATM, but little has been said about using the Internet to connect two business offices. Why were frame relay and ATM used so much, if the Internet was available and essentially free? Two main reasons, as we have already seen. First, there is no guarantee that the Internet will deliver data in a consistent and timely manner. The Internet was not designed to handle data streams with small latency values. In other words, as data packets move into routers, there is no guarantee when the packet will leave the router. Part of this is because routers handle a large amount of traffic, and the routing tables found in routers continue to get more and more complex all the time. Frame relay and particularly ATM were designed to move packets quickly through the network.

The second reason that frame relay and ATM were considered over the Internet was the perception that frame relay and ATM were more secure. While it is true that frame relay and ATM are private networks (you must have a contract with a service provider before you can transfer data over the network) there is no guarantee that all data on private networks are secure unless proper precautions are in place. This is in contrast to the Internet. Virtually anyone, anywhere in the world can get connected to the Internet.

Despite these apparent shortcomings, more and more businesses are turning away from frame relay and ATM and sending their data over the Internet. What has changed? Internet users now have two new weapons in their arsenal: MPLS and VPNs. As we saw in the previous chapter, MPLS (Multi-Protocol Label Switching) can be used to create a faster path through the Internet. And a VPN (virtual private network) can create a secure tunnel through the Internet through which a business can confidently send private data.

There are basically two ways a business can create a tunnel through the Internet. In the first technique, a user installs the necessary software at each end of the tunnel and contracts an Internet service provider to create and maintain the tunnel. This typically incurs a monthly charge from the ISP. A second technique involves the user purchasing the necessary hardware and software that is placed at each end of the tunnel. The hardware is essentially a gateway (an advanced router) that provides the necessary VPN and security software. The user often pays a one-time price for the hardware but still needs a standard Internet connection via an ISP.

Only time will tell whether the trend of MPLS/VPN tunnels replacing other data delivery services continues. Nonetheless, frame relay and ATM are not going to disappear over night.

Comparison of the Data Delivery Services

To summarize, let us compare DSL, cable modems, frame relay, ATM, and MPLS/VPNs along the following factors: maximum transmission speed, typical cost per month, if the connection is switchable or fixed, whether the service provides quality of service choices, if the service can support data or voice or both, and how available the service is within the U.S. For your reference, Table 11-1 summarizes the results of this comparison.

Note that when we compare maximum transmission speeds, ATM is one of the fastest, but this comes at a premium price. Also, ATM is the most complex technology (which usually translates into higher maintenance and higher costs), making it an unlikely candidate for small business and home connections. DSL systems that transmit around 0.5 to 1.5 Mbps compare favorably to cable modems and are roughly the same price. DSL and cable modems are less expensive than frame relay systems at comparable speeds. MPLS/VPN can be the most economical and can achieve high speeds depending upon Internet connection speeds.

If a user requires a service that is switchable, or can change—for example, one in which a different telephone number can be dialed to create a new connection—then only frame relay, ATM, and MPLS/VPN are acceptable. The other technologies are fixed—once the connection is established, the user cannot change it without contacting the service provider.

In this group, only ATM and to a lesser degree, MPLS/VPN, can offer the quality of service feature. This is clearly a strong advantage over the other technologies. In regard to transmitting data and voice, all the technologies can transmit both. Because many companies wish to use the service for supporting multiple types of applications, the capability of sending both data and voice is a distinct advantage.

The last issue worth noting is the availability of each service. Although the availability of each type of service is certain to change, all now seem to have wide availability across the country.

Table 11-1

Summary of different high-speed data delivery network technologies

	Typical Maximum Speeds	Cost per Month	Switchable or Fixed	QoS	Data or Voice	Availability
DSL-consumer	600 kbps	~$35	Fixed	No	Both	Wide
DSL-business	7 Mbps	$100s – $1000s	Fixed	No	Both	Wide
Cable modem	1.5 Mbps	~$40	Fixed	No	Both	Wide
Frame relay	45 Mbps	$100s – $1000s	Both	No	Both	Wide
ATM	622 Mbps	$100s – $1000s	Both	Yes	Both	Wide
MPLS/VPN	*	*	Both	Yes	Both	Wide

*Depends on connection to Internet

Convergence

We have seen examples of convergence in a number of earlier chapters of this book. Starting in Chapter One, we saw how the worldwide market of Internet users has converged on the TCP/IP protocol suite. In Chapter Two, we saw how pulse code modulation has become the predominant technique for converting analog data to digital signals. In Chapter Seven, we discovered how the local area network market has converged on Ethernet and has virtually wiped out token ring and FDDI networks. We have also seen how two or more technologies have converged into one. For example, we saw in Chapter Three how cell phones can be used to snap and transmit photos, as well as perform walkie-talkie-like operations and even download television signals. We have also seen how handheld personal digital assistants can perform a variety of operations, including placing cell phone calls, sending e-mail, datebook and calendaring functions, high-speed data access, and Bluetooth capabilities.

Although convergence is an important trend in many areas of data communications and computer networks, it has had a particularly noticeable impact on the telecommunications market. We are seeing a number of telephone companies merging into single entities. Recent examples of these include the merger of SBC with AT&T, and Verizon with MCI. The trend toward these large telecomm mergers is interestingly ironic when one recalls the industry concerns that led to the breakup of AT&T. At that time, separating telecommunications services so that multiple companies could offer them was thought to be in the best interest of consumers, because it would promote competition, which would yield lower prices. Now the pendulum seems to be swinging in the other direction—and the industry is favoring the bundling together of various telecommunications services.

Another important convergence issue (as we just saw) is the fact that data delivery services such as frame relay and ATM are starting to give way to other services such as Ethernet and MPLS/VPNs over IP-backbones. Ethernet clearly dominates in the local area network arena. With the expansion of Metro Ethernet into many metropolitan area markets, some in the industry feel it is only a matter of time before Ethernet becomes a common wide area network. If this happens, Ethernet will then be available on local, metropolitan, and wide area networks. Many companies will find this appealing, because they could then support all their networking needs using only a single technology—Ethernet. In other words, a business would be able to set up a seamless series of networks, all running the Ethernet protocol.

MPLS, introduced in the previous chapter, is also causing the number of frame relay systems to decline. By layering MPLS frames over the top of IP frames, service providers are capable of sending high-speed data packets over Internet backbones.

Another example of convergence in the telecommunications industry involves combining a telephone network with a local area network in order to create one network that can treat an incoming telephone call the same way that it would the arrival of an e-mail message. Let us take a look at this form of convergence—known as computer-telephony integration.

Computer-Telephony Integration

Computer-telephony integration (CTI) is an emerging field that combines more traditional voice networks with modern computer networks. CTI integrates the PBX phone switch with computer services to create modern voice and data applications that run on computer systems. Recall that the PBX is the internal telephone system many businesses use to support all their telephone operations, including in-house calling, in-house to outside calling, voice mail, call transfers, and conference calling. Traditionally, these services require a PBX switch, a set of telephones, and standard telephone wiring. In contrast, with CTI, all you need is a workstation that is running the appropriate telephony software. Because CTI combines the power of computer systems with the services of a telephone network, a user can perform typical telephone operations by clicking in a window of a program.

Consider the following scenario: A business sells a product and provides customer service to support that product. To obtain customer service, a customer calls the company's toll-free number. Because the company has a CTI system that combines the power of a local area network with modern telephone services, the caller's telephone number appears on the screen of the customer service representative's computer monitor as the telephone rings (caller ID). Before the customer service representative even answers the telephone, the CTI system uses the customer's telephone number as the key to a database query, and a summary of the customer's account also appears on-screen. The customer service representative answers the telephone and has customer information on the screen when the customer says, "Hello." As the representative talks to the customer and learns about the customer's problem, he or she can click icons on the computer screen to transfer the customer's call to another party, put the customer on hold, or retrieve additional information from the computer database. Thus, the distinction between computers and telephone systems is greatly blurred. The computer and local area network are performing telephone operations as easily as they perform data operations.

CTI can also integrate voice cabling with data cabling. Early CTI systems created a strong marriage of data and voice operations but still maintained separate wiring—one set of wires for the local area network and its workstations, and another set for the telephone lines. The PBX may have talked directly to the local area network, sending data and telephone commands back and forth, but the actual computer data and telephone data remained physically separated. In contrast, recently developed systems support computer data and voice data on the same set of wires. This innovation can save a company money in terms of cabling—often one of the costliest pieces of a network. This design, however, has a drawback, in that it places a larger demand on the single wiring system.

Using CTI has three advantages. First, it creates new voice/data business applications that can save companies time. Second, it makes optimal use of current resources. Third, it saves money. These advantages mean that businesses can realize many benefits from CTI applications. For example:

> ► Unified messaging—Users utilize a single desktop application to send and receive e-mail, voice mail, and fax.

> ► Interactive voice response—When a customer calls your company, his or her telephone number is used to extract the customer's records from a corporate database. The customer records are displayed on a service representative's workstation as the representative answers the telephone.

> ► Integrated voice recognition and response—A user calling into a company telephone system provides some form of data by speaking into the telephone, and a database query is performed using this spoken information.

> ► Fax processing and fax-back—In fax processing, a fax image that is stored on a LAN server's hard disk can be downloaded over a local area network, converted by a fax card, and sent out to a customer over a trunk. An incoming fax can be converted to a file format and stored on the local area network server. With fax-back, a user dials in to a fax server, retrieves a fax by keying in a number, and sends that fax anywhere.

> ► Text-to-speech and speech-to-text conversion—As a person speaks into a telephone, the system can digitize the voice and store it on a hard disk drive as computer data. The system can also perform the reverse operation.

> ► Third-party call control—Users have the ability to control a call—for example, set up a conference call—without being a part of it.

> ► PBX graphic user interface—Different icons on a computer screen represent common PBX functions such as call hold, call transfer, and call conferencing, making the system easier for operators to use.

> ► Call filtering—Users can specify telephone numbers that are allowed to get through. All other calls will be routed to an attendant or voice mailbox.

> ► Customized menuing systems—Organizations can build customized menuing for an automated answering system to help callers find the right

information, agent, or department. Using drag-and-drop functionality, it is possible to revise the menu system daily. The menu system can be interactive, allowing callers to respond to voice prompts by dialing different numbers. Different voice messages can be associated with each response, and voice messages can be created instantly via a PC's microphone.

As you can see, some interesting applications are emerging from the convergence of voice and data systems. Only time will tell if CTI is a passing fad or the beginning of a whole new area of computer networks.

Telecommunications Systems In Action: A Company Makes a Service Choice

Better Box Corporation manufactures cardboard boxes. Its administrative headquarters are in Chicago, Illinois, and it has regional sales offices in Seattle, San Francisco, and Dallas. Better Box is expanding its data networking capability and has asked for your help in designing its new corporate network. It wants to interconnect all four of its locations, so that the headquarters in Chicago can accept data records from each of the other three offices. Typically, about 20,000 sales records per month will be uploaded from each of the regional sales offices. Each sales record is 400,000 bytes in size. It should take no longer than 20 seconds to upload a single sales record. Although the company has enough funds to support many possible solutions, it wants to seek the most cost-effective solution for the long run.

Prices

In order to make a proper decision as to the type of data transmission service, we will consider the prices of a number of different types of services:

> ▶ A plain old telephone service (POTS) transmitting data at 56 kbps over an analog line
> ▶ An intraLATA T-1 (flat monthly rate) and an interLATA T-1 (flat monthly rate plus mileage charge)
> ▶ A series of frame relay services, which include port speed and PVC capacity
> ▶ A series of ATM services, which include port speed, path charge, and channel charge
> ▶ Using MPLS and VPN to create a secure, fast connection over the Internet

Let's examine each of these in turn.

Making the choice

To choose the right data delivery service, you need to choose the appropriate data transmission speeds for each service, keep the cost low, and meet Better Box's requirements. For each network service, let us examine the total cost per month that Better Box Corporation might pay to a telecommunications carrier to use that service to connect all four of its sites. Remember that the cost of frame relay has three components: (1) the access line charges, (2) the frame relay port charges, and (3) the PVC charges. Likewise, ATM has four cost components: (1) the access line charges, (2) the ATM port charges, (3) the PVC's channel charges, and (4) the PVC's path charges. Table 11-2 lists the costs for these components as well as charges for other types of services.

To begin, you need to reread Better Box's plans and discover the company's transmission requirements. We see the requirement that 400,000 byte records should take no longer than 20 seconds to download. Here are some simple calculations:

$$400,000 \text{ bytes} \times 8 \text{ bits/byte} = 3,200,000 \text{ bits}$$
$$n = 3,200,000 \text{ bits/20 seconds}$$
$$n = 160,000 \text{ bps}$$

Thus, Better Box needs a telecommunications line connecting to the outside world with a transmission speed of at least 160 kbps. What service will meet this requirement?

Table 11-2

Hypothetical prices of different telecommunications services

Type of Service	Speed	Per-Month Cost	Usage Fee
POTS	56 kbps	$20 + usage fee	$.10 per minute
T-1	IntraLATA T-1 (1.544 Mbps) InterLATA T-1 (1.544 Mbps)	$350 $1200 + $2.50 per mile	
Frame Relay (port speed and price)	56 kbps 128 kbps 256 kbps 512 kbps 768 kbps 1.544 Mbps	$220 $400 $495 $920 $1240 $1620	
Frame Relay PVC (CIR and price)	16 kbps 56 kbps 128 kbps 256 kbps 384 kbps 512 kbps 1024 kbps 1536 kbps	$25 $60 $110 $230 $330 $410 $1010 $1410	
ATM (CBR) port	1.544 Mbps 3 Mbps	$2750 $3400	
ATM (ABR) port	1.544 Mbps 3 Mbps	$1750 $2400	
ATM PVC Path		$2 per mile	
ATM PVC Channel		$250 each channel (no mileage charge)	
MPLS/VPN Tunnel	1.544 Mbps	$1000 initial hardware plus $375 per month ISP charge	

The best transmission speed that can be achieved using a dial-up modem with POTS is 56 kbps (actually 53 kbps). To achieve transmission speeds of at least 160 kbps, Better Box needs to use a T-1 line (1.544 Mbps), a frame relay service, an ATM service, or MPLS.

When you consider a T-1 scenario, you recognize that Better Box would need three T-1 lines: one from Seattle to Chicago (2050 miles), one from San Francisco to Chicago (2170 miles), and one from Dallas to Chicago (920 miles). Each of these lines is a long-distance, or interLATA, line. Using the mileage between these points, the three interLATA T-1 lines would cost approximately $6325 ($1200 per month + $2.50 per mile from Seattle to Chicago), $6625 ($1200 + $2.50 per mile from San Francisco to Chicago), and $3500 ($1200 + $2.50 per mile from Dallas to Chicago), respectively, for a monthly total of $16,450. Unfortunately, this scenario does not even allow Seattle to talk directly to San Francisco, or Seattle to talk directly to Dallas; it provides communication only between Chicago and each of the regional sites (Figure 11-12).

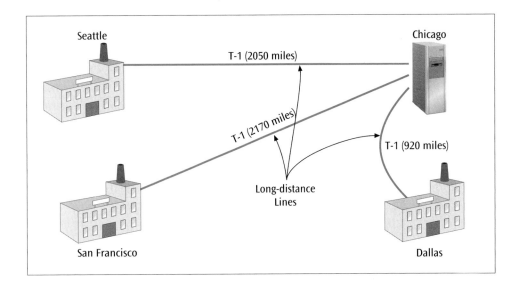

With a frame relay scenario, you do not have to worry about running a separate line between every pair of cities. You only have to make a frame relay connection into the frame relay network (Figure 11-13). Thus, four connections allow us to send data between any two cities. As you have calculated, each of these connections has to support the 160-kbps transmission speed. A 256-kbps frame relay connection is the closest size that will support the desired speed.

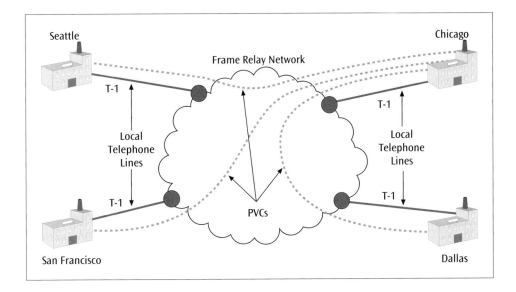

The frame relay charges would include:

▶ Three port charges at 256 kbps = 3 × $495.

▶ One port in Chicago at 768k (which is the total of three connections passing through it) = $1240.

▶ Total port charges = $2725.

▶ Three PVC charges for three 256-kbps connections (represented by the dotted lines in Figure 12-13) = 3 × $230. Thus, the total PVC charges = $690.

▶ Four IntraLATA T-1 telephone lines to connect the four cities to the frame relay network = 4 × $350. The total T-1 charge = $1400.

The total charges for frame relay would be $4815 per month, which is much less expensive than the preceding option of using three interLATA T-1 lines among the four cities.

An ATM solution would be similar to a frame relay solution in that you only have to connect each of the four cities to the ATM cloud. Once these were connected, you would create three paths with one channel over each path. This layout would be very similar to the frame relay layout. A 1.544 Mbps ABR ATM connection is the closest size that will support the desired speed. The ATM charges would include:

▸ Four port charges at the four cities, each 1.544 Mbps ABR = 4 × $1750. Thus the total port charges = $7000.

▸ Three channels to support the three connections = 3 × $250. Thus the total channel charges = $750.

▸ Three paths to support the three channels = $2 per mile × 5140 miles. The total path charges = $10,280.

▸ Four intraLATA T-1s to connect the four locations to the ATM cloud = 4 × $350. Total T-1 charges = $1400.

The total cost for an ABR ATM network would be $19,430 per month.

Our final possibility is the creation of a MPLS/VPN tunnel over the Internet. The costs associated with an Internet tunnel consist primarily of the initial hardware costs (two gateways, one on each end of the connection) plus a monthly connection between the business and the Internet service provider. We'll need the equivalent of two intraLATA T-1s, much as we did for connecting frame relay and ATM to their respective clouds. Let's assume that an ISP monthly charge for a small business is $350 per month. Plus, we'll still need a T-1 on each end. That gives us an initial cost of $2000 for the two gateways, plus a monthly charge of two T-1s ($700) and two ISP connections ($750). It appears that at $1450 per month (after the initial gateway purchases) the MPLS/VPN tunnel is the most cost-effective solution.

◆ ◆

SUMMARY

▸ The basic telephone system that covers the U.S. is called plain old telephone service (POTS) and is a mix of analog and digital circuits. Telephone channels are designed to carry voice signals with 4000-Hz bandwidth signals. The telephone line is the set of wires that runs from the central office to a house or business and transfers analog signals; it is called the local loop. A telephone trunk carries multiple telephone signals and typically runs between central offices and switching centers.

▸ The divestiture of AT&T in 1984 opened the long-distance telephone market to other long-distance providers, forced AT&T to sell off its local telephone companies, and divided the country into local access transport areas (LATAs).

▸ A PBX is an on-premise computerized telephone switch that handles all internal and outgoing telephone calls and offers a number of telephone services. A Centrex offers the same services as a PBX, but the equipment resides on the telephone company's property, and the business leases the service. Private lines or leased lines are permanent telephone lines that run between two locations and provide constant access without the need to dial a telephone number.

▶ The Telecommunications Act of 1996 opened local telephone service to new competitors and required the existing local telephone companies to provide these competitors with access to the local telephone lines. Existing local telephone companies became known as incumbent local exchange carriers (ILECs), while new local telephone companies became known as competitive local exchange carriers (CLECs).

▶ The data rate of standard modems using voice-grade telephone lines has peaked at 33,600 bits per second. The newer digital modems are capable of speeds near 56,000 bits per second, depending on line conditions.

▶ Leased lines are established by a communications service provider and serve as permanent, private connections between two locations. Many different types of leased line services are offered by a telephone company. The most popular leased line service is the all-digital 1.544 Mbps T-1.

▶ Technologies such as digital subscriber line (DSL) and cable modems have improved the data transfer rates available between homes and businesses and Internet service providers; these data transfer rates are now as high as millions of bits per second. Currently, there are many different types of DSL service around the country, but most can be classified as either splitterless (only DSL service), splitter (DSL service and a POTS telephone service on the same set of wires), asymmetric, or symmetric.

▶ Frame relay is a service that provides digital data transfer over long distances and at high data transfer rates.

▶ To use a frame relay service, a customer contacts a frame relay provider who creates a permanent virtual circuit (PVC) between the customer and the site to which the customer wants to be connected. The customer pays for the PVC, the access to the frame relay network (a port charge), and the telephone line that gives the customer access to the port.

▶ Frame relay is a layer 2 protocol, which means that it can run over a variety of physical media and supports many different types of applications at layers 3 and up. The committed information rate (CIR) is an agreement between the frame relay customer and the service provider. The customer requests a particular transmission speed, and if the customer does not exceed that rate, the service provider will guarantee accurate and timely delivery of data frames.

▶ Frame relay is more reliable than the Internet and provides a better level of security, but the Internet is less expensive and available essentially everywhere.

▶ Asynchronous Transfer Mode (ATM) is also a packet-switched service, but it supports all types of traffic and operates over LANs as well as WANs and MANs. Because ATM uses very fast switches, it can transfer ATM cells at very high rates (over 600 Mbps). To offer ATM, the ATM service provider creates a virtual path connection between two points, and the user then creates one or more virtual channel connections over that virtual path. A powerful advantage of ATM is that it provides a number of different classes of service to support a wide range of network applications, some of which may require varying transmission speeds and network throughputs. The disadvantages of the service include its complexity and cost.

▶ Frame relay and ATM are seeing a decrease in market share due to two relatively recent newcomers: MPLS and virtual private networks (VPNs). The advantages of MPLS and VPNs are lower costs and the ability to use the ubiquitous Internet.

▶ Computer-telephony integration is the convergence of data communications networks and voice systems. This convergence provides a number of new voice/computer applications, such as workstation-controlled telephone services and data retrieval services.

KEY TERMS

asymmetric connection

asymmetric digital subscriber
line (ADSL)

Asynchronous Transfer Mode (ATM)

available bit rate (ABR)

burst rate

cable modem

call filtering

central office (CO)

Centrex (central office exchange
service)

class of service

committed information rate (CIR)

competitive local exchange
carrier (CLEC)

computer-telephony integration (CTI)

constant bit rate (CBR)

customized menuing systems

digital subscriber line (DSL)

DSL Lite

fax-back

fax processing

frame relay

incumbent local exchange
carrier (ILEC)

integrated voice recognition and
response

interactive voice response

interexchange carrier (IEC, IXC)

layer 2 protocol

local access transport area (LATA)

local exchange carrier (LEC)

local loop

Modified Final Judgment

network-network interface

PBX graphic user interface

permanent virtual circuit (PVC)

plain old telephone service (POTS)

Private Branch Exchange (PBX)

private line

rate-adaptive DSL (RADSL)

splitterless DSL

symmetric connection

Telecommunications Act of 1996

text-to-speech and speech-to-text
conversion

third-party call control

tie line

trunk

unified messaging

unspecified bit rate (UBR)

user-network interface

V.90 standard

V.92

variable bit rate (VBR)

very high data rate DSL2 (VDSL)

virtual channel connection (VCC)

virtual path connection (VPC)

xDSL

REVIEW QUESTIONS

1. What is plain old telephone service (POTS)?

2. What is the typical frequency range for the human voice?

3. What two entities does a local loop connect?

4. What is a LATA?

5. How does a trunk differ from a telephone line?

6. List the most important results of the Modified Final Judgment of AT&T.

7. What is the difference between a local exchange carrier and an interexchange carrier?

8. List the most important results of the Telecommunications Act of 1996.

9. Why are the 56-kbps modems faster than the older 33,600-bps modems?

10. Why don't the 56-kbps modems transmit at 56 kbps?

11. What are the basic features and transfer speeds of DSL?

12. What are the basic features and transfer speeds of cable modems?

13. How does a leased line service differ from POTS?

14. What are the basic characteristics of a T-1 line?

15. What are the basic services of a T-3 line?

16. What are the basic characteristics of frame relay?

17. What features make frame relay so attractive?

18. How do you create a frame relay permanent virtual circuit?

19. How does a switched virtual circuit differ from a permanent virtual circuit?

20. Is it possible to have more than one PVC over a physical line? Explain.

21. What is established when a customer and a frame relay service agree on a committed information rate?
22. What happens when a user transmits data faster than the agreed upon committed information rate?
23. How does frame relay compare to sending data over the Internet?
24. What are the basic features of ATM?
25. What is the relationship between an ATM virtual channel connection and a virtual path connection?
26. What is meant by ATM having classes of service?
27. Describe some applications that incorporate computer-telephony integration.
28. What technologies are threatening the use of frame relay and ATM?

EXERCISES

1. The telephone line that connects your house or business to the central office (the local loop) carries your conversation and the conversation of the person to whom you are talking. What do you estimate is the bandwidth of a local loop?
2. If you play a CD for a friend over the telephone, will the friend hear high-quality music? If not, why not?
3. If you place a telephone call and it leaves your LATA and enters another LATA, what kind of telephone call have you placed? What kind of telephone company handles this telephone call?
4. For each of the following scenarios, state whether a telephone line or a trunk should be used:
   a. the connection from your home to the local telephone company
   b. the connection between a large company's PBX and the telephone company
   c. the connection between two central offices
5. Many within the telephone industry feel we will eventually run out of area codes and telephone numbers. How many different area codes are currently available? What would you suggest to increase the number of telephone numbers?
6. State whether each of the following was a result of the Modified Final Judgment of 1984, the Telecommunications Act of 1996, or neither:
   a. The FCC was created to watch over interstate telephone systems.
   b. AT&T had to sell off its local telephone companies.
   c. The LATA was created.
   d. Cable television companies could provide local telephone service.
   e. Customers could choose between different long-distance telephone providers.
   f. AT&T split off its technology division, which became Lucent.
7. You want to start your own local telephone company. Do you have to install your own telephone lines to each house and business? Explain.
8. What type of telephone service do you use to connect a PBX to a central office if you want to allow 40 users to dial out simultaneously?
9. If you install a 56-kbps modem in your computer and dial in to a remote network that has only 33,600-bps modems, is your modem useless? Defend your answer.
10. Why would you want to use the newer V.92 modem rather than the older V.90? Why might an online Internet service not want you to use the V.92 modem?
11. You dial in to your Internet service provider using your 56k modem. Once connected, your data rate is shown as 42,000 bps. Why not 56k bps? List all possible reasons.

12. The next day, you dial in to an Internet service provider different from the one used in the previous question. This time you get connected at 24,000 bps. Could there be different reasons for this slower connection? Explain your response.

13. Cable modems, like 56k modems, create asymmetric connections. What does this mean, and why doesn't this asymmetric connection affect the average Web user?

14. What are the basic functions of a cable modem?

15. Give an example in which someone would want to use a symmetric DSL service.

16. What is the main advantage of asymmetric DSL over symmetric DSL?

17. In a DSL service, what is the function of the DSLAM?

18. You live in a house that is 10 miles from the nearest city. Will you be able to get DSL service to your home? If the answer is no, what could the DSL service provider do to provide you with DSL?

19. If you have a frame relay service installed, can you just pick up the telephone and dial the number of the party to whom you want to connect? Explain.

20. Suppose you want to have a frame relay connection between your Chicago office and your New York office. Itemize the different charges that you will have to pay for this connection.

21. You have established a frame relay connection with a committed information rate of 256 kbps and a burst rate of 128 kbps. Several times a day, your computer systems transmit in excess of 512 kbps. What will happen to your data?

22. What exactly does frame relay do if a frame is garbled and produces a checksum error?

23. State whether frame relay or the Internet is the better transmission medium for each of the following activities:
    a. sending e-mail
    b. sending high-speed data
    c. transmitting interactive voice communications
    d. receiving a live video stream
    e. participating in a chat room or newsgroup

24. One of the disadvantages of ATM is the "5-byte cell tax." Explain what the 5-byte cell tax means.

25. You have an ATM connection that goes from your location to a network entity, on to a second network entity, and then to your desired destination. Draw a simple sketch that shows each virtual channel connection.

26. State which ATM class of service would best support each of the following applications:
    a. e-mail with image attachments
    b. interactive video
    c. simple text e-mail
    d. voice conversation

27. How can a local area network support telephone operations using CTI?

28. Describe a business or school application that would benefit from CTI.

THINKING OUTSIDE THE BOX

1 A company wants to connect two offices, located in Memphis, Tennessee, and Laramie, Wyoming. The offices need to transfer data at 512 kbps. Which is the less expensive solution: a T-1 connection or a frame relay connection? What happens if an office in Baton Rouge, Louisiana, also must be connected to both Memphis and Laramie at a rate of 512 kbps? Which solution is cheaper now? How about considering an MPLS over IP solution or a form of long-haul Ethernet?

2 You are consulting for a hospital that wants to send three-dimensional, high-resolution, color ultrasound images between the main hospital and an outpatient clinic. A patient will be receiving the ultrasound treatment at the clinic, while a doctor at the main hospital observes the images in real time and talks to the ultrasound technician and patient via a telephone connection. What type of telecommunications service presented in this chapter might support this application? Identify the telecommunications service, and be sure to list the particular details concerning the service. Explain your reasoning.

3 Your company wants to create an application that allows employees to dial in from a remote location and, using a single connection, access their voice mail, e-mail, and data files. What kind of system would allow this? Describe the necessary hardware and software components.

HANDS-ON PROJECTS

1. What kind of hardware is needed to support a T-1 connection to your business?

2. What newly emerging telecommunications technology or technologies may replace the technologies introduced in this chapter? Give a brief description of each technology.

3. Using the Better Box Corporation example, what further steps will be needed, and what will be the resulting costs of including:
   a. e-mail access for all locations
   b. Internet server capabilities for the Chicago location

4. To find out more technical details about DSL, investigate the different modulation techniques that are used to transmit DSL signals. Although these techniques are quite complex, they are an interesting study in the technological advances necessary to provide high-speed data streams over ordinary copper twisted pair wire. Using either the Internet or hard copy sources (which, by necessity, must be fairly current), look up the following four technologies. All four are being used somewhere on DSL circuits:
   - discrete multitone technology (DMT)
   - carrierless amplitude modulation (CAP)
   - multiple virtual line (MVL)
   - echo cancellation

12

Network Security

◆ ◆

IN THE LAST SEVERAL years, computer users and computer networks have been bombarded with computer viruses. One of the more destructive viruses, the Nimda virus (*admin* spelled backward), was launched in September of 2001. Internet security experts estimated that Nimda affected more than 140,000 computers in the United States alone. The virus entered a user's computer or Web server via e-mail and Web pages, created an administrator-level account, and overwrote files. The costs to clean up the Nimda virus were estimated to be in excess of $2.6 billion.

While 2002 was a relatively quiet year for viruses, 2003 started off with a bang. In January, four new viruses were launched: Lirva, SoBig, SQL Slammer, and Yaha. Three of these viruses easily bypassed already patched or protected e-mail programs and almost instantly flooded the Internet with millions of messages. Once again, clean-up costs were estimated to be in the hundreds of millions to billions of dollars.

The Netsky and Bagle viruses struck in March 2004, followed by the Sasser worm, which hit Windows systems hard in May 2004. Since 2005, computer software has improved and these kinds of viral attacks are much less common. Instead, 2006 through 2010 saw new attacks in the form of phishing, malware, and botnets.

What will the future hold? As software companies issue patches to cover holes and vulnerabilities in their software, more holes and vulnerabilities are discovered. If this trend continues, the money and time spent on protecting computers and then recovering from attacks will continue to increase at alarming rates.

What different types of data security exist today?

What different types of communications security exist today?

If the Internet is not capable of protecting itself, what, if anything, can you do to protect yourself while you are using the Internet?

Objectives ▶

After reading this chapter, you should be able to:

- ▶ Recognize the basic forms of system attacks
- ▶ Recognize the concepts underlying physical protection measures
- ▶ Cite the techniques used to control access to computers and networks
- ▶ Discuss the strengths and weaknesses of passwords
- ▶ List the techniques used to make data secure
- ▶ Explain the difference between a substitution-based cipher and a transposition-based cipher
- ▶ Outline the basic features of public key cryptography, Advanced Encryption Standard, digital signatures, and the public key infrastructure

- ▶ Cite the techniques used to secure communications
- ▶ Describe the differences between the frequency hopping spread spectrum technique and the direct sequence spread spectrum technique
- ▶ Recognize the importance of a firewall and be able to describe the two basic types of firewall protection
- ▶ Recognize the techniques used to secure wireless communications
- ▶ List the advantages to a business of having a security policy

Introduction ▶

Computer network security has reached a point at which it can best be characterized by two seemingly conflicting statements: Never has network security been better than it is today; and never have computer networks been more vulnerable than they are today. How both these statements can be true is an interesting paradox. Network security, as well as operating system security, has come a long way from the early days of computers. During the 1950s and 1960s, security amounted to trust—one computer user trusted another computer user not to destroy his or her data files. Since then, computer systems have grown in complexity, and modern systems support many simultaneous users with increasingly demanding requests, such as database retrievals and graphic-intensive downloads of Web pages. To protect users from one another, computer system security also has had to grow in complexity.

Today, the Internet allows anyone in the world to access or attempt to access any computer system that is connected to the Internet. This interconnectivity between computer systems and networks is both a boon and a bane. It allows us to download Web pages from Europe and Asia and order toys for the kids (or ourselves) from electronic stores, but it also exposes all Internet-attached systems to invasion. And the reality is that there are certain Internet users who have a single goal in mind: to access forbidden systems and steal or destroy anything they can get their "hands" on.

Internet systems are not the only systems that experience security problems. Any system with wireless capabilities is also open for vandalism, just as any corporate office center or educational facility is a potential target for someone wishing to walk in and steal or destroy computer files. Even building a 30-foot wall and a moat around your company and severing all connections to the outside world will not create a secure environment. In fact, many studies show that a majority of business thefts are committed by the employees working at the company. Carrying a flash drive home in a pocket is a very convenient (and easy) way to remove data files from a corporate computer network. In today's environment, managing computer network security is an all-encompassing, never-ending job.

This chapter's discussion of network security begins by examining the standard system attacks that are launched against computer users and their networks. We will then examine four basic areas of network security: implementing physical protection of computer networks and equipment, controlling access to computer systems, securing data, and securing communications. The chapter will then conclude with the basic principles of creating a network security policy.

Standard System Attacks

As a result of the large number of attacks on computers and networks in recent years, many studies have been performed to try to determine the standard methods of system attacks. The two leading attack methods for the last few years have exploited known vulnerabilities in operating systems and in application software. In particular, browser vulnerabilities have been one of the most common attacks. Hackers will spend long hours digging into popular operating systems and application software in an attempt to find an opening. Once this opening is found, the hacker will launch an attack that compromises the host computer or network server. Often, the company that created and supports the compromised operating system or application will issue a patch, a corrective piece of software designed to close the vulnerability. Unfortunately, not all computer owners and operators install the patch, or, even worse, the patch itself may have additional vulnerabilities. Interestingly, many hackers wait until a company announces the release of a patch. Once the vulnerability is known, the hacker creates code to take advantage of that vulnerability, knowing that many companies and users will be slow to install the patch, if they install it at all.

In both of these types of attacks, the perpetrator can gain access to the user's system in a number of ways. One very common technique is to deliver an e-mail or Web page that contains a malicious piece of code called a **mobile malicious code** or a **Trojan horse**, because it is hiding inside a harmless-looking piece of code. Once the user reads the e-mail and opens the attachment, the Trojan horse is released, and the damage occurs. Many people consider the Trojan horse to be a form of computer virus. A computer **virus** is a small program that alters the way a computer operates without the knowledge of the computer's users and often does various types of damage by deleting and corrupting data and program files, or by altering operating system components, so that computer operation is impaired or even halted. More recently, malicious programs that take over operations on a compromised computer are termed **botnets**.

Some other types of viruses include:

> ► Macro virus—A common type of virus that is programmed into an attached macro file. Macros are often found in spreadsheet, database, and word-processing documents. A macro virus hides within an application's macro and is activated when the macro is executed.

> ► Boot sector virus—One of the original forms of viruses, a boot sector virus is usually stored on some form of removable media. When the removable media is connected to a new machine and is used to boot the machine, the virus moves from the media into the host system. Newer versions do not need to be booted; they simply have to reside on the removable media, such as on a flash drive.

> ► Polymorphic virus—These viruses mutate with every infection, thus making them difficult to locate.

> ► File infector virus—A virus that infects a piece of executable code such as an .exe or .com file. When the program is executed, either from a disk or over the network, the host computer becomes infected.

Another form of computer virus is a worm. A computer **worm** is a program that copies itself from one system to another over a network, without the assistance of a human being. Worms usually propagate themselves by transferring from computer to computer via e-mail. Typically, a virus or a worm is transported as a Trojan horse—in other words, hiding inside a harmless-looking piece of code such as an e-mail or an application macro.

Some system attacks do not even need a user to open an e-mail or Web page at all. All a user has to do to become infected is connect to the Internet. While connected, a user's computer is constantly vulnerable to malicious software programs on the Internet that are scanning for unprotected computers (open TCP ports) and trying to exploit known operating system and application vulnerabilities. Unless the user has up-to-date virus protection software installed on his or her computer, the computer is likely to be compromised.

Another common category of system attacks that was more popular at the end of the twentieth century was the denial of service. **Denial of service attacks** (or distributed denial of service attacks) bombard a computer site with so many messages that the site is incapable of performing its normal duties. Some common types of denial of service include e-mail bombing, smurfing, and ping storm. In

e-mail bombing, a perpetrator sends an excessive amount of unwanted e-mail messages to someone. If these e-mail messages have a return address of someone other than the person actually sending the e-mail, then the sender is **spoofing**.

Smurfing is the name of a particularly nasty automated program that attacks a network by exploiting Internet Protocol (IP) broadcast addressing and other aspects of Internet operation. Simply stated, an attacker sends a packet to an innocent third party—the "amplifier"—who then unknowingly multiplies the packet hundreds or thousands of times and sends these copies on to the intended victim.

A ping is a common Internet tool used to verify if a particular IP address of a host exists and to see if the particular host is currently available. Pings are used most commonly by Unix-based systems, which are extensively used to support Internet Web servers. A **ping storm** is a condition in which the Internet ping program is used to send a flood of packets to a server to make the server inoperable.

For an example of spoofing and ping storms, let us consider the denial of service attacks that hit a number of commercial Web sites in early February 2000. The sites attacked included popular ones such as Yahoo!, eBay, Amazon, CNN, and e*Trade. In one type of attack, a hacker somehow took control of a number of servers on the Internet and instructed them to make contact with a particular Web server (the server that the hacker wished to disable). The compromised servers sent TCP/IP SYN (synchronize/initialization) packets with bogus source IP addresses (spoofing) to the intended target. As each SYN packet arrived, the target server tried to return a legitimate response but could not because the IP address was bogus. While the target server waited for a response from the bogus IP address, its resources were consumed as more bogus SYN commands arrived.

To see how a denial of service attack works, consider the following example. The hacker instructs a number of compromised Web servers (zombie servers) to send packets to a second set of servers (see Figure 12-1). These packets contain the destination address of the target Web site. The second set of servers, called bounce sites, receives multiple spoofed requests and responds by sending multiple packets to the target Web site at the same time. The target site is then overwhelmed and essentially crippled by the incoming flood of illegitimate packets, which leaves no room for legitimate packets.

Figure 12-1

An example of smurfing intended to cripple a Web server

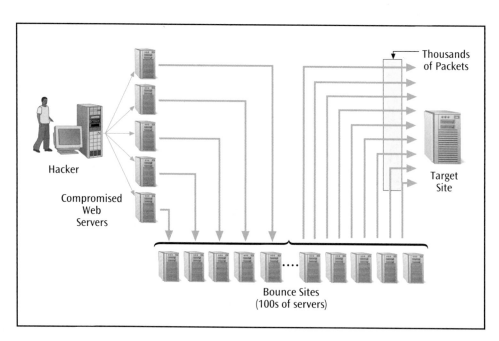

Another popular attack method is to abuse or take advantage of valid user accounts and the permissions associated with those accounts. For example, a user at a company or college who has a valid account will try to access forbidden documents, upload unauthorized files and datasets, use the company network as a site for illegal Web and e-mail attacks, or simply try to circumnavigate the system security features in an attempt to access forbidden services.

One more common attack method is to try to guess or intercept valid IDs and passwords from authorized users. Hackers will try to steal passwords by guessing simple combinations or eavesdropping on transmissions in which a password is being transmitted. Some hackers will even go so far as to create an application that appears to be legitimate and prompts users for an ID and password. Once an individual enters his or her ID and password, the software displays a message that gives the appearance of a system failure. The user moves on, not knowing that his or her ID and password have just been passed to a bogus program. In the last few years, there have been a number of attempts to steal IDs and passwords. In several instances, hackers created an e-mail that seemed to be a legitimate request coming from a well-known company. The e-mail even included the official-looking corporate logos and emblems. In the e-mail, the unsuspecting user was asked to provide private information such as a Social Security number or credit card number. The hacker would collect this personal information and use it illegally to purchase items or, even worse, commit identity theft. This type of attack is termed phishing.

Another type of attack that involves tricking the user into supplying confidential information is called pharming. In this attack, a Web user seeking to visit a particular company's Web site is unknowingly redirected to a bogus Web site that looks exactly like that company's official Web site. Not knowing that he or she is visiting a bogus Web site, the user may enter confidential information to register for a service or make a purchase, which can then be stolen.

Two more tools that are used by malicious users are root kits and keyloggers. A root kit (or simply rootkit) is a program or programs that have been installed (usually unknowingly) into a user's operating system. They are installed so deep within an operating system that normal protection software does not even notice the root kit, thus making them a highly manipulative form of attack. With a root kit in place, a user's computer can essentially be taken over by a remote user. While some root kits are actually useful and can help with problem computers, most root kits are meant to be destructive. A keylogger is a software system that runs in a computer and captures and records all the keystrokes made at the keyboard. This can be a useful program for someone who wants to monitor a computer user's progress on a certain task. Unfortunately, it can also be used to capture someone's user ID and password, or other private information. Some companies also use keyloggers to monitor their employee productivity or habits. This kind of use can be either good or bad depending upon your frame of reference.

Professionals who support computer networks, as well as individual computer users who have a computer at home or work, need to be aware of these common attacks, so they can decide how to best protect their systems from vandalism and intrusion.

Now that we understand several of the more common system attacks, let us investigate what we can do to prevent them. We will begin by examining what we can do to protect our equipment and our data physically.

Physical Protection

The physical protection of a computer system or a computer network consists of protecting the equipment from physical damage. Causes of physical damage include fire, floods, earthquakes, power surges, and vandalism. In many cases, the techniques to prevent damage are a matter of common sense. For example, rooms containing computer equipment should always be locked, and unauthorized persons should not be allowed to enter them. If at all possible, cabling, and the devices that cables plug into, should not be left exposed.

Some equipment, obviously, has to be in the open for public access. In this case, the equipment should be locked down. There are many kinds of antitheft devices for locking cabinets, locking cables to cabinets, and locking down keyboards and other peripheral devices. For example, one such device transmits a wireless signal to a pager whenever a computer cabinet is opened. This way, the person carrying the pager will know immediately when a cabinet is being opened, and which one it is; he or she can then direct security personnel to the appropriate location.

Another matter of common sense is that expensive computer systems should not be placed in the basements of buildings. Basements can flood and are often high-humidity locations. Rooms with a large number of external windows are also not advisable. Windows can let in sunshine, which can increase the temperature of a room. Computer equipment itself is known to heat up rooms. With the addition of sunlight, the increase in temperature may strain the capacity of the air-conditioning system (if there is one). As temperatures rise, the life expectancy of computer circuits decreases. Rooms with many external windows should also be avoided because they are more vulnerable to break-ins—and thus vandalism or theft.

To prevent electrical damage to computing equipment, high-quality surge protectors should be used on all devices that require an electrical current. The electrical circuits that provide power to devices should be large enough to adequately support them without placing a strain on the electrical system. In addition, computer devices should not be on the same circuits as electrical devices that power up and down and cause power fluctuations, such as large motors. Finally, devices that are susceptible to damage from static electricity discharges, such as memory cards and printed circuit boards, should be properly grounded.

Surveillance may also be considered a form of physical protection. Although many employees feel surveillance is an intrusion into their privacy, many network administrators consider it a good deterrent of computer vandalism and theft. The placement of video cameras in key locations can both deter criminals and be used to identify criminals in the event of vandalism or theft.

Other forms of surveillance can be used, beyond the capturing of live video with a camera. For example, companies that accept merchandise orders over the telephone often monitor each telephone call. Companies claim this form of surveillance can improve the quality of customer service and help settle future disputes. In another example, some companies use a form of surveillance called intrusion detection, or an intrusion detection system, which involves electronically monitoring data flow and system requests into and out of their systems. If unusual activity is noticed, protective action can be taken immediately. Intrusion detection is a growing field of study in network security. Both intrusion detection and surveillance are important areas of physical protection.

In addition to video surveillance and intrusion detection, there is an interesting surveillance technique built on the concept of a honeypot. In computer terminology, a honeypot is a trap that is set by network personnel in order to detect unauthorized use of a network resource. Typically, a honeypot is a network location that appears to be part of the company's resources and contains data that might be interesting or valuable to a hacker. In reality, this data is bogus and meant to tempt hackers into committing a break-in. Companies that implement this technique usually maintain surveillance of the honeypot, and thus are able to observe the actions of would-be hackers. Basically, the honeypot can serve as an early warning tool, but companies should be aware that using a honeypot for other purposes (such as luring a potential hacker into committing a crime) is oftentimes questionably ethical and possibly even illegal.

Controlling Access

Controlling access to a computer network involves deciding and then limiting who can use the system and when the system can be used. Consider a large corporation in which there are many levels of employees with varying job descriptions. Employees who do not need to come in contact with sensitive data should not have access to that level of data. For example, if an employee simply performs data-entry operations, he or she should probably not be allowed access to payroll information. Likewise, employees working in payroll need access to the payroll database, but, more than likely, they do not need access to information regarding corporate research programs. A manager of an area would probably have access to much of the information in his or her department, but the manager's access to information in other departments should probably be limited. Finally, top-level executives often have access to a wide range of information within a company. But there are many companies that wisely limit the information access capabilities of even top-level management.

Local area networks and database systems provide much flexibility in the assigning of access rights to individuals or groups of individuals, as you will see shortly. Access rights define the network resources that a user or set of users can access. A company's computer network specialists, along with database administrators and someone at the top levels of management, such as the Chief Information Officer (CIO), often work together to decide how the company should be broken up into information access groups. Then they resolve each group's access rights and determine who should be in each group. As you may recall from Chapter Eight, network operating systems, such as Windows 2008, can be very useful in the task of creating such workgroups and assigning access rights.

It is also possible to limit access to a system by the time of day or the day of the week. If the primary activity in one part of your business is accessing personnel records, and this activity is performed only during working hours by employees in the personnel or human resources department, then it might be reasonable to disable access to personnel records after working hours, for example, from 5:30 p.m. until 7:00 a.m. the next morning. The network administrator could also deny access to this system on weekends. Figure 12-2 shows an example of a network operating system's dialog box that an administrator would use to set user access limits within a network.

Figure 12-2
*Sample dialog box
from a network
operating system for
setting time of day
restrictions*

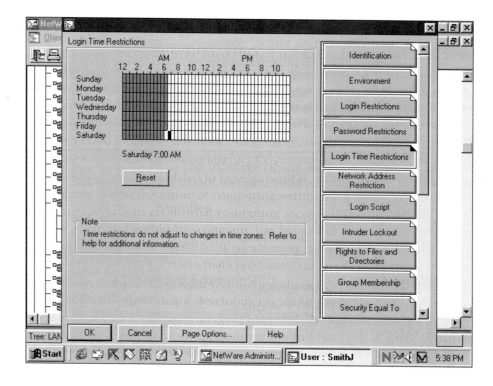

It may also be wise to limit remote access to a system during certain times of the day or week. One reason for this is to prevent someone from dialing in at 2:30 a.m. to engage in illegal activities, such as the transferring of funds from one account to another. Given such possibilities, it is reasonable for companies to protect themselves by deciding that corporate fund transfers can occur only during typical business hours and to restrict dial-in activity to these hours.

One of the most common ways to control access to a system is to require user IDs and passwords, even though this technique has a number of shortcomings. Let us examine passwords and other ID systems in more detail.

Passwords and ID systems

These days, almost every system that stores sensitive or confidential data requires an authorized user to enter a password, personal identification number (PIN), or some other form of ID before gaining access to the system. Typically, this password or ID is either a string of characters that the user must remember or a physical feature of a user, such as a fingerprint. The technology in the area of identity management is improving rapidly as companies try to incorporate systems that are less vulnerable to fraud.

Perhaps the most common form of protection from unauthorized use of a computer system is the **password**. Anyone accessing a computer system, banking system, or voice mail/e-mail system is required to enter a password or personal identification number (PIN). Although the password is the most common form of identification, it is also one of the weakest forms of protection. Too often, passwords become known, or "misplaced," and fall into the wrong hands. Occasionally, a password is written on paper, and the paper is discovered by the wrong people. More often, however, the password is too simple, and an intruder guesses it. The standard rules that an individual should follow when creating or changing a password include:

▶ Change your password often.

▶ Pick a good password by using at least eight characters, mixing uppercase and lowercase if the computer system is case-sensitive, and mixing letters with numbers.

▶ Do not choose passwords that are similar to your first or last names, pet names, car names, or other choices that can be easily guessed.

▶ Do not share your password with others; doing so invites trouble and misuse.

Figure 12-3 shows how a typical network operating system enables the network administrator to require a user to use a password, gives the user the ability to change his or her own password, and require the user to select a password of a particular size.

Figure 12-3
Controlling a user password with a typical network operating system

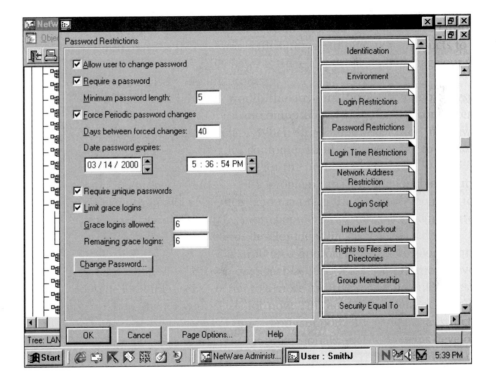

Some computer systems generate random passwords that are very difficult to guess, but are also hard to remember. Often, the user who is given a randomly generated password either changes it to something simpler, which makes the password easier to guess, or writes the password down on a piece of paper, which defeats the whole purpose of having a secret password. Some systems disallow obvious passwords or already used passwords, thus requiring the user to be creative and to select a password that is difficult to guess.

A common fallacy among computer system users is that the internal operating system file that stores the login IDs and passwords is susceptible to intrusion. Interestingly, most computer systems store passwords in an encrypted form for which there is no known decryption. How then does the system know when you have entered the correct password? When a user enters his or her login ID and password, the password is encrypted and compared with the entry in the encrypted password file. If the two encrypted passwords match, the login is allowed. Anyone who gets access to this encrypted password file will discover only unreadable text. This encryption technique is the reason that, when you forget your password, a computer operator cannot simply read a file and tell you what it is. The computer operator can only reset the password to something new.

Because there are so many weaknesses to the password, other forms of identification have emerged. **Biometric techniques** that observe and record some aspect of the user, such as voiceprints, fingerprints, eyeprints, and faceprints, appear to be the wave of the future. For example, England has a large database of earprints. Research shows that no two ears are the same; thus, an earprint is useful in helping to identify an individual. Fingerprints have, of course, long been used to distinguish one individual from another. Now, some laptop computers have built-in fingerprint scanners that can scan a user's thumbprint and allow or disallow the user access to a computer system. Systems also are available that can record and digitize your voice. The digital voice pattern is compared to a stored sample, and the software determines if the match is close enough for validation. Retinal scans (the retina is the inside back lining of the eye) have been commonly featured in movies and do exist in the real world as a security technique, although to a small degree due to complexity and cost. An additional impediment has been that people are squeamish when told a laser is going to scan the inside of their eyeballs. Another part of the eye that is unique among all individuals is the iris, or the colored portion of the eye. Some security devices use the iris to identify people who are allowed to access a given system. Because the technology needed to perform iris scans is less expensive than retinal scan technology, it is more likely that we will see iris scans performed in the near future. Other research in the field of biometric techniques has been aimed at digitizing the features of the entire face and comparing this digital representation with a stored image. Companies that manufacture 24-hour automatic teller machines are interested in replacing the credit-card-sized automatic teller machine (ATM) card and corresponding PIN with something that cannot be as easily stolen, such as a fingerprint, faceprint, or eyeprint. To enhance their security, many companies require combinations of forms of identification, such as a password and a thumbprint.

Access rights

Modern computer systems and computer networks allow multiple users to access resources such as files, tapes, printers, and other peripheral devices. Many times, however, the various resources of an organization are not supposed to be shared, or they should be shared only by a select group. If resource sharing is to be restricted, then a user or network administrator should set the appropriate access rights for a particular resource. Most access rights have two basic parameters: *who* and *how*. The *who* parameter lists who has access rights to the resource. Typical examples of *who* include the owner, a select group of users, and the entire user population. The *how* parameter can specify how a user may access the resource, and rights are listed as RWX, for read, write, and execute. More precisely, the R includes read and print privileges; W includes write, edit, append, and delete rights; and X stands for execute, or run. When a user creates a file, the usual system defaults may grant the user full read, write, and execute access (or ALL), so that he or she can modify or delete the file at any time. This user, or owner, may allow other users to access the file, but can limit them to having read access rights only. Figure 12-4 shows an example of how a network operating system assigns access rights to a resource. As you can see in the figure, a network administrator can use this application to assign supervisor, read, write, create, erase, modify, file scan, and access control rights to a particular user.

Figure 12-4
A network operating system assigning access rights to a resource

Modern network operating systems allow network administrators to create workgroups. These workgroups are defined by the network administrator and can contain any form of user grouping desired. For example, one workgroup might be all the employees from marketing and engineering who are currently working on a particular project. Once the workgroup is defined, it is then possible to assign this workgroup a unique set of access rights.

Auditing

Auditing a computer system is often a good way to deter crime and can also be useful in apprehending a criminal after a crime has been committed. Computer auditing usually involves having a software program that monitors every transaction within a system. As each transaction occurs, it is recorded into an electronic log along with the date, time, and "owner" of the transaction. If a transaction is suspected of being inappropriate, the electronic log is scanned and this information is retrieved. In a classic case of a computer crime that was thwarted by auditing, a New York man discovered that when invoices for amounts under $500 were sent to local government agencies, they would be routinely paid without any request for further details. Armed with this knowledge, he created his own invoices and sent them to these government agencies. True to form, the agencies paid the invoices, and his savings account grew. But several months later a lawyer examining the computer audit trail of payments noticed that there was a pattern of checks (all under $500) that were sent to the same individual.

Figure 12-5 shows an example of an audit log from the Windows operating system's Event Viewer window. Note that for each transaction (or event), the date, time, and source of the event are recorded. Typical recorded events include the failure of a driver or other system components to load during system startup, any possible breaches in security, and any program that might record an error while trying to perform a file operation.

Figure 12-5
Windows Event Viewer example

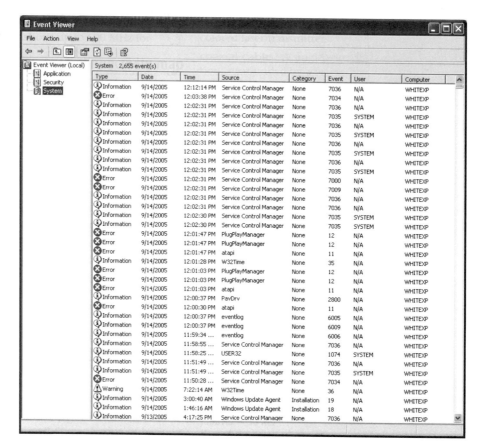

Many good computer programs are available that can audit all the transactions that take place on a computer system. The money spent on purchasing, installing, and supporting an audit program may be well spent if the program helps catch a person performing unauthorized transactions.

Securing Data

Computer systems store and manipulate data, while communication systems transfer data. What precautions can we take to ensure that this data is not corrupted or intercepted by the wrong people? Let us examine several techniques that can be used to make data more secure. We will begin by introducing the basic techniques used to encrypt and decrypt data. Then we will look at some of the more recent techniques used to secure data, such as the Advanced Encryption Standard, digital signatures, public key infrastructure, and steganography.

Basic encryption and decryption techniques

When users transfer data from one point to another in a computer network, it is often necessary to ensure that the transmission is secure from anyone who might be eavesdropping on the line. The term "secure" means two things. First, it should not be possible for someone to intercept and copy an existing transmission. Second, it should not be possible for someone to insert false information into an existing transmission. Financial transactions and military communications are two good examples of data that should be made secure during transmission.

Fiber-optic cable represented a major improvement in the ability of transmission media to secure sensitive data. As you may recall from Chapter Three, metal media such as twisted pair and coaxial cable conduct electrical signals and are therefore relatively easy to tap, but fiber-optic wires are much more difficult to tap because they transmit pulses of light, which are not electromagnetic. To tap a fiber-optic system illegally, you must either physically break the fiber-optic cable—an intrusion that would be noticed immediately—or you must gain access to the fiber-optic junction box, which is usually in a locked location.

Regardless of how secure a given computer's medium is, the data in the system is still vulnerable because sometime in its life it will probably be moved through other (less secure) computers, stored on unknown/unprotected hard disk drives, and/or transmitted over standard telephone systems—in other words, data can be intercepted or stolen from other systems. Considering these risks, sensitive data requires additional security measures. One such additional measure is to use encryption software to encrypt the data before it is transmitted, transmit the encrypted data over secure media, then decrypt the received data to obtain the original information. **Cryptography** is the study of creating and using encryption and decryption techniques. Many volumes could be filled on cryptographic techniques, but an introduction to the basic principles will give you a sufficient understanding of the encryption techniques used today.

Delving into the topic of cryptography requires you to understand a few basic terms. **Plaintext** (which will always be shown in lowercase characters in the examples that follow) is data before any encryption has been performed (Figure 12-6). An **encryption algorithm** is the computer program that converts plaintext into an enciphered form. The **ciphertext** (shown in uppercase characters) is the data after the encryption algorithm has been applied. A key is the unique piece of information that is used to create ciphertext and then decrypt the ciphertext back into plaintext. After the ciphertext is created, it is transmitted to the receiver, where the ciphertext data is decrypted.

Figure 12-6
Basic encryption and decryption procedure

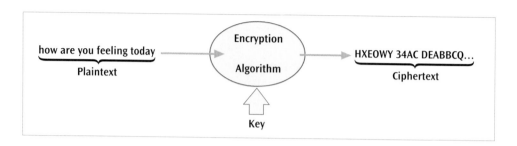

Early cryptography algorithms used the same key for both encryption and decryption. The use of a single key greatly concerned many experts. To allow local and remote locations to send and receive encrypted data, the key had to be given to both local and remote parties. If this key were intercepted and fell into the wrong hands, not only could encrypted data be decrypted, but false data could be encrypted and sent to either party. Newer techniques, as you will soon see, solve this problem by allowing the use of two different, but mathematically related, keys. One key is for encrypting the data, and the second key is for decrypting the data. Let us begin our discussion of encryption techniques by looking at one of the simplest: monoalphabetic substitution-based ciphers.

Monoalphabetic Substitution-based Ciphers

Despite its daunting name, the monoalphabetic substitution-based cipher is actually a fairly simple encryption technique. A monoalphabetic substitution-based cipher replaces a character or group of characters with a different character or group of characters. Consider the following simple example. Each letter in the plaintext row maps onto the letter below it in the ciphertext row.

```
Plaintext:   a b c d e f g h i j k l m n o p q r s t u v w x y z
Ciphertext:  P O I U Y T R E W Q L K J H G F D S A M N Z V C X B
```

This ciphertext simply corresponds to the letters on a keyboard, scanning right to left, top to bottom. To send a message using this encoding scheme, each plaintext letter of the message is replaced with the ciphertext character directly below it. Thus, the message

```
how about lunch at noon
```

would encode to

```
EGVPO GNMKN HIEPM HGGH
```

A space has been placed after every five ciphertext characters to help disguise obvious patterns. This example is monoalphabetic, because one alphabetic string was used to encode the plaintext. It is a substitution-based cipher, because one character of ciphertext was substituted for one character of plaintext.

Polyalphabetic Substitution-based Cipher

The polyalphabetic substitution-based cipher is similar to the monoalphabetic cipher, but it uses multiple alphabetic strings to encode the plaintext, rather than one alphabetic string. Possibly the earliest example of a polyalphabetic cipher is the Vigenére cipher, devised by Blaise de Vigenére in 1586. For the Vigenére cipher, a 26 × 26 matrix of characters is created, as shown in Table 12-1.

Table 12-1

An example of a Vigenére 26 × 26 ciphertext character matrix

Key Character	Plaintext Letters																									
	A	B	C	D	E	F	G	H	I	J	K	L	M	N	O	P	Q	R	S	T	U	V	W	X	Y	Z
A	A	B	C	D	E	F	G	H	I	J	K	L	M	N	O	P	Q	R	S	T	U	V	W	X	Y	Z
B	B	C	D	E	F	G	H	I	J	K	L	M	N	O	P	Q	R	S	T	U	V	W	X	Y	Z	A
C	C	D	E	F	G	H	I	J	K	L	M	N	O	P	Q	R	S	T	U	V	W	X	Y	Z	A	B
...	...																									
Z	Z	A	B	C	D	E	F	G	H	I	J	K	L	M	N	O	P	Q	R	S	T	U	V	W	X	Y

To perform this cipher, you choose a key, such as COMPUTER SCIENCE, which you repeatedly place over the plaintext message. For example:

Key: COMPUTERSCIENCECOMPUTERSCIENCECOMPUTERSCIENCECO
Plaintext: thisclassondatacommunicationsisthebestclassever

To encode the message, you look at the first letter of the plaintext, *t*, and the corresponding key character immediately above it, *C*. The C tells you to use row C of the 26 × 26 matrix to perform the alphabetic substitution for the plaintext character *t*. You then go to column T in row C and find the ciphertext character V. This process is repeated for every character of the plaintext. The key, COMPUTER SCIENCE, must be kept secret between the encoder and decoder.

To make matters more difficult for an intruder, the standard 26 × 26 matrix with row A, row B, row C, and so on does not have to be used. Instead, the encoding and decoding can be done using a unique matrix. In this case, both the matrix and the key must remain a secret.

Transposition-based Ciphers

A transposition-based cipher is different from a substitution-based cipher in that the order of the plaintext is not preserved. Rearranging the order of the plaintext characters makes common patterns unclear and the code much more difficult to break. Let us consider a simple example of a transposition cipher. Choose a keyword that contains no duplicate letters, such as COMPUTER. Over each letter in the keyword, write the number that corresponds to the order in which that letter appears in the alphabet when compared to the other letters in the keyword. For the keyword COMPUTER, C appears first in the alphabet, E is second, M is third, O is fourth, and so on.

14358726
COMPUTER

Take a plaintext message such as "this is the best class i have ever taken" and write it under the keyword in consecutive rows going from left to right.

```
14358726
COMPUTER
thisisth
ebestcla
ssihavee
vertaken
```

To encode the message, read down each column starting with the column numbered 1 and proceeding through to the column numbered 8. Reading column 1 gives us TESV, and column 2 gives us TLEE. Encoding all eight columns gives us the following message:

TESVTLEEIEIRHBSESSHTHAENSCVKITAA

Two interesting observations can be made about this example. First, the choice of the keyword is once again very important, and care must be taken to make sure the keyword does not fall into the wrong hands. Second, you could make the encryption even more difficult by performing an additional substitution-based cipher on the result of the transposition cipher. In fact, why stop there? You could create a very difficult code if you repeated various patterns of substitution- and transposition-based ciphers, one after another.

Public Key Cryptography

All the encoding and decoding techniques shown thus far depend on protecting the key and keeping it from falling into the hands of an intruder. Given how important key secrecy is, it is surprising how often keyword or password security is lax or altogether nonexistent. Consider the episode of Stanley Mark Rifkin and the Security National Bank. In 1978, Rifkin posed as a bank employee, gained access to the wire funds transfer room, found the password to transfer electronic funds taped to the wall above a computer terminal, and transferred $12 million to his personal account. When he was apprehended, he was bragging about the feat in a bar.

One technique for protecting a key from an intruder is often seen in late-night, black-and-white spy movies: Break the key into multiple pieces and assign each piece to a different individual. In real life, rather than simply assigning one or two characters to each person, cryptographers use mathematical techniques, such as simultaneous linear equations, to divide the key. Other techniques that require manipulation of the key, or masking the key so it is not obvious, can be tedious and sometimes do not add any real security to the key.

One of the inherent problems with protecting a single key is that it means that only one key is used to both encode and decode the message. The more people who have possession of the key, the more chances there are for someone to get sloppy and let the key become known to unauthorized personnel. But what if two keys are involved—one public and one private? Data encrypted with the public key can be decoded only with the private key, and data encrypted with the private key can be decoded only with the public key. This concept of two keys, public and private, is called **public key cryptography**. It is also called **asymmetric encryption**. (The opposite of asymmetric encryption is symmetric encryption, in which one key is used to encrypt and decrypt.) When one key encrypts the plaintext and another key decrypts the ciphertext, it is nearly impossible—even if you have access to one of the keys—to deduce the second key from the first key. Furthermore, the encrypted data is also extremely difficult to decode without the private key—even for the experts.

How does public key cryptography work? Consider a business that has its headquarters in New York. A branch office in Atlanta wishes to send secure data to the New York office. The New York office sends the public key to Atlanta and keeps the private key locked up safe in New York. The Atlanta office uses the public key to encrypt the data and sends the encrypted data to New York. Only the New York office can decode the data, because it is the only one possessing the private key. Even if other parties intercept the transmission of the public key to Atlanta, nothing will be gained, because it is not possible to deduce the private key from the public key. Likewise, interception of the encrypted data will lead to nothing, because the data can be decoded only with the private key.

For a more familiar example, consider a situation in which a person browsing the Web wishes to send secure information (such as a credit card number) to a Web server. The user at a workstation clicks on a secured Web page and sends the appropriate request to the server. The server returns a "certificate," which includes the server's public key, and a number of preferred cryptographic algorithms. The user's workstation selects one of the algorithms, generates a set of public and private keys, keeps the private key, and sends the public key back to the server. Now both sides have their own private keys, and they both have each other's public key. Data can now be sent between the two endpoints in a secure fashion.

This technique is a part of the Secure Sockets Layer and can be used by most Web browsers and servers when it is necessary to transmit secure data. Secure Sockets Layer (SSL) is an additional layer of software added between the application layer and the transport (TCP) layer that creates a secure connection between sender and receiver. Because Unix operating systems create connections between endpoints in a network via sockets, the software has the title of Secure Sockets Layer. SSL's successor is Transport Layer Security (TLS), which is based upon SSL version 3 and contains a few technical improvements over SSL. A second technique similar to SSL is Secure-HTTP or S-HTTP, but it has not experienced the widespread success that SSL has. Both SSL and S-HTTP may eventually give way to IPsec, the newer security standard created by the Internet Engineering Task Force.

IPsec (IP security) is a set of protocols created to support the secure transfer of data at the IP layer. One popular application of IPsec is in the support of virtual private networks. Recall that virtual private networks create a "tunnel" between a user and a remote destination via the Internet. IPsec uses encryption to protect the packet as it is transmitted through this tunnel. More precisely, the tunnel endpoints exchange a public key/private key pair.

Data Encryption Standard and Advanced Encryption Standard

The Data Encryption Standard (DES) is a commonly employed encryption method used by businesses to send and receive secure transactions. The standard came into effect in 1977 and was reapproved in 1983, 1988, and 1993. The basic algorithm behind the standard is shown in Figure 12-7. The DES algorithm works with 64-bit blocks of data, and subjects each block to 16 levels, or rounds, of encryption. The encryption techniques are based upon substitution- and transposition-based ciphers. Each of the 16 levels of encryption can perform a different operation based on the contents of a 56-bit key. The 56-bit key is applied to the DES algorithm, the 64-bit block of data is encrypted in a unique way, and the encrypted data is transmitted to the intended receiver. It is the responsibility of the parties involved to keep this 56-bit key secret and prevent it from falling into the hands of other parties.

Figure 12-7
The basic operations of the Data Encryption Standard

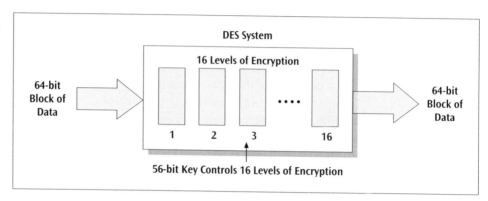

Even though a 56-bit key provides for more than 72 quadrillion possible keys, the Data Encryption Standard was criticized from its inception for being too weak. Businesses and researchers felt a much larger key was needed to actually make the encryption virtually impossible to break. The government, however, felt that 56 bits were sufficient and that no one would spend the time or resources necessary to try to break the key. Why was the government reluctant to support businesses' concerns about their data security? Perhaps the government's own concern was that criminals armed with a key larger than 56 bits could send encrypted materials that no one, including the government itself, could decrypt, which would create a disadvantage that could compromise national security. In July 1998, the government's side of the debate suffered a setback when a group called the Electronic Frontier Foundation claimed that it had cracked a 56-bit key in 56 hours, using a single customized personal computer it called the DES Cracker.

Attempts to design a stronger encryption technique led to the creation of triple-DES. With **triple-DES**, the data is encrypted using DES three times—the first time by a first key, the second time by a second key, and the third time by the first key again. While this technique creates an encryption system that is virtually unbreakable, it is CPU-intensive; that is, the use of triple-DES encryption requires a large amount of processing time. A CPU-intensive process is not desirable on small devices such as cell phones, smart cards, and personal digital assistants. Thus, what is needed is an encryption technique that is as powerful as triple-DES but also fast and compact.

The Advanced Encryption Standard answers this need. The **Advanced Encryption Standard (AES)** was selected by the U.S. government to replace DES. More precisely, the National Institute of Standards and Technology selected the algorithm Rijndael (pronounced rain-doll) in October 2000 as the basis for AES. The Rijndael algorithm involves very elegant mathematical formulas, requires only one pass (not three passes like triple-DES), computes very quickly, is virtually unbreakable, and operates on even the smallest computing devices. With AES, one has the ability to select a key size of 128, 192, or 256 bits.

Digital Signatures

Basic encryption and decryption techniques have come a long way since their inception. While today's techniques often employ state-of-the-art encryption and decryption methods, many of these methods have foundations based on the earlier basic techniques. Let us examine a number of the more recent advances in cryptography by beginning with the concept of a digital signature.

When participating in financial or legal transactions, people often identify themselves through a handwritten signature or by entering a PIN. For example, when you use your debit card to purchase something, you have to enter your PIN into an electronic keypad and your PIN is compared to the one stored in the system. Assuming your card and PIN have not been stolen and someone is not posing as you, this system is fairly safe. But what happens when you want to "sign" an electronic document so that later you will be able to prove that it is your document and no one else's? To authenticate electronic documents as yours, you need to create a digital signature. A **digital signature** is a security procedure that uses public key cryptography to assign to a document a code for which you alone have the key.

Digitally signing an electronic document involves sending the document through a complex mathematical computation that generates a large prime number called a hash. The original document and the hash are inextricably tied together. If either the document or the hash is altered, they will not match, and it will not be possible to decode the document.

When a user digitally signs a document, a hash (the original hash) is generated from the document and then encoded with the user's private key. This encoded hash becomes the digital signature and is either stored with the document or transmitted with the document. Later, if someone wants to verify that this document belongs to this user, a new hash is created from the document. The original hash, which has been encrypted with the owner's private key, is decrypted with the owner's public key, and the two hashes are compared. If the two hashes agree, the data was not tampered with, and the user's digital signature is valid. Someone could argue that he or she did not *see* the user "sign" the original document, and therefore does not truly know that the document is legitimate. This argument is resolved by the fact that only this one user should have this private key. Thus, this user is the only person who could have created this particular encoded document.

One drawback to this system is that if someone discovers the user's private key, a digital signature could be forged. Creating a system that assigns hashes and keys and keeps them secret requires a public key *infrastructure*, a topic we will discuss shortly.

In 2000, the U.S. government approved the Electronic Signatures in Global and National Commerce Act. This act grants digital signatures in electronic documents the same legal standing as that of handwritten signatures on pieces of paper. It is important to note, however, that not all documents may be legally signed by a digital signature. For example, although home loans and mortgages may be digitally signed, documents such as divorce agreements, wills, and adoption contracts will still need the old-fashioned pen-and-paper signatures.

Pretty Good Privacy (PGP)

In an effort to create an encryption scheme that could be used by the average person, an entrepreneur named Philip Zimmermann created encryption software called **Pretty Good Privacy (PGP)**. PGP is high-quality encryption software that has become quite popular for creating secure e-mail messages and encrypting other types of data files. PGP employs some of the latest techniques of encryption, including public key cryptography and digital signatures, and is available to anyone in the United States for free. Web sites, FTP sites, and bulletin boards offer free copies for individual use, and commercial versions can be purchased for commercial installations. You should be aware, however, that even though the use of PGP and other encryption software is legal in the U.S., it may not be legal in other countries. Also, it is illegal to send PGP from the U.S. to another country. Zimmermann himself ran into much legal trouble because he allowed his PGP software to leave the United States, which violated federal export laws banning the external transportation of encryption software.

If you wish to send a secure transmission, you can apply PGP to your document, be it a Web page or an e-mail message. Note, however, that because PGP is based on public key encryption techniques and therefore public and private keys are necessary, the receiving workstation must possess the same PGP software as your sending workstation in order to decrypt your message. Because there are many different versions of PGP and various restrictions on their use, you should consult one or more of the popular Web sites devoted to supporting PGP to learn more specific details.

Kerberos

Kerberos is an authentication protocol designed to work on client/server networks that use secret or symmetric cryptography. Unlike public key cryptography, in which two keys are used—a public key and a private key—Kerberos uses one key for both encryption and decryption.

One of the primary functions of Kerberos is to authenticate users. For example, if a client workstation requests a service from a Web server, and the Web server wants assurance that the client is who he or she claims to be, Kerberos can provide the authentication. To ensure this level of authentication, a client presents a *ticket* that was originally issued by the Kerberos authentication server. In this ticket is a password that was selected by the client at an earlier time. The server accepts the ticket along with the transaction request, examines the ticket, and verifies that the user is who he or she claims to be. Many application packages found on client/server systems support the use of Kerberos.

Now that we have learned the basic techniques for keeping data safe, let us look at a technology that combines all these techniques.

Public Key Infrastructure

Suppose you are working for a company that wants to open its internal local area network to the Internet, so as to allow its employees access to corporate computing resources—such as e-mail and corporate databases—from remote locations, to allow corporate customers or suppliers access to company records, or to allow retail customers to place orders or inquire about previous orders. In each of these transactions, the company wants to be sure that the person conducting the transaction is a legitimate employee or customer, not a hacker trying to compromise its system. A technology that assists a company to achieve this goal is public key infrastructure.

Public key infrastructure (PKI) is the combination of encryption techniques, software, and services that involves all the necessary pieces to support digital certificates, certificate authorities, and public key generation, storage, and management. A company that adheres to the principles of PKI issues digital certificates to legitimate users and network servers, supplies enrollment software to end users, and provides network servers with the tools necessary to manage, renew, and revoke certificates.

A digital certificate, or simply a **certificate**, is an electronic document, similar to a passport, that establishes your credentials when you are performing transactions on the World Wide Web. It can contain your name, a serial number, expiration dates, a copy of your public key, and the digital signature of the certificate-issuing authority (to allow you to verify that the certificate is legitimate). Certificates are usually kept in a registry, so that other users may look up a particular user's public key information.

Many certificates conform to the X.509 standard. Created and supported by the International Telecommunication Union-Telecommunication Standardization Sector (ITU-T), the X.509 standard defines what information can go into a certificate. All X.509 certificates contain the following pieces of information:

- ► Version—Identifies the version of the X.509 standard that applies to this certificate
- ► Serial number—A unique value that identifies a particular certificate; when a certificate is revoked, its serial number is placed in a certificate revocation list (CRL)
- ► Signature algorithm identifier—Identifies the algorithm used by the certificate authority (an entity that will be defined shortly) to sign the certificate
- ► Issuer name—The name of the entity, normally a certificate authority, that signed the certificate (in some instances, the issuer signs his or her own name)
- ► Validity period—The limited amount of time for which each certificate is valid (a period that can be as short as a few seconds or as long as a century), denoted by a start date and time and an end date and time
- ► Subject name—The name of the entity whose public key this certificate identifies
- ► Subject public key information—The public key of the entity being named, together with an algorithm identifier that specifies to which public key encryption system this key belongs
- ► Digital signature—The signature of the certificate authority that will be used to verify a legitimate certificate

All certificates are issued by a certificate authority. A **certificate authority** (CA) is either specialized software on a network or a trusted third-party organization or business that issues and manages certificates. One such third-party business is VeriSign, which is a fully integrated PKI service provider, designed to provide certificates to a company that wants to incorporate PKI but does not want to deal with the elaborate hardware and software necessary to create and manage its own certificates.

Consider a scenario in which a user wants to order some products from a Web site. When the user is ready to place the order, the Web site tries to make sure the user really is who he or she claims to be, by requesting that the user sign the order with his or her private key. This private key should have been already issued to the user by a certificate authority, such as VeriSign. The package consisting of the user's order and digital signature is sent to the server. The server then requests both the user's certificate and VeriSign's certificate. The server validates the user's certificate by verifying VeriSign's signature, and then uses the user's certificate to validate the signature on the order. If all the signatures match, confirming that the user is who he or she claims to be, then the order is processed. Typically all of this happens without the user's knowledge. Even though most Web retailers do not use PKI to verify their customers—relying instead on Secure Sockets Layer—there are indications that more of them will do so in the near future.

A certificate revocation list (CRL) is a list of certificates that have been revoked before their originally scheduled expiration date. Why might a certificate be revoked? A certificate will be revoked if the key specified in the certificate has been compromised and is no longer valid, or the user specified in the certificate no longer has the authority to use the key. For example, if an employee has been assigned a certificate by a company, then later quits or is fired, the certificate authority will revoke the certificate and place the certificate ID on the CRL. The employee will no longer be able to send secure documents using this certificate. Actually, this will be true only if the software takes the time to *check* each certificate against the CRL. The decision as to whether taking the time to perform this check is worth the effort might depend on the importance of the signed document.

Companies that need to conduct secure transactions will often invest in their own PKI systems. Because PKI systems are specialized and relatively involved, they are typically very expensive. Before purchasing a PKI system, a company needs to consider whether the transactions that need to be secure remain internal (within the network), or whether they need to traverse external networks. External transactions, such as electronic commerce transactions, require a system that can interoperate with the systems of other companies. Another factor that the company needs to consider when choosing a PKI system is the size of the application requiring secure transactions. If you have a small number of employees (roughly, fewer than 100), it might be more economical to purchase a PKI system that is embedded within an application. If the number of employees is larger (more than 100), then a PKI system that supports a complete system of key and certificate storage and management might be more useful.

What kind of applications or transactions would benefit from public key infrastructure? Here is a brief sample:

- World Wide Web access—Ordering products over the Internet is a common activity that benefits from PKI.
- Virtual private networks—A PKI system can assist in creating a secure VPN tunnel (described in Chapter Ten) through the insecure Internet.
- Electronic mail—PKI can be used to secure messages when users wish to send secure e-mail across an insecure network such as the Internet.
- Client/server applications—Client/server systems, such as interactive database systems, often involve the transfer of secure data. PKI can be used to ensure that the data remains secure as it travels from client to server and back.
- Banking transactions—Banking transactions that traverse external networks need to be secure. PKI is commonly employed in banking systems to create public/private key pairs and digital signatures.

To summarize, as good as PKI sounds, it does have its problems. Currently, it is, as stated earlier, expensive. Typical PKI systems cost tens of thousands to hundreds of thousands of dollars for medium-sized businesses. Moreover, many PKI systems are proprietary and do not interact well with other PKI systems. On the positive side, the price of PKI technology, like that of most technologies, is constantly dropping. In addition, the Internet Engineering Task Force continues to work on standards to which future PKI systems can adhere. It is hoped that these new standards will increase the possibility that PKI systems will interoperate.

Steganography

Steganography is the art and science of hiding information inside other, seemingly ordinary messages or documents. Unlike the sending of an encrypted message, steganography involves *hidden* messages; thus, you can tell when an encrypted message has been sent, but you cannot tell when steganography is hiding a secret message within a document. While there are many branches of steganography, involving such means of concealment as invisible inks and microdots, the branch of steganography that has drawn the most attention recently is the concealment of secret messages within digital images.

When stored in a computer, digital images are essentially large arrays of numbers. Each number represents the presence of a pixel, a colored or a black-and-white dot. It is common to have 8-bit pixels that represent 256 different colors, and typical image sizes are 640 × 480 pixels, or 2,457,600 bits (640 × 480 × 8). The idea behind steganography is that secret messages can be hidden within these 2.4-million-plus bits. Currently, there are three ways to conceal the bits of a secret message within a digital image: by putting them in the least significant bit (rightmost bit) of various pixels, by marking the image with a process (known as masking and filtering) that is similar to creating a watermark on paper, and by using algorithms and transformations. Readers intrigued by this unique form of securing data are encouraged to search the Web for further information.

Now that we have covered the various encryption technologies, let us examine another type of defense system that can be used to keep resources secure.

Securing Communications

So far in the chapter we have examined some of the more common attack techniques used by hackers, various ways to make computer and network equipment physically secure, techniques for controlling access, and techniques for securing data. Now let us turn our attention to examining ways to make communications more secure. We will start by discussing ways to protect the transmission of secure signals over short distances using spread spectrum technology.

Spread Spectrum Technology

If you want to listen to a particular station (either AM or FM) on the radio, you have to tune your radio (the receiver) to the same frequency on which that station's transmitter (the sender) is broadcasting. Anyone within listening range of the transmitter can tune to the same frequency and listen. This works well for commercial radio, but has always been a serious problem for the military. When one side transmits military information, it generally does not want the enemy to be able to hear it. In such cases, the *transmission* of information has to be made secure somehow. Julius Caesar was perhaps one of the first military leaders to use a coding technique in which letters of a message were scrambled such that, if the message courier was captured, the enemy still could not read Caesar's commands. When radios became commonplace telecommunication devices for the military, some form of encryption was needed so that secure transmissions could be sent and received. Shortly after World War II such a technology became available—it was called spread spectrum technology. Spread spectrum technology essentially takes the data to be transmitted and rather than transmitting it in a fixed bandwidth spreads it over a wider bandwidth. By spreading the data, a level of security is incorporated, thus making the eventual signal more resistant to eavesdropping or wire-tapping. One of the more common commercial applications of spread spectrum technology is wireless communications, such as those found in cordless telephones. Cordless telephones that incorporate spread spectrum technology are impervious to intruder eavesdropping.

Two basic spread spectrum techniques are commonly used in the communications industry today: frequency hopping spread spectrum and direct sequence spread spectrum. The idea behind frequency hopping spread spectrum transmission is to bounce the signal around on random frequencies rather than transmit it on one fixed frequency. Anyone trying to eavesdrop will not be able to listen because the transmission frequencies are constantly changing. How does the intended receiver follow this random bouncing around of frequencies? It turns out that the signal does not actually bounce around on random frequencies; it only *seems* to do so. The transmitter actually follows a *pseudorandom* sequence of frequencies, and the intended receiver possesses the hardware and software knowledge to follow this pseudorandom sequence of frequencies.

Figure 12-8 demonstrates the basic operation of a frequency hopping spread spectrum receiver and transmitter system. The input data enters a channel encoder, which is a device that produces an analog signal with a narrow bandwidth centered around a particular frequency. This analog signal is then modulated onto a seemingly random pattern of frequencies, using a pseudorandom number sequence as the guide. The pseudorandomly modulated signal is then transmitted to a friendly receiver. The first operation the friendly receiver performs is to "unscramble" the modulated signal, using the same pseudorandom sequence that the transmitter used to encode the signal. The unmodulated signal is then sent to the channel decoder, which performs the opposite operation of the channel encoder. The result is the original data.

The second technique for creating a spread spectrum signal to secure communications is direct sequence spread spectrum. Direct sequence spread spectrum spreads the transmission of a signal over a wide range of frequencies using mathematical values. Figure 12-9 shows that as the original data is input into a direct sequence modulator, it is exclusive-ORed with a pseudorandom bit stream. (The exclusive-OR is a logical operation that combines two bits and produces a single bit as the result. More precisely, if you exclusive-OR a 0 and a 0, the result is 0; 0 exclusive-OR 1 equals 1, and 1 exclusive-OR 0 equals 1; when you exclusive-OR a 1 and a 1, the result is again 0.) Thus, the output of the direct sequence modulator is the result of the exclusive OR between the input data and the pseudorandom bit sequence. When the data arrives at the intended receiver, the spread spectrum signal is again exclusive-ORed with the same pseudorandom bit stream that was used during the transmission of the signal. The result of this exclusive-OR at the receiving end is the original data. Code Division Multiple Access (CDMA), which is commonly used in modern cellular telephones, is a form of direct sequence spread spectrum technology.

Figure 12-8

Basic operation of a frequency hopping spread spectrum receiver and transmitter system

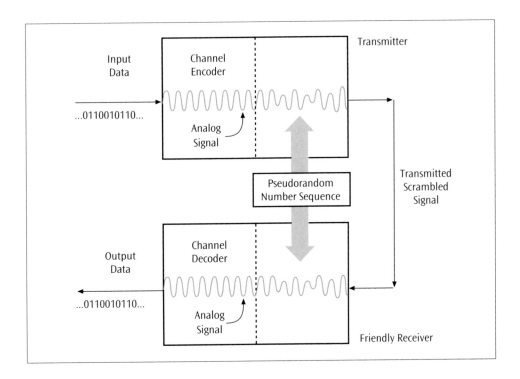

Figure 12-9

Example of binary data as it is converted into a direct sequence spread spectrum and back

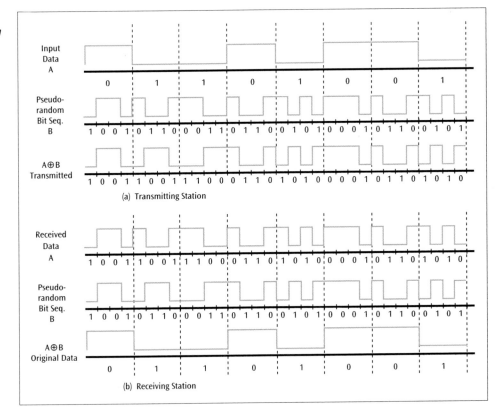

In conclusion, spread spectrum technologies offer effective means of securing the transmission of data between sender and receiver. Let us now turn our attention to other ways to secure communications. In particular, let us examine how secure communications can be created between a user computer and the Internet.

Guarding against viruses

As the introduction to this chapter stated, despite the higher levels of computer security available today, never before have users been so susceptible to the malicious intents of other computer users. One of the most common forms of invasion delivered over communications channels is the computer virus. Because a virus is a computer program, it has a binary pattern that is recognizable. Early models of virus scanners, from the 1980s, looked for the unique bit pattern of a virus. More recent virus scanners watch for the *actions* of a virus, such as unusual file changes or directory activity. To guard against viruses, you can purchase antivirus software that checks all of your files periodically and removes any viruses that are found. As you download files and e-mail and open applications with this software, you may get a message warning you of a new virus. To more effectively protect your computer against virus attacks, you need to use antivirus software that includes signature-based scanning, terminate-and-stay-resident monitoring, and integrity checking.

Signature-based scanning works by recognizing the unique pattern of a virus. All viruses have a unique bit pattern, much as a strand of DNA has a unique sequence. Antivirus product developers and virus researchers catalog the known viruses and their signatures. Signature-based scanning then uses these catalog listings to search for *known* viruses on a user's computer system. Because new viruses are created daily, it is necessary for a user to update the known virus catalog frequently. This updating is usually accomplished by downloading new virus information from the Internet.

In order to provide their computers with a constant level of protection from viruses, most users employ antivirus software that continuously watches for viruses—a technique known as terminate-and-stay-resident monitoring. Antivirus software with terminate-and-stay-resident monitoring runs in the background while an application that the user is executing runs in the foreground. Terminate-and-stay-resident programs can provide a combination of protective services, including real-time monitoring of disk drives and files, intelligent analysis of virus-like behavior, and detection of polymorphic viruses. An advantage of antivirus software with terminate-and-stay-resident monitoring is its automatic nature—a user does not have to activate the software each time a new file is opened or downloaded. One disadvantage of this software is that, because the virus checker is always running, it consumes memory and processing resources. An additional disadvantage is that this type of antivirus software can create false alarms. After being interrupted by a number of false alarms, users often disable the antivirus software completely, leaving their systems open to attack.

An antivirus technique that is used in conjunction with signature-based scanning and terminate-and-stay-resident monitoring is integrity checking. This technique is a combination of antivirus techniques such as intelligent checksum analysis and expert system virus analysis. Intelligent checksum analysis calculates and applies a checksum to a file and dataset at two major times in a file's lifetime—at the beginning, when a file is new (or in a known safe state), and at a later time, after the file has existed for a while. As in cyclic checksum, the checksum from the file when it's new is compared with the later checksum to determine if the file has changed due to the actions of a virus. Expert system virus analysis involves a series of proprietary algorithms that perform millions of tests on your software and examine the flow of program code and other software functions. Based on the results of these tests, this antivirus technique assigns a number of points to the software in question and indicates the existence of a virus if the point score reaches a certain level.

Users can simplify the work of antivirus software by not even letting the virus attack or infected e-mail enter their systems. They can do this with a firewall, which, if properly installed, can be just the thing for keeping the bad guys out of their computers. Let us examine the firewall more closely, and see how it can provide users with another level of protection.

Firewalls

A **firewall** is a system or combination of systems that supports an access control policy between two networks. The two networks are usually an internal corporate network and an external network, such as the Internet. A firewall can limit users on the Internet from accessing certain portions of a corporate network and can limit internal users from accessing various portions of the Internet. Figure 12-10 demonstrates how a firewall blocks both internal and external requests.

Figure 12-10
A firewall as it stops certain internal and external transactions

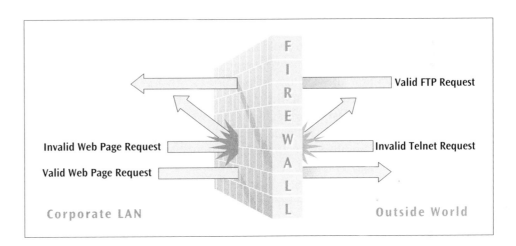

In a corporate environment in which a local area network is connected to the Internet, the company may want to allow external e-mail messages to enter the network but may not want to allow remote logins to the corporate network, because an intruder who learned how to log in to a corporate application could vandalize the system. Alternately, the company may want to outlaw file transfers out of the corporate network for fear of losing corporate data or trade secrets. In any case, a firewall, as you will see, involves more than just programming a device to accept or reject certain types of transactions.

Firewall Efficacy

What types of transactions will a firewall stop, and what types of transactions will it not stop? It is possible for a firewall system to stop remote logins as well as inbound or outbound e-mails and file transfers. It is also possible for a firewall to limit inbound or outbound Web page requests. Once a company has created a security policy—which means it has decided which types of transactions should be restricted and which should be allowed—it can create a firewall system that allows or disallows a fairly wide range of these transactions.

TCP port numbers indicate which application service is desired. When a transaction destined for a particular application enters a system, the appropriate port number is included as part of the transaction. Some typical port number assignments are:

- Telnet, or remote login service, is usually port 23.
- Finger, a service for finding a user ID, is often port 79.
- Usenet, the newsgroup service, is often port 119.
- E-mail is usually port 25.
- Web browsing, which uses HTTP, is port 80.

If a company wanted to block outsiders from trying to Telnet in to its network, it could configure its firewall to stop all incoming transactions requesting port 23. Or a company could stop employees from spending time on Usenet by configuring the firewall to block all outgoing transactions with port number 119.

Firewalls, unfortunately, do not protect a network from all possible forms of attack. Because a virus can hide within a document, it will probably not be detected by a firewall if its host document is allowed into the system. Some firewall systems, however, do boast that they can detect viruses within documents. But these systems are rare and may not detect all current viruses.

A firewall also will not protect a computer or network properly if it is possible for an intruder to avoid the firewall and enter the system through an alternate route. For example, if a network system has one or more dial-up modems that are not part of the firewall system, a perpetrator can bypass the firewall and gain illegal access. A fax machine is another route through which sensitive information can leave a company. The best firewall system in the world will not stop a determined employee from faxing a confidential document over a standard telephone line to an outside recipient. For a firewall system to work, a comprehensive security policy must be created, installed, and enforced. This security policy must cover the management of all possible entrances to and exits from the corporate environment, from computer, telephone, and fax systems to persons physically carrying media into and out of the building.

Basic Firewall Types
Firewalls come in two basic types:

- Packet filter
- Proxy servers

Let us examine both of these types of firewalls and review their advantages and disadvantages.

The **packet filter** firewall is essentially a router that has been programmed to filter out certain IP addresses or TCP port numbers. These types of routers perform a static examination of the IP addresses and TCP port numbers, then either deny a transaction or allow it to pass, on the basis of information stored in their tables. These types of routers are relatively simple in design. Also, they act very quickly, but they are a little too simple to provide a high level of security. For example, it is relatively easy for an external user bent on causing internal network problems to spoof a static packet filter into thinking he or she is a friendly agent.

Routers (and firewalls) have become increasingly intelligent over the years, and modern models can even follow the flow of conversation between an internal entity and an external entity. This capability allows modern routers to detect spoofing. Modern routers are also beginning to scan for viruses on incoming transactions.

The proxy server is a more complex firewall device. A proxy server is a computer running proxy server software, whose function is much like that of a librarian who controls access to books in a library's rare books room. To keep costly books from getting damaged by vandalism or careless handling, many libraries do not allow patrons to enter their rare books room. Instead, a patron fills out an information request slip and hands it to a librarian. The librarian enters the rare books room and retrieves the requested volume. The librarian then photocopies the requested information from the book and gives the photocopies to the patron. Thus, the patron never comes in contact with the actual rare book. Similarly, a transaction making a request of a company's proxy server firewall never comes into contact with the company's network. Any external transactions that request something from the corporate network must first approach the proxy server. The proxy server creates an application called a proxy that goes into the corporate network to retrieve the requested information. Because all external transactions must go through the proxy server, it provides a very good way to create an audit log.

A proxy server that supports FTP requests provides a realistic example. Recall from Chapter Ten that the FTP protocol uses two network connections—one for control information and one for data transfer. When an FTP request comes in, the FTP proxy examines the control connection at the application layer, decides which commands are allowed or denied, and then logs individual commands. When the proxy encounters a command requesting the second connection—the data transfer connection—it uses company-programmed security policy information to decide whether to filter the data packets.

Because a proxy server sits outside a company's corporate network and its various security barriers, the proxy server itself is wide open to vandalism (Figure 12-11). To protect the proxy server, the company should make sure its proxy server is a stripped-down version of a network computer. Thus, a vandal can do little to harm the proxy server.

Figure 12-11
The proxy server sitting outside the protection of the corporate network

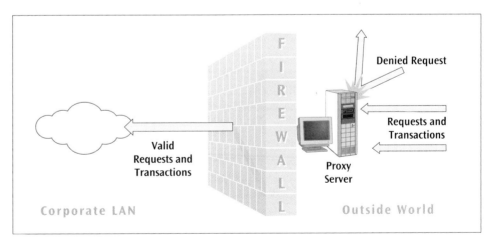

Lastly, although the proxy server provides a higher level of security, it is slower than a packet filter, because a proxy has to be created for each type of transaction that may request data from inside the corporate network. Like packet filter technology, proxy technology is improving, as you might expect. Router-like devices that can efficiently and effectively create a proxy that enters an internal network to retrieve the requested information now appear on the market.

Wireless security

Local area networks have been in existence for almost 35 years. During these years, we have seen them evolve from simple bus-based, single-server systems to complex networks with layer two and layer three switches, advanced network operating systems, and multiple servers. While local area networks seem to have matured past their growing pains, we are now experiencing the same pains with a relatively newer technology—wireless local area networks. One of the biggest growing pains of wireless local area networks is security. Because virtually anyone can have a wireless laptop or other wireless device, it is almost impossible to control who can connect into a wireless local area network and from where. Thus, wireless security is essential but also complex.

The first security protocol for wireless LANs was Wired Equivalency Protocol (WEP). Although WEP was a step in the right direction, it suffered two serious drawbacks. First, WEP used weak encryption keys that were only 40 bits in length. Second, the keys were static, not dynamic. Static keys are easier to crack because they are not likely to be changed. In order for a network administrator to change a WEP key, the administrator has to physically visit the machine. Many administrators thought WEP was so inadequate and difficult to work with that they preferred to have no security mechanism in place. Fortunately, WEP has been replaced by a new standard, Wi-Fi Protected Access (WPA). WPA keeps the 40-bit-sized encryption keys of WEP but has one significant improvement: the inclusion of the Temporal Key Integrity Protocol (TKIP) and IEEE 802.1x features, which together provide dynamic key encryption and mutual authentication for wireless clients. Thus, although the 40-bit key is still not very powerful, the keys are now being dynamically assigned, and wireless clients must prove they are legitimate users before they can connect.

In any case, WPA was a temporary standard that was used until a more powerful security standard was approved and implemented: that standard was IEEE 802.11i. IEEE 802.11i (also called WPA2) addresses *both* weaknesses of WEP by allowing the keys, encryption algorithms, and negotiation to be dynamically assigned, and by adopting the AES encryption based on the Rijndael algorithm with 128-, 192-, or 256-bit keys. IEEE 802.11i is based upon WPA and the robust security network (RSN), software that performs a robust negotiation of the encryption and authentication algorithms used between access points and wireless agents.

Security Policy Design Issues

When a company is designing a firewall system and its corresponding security policy, a number of questions should be answered. The first question involves the company's expected level of security. Is the company trying to allow access only to services deemed essential to its business? Or does the company wish to allow all or most types of transactions, and thus need its firewall system only for auditing transactions and creating an orderly request for transactions? Allowing access only to services deemed essential requires a more elaborate firewall system and thus more work and expense. Allowing most types of transactions to take place requires a simpler firewall system that only performs queue management operations and creates an audit trail. To determine the appropriate level of security it needs, a company must assess the nature of its data. More precisely, the company should determine how secure its data ought to be. Also, it must decide whether all this data or only certain parts of it should be secured. This step of examining data and setting the security level for a company and its operations is typically time-consuming and costly.

The decision about level of security leads to the second question: How much money is the company willing to invest in a firewall system? Commercially purchased firewall systems can be powerful, complex, and expensive. It is possible, however, to construct a firewall system in-house that takes advantage of the capabilities of existing resources, such as network operating systems and routers. As we learned earlier, it is possible to use a network operating system to restrict access to a system according to time of day, day of week, and location. It is also possible to use existing software to create an audit trail of all incoming and outgoing transactions. Depending on the level of auditing detail required, a company can purchase and install additional software that will work in concert with network operating system software to provide any desired level of audits.

Like network operating systems, many routers can be programmed to restrict access to certain kinds of traffic. A router can be programmed to accept and reject requests with specific IP addresses or a range of IP addresses. Routers can also be programmed to deny access to certain port addresses at the TCP level.

A third question that the company must address relates to its commitment to security. Is the company sufficiently serious about restricting access to the corporate network that it not only secures entrances to the network that are Internet-related but also supports the security of any and all other links into the corporate network environment? In other words, when making security decisions, the company should also consider access to dial-up modems, wireless networks, and other telecommunications links. Fax machines, both stand-alone and computer-based, as well as removable disk media, are two more examples of ways that data may enter or leave a corporation. A security policy must take these entrance and exit points, as well as the Internet, into consideration.

Having a well-designed security policy in place will make the jobs of network support staff clearer. A well-designed security policy will make enforcement more straightforward, and it will allow the staff to react properly to specific security requests. The staff employees will know what the network users can and cannot access, and where they can and cannot go. The policy will also make clear the goals and duties of network employees when they must enforce security with respect to requests from the outside.

If there is a good security policy, the corporate users themselves will have a better understanding of what they can and cannot do. This understanding will, one hopes, assist the network staff members in conducting their jobs and will allow the company to maintain security in an increasingly less secure world.

Perhaps because a lot of companies have well-designed security policies in place, many people who use the Internet to purchase items online are fairly confident that when they transfer credit card information during a secure session, their data is safe from hackers and other eavesdroppers. (This is despite the fact that the storage of credit card information in company databases has been a serious concern, with a number of well-publicized database break-ins.) This sense of security may change, however, as the Internet Engineering Task Force is considering a proposal that allows the creation of a backdoor entry into all Internet traffic. This backdoor entry would allow authorized persons to intercept any data traffic on the Internet. Because the use of a backdoor would clearly constitute a violation of privacy, why would anyone want to propose such a thing?

At the core of the argument is the fact that standard telephone systems currently allow agencies of the U.S. government to wiretap communications. This wiretap occurs at the telephone central office and is built into central office telephone switches. The act that allows wiretapping, the Communications Assistance for Law Enforcement Act, has been in existence since 1994. The proponents of the backdoor entry proposal reason that, now that the Internet is beginning to carry voice traffic, the government's authority to wiretap voice transactions should extend to the Internet. But as critics of the proposal warn, if the government can tap voice, then it can also tap data. Critics also point out that if the designers of the Internet were to create such a backdoor, this backdoor access could fall into the wrong hands and be used for criminal purposes.

This issue is further complicated by the fact that many businesses presently encrypt all data leaving their corporate networks. Most encryption techniques used by businesses are so effective that virtually no one, including the government, can crack them. In cases where this encryption is performed by the network just before the data leaves corporate boundaries, it would be the responsibility of the corporate network's support personnel to provide the U.S. government, if asked, with unencrypted data. If, on the other hand, the encryption is applied at the user workstation before the data is inserted into the corporate network, it is not clear who would have to supply the government with the unencrypted data. Clearly, this issue will be hotly debated for some time to come.

Although companies may have well-designed security policies in place, external events are making this area more complex all the time.

Network Security In Action: Making Wireless LANs Secure ▶

The last time we saw Hannah was in Chapter Eight, when she was examining possible candidates to choose a network operating system for her company. Now Hannah's company is considering offering wireless local area network service to its employees and administration staff. The question is: How can the wireless local area network be made secure from unauthorized users? This is a problem that universities, businesses, and even homeowners are also encountering. More precisely, the problem is: How do you keep your wireless network secure from war drivers? War drivers are unauthorized users, who, using their wireless devices, try to connect to someone's wireless network either to gain free access to the Internet or to steal and vandalize files.

As we have already seen, Wired Equivalency Protocol (WEP) is too weak to be a serious wireless security protocol. It is based on an algorithm called RC4, which is apparently fairly easy to crack. Even though WEP upgraded from using 40-bit keys to 128-bit keys, individuals were able to crack the larger key with software and persistence. An additional disadvantage of WEP is that the keys are static, meaning that an IT person has to visit each machine personally to change the key, which is not practical, even in a small company such as Hannah's.

Hannah could consider using Cisco's LEAP (Lightweight Extensible Authentication Protocol) as a wireless protocol. Unfortunately, LEAP is not a standard, but proprietary software, and it has interoperability issues with other devices. Hannah definitely wants to use software that has been accepted by a standards-making organization. WPA (Wi-Fi Protected Access) was created by the WiFi Alliance. Acceptance of WPA has been accelerated by the fact that the WiFi Alliance has demanded that the WPA security mechanism be required for all new WiFi certifications.

Although WPA uses the same RC4 encryption as WEP, it includes TKIP (Temporal Key Integrity Protocol), which rotates the keys and thus strengthens the encryption process. But WPA is not without its problems. Hannah has discovered that it might not be compatible with older network operating systems—which is not actually a problem for her, because she just installed the latest version of a popular network operating system. She has also learned that some personal digital assistants may not have enough processing power for WPA, but because her company has not issued PDAs to its employees, this is also not a problem for Hannah—at least not yet. What could be a potential problem for Hannah is that WPA degrades device performance. After all her efforts to make the best selections for her company's local area network system, she does not relish dealing with performance slowdowns.

Fortunately, IEEE 802.11i, or WPA2, is now available for wireless LANs. With its stronger encryption methods and faster algorithms, it is hard to beat. This clearly looks to be the best encryption software for a wireless local area network. Hopefully she will not encounter too many compatibility problems as the standard has now been in existence for a number of years.

Thus, Hannah is considering installing WPA2 on her company's network. The next question is, in what devices should WPA2 be installed? For a wireless network to be properly secured, the security software has to operate at three different points within the system: at the user's wireless laptop, at the wireless access point (the connection between the wireless user and the wired network), and at the network server. Fortunately, the software that supports all three points in the system is now available.

Hannah purchases and installs her WPA2 wireless system. Because the passwords can be dynamically assigned, she is fairly confident that the corporate network will be secure from war drivers outside the building. The employees enjoy their newfound freedom of wirelessly moving their laptop computers anywhere within the building.

◆ ◆

SUMMARY

- ► Network security continues to be an increasingly important topic, particularly with the increase in network interconnectivity. The Internet has helped to open the door to vandals all over the world, making any system connected to it vulnerable to attack.

- ► Three common system attacks are (1) attacking known operating system and application software vulnerabilities, (2) bombarding a system so that it is incapable of performing its normal duties (denial of service attacks), and (3) using valid user accounts for unauthorized purposes.

- ► Network personnel and users must be aware of (and take) physical protection measures, such as placing equipment in appropriate locations, to protect their systems from potential flooding, vandalism, and other detrimental environmental factors. Other physical protection measures such as surveillance systems can also deter vandalism and equipment theft effectively.

- ► Controlling access to a computer system and its network is an essential aspect of network security. Network security measures exist that allow a network administrator to restrict the time of day, day of week, and location from which someone can log on to a computer system.

- ► Passwords and other ID systems are very common access-controlling security techniques, but passwords can be stolen and used by unscrupulous parties. New biometric techniques that use some part of the body to identify an individual are more secure than password systems.

- ► Most computer systems apply access rights to the resources of the system and the users. By properly setting access rights, network administrators can make computer resources more secure.

▶ Software that conducts a continuous audit of network transactions creates an electronic trail that companies can use when they are trying to catch malicious users.

▶ Providing security for system data is just as important as securing the system itself. Encryption technology provides an important set of techniques that can aid in the fight against computer fraud. Substitution-based ciphers replace one letter or series of letters in a message with a second letter or series of letters. Transposition-based ciphers rearrange the order of letters in a message.

▶ Public key cryptography uses two keys—one key to encode messages and a second key to decode messages. The Secure Sockets Layer is used to encrypt data that travels back and forth between Web servers and Web browsers.

▶ The Data Encryption Standard was created in 1977 and uses a 56-bit key to encrypt data transmitted between two business locations. The Advanced Encryption Standard has replaced the Data Encryption Standard and provides a much higher level of protection.

▶ Digital signatures use public key cryptography and can be used to verify that a given document belongs to a given person.

▶ Pretty Good Privacy is free encryption software that allows regular users as well as commercial users to encrypt and decrypt everyday transmissions.

▶ Kerberos is a secret key encryption technique that can be used by commercial application programs to verify that a user is who he or she claims to be.

▶ Public key infrastructure uses public key cryptography, digital signatures, and digital certificates to enable the secure passage of data over unsecured networks.

▶ Steganography is the study of hiding secret data within an unrelated document, for example, hiding the bits of a message within the pixels of an image.

▶ Along with securing network data, it is imperative to secure network communications. Spread spectrum technology is the technique of spreading data and their signals over a wider range of frequencies in order to make the transmission secure. When a signal is transmitted using frequency hopping spread spectrum techniques, the signal continuously hops from one frequency to another to prevent eavesdropping, disruption of the transmission, or other malicious intervention. When a signal is transmitted using direct sequence spread spectrum techniques, the 1s and 0s of the original data are converted to longer bit sequences.

▶ In order to secure communications, network administrators and users must be aware of standard computer attacks and viruses that can damage computer systems. They must also be aware of software and hardware that can help to protect a system and its users from computer attacks and viruses. Virus scanners have three basic forms: signature-based scanning, terminate-and-stay-resident monitoring, and integrity checking.

▶ Another means of securing communications is a firewall, a system or combination of systems that supports an access control policy between two networks. Firewalls come in two basic types: (1) packet filters, which examine all incoming and outgoing transmissions and filter out those transmissions that have been deemed illegal, and (2) proxy servers, which are computers running at the entrance to a computer network and acting as gatekeepers into the corporate network.

▶ Securing wireless networks is a new and exciting field of study. Standards such as IEEE 802.11i are powerful and robust at providing dynamic key protection and user authentication.

▶ A proper network security design helps the corporate network staff by clearly delineating which of the network transactions submitted by internal employees and by external users are acceptable.

KEY TERMS

access rights
Advanced Encryption Standard (AES)
asymmetric encryption
biometric techniques
botnet
certificate
certificate authority (CA)
certificate revocation list (CRL)
ciphertext
computer auditing
cryptography
Data Encryption Standard (DES)
denial of service attacks
digital signature
direct sequence spread spectrum
e-mail bombing
encryption algorithm
firewall
frequency hopping spread spectrum
honeypot
IEEE 802.11i

integrity checking
intrusion detection
IPsec
Kerberos
key
keylogger
mobile malicious code
monoalphabetic substitution-based
  cipher
packet filter
password
pharming
phishing
ping storm
plaintext
polyalphabetic substitution-based
  cipher
Pretty Good Privacy (PGP)
proxy server
public key cryptography
public key infrastructure (PKI)

root kit
Secure Sockets Layer (SSL)
signature-based scanning
smurfing
spoofing
spread spectrum technology
steganography
surveillance
symmetric encryption
terminate-and-stay-resident
  monitoring
Transport Layer Security (TLS)
transposition-based cipher
triple-DES
Trojan horse
Vigenére cipher
virus
war driver
Wi-Fi Protected Access (WPA)
Wired Equivalency Protocol (WEP)
worm

REVIEW QUESTIONS.

1. How do hackers exploit operating system vulnerabilities?

2. What is a Trojan horse?

3. How does a denial of service attack work?

4. What is spoofing, and how does it apply to a denial of service attack?

5. What is a ping storm, and how does it apply to a denial of service attack?

6. List three forms of physical protection.

7. How can surveillance be used to improve network security?

8. How does an intrusion detection system work?

9. What is the major weakness of the password system? What is its major strength?

10. What are the most common types of access rights?

11. How can auditing be used to protect a computer system from fraudulent use?

12. Describe a simple example of a substitution-based cipher.

13. Describe a simple example of a transposition-based cipher.

14. How can public key cryptography make systems safer?

15. Give a common example of an application that uses Secure Sockets Layer.

16. What is the Data Encryption Standard?

17. How is the Advanced Encryption Standard different from the Data Encryption Standard?

18. What is a digital signature?

19. What kind of applications can benefit from Pretty Good Privacy?

20. Is Kerberos a public key or a private key encryption technique? Explain.

21. List the basic elements of public key infrastructure.

22. What kind of applications can benefit from public key infrastructure?

23. What kind of entity issues a certificate?

24. Under what circumstances might a certificate be revoked?

25. How is steganography used to hide secret messages?

26. What are the two different forms of spread spectrum technology?

27. What is a computer virus, and what are the major types of computer viruses?

28. What are the different techniques used to locate and stop viruses?

29. What is the primary responsibility of a firewall?

30. What are the two basic types of firewalls?

31. What are the advantages of having a security policy in place?

EXERCISES

1. A major university in Illinois used to place the computer output from student jobs on a table in the computer room. This room was the same computer room that housed all the campus' mainframe computers and supporting devices. Students would enter the room, pick up their jobs, and leave. What kinds of security problems might computer services encounter with a system such as this?

2. You have forgotten your password, so you call the help desk and ask the representative to retrieve your password. After a few moments, the help desk representative tells you your forgotten password. What has just happened, and what is its significance?

3. Create (on paper) a simple example of a substitution-based cipher.

4. Create (on paper) a simple example of a transposition-based cipher.

5. Using the Vigenére cipher and the key NETWORK, encode the phrase "this is an interesting class".

6. Using the sample transposition-based cipher described in this chapter and the same key, COMPUTER, encode the phrase "Birthdays should only come once a year".

7. You are using a Web browser and want to purchase a music CD from an electronic retailer. The retailer asks for your credit card number. Before you transfer your credit card number, the browser enters a secure connection. What sequence of events created the secure connection?

8. You want to write a song and apply a digital signature to it, so that you can later prove that it is your song. How do you apply the signature, and later on, how would you use it to prove that the song is yours?

9. List three examples (other than those listed in the chapter) of everyday actions that might benefit from applying PKI.

10. Can a firewall filter out requests to a particular IP address, a port address, or both? What is the difference?

11. One feature of a firewall is its ability to stop an outgoing IP packet, remove the real IP address, insert a "fake" IP address, and send the packet on its way. How does this feature work? Do you think it would be effective?

12. How does the size of a key affect the strengths and weaknesses of an encryption technique? Consider both a friendly use of the key and an unfriendly use of the key.

13. Assume a key is 56 bits. If it takes a computer 0.00024 seconds to try each key, how long will it take to try all possible keys? What if 10,000 computers are working together to try all keys?

14. What are the answers to the questions in Exercise 13 if the key is 128 bits in length?

15. You want to hide a secret message inside an image file using steganography. You have decided to place one bit at a time from the message into the image's pixels. How are you going to select the pixels? Will they be random or all in a row? And once a pixel is chosen, which bit are you going to replace with the bit from the secret message? Why?

16. Why can't a truly random sequence be used in a frequency hopping spread spectrum system?

THINKING OUTSIDE THE BOX

1 Create an encoding scheme that is composed of either a substitution-based cipher, a transposition-based cipher, or both. Encode the message "Meet me in front of the zoo at midnight". Explain your encoding technique(s).

2 Create a hypothetical business with approximately 50 to 100 employees. Place the employees in two or three different departments. Assign to each department a title and basic job duties. All employees in all departments use personal computers for numerous activities. Identify the computer activities for the employees of each department. Then create a security policy for the employees of each department. Be sure to address when and where the employees have access to computer resources, and if any types of transactions should be restricted.

3 You are working for a company that allows its employees to access computing resources from remote locations and allows suppliers to send and receive order transactions online. Your company is considering incorporating PKI. How would you recommend that PKI be implemented to support these two application areas?

4 You have a computer at home with a wireless NIC and wireless router. List all the security measures that should be employed so that your home network is secure.

5 Your supervisor has asked you to explore the concept of ID management for the company. What is involved? How does it pertain to the topic of security? Is it reasonable for the company to consider implementing ID management, or is this concept too new (and therefore too risky) for a functioning business to consider?

6 Some companies use the prevent/detect/correct model for establishing corporate security. How does this model operate? Can you give examples of each of the three categories? If you are currently employed, how do you think your company fits this model? If you are a student taking a college course, how does your college or university fit this model?

HANDS-ON PROJECTS

1. Find and report on three different security techniques that use some part of a person's physical characteristics for verification.

2. Intrusion detection is a popular area of network security. A number of companies offer software systems that perform intrusion detection. Write a one- to two-page paper that summarizes how intrusion detection systems operate.

3. What are the allowable access rights for protecting files and other objects on the computer system at your school or place of employment? List the options for "who" and "how."

4. Create a short list of some of the recent viruses that travel inside application macros, specifically the macros that work with Microsoft Office applications.

5. Report on the current status of encryption standards for businesses that send data outside the country. Do businesses want stronger restrictions on sending encrypted material outside the country, or weaker restrictions? What is the government's position? What types of encrypted documents will the government currently allow businesses to export?

6. Pretty Good Privacy (PGP) can be used as a form of public key infrastructure (PKI). Explain how PGP and PKI relate.

7. What is the *clipper chip*? Did it ever come to fruition? Explain.

8. Occasionally you hear of a virus that is not a virus but instead is a hoax. Find a Web site that has a list of hoax viruses and report on some of the more interesting ones.

9. What are some of the other spread spectrum technologies besides frequency hopping and direct sequence? Be sure to list your sources.

13
Network Design and Management

◆ ◆

SUPPOSE YOU ARE A NETWORK administrator or systems analyst for your company and need to ask your boss for money to purchase a new network system or a new piece of equipment. *Network Computing* magazine offers seven tips on how to make an effective pitch in such cases.

Know the people who hold the purse strings: If management is likely to demand a detailed ROI (return on investment) model for your request, then create one. If management is more flexible, you should still consider supporting your request by preparing a less formal business case that shows the subjective benefits and costs of the new purchase.

Quantify where possible: Because creating a valid ROI can be difficult, you should, if you need to create one, be sure to concentrate on costs and benefits that can be clearly and legitimately quantified.

Get the finance department involved: If you can get a company accountant on your side early in the game (to validate your calculations and support your request), it will help your position.

Do your research: Talk to anyone in the company who might be involved with or benefit from the acquisition of the new computer system or equipment. Be sure to use this information to build your case.

Even the playing field: Use the appropriate value assessment for the type of system being requested. For example, for a security-related product or system, you could use risk-based assessment models.

Produce qualitative evidence: Gather testimonials in support of the purchase from different sources, such as other employees who might benefit from using the system and any corporate managers who may already be using the new product. To support your case, you might also consider investigating whether other companies (especially competitors) are using the product.

How difficult is it to calculate an ROI for a new product?

Do professionals follow a procedure when they propose and design computer systems?

Source: Wilson, T., "What's It Worth to You?" *Network Computing*, March 17, 2005, pp. 79–86.

Objectives ▶

After reading this chapter, you should be able to:

▶ Recognize the systems development life cycle and define each of its phases

▶ Explain the importance of creating one or more connectivity maps

▶ Outline the differences among technical, financial, operational, and time feasibility

▶ Create a cost-benefit analysis incorporating the time value of money

▶ Explain why performing capacity planning and traffic analysis is difficult

▶ Describe the steps involved in performing a baseline study

▶ Discuss the importance of a network administrator and the skills required for that position

▶ Calculate component and system reliability and availability

▶ Recognize the basic hardware and software network diagnostic tools

▶ Describe the importance of a help desk with respect to managing network operations

▶ List the main features of the Simple Network Management Protocol (SNMP) and distinguish between a manager and an agent

▶ Describe the use of the Remote Network Monitoring (RMON) protocol and its relationship to SNMP

Introduction

For a computer network to be successful, it has to be able to support both the current and future amounts of traffic, pay for itself within an acceptable period of time, and provide the services necessary to support users of the system. All these goals are very difficult to achieve. Why? First, computer networks are constantly increasing in complexity. In many business environments, it is extremely difficult for one person to completely understand every component, protocol, and network application. Thus, network management is becoming increasingly challenging.

A second reason is related to how difficult it is for an individual or a business to properly define the future of computing within a company. Each company has its own expectations about what computing services it should provide. In addition to this, each user within the company has his or her own idea of the computing services that should be available. It is extremely difficult, therefore, to determine a single service or set of services that could serve the needs of an entire company.

Finally, computer network technology changes at breakneck speed. In some areas, you can expect major developments about every six months, and in other areas new technologies are emerging almost daily. Keeping abreast of new hardware, software, and network applications is a full-time job in itself. Incorporating the new technology into existing technology, while trying to guess the needs of users, is exhausting. This ever-changing environment is one of the main reasons that designing new systems and updating current systems are areas filled with dangerous pitfalls. With one wrong decision, much time and money can be wasted.

Most of the topics introduced in this chapter come from very large areas of study. The area of planning, analyzing, designing, and implementing computer-based solutions, for example, is often the subject of an entire college course or a two-term course sequence. Thus, the chapter will not be able to cover these topics in great depth. Nonetheless, we will attempt to survey the field in order to give you an understanding of the basic concepts involved in creating computer-based solutions.

Why is it important to understand the basic concepts that are used to develop computer systems? These concepts are important because, if you pursue a career that involves computer networks, there is a good chance that sometime in the future you will either be designing or updating a network system, or assisting one or more persons who are designing or updating a network system. If you will be performing this task yourself, you will need to know how to attack the problem logically and set up a proper progression of steps. If you will be working with one or more network professionals, you will need to be cognizant of the steps involved and how you might participate in them. In addition to understanding the basic concepts of network design, a network manager needs to learn other complex skills, such as how to conduct feasibility studies, do capacity planning, perform traffic analysis, and create a baseline. This chapter will introduce each of these activities and present relatively simple examples that demonstrate their use. The chapter will conclude with a description of the personal skills required of a network administrator and a look at some of the tools available for properly supporting a network.

Systems Development Life Cycle

Every company, whether it is a for-profit or nonprofit organization, usually has a number of major goals, some of which may include:

- ▶ Increasing the company's customer base
- ▶ Keeping customers happy by providing the company's services as well as possible
- ▶ Increasing the company's profit level, or, in a nonprofit organization, acquiring the funds necessary to meet the organization's goals and objectives
- ▶ Conducting business more efficiently and effectively

From these major goals, systems planners and management personnel within a company try to generate a set of questions that, when satisfactorily answered, will assist their organization in achieving its goals and move the organization forward. For example, someone in management might ask: Is there a way to streamline the order system to allow the company to conduct business more efficiently and effectively? Can we automate the customer renewal system, to better serve customers and keep them happy? Is there a more efficient way to offer new products to help the company increase the customer base? Is there a better way to manage our warehouse system to increase company profits?

Very often in today's business world, the answers to these types of questions involve solutions that require the use of computer systems, and the computer systems used are often large and complex. When large amounts of time, money, and resources are involved in a particular computer solution, you want to make sure that this solution is the best possible one for the problem.

To be able to properly understand a problem, analyze all possible solutions, select the best solution, and implement and maintain the solution, you need to follow a well-defined plan. One of the most popular and successful plans currently used by businesses today is the systems development life cycle (SDLC). The systems development life cycle (SDLC) is a structured approach to the development of a business system. While newer techniques such as RAD (Rapid Application Development) and Agile lend themselves better to a more iterative development process, for simplicity's sake we'll stick with SDLC for now. This approach often includes planning, analysis, design, implementation, and support. In general, a structured approach is a series of steps and tasks that professionals, such as systems developers, can follow to build high-quality systems faster, with fewer risks, and at lower costs. Although virtually every company that uses SDLC and every textbook that teaches SDLC has its own slightly different variation of the methodology, most agree that the SDLC includes at least the following phases:

- ▶ Planning—Identify problems, opportunities, and objectives.
- ▶ Analysis—Determine information requirements, analyze system needs, and prepare a written systems proposal.
- ▶ Design—Design and build the system recommended at the end of the analysis phase and create the documentation to accompany the system.
- ▶ Implementation—Install the system and prepare to move from the old system to the new system; train the users.
- ▶ Maintenance—Correct and update the installed system as necessary.

The idea of *phases* is critical to the SDLC concept. The intent of SDLC is for phases not to be disjointed steps in a big plan, but overlapping layers of activity. It is quite common for two and three phases of a single project to be going on at the same time. For example, the design of one component of a system can be in progress, while the implementation of another component is being performed.

A second critical concept is that of the *cycle*. After a system has been maintained for a period of time, it is relatively common to restart the planning phase—hence, another cycle—in an attempt to seek a better solution to the problem. Thus, the systems development life cycle is a never-ending process (see Figure 13-1).

Figure 13-1
The cyclic nature of the phases of the systems development life cycle

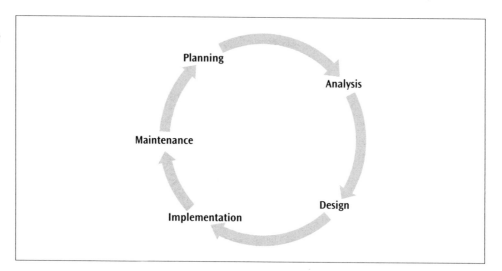

Who is responsible for initiating the phases and shepherding a project through the cycle? A professional called a systems analyst typically is responsible for managing a project and following the SDLC phases, particularly the analysis and design phases. Although becoming a systems analyst requires specialized professional training, many individuals who work with computer networks can learn systems analysis and design methods and incorporate them into designing computer networks. It is also possible that a person who supports computer networks may be called upon by a systems analyst to provide input into a new network solution. In this instance, the network professional should be aware of the SDLC and be prepared to provide any requested information or design materials.

Network Modeling

When a systems analyst or a person acting in the role of systems analyst is asked to design a new computer system, he or she will typically create a set of models for both the existing system (if there is one) and the proposed system. These models are usually designed to show the flow of data through the system and the flow of procedures within the system, and thus they help the analyst and other professionals visualize the current and proposed systems. One very important part of most computer systems today is the network. Most businesses have at least one internal local area network, and one or more connections to external wide area networks such as the Internet. Many businesses have multiple office locations scattered over a geographic location. The models created for a network design can either depict the current state of the network or illustrate the desired computer network.

The network model does not have to be an elaborate creation. Oftentimes, it is only a hand-drawn model that depicts the proposed design of the network. One technique used to model a corporation's network environment is to create connectivity maps. More precisely, three different modeling techniques can be used, depending on what type of network you are modeling: wide area connectivity maps, metropolitan area connectivity maps, and local area connectivity maps. Not all analysis and design projects require all three network connectivity maps. For example, your company may not use a metropolitan area network to connect its local area network to the outside world, and thus will not need to create a metropolitan area connectivity map. Nonetheless, all three maps will be described here, as each type has slightly different but important characteristics. Let us begin with the big-picture perspective of the wide area connectivity map and work our way down to the smaller details shown in the local area connectivity map.

Wide area connectivity map

In order to create a wide area connectivity map, the modeler begins by identifying each site or location in which the company has an office. Each fixed site is denoted by a circle; mobile or wireless sites are indicated by circles containing the letter M; and external sites, such as suppliers or external agents, are denoted by circles containing the letter E. A solid line between two sites indicates a desired path for data (or voice) transmission. Figure 13-2 shows four sites, Chicago, Seattle, Los Angeles, and San Antonio, in such a map. The user in San Antonio is actually a government office and thus is shown as a site with an E. Also, there are wireless users in the Chicago office, and they are shown all together with a single circled M.

Figure 13-2
Example of a wide area connectivity map for sites in Chicago, Seattle, Los Angeles, and San Antonio

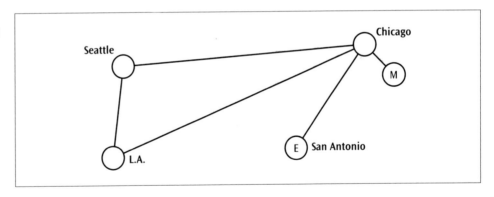

To identify the connections between sites, the following link characteristics can be applied to each connection (in any order):

- d = Distance of the connection (usually shown in either miles or kilometers)
- s = Security level (high, medium, low, or none)
- du = Duplexity (full duplex, half duplex, or simplex)
- dr = Data rate desired (in bps)
- l = Latency, or acceptable delay time across the network (usually in milliseconds, or ms)
- QoS = Quality of service (CBR = constant bit rate, VBR = variable bit rate, ABR = available bit rate, UBR = unreliable bit rate, or none)
- de = Delivery rate (sometimes called throughput percentage)

Figure 13-3 shows how these link characteristics can be used to more fully define a connection—in this case, the one between Chicago and Los Angeles. The distance between these two cities is 2250 miles; the level of security for the connection is medium; the duplexity is full; the desired data rate is 256 kbps; the latency across the network is 200 ms; the quality of service is ABR; and the delivery rate is 99.9 percent.

Figure 13-3

Detailed link characteristics for the connection between Chicago and Los Angeles

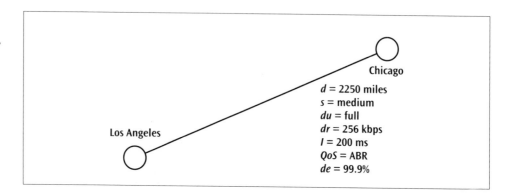

After identifying the sites, the modeler will then define the characteristics for each link, creating a map that shows the wide area network interconnections for the entire business. Now that the wide area network has been mapped, the modeler can zoom in on the metropolitan area network connections.

Metropolitan area connectivity map

If a company desires a metropolitan area network connection between one of its offices and another business, such as an Internet service provider (ISP), it can use a metropolitan area connectivity map to outline this connection and define the desired network characteristics. A metropolitan area connectivity map shares some of the characteristics of wide area maps and some of the characteristics of local area maps. Data rate, quality of service, and security are still important parameters at the metropolitan layer, but distance is probably not as important as it was in the wide area map. A new parameter that might have an impact on metropolitan area design is the failover time. Recall from Chapter Nine that failover time is the amount of time necessary for the metropolitan area network to reconfigure itself or reroute a packet, should a given link fail. An example of a metropolitan area connectivity map is shown in Figure 13-4.

Figure 13-4

Two nodes and the connecting link in a metropolitan area connectivity map

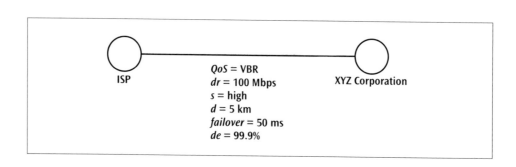

The metropolitan area connectivity map can stand apart from the wide area connectivity map or can be incorporated into the wide area map. The modeler would do the latter by including the metropolitan area link and characteristics at the appropriate wide area connectivity site. Once the metropolitan area connectivity map is complete (or if there was no need to create a metropolitan area map because there are no metropolitan networks), the modeler then proceeds to the final set of maps—the local area connectivity map.

Local area connectivity map

To examine the nodes in a wide area connectivity map in more detail, an analyst can expand each individual site into a local area connectivity map. The local area network design can then be performed in one or two stages, depending upon the level of detail desired. If only an overview of a local network is desired, then the analyst can create a local area overview connectivity map. In this stage, entire logical or physical groups, such as clusters of users and workstations, are denoted as a single node. The links between such nodes are defined by factors such as distance, security, data rate, QoS, duplexity, and throughput (or thru), which is the percentage of actual data transmitted in a given time period. Latency, delivery rate, and failover are usually not significant enough factors to be included at the local area level. Figure 13-5 shows an example of a local area overview connectivity map. Note that the overview map does not include any connection points, such as hubs, switches, or routers.

Figure 13-5
Example of a local area overview connectivity map

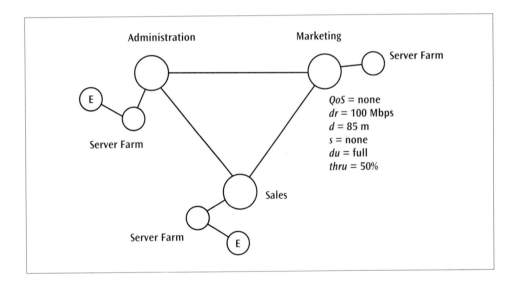

If more detail is desired, the analyst can create a local area detailed connectivity map. A detailed map can show how individual workstations or groups of workstations are clustered with switches, routers, hubs, and server farms. For example, the local area detailed connectivity map in Figure 13-6 has zoomed in on the Marketing node from the overview map in Figure 13-5 to show this node's workstations and their interconnections with the node's switches. The level of detail shown in a local area detailed connectivity map depends on the needs of a given project. Some projects require that all interconnections among the components be shown, while others can work from a map that shows only the major interconnections. As you can imagine, local area detailed connectivity maps generally capture a fair amount of information.

Figure 13-6
Local area detailed connectivity map for Marketing

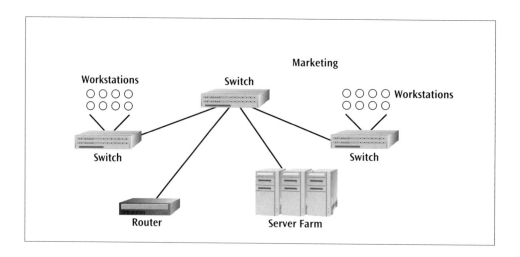

Now that we have learned about network modeling, let us examine another tool used in the SDLC model—the feasibility study.

Feasibility Studies

Analyzing and designing a new computer system can be time-consuming and expensive. While the project is in the analysis phase, and before a system is designed and installed, a feasible solution must be found. The term "feasible" has several meanings when it is applied to computer-based projects. The proposed system must be technically feasible. The technical feasibility of a system is the extent to which the system can be created and implemented using currently existing technology. Does the technology specified in the system proposal exist, and can it be incorporated into a working solution? If technology from two or more vendors is suggested, will the differing technologies work together? When a vendor claims that a particular hardware or software system will do what is desired, are the vendor's claims accurate or misleading? Does your company have the technical expertise to build, install, or maintain the proposed system?

The proposed system must also be financially feasible. A system's financial feasibility is the extent to which the system can be created, given the company's current finances. Can the proposed system both solve the company's current problem and stay within budget? Will the proposed system return a profit? If so, how long will it take for this to happen?

In addition, the proposed system must be operationally feasible. When a system demonstrates operational feasibility, it operates as designed and implemented. Thus, the company must ask: Will the proposed system produce the expected results? Will users be able to use the proposed system, or is there a chance that it will be so difficult or inconvenient to use that users will not adopt it?

Finally, the proposed system must be time feasible. A system's time feasibility is the extent to which the system can be installed in a timely fashion and meets organizational needs. Can the proposed system be designed, built, tested, and installed in an amount of time that all parties find reasonable and to which they can agree?

All of these feasibility questions are difficult to answer, but they must be answered. Technical, operational, financial, and time feasibility are best determined when the studies are based on a sound knowledge of computer systems, an understanding of the state of the current market and its products, and experience.

Individuals embarking on designing and installing a new computer network will perform better if they also understand analysis and design techniques, project and time management techniques, and financial analysis techniques. These techniques are an integral part of feasibility studies.

For an example of how these techniques are used, let us consider a common financial analysis technique that involves determining a proposed system's costs and benefits: payback analysis. Payback analysis charts the initial costs and yearly recurring costs of a proposed system against the projected yearly income (benefits) derived from the proposed system. Systems analysts and middle and upper management use payback analysis, along with other financial techniques, to determine the financial feasibility of a project.

Before you see how to perform a payback analysis calculation, you need to review a few common financial concepts as they apply to computer systems. To determine the cost of a system, it is necessary to include all possible costs. To do this, you will first need to consider all one-time costs such as:

- ▸ Personnel costs related to those individuals specifically hired to work on developing the system; these may include the salaries of analysts, designers, programmers, consultants, specialists, operators, secretaries, and so on
- ▸ Computer usage costs, which reflect the computing needed to perform the analysis and feasibility studies
- ▸ Costs of hardware and software for the proposed system
- ▸ Costs to train the users, support personnel, and management to use the proposed system
- ▸ Supplies, duplication, and furniture costs for the personnel creating the proposed system

But to get a comprehensive understanding of the cost of the system, you must also calculate the recurring costs of the proposed system. These include:

- ▸ Lease payments on computer hardware or other equipment
- ▸ Recurring license costs for software purchased
- ▸ Salaries and wages of personnel who will support the system
- ▸ Ongoing supplies that will keep the proposed system working
- ▸ Heating, cooling, and electrical costs to support the proposed system
- ▸ Planned replacement costs to replace pieces of the system as they fail or become obsolete

Once the one-time and recurring costs have been established, it is time to determine the benefits that will result from the proposed system. When calculating benefits, you will need to include both tangible benefits and intangible benefits. With respect to tangible benefits, the most common measurement is the monthly or annual savings that will result from the use of the proposed system. Intangible benefits are ones for which assigning a dollar amount is difficult; they include customer goodwill and employee morale.

Now that the costs and benefits have been determined, you can apply them to a payback analysis. When performing a payback analysis calculation, you should show all dollar amounts using the time value of money. The time value of money is a concept that states that one dollar today is worth more than one dollar promised a year from now, because today's dollar can be invested now and therefore accumulate interest. This also means that if something is going to cost one dollar one year from now, you need to put away less than a dollar today to pay for it. How much less depends on what assumptions you make about the discount rate. The discount rate is the opportunity cost of being able to invest money in other projects, such as stocks and bonds. The value of the discount rate is often set by a company's chief financial officer. Assuming a discount rate of, for example, 8 percent, means that you would need to put away only 92.6 cents today to pay for something that is going to cost a dollar in one year. Thus, in this case, the value 92.6 cents is the present value of one dollar one year from now. Table 13-1 shows the present value of a dollar over a period of seven years, given four different discount rates.

Table 13-1
The present value of a dollar over time, given discount rates of 4, 6, 8, and 10 percent

Present Value of a Dollar for Different Discount Rates				
Year	4%	6%	8%	10%
1	0.961	0.943	0.926	0.909
2	0.924	0.889	0.857	0.826
3	0.888	0.839	0.794	0.751
4	0.854	0.792	0.735	0.683
5	0.821	0.747	0.681	0.621
6	0.790	0.704	0.630	0.564
7	0.759	0.665	0.583	0.513

Figure 13-7 shows an example of a payback analysis. When examining the figure, note that *Development Cost* is a one-time cost that occurs in Year 0, and that recurring costs are listed as *Operation and Maintenance Costs* in Years 1–6. (Negative values are denoted with parentheses.) Together, Development Cost and Operation and Maintenance Costs total the costs of the project over its intended lifetime. Assuming a discount rate of 6 percent, the *Time-Adjusted Costs* row reflects the total costs for each year times the present value of a dollar for that year. The *Cumulative Time-Adjusted Costs* are simply the running sums of the Time-Adjusted Costs over all the years. *Benefits Derived* values are the benefits, or income amounts, that are expected each year. Still assuming a discount rate of 6 percent, *Time-Adjusted Benefits* are the Benefits multiplied by the present value of a dollar for each year. The *Cumulative Time-Adjusted Benefits* values are the running sums of the Time-Adjusted Benefits over all the years. *Cumulative Lifetime Time-Adjusted Costs* are the Cumulative Time-Adjusted Costs plus the Cumulative Time-Adjusted Benefits for each year.

Figure 13-7
Payback analysis calculation for a proposed project

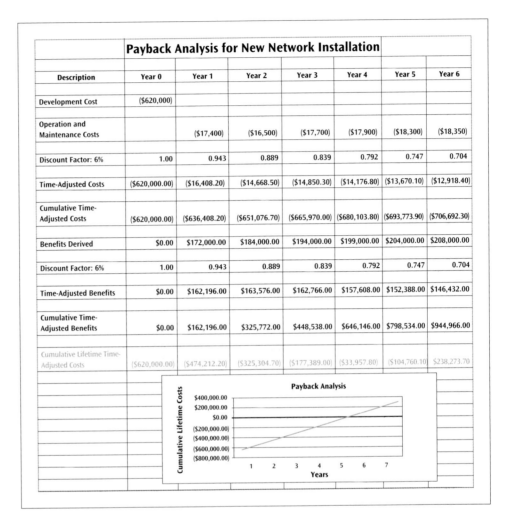

Description	Year 0	Year 1	Year 2	Year 3	Year 4	Year 5	Year 6
Payback Analysis for New Network Installation							
Development Cost	($620,000)						
Operation and Maintenance Costs		($17,400)	($16,500)	($17,700)	($17,900)	($18,300)	($18,350)
Discount Factor: 6%	1.00	0.943	0.889	0.839	0.792	0.747	0.704
Time-Adjusted Costs	($620,000.00)	($16,408.20)	($14,668.50)	($14,850.30)	($14,176.80)	($13,670.10)	($12,918.40)
Cumulative Time-Adjusted Costs	($620,000.00)	($636,408.20)	($651,076.70)	($665,970.00)	($680,103.80)	($693,773.90)	($706,692.30)
Benefits Derived	$0.00	$172,000.00	$184,000.00	$194,000.00	$199,000.00	$204,000.00	$208,000.00
Discount Factor: 6%	1.00	0.943	0.889	0.839	0.792	0.747	0.704
Time-Adjusted Benefits	$0.00	$162,196.00	$163,576.00	$162,766.00	$157,608.00	$152,388.00	$146,432.00
Cumulative Time-Adjusted Benefits	$0.00	$162,196.00	$325,772.00	$448,538.00	$646,146.00	$798,534.00	$944,966.00
Cumulative Lifetime Time-Adjusted Costs	($620,000.00)	($474,212.20)	($325,304.70)	($177,389.00)	($33,957.80)	($104,760.10	$238,273.70

As you can see in Figure 13-7, the values in the *Cumulative Lifetime Time-Adjusted Costs* row become positive in the project's sixth year—which means the project finally turns a profit at this time. Thus, in this case, it takes approximately five and one-half years for a payback, or **return on investment (ROI)**, to occur. Most companies establish their own acceptable ROI period. If the preceding payback analysis had been done for a company whose ROI period was six to seven years, its results would demonstrate to that company that this proposed project might be financially feasible.

Now let us turn our attention to additional techniques that should be used when developing a system solution.

Capacity Planning

Computer networks are mission-critical systems, and designing a new computer network or increasing the capacity of a current system requires careful planning. If you design a system for a company, and the system is not capable of supporting the traffic generated within the company, response times will be sluggish, and users may not be able to complete their work on time. This inability to perform work duties will lead to missed deadlines, projects backing up or even failing, and low employee morale. If this company is selling a product that can be purchased electronically, sluggish response times will also lead to dissatisfied customers. In the Internet age, when a competitor's Web site is a simple click away, dissatisfied customers will quickly turn elsewhere.

At the other end of the spectrum, if you over-design a system, you will have spent money unnecessarily to create a system that may never reach its capacity. It can be argued that if you are going to err in design, it is probably best to err in the direction of over-designing, especially because it is difficult to predict the growth rate of new users and applications. But the reality is that a large percentage of systems designed during the last 10 years are probably too small and will not be able to—or already cannot—support the demands placed on them. A smaller percentage of network systems are over-designed and remaining relatively idle.

Capacity planning involves trying to determine the amount of network bandwidth necessary to support an application or a set of applications. Capacity planning is a fairly difficult and time-consuming operation. But if capacity planning is

---

Details ▶

ROI for Wireless LANs and Network Security

When a company calculates the ROI (return on investment) for a proposed project, it compares the revenue generated from the project with the expenses incurred. When the revenue becomes greater than the expenses, the company has achieved a return on its investment. What happens if the project does not generate revenue—that is, if it is not even designed to generate revenue? How then do you measure the return on your company's investment? Also, if you cannot calculate ROI, how do you "sell" the importance of this new project to management?

Two technologies in computer networking that often do not generate revenue are wireless LANs and network security. Or perhaps we should say that these two do not generate revenue in terms of conventional ROI calculations. To determine how his or her company could benefit from investments in wireless LANs and network security, a network administrator has to look beyond the conventional techniques that simply account for profit increases and cost reductions, and consider qualitative factors instead.

For example, if a company is considering installing a wireless local area network, the typical expenses incurred are the costs associated with:

▶ Hardware (including the cost of access points and wireless NICs)

▶ Software

▶ Installation

▶ Training and documentation

▶ Ongoing maintenance

What about the benefits? The benefits of installing a wireless LAN include:

▶ No need for network cabling for the wireless devices

▶ Lower maintenance costs, because a network jack does not have to be moved every time an employee moves his or her desk or work location

not done well, it can be disastrously easy to plan poorly and thus design a system that will not support the intended applications. A number of techniques exist for performing capacity planning, including linear projection, computer simulation, benchmarking, and analytical modeling.

Linear projection involves predicting one or more network capacities based on the current network parameters and multiplying those capacities by some constant. Suppose you currently have a network of, say, 10 nodes, and the network has a response time of x. Using a linear projection, you might conclude that a system of 20 nodes would have a response time of $2x$. Some systems, however, do not follow a linear projection. If you apply a linear projection to these systems, you may produce inaccurate predictions. In these cases, an alternate strategy is required.

A **computer simulation** involves modeling an existing system or a proposed system using a computer-based simulation tool and subjecting the model to varying degrees of user demand (called load). The advantage of using a computer simulation is that it can mimic conditions that would be extremely difficult, if not impossible, to create on a real network. On the negative side, computer simulations are difficult to create, mainly because it is easy to make mistakes in the modeling process and difficult to discover them. Plus, just one mistake in a simulation can produce false results. Thus, extreme care must be taken when creating a simulation.

Benchmarking involves generating system statistics under a controlled environment and then comparing those statistics against known measurements. A number of network benchmark tests exist that can be used to evaluate the performance of a network or its components. Compared to simulation, benchmarking is

---

► Increased employee productivity (and morale) gained from giving employees benefits such as the freedom to work in any location as well as the ability to, say, directly access their notes during meetings or immediately retrieve product specifications from the factory floor

Note that although the financial consequences of the third benefit—increasing employee productivity—are less directly quantifiable than those of the first two, they are no less significant.

Now let us consider security. What are the typical costs of adding or increasing network security? Usually, these costs are associated with:

► Hardware (including the cost of firewall routers and other devices)

► Software

► Ongoing maintenance and support

The benefits for adding security are the following:

► Decreased system downtime, because the reduction of viruses should help minimize system failure and system slowdowns

► Increased employee productivity, because employees will spend less time recovering from viruses

► Decreased storage requirements, because fewer viruses will be using or destroying hard disk space

As more articles about the high costs associated with virus infections are published, more and more network administrators are discovering that it is getting easier to document the productivity time that is lost as a result of viruses.

In conclusion, it is not impossible to calculate an ROI for wireless and security networking projects. But to do this, a network administrator has to look beyond the normal definitions of revenue and consider more creative ways to define benefits. Many times, this involves calculating the costs associated with various "what if" scenarios—the main one being: What if we do not implement the proposed improvements? What will this cost us in terms of lost productivity and downtime?

a relatively straightforward technique and can provide useful information when it is used to analyze a network. But setting up a benchmark test can be quite time-consuming. Unfortunately, like simulation, this process can also suffer from possible errors. In addition, if all the variables in the test environment are not the same as all the variables in the benchmark environment, inaccurate comparisons will result. For example, if the test environment of the network uses one brand of router or switch and the benchmark network uses a different brand, comparing the results of the two networks may not be valid.

Analytical modeling involves the creation of mathematical equations to calculate various network values. For example, to calculate the utilization (the percentage of time that the line is being used) of a single communications line within a network, you can use the following equation:

$$U = t_{frame} / (2t_{prop} + t_{frame})$$

in which t_{frame} is the time to transmit a frame of data, and t_{prop} is the propagation time, the time it takes for a signal to be transferred down a wire or over the airwaves.

Many experts feel analytical modeling is a good way to determine network capacity. As in the computer simulation technique, you can create analytical models representing network systems that are difficult to create in the real world. Unfortunately, it is easy to create inaccurate analytical models and thus generate results that are invalid.

Because most networks support multiple applications, a person who performs capacity planning for a network has to calculate the capacity of each application on the network. Once the capacity of each application has been determined, it should be possible to determine the capacity for the entire network. You can calculate the individual capacities using analytical methods and then estimate the total network capacity using a linear projection. Let us consider two simple examples of capacity planning for single applications on a computer network.

The first example is based on the scenario featured in Chapter Eleven's In Action section. It is a simple mathematical calculation that involves selecting an appropriate telecommunications technology that will support the transfer of very large sales records. Each sales record is 400,000 bytes and has to download in 20 seconds or less. The following equations calculate the capacity (or data transfer rate, *n*) needed for one user downloading one file:

$$400,000 \text{ bytes} \times 8 \text{ bits} / \text{byte} = 3,200,000 \text{ bits}$$
$$n \text{ bps} = 3,200,000 \text{ bits} / 20 \text{ seconds} = 160,000 \text{ bps}$$

Thus, this application requires a telecommunications line capable of supporting a data rate that is at least 160,000 bps.

For another example of capacity planning—this time using a linear projection—suppose you work for a company that allows its employees to access the Internet. Informal studies performed in the late 1990s showed that the average Internet user requires a transmission link of 50 kbps. As a precaution, we will double that figure to 100 kbps to allow room for growth. Additional studies have shown that the peak hour for Internet access is around 11:00 a.m., at which time about 40 percent of the potential users are on the system. If your system has 1000 potential users, your peak capacity would correspond to 400 *concurrent* users, each of whom requires a 100-kbps connection. To satisfy this demand, network capacity would need to be

40 Mbps (400 × 100 kbps). Can your local area network support this much traffic? In this example, capacity planning indicates that if your company is not willing to install communications links totaling 40 Mbps, it may have to apply restrictions on how its employees use the Internet. For example, the company may have to limit the number of simultaneous users or not give the full 100-kbps capacity to each user. Alternatively, the company can simply hope that no more than 40 percent of all its users will access the Internet at the same moment in time.

As you can see, capacity planning is a difficult and nontrivial matter. Once capacity planning is properly done, however, a network administrator can determine if the company's local area networks and wide area network connections can support the intended applications. But knowing the capacity that is needed is not the whole story. Still other important questions must be resolved. For example: How do you know whether the current network can or cannot handle the intended applications? Once the needed capacity is determined, you need to closely examine the current network to determine its actual capacity. One of the best techniques for determining current network capacities is creating a baseline.

Creating a Baseline

Creating a baseline for an existing computer network involves measuring and recording a network's state of operation over a given period of time. As you will learn shortly, creating a baseline actually involves capturing many network measurements over all segments of a network, including numerous measurements on workstations, user applications, bridges, routers, and switches. Because collecting this information appears to be such a large undertaking, why would you want to create a baseline? Network personnel create a baseline to determine the normal and current operating conditions of the network. Once a baseline is created, the results can be used to identify network weaknesses and strengths, which can then be used to intelligently upgrade the network.

Very often, network administrators feel pressure from users and network owners to increase the bandwidth of a network. Without a thorough understanding of whether network problems actually exist—or, if they do, where they exist—a network administrator trying to improve network operation may inadvertently "fix" the wrong problem. Improving a network's operation may involve increasing network bandwidth, but it could just as easily involve something less expensive, such as upgrading some older equipment or segmenting a network by means of a switch. By conducting a baseline study (preferably one that is *continuing*), a network administrator can gain a better understanding of the network and can improve its overall quality more effectively.

Baseline studies can be started at any time but are most effective when they are initiated during a time when the network is not experiencing severe problems, such as a node failure or a jabber (a network interface card that is transmitting non-stop). Therefore, before you begin a baseline study, you must extinguish all immediate fires and try to get the network into fairly normal operation. Because you will be generating a large number of statistics, you will want to have access to a good database or spreadsheet application to keep the data organized. Once the database or spreadsheet has been set up, you are ready to begin your baseline study.

The next question is: On what items are you going to collect baseline information? You may find it useful to collect information on items such as system users, system nodes, operational protocols, network applications, and network utilization levels. Collecting information on system users involves determining the maximum number of users, the average number of users, and the peak number of users.

To collect baseline information on system nodes, you create a list of the number and types of system nodes in the network. These may include computer workstations, routers, bridges, switches, hubs, and servers. It is a good practice to have up-to-date drawings of the locations of all nodes, along with their model numbers, serial numbers, and any address information such as Ethernet and IP addresses. This information should also include the name and telephone number(s) of each product's vendor, in case technical assistance is ever needed.

Collecting baseline information on operational protocols involves listing the types of operational protocols used throughout the system. Most networks support multiple protocols, such as TCP, IP, NetBIOS, and more. The more protocols supported, the more processing time necessary to convert from one protocol to another. If a baseline study discovers an older protocol that can be replaced, it could contribute to improving network efficiency.

During the baseline study, you should also list all network applications found on the network, including the number, type, and utilization level of each application. Having a comprehensive list of the applications on the network will help you to identify old applications that should no longer be supported and thus should be deleted from the system. A list of applications will also help you to identify the number of copies of a particular application and thus help the company avoid violating its software licenses. When creating a comprehensive list of network applications, do not forget to include the applications stored on both individual user workstations and network servers.

Assembling information on network utilization levels requires that you create a fairly extensive list of statistics. These statistics include many of the following values:

- Average network utilization (%)
- Peak network utilization (%)
- Average frame size
- Peak frame size
- Average frames per second
- Peak frames per second
- Total network collisions
- Network collisions per second
- Total runts
- Total jabbers
- Total cyclic redundancy checksum (CRC) errors
- Node(s) with the highest percentage of utilization and corresponding amount of traffic

Once you have collected and analyzed network utilization data, you can make several important observations. First, you can detect when a network may be reaching saturation. Typically, a network reaches saturation when its network utilization is at 100 percent, which means 100 percent of the usable transmission space on the network is consumed by valid data. In certain cases, however, networks may reach saturation at lower levels of network utilization. Take a shared-segment CSMA/CD LAN, for example. Because shared-segment CSMA/CD networks are contention-based, they suffer from collisions. As the number of transmitting stations increases, the number of collisions increases, which *lowers* the network utilization. Thus, CSMA/CD networks that are consistently experiencing network utilization in the 40 to 50 percent range are probably experiencing a high number of collisions and might need to be segmented with switches.

A second observation you can make is when peak periods of network use occur. Making observations about peak periods of network use is easiest when you graph network activity data. Consider the hypothetical example shown in Figure 13-8. Peak periods occur at approximately 8:30 a.m., 11:30 a.m., 1:00 p.m., and 4:00 p.m. The most likely reason for these peaks would be users logging in and checking e-mail at 8:30 a.m. and 1:00 p.m., and users finishing work before going to lunch (11:30 a.m.) or going home (4:00 p.m.). Once you identify regular peak periods and their underlying causes, you can recognize when a peak period occurs at an unusual time. Plus, if you know when peak periods occur and why they occur, you can rearrange network resources to help lessen the load during these periods.

Figure 13-8
Peak periods of network activity in a typical day

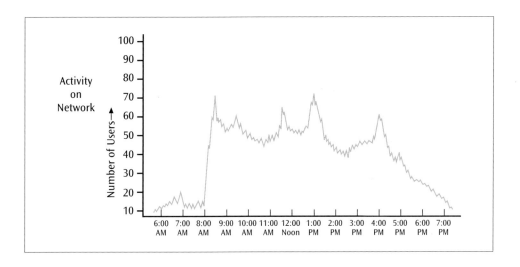

Examining the amount of traffic on each node also yields valuable information about network performance. Typically, a small percentage of network nodes generates a large percentage of network traffic. It is not unusual to encounter a node, such as a router or server, that is at the center of a great deal of traffic. A user workstation that generates a high amount of traffic, however, is suspect and should be examined more closely. A common example today is that of a user downloading music or video over the corporate or school network. Music and video downloads consume a great deal of bandwidth and tie up network resources, causing delays for employees and students with legitimate network requests. Network administrators can and should detect these downloads, which are often unauthorized, and ask the user to discontinue performing them on network workstations.

Once you have performed the baseline study, do not stop observing the network. For a baseline study to be really effective, you need to maintain it. An ongoing baseline study gives a systems manager an effective tool for identifying network problems, repairing the network, responding to complaints, improving the weak spots, and requesting additional funding.

Now that we have studied all the tools and steps necessary for developing a network, let us examine the types of skills a network administrator needs to keep the system operating successfully.

Network Administrator Skills

Once the analysis and design phases of network development are completed and the computer network is in place and operating, it is the network administrator's responsibility to keep it running. Keeping a network running involves making repairs on failed components, installing new applications and updating the existing ones, keeping the system's existing users up to date, and looking for new ways to improve the overall system and service level. It is not an easy job. With the complexity of today's networks and businesses' dependence on their applications, network administrators are highly valuable, visible, and always on the move.

Because many network administrators are dealing with both computers and people, they need the skills necessary to work with both. A checklist of skills for the network administrator would include a wide platform of technology skills, including, but not limited to, knowledge of local area networks, wide area networks, voice telecommunications systems, data transmission systems, video transmission, basic hardware concepts, and basic software skills. A network administrator should also have interpersonal skills, including the ability to talk to users in order to service problems and explore new applications. Along with interpersonal skills, a network administrator also needs training skills, which involve the ability to train users or other network support personnel.

To make effective use of limited resources, a network administrator should also possess a number of common management skills. For one, the network administrator should have budget management skills, which include knowing how to prepare a budget to justify continuing funds or to request additional funds. Along with those skills, a network administrator needs basic statistical skills, which means that he or she must know how to collect and use system statistics to justify the performance of existing systems or to validate the addition of new ones. Time management skills are also a necessity. These include the ability to manage not only one's own time, but also that of projects and any information technology workers who may be working for the administrator. Just as valuable as time management skills are project management skills, which center on the ability to keep a project on schedule and to use project-estimating tools, project-scheduling tools, and other methods for continuous project assessment. Finally, a network administrator should possess policy creation and enforcement skills, which include the ability to create policies concerning the use of the computer systems, access to facilities, password protection, access to applications, access to databases, distribution of hardware and software, replacement of hardware and software, and the handling of service requests.

To learn new skills and demonstrate proficiency within a particular area, the network administrator can obtain certification. It is possible to become certified on a particular type of network operating system, such as Windows 2008, or on a particular brand of network equipment, such as Cisco routers or Nortel Networks. The following is a list of some of the more popular certification programs:

- ► Microsoft's Certified Network Systems Engineer (CNSE)—This certificate addresses the design, installation, and support of the Windows network operating system.
- ► Cisco Certified Network Associate (CCNA)—This certificate covers the topics of installing, configuring, operating, and troubleshooting enterprise level router and switched networks.
- ► Nortel Networks Certification—A certificate in Nortel Networks provides an in-depth study of Nortel network systems.
- ► IBM Certified Systems Expert (CSE) and Certified Administrator (CA)—These certificates demonstrate the ability to successfully plan for, install, and support IBM's Networking LAN products.

The position of network administrator is demanding, challenging, and always changing. Being a successful network administrator requires a wide range of technical, management, and interpersonal skills. A good network administrator is constantly learning new skills and trying to keep abreast of the rapidly evolving technology. A computer network system could not survive without the network administrator. The network administrator has to remember, however, that the system would also not survive without users.

Let us look at an additional tool that network administrators can use for supporting or improving a system—statistics.

Generating Usable Statistics

Computer networks are in a constant state of change. New users and applications are added, while former users and unwanted applications are deleted. The age of a network (and its underlying technology) is often based on Internet-years, which many experts equate to approximately 90 calendar days. Because the technology changes so quickly and networks are constantly being called upon to support new and computationally intensive applications, a network administrator is constantly working on improving the data transfer speed and throughput of network applications.

To support changes to a network, a network administrator needs funding. Management, unfortunately, is not always receptive to investing more funds in technology. Often, management needs to be persuaded that services are suffering or response time is not what it needs to be. Statistics on computer network systems can be a very useful tool for demonstrating the need to invest in technology. If properly generated, statistics can be used to support the request for a new system or modifications to an existing system.

Four statistics, or measures, that are useful in evaluating networks are mean time between failures, mean time to repair, availability, and reliability. Mean time between failures (MTBF) is the average time a device or system will operate before it will fail. This value is sometimes generated by the manufacturer of the equipment and passed along to the purchaser. But often this value is not available, and the owner of the equipment has to generate an MTBF value from the equipment's past performance. Although every device is different, the longer the mean time between failures, the better.

Mean time to repair (MTTR) is the average time necessary to repair a failure within the computer network. This time includes the time necessary to isolate the failure. It also includes the time required to either swap the defective component with a working component or repair a component—either on-site or by removing the component and sending it to a repair center. Finally, mean time to repair includes the time needed to bring the system back up to normal operation. The value of mean time to repair depends on each installation and, within an installation, on each type of component. For example, a network server with hot-swappable devices should have a shorter mean time to repair than a device that has to be shut off, opened, repaired, and then rebooted.

The third statistic, availability, is the probability that a particular component or system will be available during a fixed time period. A component or network with a high availability (near 1.0) is almost always operational. Software that generates statistics often calculates the value for availability based on mean time to repair and mean time between failure values. Components with a small MTTR and a large MTBF will produce availability values very near to 1.0. For simplicity, however, we will calculate availability by simply subtracting the downtime from the total available time and then dividing by the total available time:

Availability % = (Total available time – Downtime) / Total available time

Suppose we want to calculate the availability of a printer for one month (24 hours per day for 30 days, or 720 hours), knowing that the printer will be down (inoperable) for 2 hours during that period.

Availability % = (720 – 2) / 720 = 0.997

Because the availability is near 1.0, there is a very high probability that the printer will be available during that one-month period.

To calculate the availability of a system of components, you should calculate the availability of each component and find the product of all availabilities. For example, if a network has three devices with availabilities of 0.992, 0.894, and 0.999, the availability of the network is the product of 0.992 × 0.894 × 0.999, or .886. Companies typically like to see availability values in the "nines," with the more nines the better. For example, 0.9999 is better than 0.999.

The fourth statistic, reliability, calculates the probability that a component or system will be operational for the duration of a transaction of time *t*. Reliability is defined by the equation:

$$R(t) = e^{-bt}$$

in which: $b = 1/\text{MTBF}$
t = the time interval of the transaction

What is the reliability of a router if the MTBF is 3000 hours and a transaction takes 20 minutes, or ⅓ of an hour (0.333 hours)?

$$R(0.333 \text{ hours}) = e^{-(1/3000)(0.333)} = e^{-0.000111} = 0.99989$$

The reliability of the router is very near to 1.0. A reliability of exactly 1.0 means the network or device is reliable 100 percent of the time.

What if the reliability of a second device was calculated and found to be 0.995? Although this value also appears to be near 1.0, there is a difference between the two reliabilities: 0.00489. What this difference basically means is that in 1000 repetitions of a trial, a particular failure may occur five times in the second device. On a network, many events occur or repeat thousands of times, so you would not want to experience five device failures in a year, especially during the transmission of data. Therefore, many network administrators strive to maintain system availability and reliability values of 0.9999 to 0.99999.

Let us now look at some other tools that a network administrator can use, along with these usable statistics, to monitor and support a system.

Network Diagnostic Tools

To support a computer network and all of its workstations, nodes, wiring, applications, and protocols, network administrators and their support staff need an arsenal of diagnostic tools. The arsenal of possible diagnostic tools continues to grow, with more powerful and helpful tools becoming available every day. Diagnostic tools can be grouped into two categories: tools that test and debug the network hardware, and tools that analyze the data transmitted over the network. Finally, the command center and the help desk should be considered. Although not exactly tools in the traditional sense, the command center and the help desk are valuable additions in a network support staff's arsenal. We will examine the tools that test the network hardware first.

Tools that test and debug network hardware

Tools that test and debug network hardware range from very simple devices to more elaborate, complex devices. Three common testing devices are electrical testers (the simplest), cable testers, and local area network testers (the most elaborate).

Electrical testers measure AC and DC volts, resistance, and continuity. An electrical tester will show if there is voltage on a line, and if so, how much voltage. If two bare wires are touching each other, they will create a short, and the electrical tester will show zero resistance. The continuity tester is a handy device that shows whether two wires are grounded to each other. Electrical and continuity testers are used to determine if the wires themselves are experiencing simple electrical problems.

Cable testers are slightly more elaborate devices. They can verify connectivity and test for line faults, such as open circuits, short circuits, reversed circuits, and crossed circuits. Certain kinds of handheld cable testers can also test fiber-optic lines, Asynchronous Transfer Mode networks, and T-1 circuits. For example, if a connector hidden in some wiring closet contains two wires that are switched, a cable tester will detect the problem and point to the source of the problem.

One of the most elaborate devices is the local area network tester. These testers can operate on Ethernet and token ring networks, whether they have switches or not. Some local area network testers have a display that graphically shows a network segment and all of the devices attached to it. When plugged into an available network jack, these testers can troubleshoot the network and suggest possible corrections. A common problem solved by these devices is the identification and location of a network interface card (NIC) that is transmitting continuously but not sending valid data (a jabber). The tester will pinpoint the precise NIC by indicating the 48-bit NIC address. A network administrator can then simply look up the particular NIC address in the system documentation and map it to a unique machine in a unique office.

Network sniffers

The second category of diagnostic tools covers tools that analyze data transmitted over the network. These tools include protocol analyzers and devices or software that emulate protocols and applications. One of the most common of these tools is the traffic analyzer or protocol analyzer. A protocol analyzer, or sniffer, monitors a network 24 hours a day, seven days a week, and captures and records all transmitted packets. Each packet's protocol is analyzed, and statistics are generated that show which devices are talking to each other and which applications are being used. This information can then be used to update the network, so that it operates more effectively. For example, if a protocol analyzer indicates that a particular application is being used a great deal and is placing a strain on network resources, a network administrator can consider alternatives such as replacing the application with a more efficient one or redistributing the application to the locations where it is used the most.

A very popular sniffer that can be used on Unix and Windows networks is Ethereal (or Wireshark). Ethereal allows you to examine the data collected from a live network or from a capture file on disk. You can interactively browse the captured data and view either detailed information for each packet or summary information for the entire network. What is perhaps even more interesting about Ethereal is that it's free.

Even a company with the latest tools and techniques may have trouble properly supporting its users. Another "tool" that is necessary is the control center and help desk. With a control center and help desk, the network support team can present a friendly "face" to the user, someone the user can turn to in times of trouble.

Managing operations

To assist network administrators and information technologists in doing their jobs, businesses have control centers for their computing services. The control center is the heart of all network operations. It contains, in one easily accessible place, all the network documentation, including network resource manuals, training manuals, baseline studies, all equipment documentation, user manuals, vendor names and telephone numbers, procedure manuals, and forms necessary to request services or equipment. The control center can also contain a training center to assist users and other information technologists. In addition, the control center contains all hardware and software necessary to control and monitor the network and its operations.

One of the more important elements of a control center is the help desk. A help desk answers all telephone calls and walk-in questions regarding computer services within the company. Whether it is called upon to address hardware problems, answer questions about running a particular software package, or introduce the company's users to new computing services, the help desk is the gateway between the user and computing and network services. Fortunately, good help desk software packages are available that the operations staff supporting computing services can use to track and identify problem areas within the system.

A well-designed control center and help desk can make an enormous impact on the users within a business. When users know there is a friendly, available person they can turn to for any computing problems, there is much less computer system/computer user friction.

Now that we have examined the physical tools that can assist a network administrator to support a company's networks, we will examine some of the software tools that can ease the burden of network administration.

Simple Network Management Protocol (SNMP)

Imagine you are the network administrator of a network that is composed of many different types of devices, including workstations, routers, bridges, switches, and hubs. The operation of the network has been smooth for some time, but then suddenly, the network begins to experience problems. It becomes sluggish, and users start calling you complaining of poor network response times. Is there something you can do to monitor or analyze the network without leaving your office and running from room to room or building to building? There is—if all or most of the devices on the network support a network management protocol. A network management protocol facilitates the exchange of management information between network devices. This information can be used to monitor network performance, find network problems, and then solve those problems—all without having any network personnel physically touch the affected device.

Although a number of different protocols exist to support network management, one protocol stands out as the simplest to operate, easiest to implement, and most widely used—Simple Network Management Protocol. Simple Network Management Protocol (SNMP) is an industry standard created by the Internet Engineering Task Force and designed originally to manage Internet components; it is now also used to manage wide area network and telecommunications systems. Currently, SNMP is in its third version.

All three versions of SNMP are based on the following set of principles. Network objects consist of network elements such as servers, mainframe computers, printers, hubs, bridges, routers, and switches. Each of these elements can be classified as either managed or unmanaged. A managed element has management software, called an agent, running in it and is more elaborate and expensive than an unmanaged element. A second type of object—the SNMP manager software—controls the operations of a managed element and maintains a database of information about all managed elements. An SNMP manager can query each agent and receive management data, which it then stores in the database. An agent can send unsolicited information to the manager in the form of an alarm. Finally, an SNMP manager itself can also act as an agent if a higher-level SNMP manager calls upon the manager to provide information for a higher-level database. All this management and passing of information can be done either locally or remotely, for example, from across the country in cases where the information is transmitted over the Internet.

The database that holds the information about all managed devices is called the Management Information Base (MIB). The information stored in the MIB can be used to repair or manage the network, or simply to observe the operation of the network. A manager can query a managed element (agent), asking for the particular details of that element's operation at that moment in time. This information is then sent from the managed element to the MIB for storage. For example, a manager might ask a router how many packets have entered the router, how many packets have exited the router, and how many packets were discarded due to insufficient buffer space. This information can later be used by a management program, which might, after looking at the information in the MIB, conclude that that particular element is not performing properly.

SNMP can also perform an autodiscovery type of operation. This operation is used to discover new elements that have been added to the network. When SNMP discovers a newly added element, the information about the element is added to the MIB. Thus, SNMP is a dynamic protocol that can automatically adapt to a changing network. This adaptation does not require human intervention (except for the intervention involved in connecting the new element to the network).

Managed elements are monitored and controlled using three basic SNMP commands: read, write, and trap. The read command is issued by a manager to retrieve information from the agent in a managed element. The write command is also issued by a manager but is used to control the agent in a managed element. By using the write command, a manager can change the settings in an agent, thus making the managed element perform differently.

A weakness of the first two versions of SNMP was the lack of security in this write command. Anyone posing as an SNMP manager could send bogus write commands to managed elements and thereby cause potential damage to the network. SNMP version addresses the issue of security, so that bogus managers cannot send malicious write commands.

The third command—the trap—is used by a managed element to send reports to the manager. When certain types of events, such as a buffer overflow, occur, a managed element can send a trap to report the event.

More often than not, the SNMP manager requests information directly from a managed element on the same network. But what if a manager wants to collect information from a remote network? Remote Network Monitoring (RMON) is a protocol that allows a network administrator to monitor, analyze, and troubleshoot a group of remotely managed elements. RMON is defined as an extension of SNMP, and the most popular recent version is RMON version 2 (often referred to as RMON2). RMON can be supported by hardware monitoring devices (known as probes), through software, or through a combination of hardware and software. A number of vendors provide networking products with RMON support. For example, Cisco's series of local area network switches includes switches with software that can record information as traffic passes through the switch and store this information in the switch's MIB. RMON can collect several basic kinds of information, such as number of packets sent, number of bytes sent, number of packets dropped, host statistics, and certain kinds of events that have occurred. A network administrator can find out how much bandwidth or traffic each user is imposing on the network and can set alarms in order to be alerted of impending problems.

To complete our discussion of network design and management, we will examine the Better Box company to see how it might increase its computing capacity.

Capacity Planning and Network Design
In Action: Better Box Corporation ▶

To see how capacity planning and network design work in a realistic setting, let us return to the Better Box Corporation from Chapter Eleven. Recall that Better Box's administrative headquarters is in Chicago, and it has regional sales offices in Seattle, San Francisco, and Dallas. The marketing group in Chicago runs applications on 25 stand-alone workstations. The other sites currently have only a few workstations. In Chapter Eleven, we solved Better Box's primary networking problem by finding a way to provide its regional sales offices with network access so they could upload sales records to the company's headquarters. That solution, as you may recall, involved establishing MPLS/virtual LAN connections between Seattle and Chicago, San Francisco and Chicago, and Dallas and Chicago.

Better Box is expanding its data networking capability now and has asked you to help design its new corporate network. It wants to add new hardware and software and provide new data applications. Although the company has enough funds to support the new applications, it wants to seek the most cost-effective solution for the long run. Better Box plans to add 10 additional workstations to the Chicago office and 35 workstations to each of the other offices, in Seattle, San Francisco, and Dallas.

In its new network, the company will also install the following six servers:

▶ Web server (Chicago)—An HTTP server to store Web pages for public access to corporate marketing information

▶ Inventory server (Chicago)—A database server that stores information about available product inventory

▶ Site servers (Chicago, Seattle, San Francisco, Dallas)—Four servers, one at each site, that provide e-mail services to enable employees to communicate about important projects, store site-specific management files and sales files, and store staff software and information for clerical workers

Once the new workstations, servers, and network equipment are in place, Better Box would like to use the network to support local site server access, e-mail services, and Internet access, and to continue to support the network access to sales records. Each employee will need to access data files on the site server at his or her location. In addition, each site server needs to be able to exchange e-mail messages with any other site server. Also, the Web server needs to be connected to the Internet, so current and potential customers around the world can use Web browsers to access company information.

Now let us find a solution for the complete problem—for setting up local site server access, e-mail access, Internet access, and internal local area networks, and also for connecting the recently installed network access to sales records.

To be able to support e-mail access and Internet Web page access, each site needs a high-speed connection to a local Internet service provider. Each Internet user will be given a 100-kbps-capacity connection (based on a calculation that was performed earlier in this chapter, in the section on capacity planning). Assuming that one-third of the 35 users (say, 12 users) at a site will be accessing the Internet at one time, a linear projection of 12 times 100 kbps per link produces a need for a 1.2-Mbps connection. Because the connection from a site to an Internet service provider should be a local connection, a local T-1 line could be added to each site to support Internet access. Recall that T-1 lines are capable of supporting a continuous 1.544-Mbps data stream between two locations. Because e-mail access also goes through an Internet service provider and requires very low capacity compared to Web page browsing, these additional T-1 lines should also be fine for supporting e-mail. Enabling Better Box's Chicago office to host a

Web server will require another connection to a local Internet service provider and the use of at least one more T-1 line to provide 1.544-Mbps access to the Web server.

To help us see the overall physical layout of the four geographic locations, we will create a simple wide area connectivity map (Figure 13-9). The connectivity map should include the four locations: Chicago, Seattle, San Francisco, and Dallas. The networking need that involves downloading sales records from the Chicago office requires us to create a connection between Seattle and Chicago, San Francisco and Chicago, and Dallas and Chicago. Note in Figure 13-9 that the mileages between the sites are included beside their respective connections. To provide Internet e-mail access and Web access to its users, each site requires a local connection to an Internet service provider. Because Internet service providers are external to the Better Box corporation, each ISP is shown as a circle with an E drawn through it.

Figure 13-9

Simple wide area connectivity map showing Seattle, San Francisco, Dallas, and Chicago sites

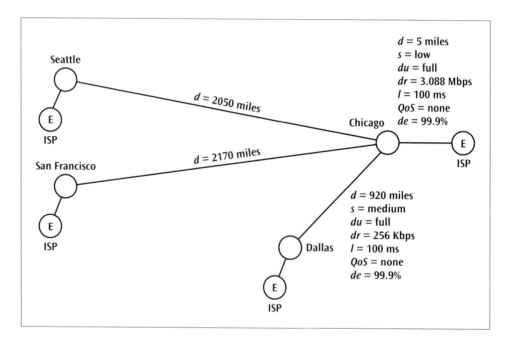

To support local site server access (which includes e-mail services for allowing employees to communicate about important projects, storage services for site-specific management files and sales files, and storage services for staff software and information for clerical workers), each site will need to create a local area network solution. Figure 13-10 shows a possible local area detailed connectivity map. Currently, each employee accesses enough local files that at least 1 Mbps of bandwidth is needed for each workstation on any shared LAN. If each site has 35 workstations, the local area network will need to support at least a 35-Mbps total capacity. Thus, a 100-Mbps CSMA/CD LAN would be prudent. CSMA/CD is also a good choice because it is the most popular local area network, which should reduce its overall costs and improves the chances that quality support is available. A 100-Mbps version has a very reasonable cost, and it should be sufficient to support a total of 35 users per site. It might be worthwhile to use high-speed switches in this local area network instead of hubs to support the 35 workstations; switches provide better network segmentation and decrease the likelihood of collisions, thus increasing overall network throughput. In addition, the cost of switches continues to decline, which makes them extremely cost-effective, given the high degree of network segmentation they provide.

Figure 13-10
Local area detailed connectivity map for one of the Better Box sites

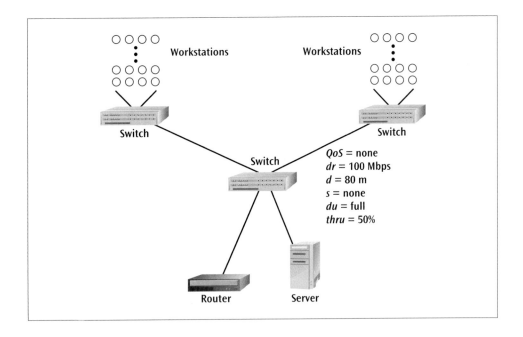

In conclusion, Better Box will need a variety of network devices to support its increased computing requirements. MPLS/virtual LAN connections from a national provider will create the virtual connections between each of the regional sales offices and the administrative office in Chicago. T-1 connections will connect each office to the virtual LAN provider using local distance services. Each office will also require additional local T-1 connections to Internet service providers. These connections will provide Internet e-mail access and World Wide Web access. Finally, CSMA/CD local area networks will provide the on-site glue that allows each workstation to access local servers as well as remote servers via the Internet.

◆◆◆

SUMMARY

▸ When creating a new network or adding to an existing network, there are many potential pitfalls and opportunities for inaccurate and incomplete assessments.

▸ The systems development life cycle (SDLC) is one of the most popular techniques used to guide analysts through the difficult decision-making process of network design. Although many versions of the SDLC model exist, most models consist of the following phases: planning, analysis, design, implementation, and maintenance.

▸ Persons designing a new network or upgrading an existing network may want to create one or more network models to help them visualize the system. The connectivity map is one of the network models used to depict internal and external network structures.

▸ An important part of SDLC is the conducting of one or more feasibility studies. Feasibility studies can be conducted during the planning phase, the analysis phase, and the design phase. But feasibility studies can also stand alone (in other words, they do not have to be conducted in conjunction with SDLC). The four basic types of feasibility studies are technical, financial, operational, and time.

▸ Payback analysis is one possible financial analysis technique that can be applied to a proposed computer network system to determine the system's costs and benefits.

▶ Capacity planning is a necessary technique that enables a network administrator to determine the network bandwidth that is needed to support one or more applications within a business. A number of techniques exist for performing capacity planning, including linear projection, computer simulation, benchmarking, and analytical modeling.

▶ A baseline study involves the measuring and recording of a network's state of operation over a given period of time. Many network administrators feel a baseline study is a must for all network operations, regardless of whether more network resources are currently being requested. The baseline study can serve a number of purposes, such as providing an understanding of the current system, helping to isolate and identify network problems, and providing evidence that more computing resources will be needed in the near future.

▶ Once a network is in operation, good network management is necessary to keep the network operating at peak efficiency. The network administrator is responsible for making sure that the network operates at peak efficiency. A network administrator should possess a number of skills, including hardware and software knowledge; people management, time management, budget management, and problem-solving skills; and a knowledge of statistics. To develop the technical skills of the manager, the more popular vendors of network hardware and software systems offer a number of certificate programs.

▶ A network administrator should be able to create and use basic statistics, such as mean time between failures, mean time to repair, availability, and reliability. These statistics can be used to justify current network resources or to validate the need for additional network resources.

▶ A large variety of diagnostic tools are available that can aid network personnel in troubleshooting and maintaining the complex computer networks that are commonly found today. The more common diagnostic tools include electrical testers, continuity testers, cable testers, local area network testers, protocol analyzers, and devices and programs that can emulate protocols and applications.

▶ All networks need a command center. It is in this command center that you find help desks, documentation, training centers, and the central nervous system of network operations.

▶ The Simple Network Management Protocol (SNMP) helps network support personnel monitor network performance, find network problems, and then solve those problems without physically touching the affected device. With SNMP, there are managed and unmanaged devices or elements. Managed elements have software agents running on them from which SNMP software managers can request information. SNMP managers can also send information to managed elements in order to control operations on the network. The database that holds the information about each managed device is called the Management Information Base (MIB). Remote Network Monitoring (RMON) is a protocol that allows a network administrator to monitor, analyze, and troubleshoot a group of remotely managed elements. RMON is an extension of SNMP and is currently in its second version.

KEY TERMS

agent
analytical modeling
availability
baseline
benchmarking
capacity planning
computer simulation
connectivity maps
    local area detailed connectivity map
    local area overview connectivity map
    metropolitan area connectivity map
    wide area connectivity map

financial feasibility
linear projection
Management Information Base (MIB)
mean time between failures (MTBF)
mean time to repair (MTTR)
network management protocol
operational feasibility
payback analysis
protocol analyzer
reliability
Remote Network Monitoring (RMON)
return on investment (ROI)

Simple Network Management
    Protocol (SNMP)
sniffer
SNMP manager
systems analyst
systems development life cycle (SDLC)
technical feasibility
time feasibility
time value of money

REVIEW QUESTIONS

1. Describe each phase of the systems development life cycle.
2. What is the primary goal of the planning phase of SDLC?
3. What is the primary goal of the analysis phase of SDLC?
4. What is the primary goal of the design phase of SDLC?
5. What is a connectivity map, and how can it assist you in designing a network?
6. Describe the four different types of feasibility studies.
7. What is meant by the time value of money?
8. Describe the four different ways to perform capacity planning.
9. For what reasons might someone perform a baseline study?
10. List the three most important skills a network administrator should possess.
11. What is the difference between mean time between failures and mean time to repair?
12. What is meant by the statistical term "availability"?
13. What is meant by the term "utilization"?
14. What is meant by the statistical term "reliability"?
15. What basic diagnostic tools are used to support a computer network?
16. What should be found in the control center for a network operation?
17. What is the function of the Simple Network Management Protocol?
18. What is the difference between a manager and an agent in SNMP?
19. How can Remote Network Monitoring be used to assist SNMP?

EXERCISES

1. State during which phase or phases of the systems development life cycle the following actions are performed:
   a. Installing the system
   b. Training users
   c. Writing documentation
   d. Performing feasibility studies
   e. Testing the system
   f. Updating the system

2. Using the following data, calculate the payback period.

   Development cost: $418,040

   Operation and maintenance costs for year 0 to year 6, respectively: 0; $15,045; $16,000; $17,000; $18,000; $19,000; $20,000

   Discount rate: 6%

   Benefits (year 0–year 6): $0; $150,000; $170,000; $190,000; $210,000; $230,000; $250,000

3. Create a simple analytical model that includes two formulas for calculating the approximate total time t for n terminals to perform roll-call polling and hub polling. Use TD = time to transmit data, TRP = time to transmit a roll-call poll, and THP = time to transmit a hub poll.

4. You are performing a baseline study for your company, which is located on the east coast. Your company does a lot of work with businesses on the west coast. You note that peak network utilization occurs at approximately noon, when most of your employees are on lunch break. What could be causing this peak activity?

5. During a baseline study, a high number of cyclic checksum errors were discovered but no runts. Explain precisely what this information has to do with network segment length.

6. A component has been operating continuously for three months. During that time, it has failed twice, resulting in 4.5 hours of downtime. Calculate the availability of the component during this three-month period.

7. If a component has an MTBF = 10 hours and a transaction takes 20 minutes, calculate the reliability of the component.

8. If a component has an MTBF = 500 hours and a transaction takes 4 seconds, calculate the reliability of the component.

9. If a network has four devices with the availabilities of 0.994, 0.778, 0.883, and 0.5, what is the availability of the entire network?

10. Is it possible for an SNMP agent in a managed device to also serve as a manager? Explain how this situation might work.

11. What are the differences between network line continuity testers and network cable testers?

12. You are working for a small company that has a local area network with two hubs. The communications line between the hubs has just been cut, but you do not know that yet. How can you determine what has happened?

THINKING OUTSIDE THE BOX

1 You have been asked to create a help desk for the computer support division of your company. What services will your help desk provide? How will you provide those services? What type of employees will you hire to work at the help desk?

2 Your company wants to create a Web server to promote its business. One of the features of the Web server is that it allows remote users to download service bulletins and repair manuals. These bulletins and repair manuals are approximately 240,000 bytes in size. You anticipate that approximately 30 users per hour will want to download these documents. What speed communications line do you need to support this demand?

3 The company you are working for sells dolls and their clothing outfits through a retail outlet and through mail-order catalogs. It is now considering selling merchandise via the Web. Create an SDLC plan for adding a Web merchandising system to the business that shows each step involved in the analysis, design, and implementation phases. The plan should have the appearance of an outline. For example, the analysis phase could begin with the following:

I. Analysis Phase

   A. Interview upper management

   B. Create a questionnaire and present to current employees

   C. Create a model showing the current data flow of the mail-order catalog business

   D. Etc.

HANDS-ON PROJECTS

1. Create a series of connectivity maps for either your place of work or your school. Try to include as many different external locations as possible. For one of these external locations, create one wide area connectivity map and one or more local area connectivity maps.

2. Perform a baseline study for the network at your workplace or school. Create a list of all network devices (servers, routers, bridges, switches, hubs, and so on). What protocols are supported by the network? What applications are supported by the network? Try to collect some statistics for one type of device on the network. If it is not possible to get actual values, simply create a list of the statistics that would be valuable for a particular device.

3. What certification programs, other than those listed in this chapter, exist for network administrators? For what type of systems are these certificates designed?

4. What are the various benchmark tests that are available for testing a computer system or computer network? Write a paragraph about each benchmark test, stating what the benchmark test actually measures.

5. Scan your local newspaper's classified job listings. What percentage of advertisements seeking network support personnel require a certification degree?

6. What are the current certifications that companies are looking for with respect to network administrators?

7. Lately there has been talk of a new type of position in the networking industry: IT architect. What is an IT architect, and what are the typical job duties associated with this position?

8. SNMP is only one of several management protocols. What other management protocols exist? Who created them? What types of situations do they manage?

9. How does SNMP version 3 handle security? Write a one- or two-page report that summarizes the main features of version 3.

10. How does the autodiscovery feature of SNMP operate?

Glossary

1-persistent algorithm   A CSMA/CD (carrier sense multiple access with collision detection) algorithm that governs a workstation to listen continuously until the transmitting medium is free, then transmit immediately.

1Base5   An 802.3 standard, created by IEEE for Ethernet (or CSMA/CD local area networks), that incorporates 1 Mbps baseband (digital) signaling for transmitting data over twisted pair for a maximum segment length of 250 meters.

1xEV (1 x Enhanced Version)   A third generation cellular telephone technology that is capable of supporting download rates of 300 kbps to 500 kbps.

4B/5B   A digital encoding scheme that takes four bits of data, converts the four bits into a unique five bit sequence, and encodes the five bits using NRZI.

5-4-3 rule   A rule for creating a LAN that states that between any two nodes on a LAN, there can only be a maximum of five segments, connected through four repeaters, and only three of the five segments may contain user connections.

8.3/125 cable   A single-mode (thinner) fiber-optic cable whose core is 8.3 microns wide, and whose cladding (the material surrounding the fiber) is 125 microns wide; this cable has fewer reflections and refractions than multimode fiber-optic cable (see 62.5/125 cable).

10Base2   An 802.3 standard, created by IEEE for Ethernet (or CSMA/CD local area networks), that incorporates 10 Mbps baseband (digital) signaling for transmitting data over coaxial cable for a maximum segment length of 185 or 200 meters.

10Base5   An 802.3 standard, created by IEEE for Ethernet (or CSMA/CD local area networks), that incorporates 10 Mbps baseband (digital) signaling for transmitting data over coaxial cable for a maximum segment length of 500 meters.

10BaseT   An 802.3 standard, created by IEEE for Ethernet (or CSMA/CD local area networks), that incorporates 10 Mbps baseband (digital) signaling for transmitting data over twisted pair for a maximum segment length of 100 meters.

10Broad36   An 802.3 standard, created by IEEE for Ethernet (or CSMA/CD local area networks), that incorporates 10 Mbps broadband (analog) signaling for transmitting data over coaxial cable for a maximum segment length of 3600 meters.

10GBase-fiber   An 802.3 standard, created by IEEE for Ethernet (or CSMA/CD local area networks), that incorporates 10 Gbps baseband (digital) signaling for transmitting data over fiber optic cable.

10GBase-T   An 802.3 standard, created by IEEE for Ethernet (or CSMA/CD local area networks), that incorporates 10 Gbps baseband (digital) signaling for transmitting data over twisted pair.

10GBase-CX   An 802.3 standard, created by IEEE for Ethernet (or CSMA/CD local area networks), that incorporates 10 Gbps baseband (digital) signaling for transmitting data over twin-axial cable.

10 Gbps Ethernet   The general term to represent 10 Gbps Ethernet local area networks.

56k leased line   A leased-line service provided by a telephone company that is fixed between two locations and provides data and voice transfer at rates up to 56 kbps.

62.5/125 cable   A multimode (thicker) fiber-optic cable whose core is 62.5 microns wide, and whose cladding (the material surrounding the fiber) is 125 microns wide; this cable has more reflections and refractions than single-mode fiber-optic cable (See 8.3/125 cable).

100BaseFX   An 802.3 standard, created by IEEE for Ethernet (or CSMA/CD local area networks), that incorporates 100 Mbps baseband (digital) signaling for transmitting data over two-pair Category 5 twisted pair for a maximum segment length of 100 meters.

100BaseT4   An 802.3 standard, created by IEEE for Ethernet (or CSMA/CD local area networks), that incorporates 100 Mbps baseband (digital) signaling for transmitting data over four pairs of Category 3 or higher twisted pair for a maximum segment length of 100 meters.

100BaseTX   An 802.3 standard, created by IEEE for Ethernet (or CSMA/CD local area networks), that incorporates 100 Mbps baseband (digital) signaling for transmitting data over two pairs of Category 5 or higher twisted pair for a maximum segment length of 100 meters.

1000BaseCX   An 802.3 standard, created by IEEE for Ethernet (or CSMA/CD local area networks), that incorporates 1000 Mbps baseband (digital) signaling for transmitting data over short-distances (in the 0.1 to 25 meter range) using copper jumper cables.

1000BaseLX   An 802.3 standard, created by IEEE for Ethernet (or CSMA/CD local area networks), that incorporates 1000 Mbps baseband (digital) signaling for transmitting data over either single-mode or multimode fiber-optic cable that is being used to support longer distance cabling within a single building.

1000BaseSX   An 802.3 standard, created by IEEE for Ethernet (or CSMA/CD local area networks), that incorporates 1000 Mbps baseband (digital) signaling for transmitting data over multimode fiber-optic cable that is being used to support relatively close clusters of workstations and devices.

1000BaseT   An 802.3 standard, created by IEEE for Ethernet (or CSMA/CD local area networks), that incorporates 1000 Mbps baseband (digital) signaling for transmitting data over Category 5e or higher twisted pair (UTP) wire for a maximum segment length of 100 meters.

access point The interconnecting bridge between a wireless local area network workstation and the wired local area network.

access rights Permissions assigned to a file or device; determines how a user or group or users may access the file or device.

Active Directory Network directory structure for Windows 2000/2003 operating system; a hierarchical structure that stores information about all the objects and resources in a network and makes this information available to users, network administrators, and application programs.

adaptive routing A dynamic system in which routing tables react to network fluctuations, such as congestion and node/link failure.

Address Resolution Protocol (ARP) An Internet protocol that takes an IP address in an IP datagram and translates it into the appropriate CSMA/CD address for delivery on a local area network.

Advanced Encryption Standard (AES) A new encryption technique selected by the U.S. government to replace the aging Data Encryption Standard (DES). AES is based on the Rijndael algorithm and uses 128, 192, or 256 bit keys.

Advanced Mobile Phone Service (AMPS) The oldest analog mobile telephone system; it covered almost all of North America and is found in more than 35 other countries.

agent The software (or management software) that runs in an element; an element that has an agent is considered a managed element and can react to SNMP commands and requests.

American National Standards Institute (ANSI) An organization that creates standards for a variety of products.

amplification The gain of the signal strength (power) of an analog signal.

amplitude The height of the wave above (or below) a given reference point.

amplitude shift keying A modulation technique for encoding digital data using various amplitude levels of an analog signal.

analog data Data that is represented by continuous waveforms, which can be at an infinite number of points between some given minimum and maximum.

analog signals Signals that are represented by continuous waveforms, which can be at an infinite number of points between some given minimum and maximum.

analytical modeling The creation of mathematical equations to calculate various network values during network analysis.

anti-spam software Software designed to detect and block spam that has been sent to a computer.

anti-spyware software Software designed to detect and remove malicious programs known as spyware (see spyware).

antivirus software Software designed to detect and remove viruses that have infected a computer's memory, disks, or operating system.

application layer The top-most layer in the OSI model and TCP/IP protocol suite; where the network application resides.

application program interface (API) A software module that acts as an interface between application programs and technical entities, such as telephone switching systems.

arithmetic checksum An error detection technique in which the ASCII values of the characters to be transmitted are summed and included on the end of the message.

ARPANET One of the country's first wide area packet-switched networks; the precursor of the modern Internet; interconnected research universities, research labs, and select government installations.

ASCII A seven-bit code that is used to represent all the printable characters on a keyboard plus many non-printable control characters.

asymmetric connection A connection in which data flows in one direction at a faster transmission rate than the data flowing in the opposite direction; for example, there are numerous systems that have a faster downstream connection (such as from the Internet) and a slower upstream connection.

asymmetric digital subscriber line (ADSL) A popular form of digital subscriber line that transmits the downstream data at a faster rate than the upstream data.

asymmetric encryption An encryption technique in which two keys are used, one to encrypt and one to decrypt; often known as public key cryptography.

asynchronous connection One of the simplest examples of a data link protocol and is found primarily in microcomputer to modem connections. Adds one start bit, one stop bit, and an optional parity bit to every character transmitted.

Asynchronous Transfer Mode (ATM) A high-speed, packet-switched service, similar to frame relay, that supports various classes of service.

attenuation The continuous loss of strength (power) that a signal experiences as it travels though a medium.

availability The probability that a particular component or system will be available during a fixed time period.

available bit rate (ABR) A class of service supported by Asynchronous Transfer Mode that is used for traffic that may experience bursts of data (called "bursty" traffic) and whose bandwidth range is roughly known, such as the traffic found in a corporation's collection of leased lines.

backbone   The main connecting cable that runs from one end of the installation to another, or from one end of a network to another.

backplane   The main hardware of a device (such as a LAN switch) into which all supporting printed circuit cards connect.

backup software   Software that allows network administrators to back up the data files currently stored on the network server's hard disk drive.

backward explicit congestion notification (BECN)   A congestion control technique in which congestion notification is sent along the return path to the transmitter.

backward learning   A technique in which a bridge or switch creates its routing tables by watching the current flow of traffic.

bandwidth   The absolute value of the difference between the lowest and highest frequencies.

bandwidth profile   A document that describes various characteristics about a data-service connection, such as basic data transfer rates and burst transfer rates.

baseband coaxial   A type of coaxial cable system that uses digital signaling/transmission technique in which the cable carries only one channel of digital data.

baseline   One of the best techniques for determining a network's current capacities.

Basic Service Set (BSS)   The transmission area surrounding a single access point in a wireless local area network; it resembles a cell in a cellular telephone network.

baud rate   The number of signal element or signal level changes per second.

benchmarking   Involves generating system statistics under a controlled environment and then comparing those statistics against known measurements.

bidirectional   A kind of transmission in which a signal transmitted from a given workstation propagates in both directions on the cable away from the source.

Bindery   A database that contains information on every resource connected to this one server, including users, groups of users, printers, and data sets.

biometric techniques   Security techniques that use parts of the body, such as fingerprints or iris prints, for verification.

bipolar-AMI   A digital encoding technique with no DC-component; logic 0s are denoted by zero voltage, and logic 1s are denoted with alternating positive and negative voltages.

bits per second (bps)   The number of bits that are transmitted across a medium in a given second.

blog   Short for Web log. An online website in the form of a journal.

Bluetooth   A wireless technology that transfers signals over short distances (of 10 meters or less) through non-metallic objects.

botnet   Malicious programs that take over operations on a compromised computer.

braided coaxial cable   Coaxial cable in which the center wire actually consists of many, smaller wires braided together.

bridge   A device that interconnects two local area networks that both have a medium access control sublayer.

broadband coaxial   A coaxial cable technology that transmits analog signals and is capable of supporting multiple channels of data simultaneously.

broadband wireless system   One of the latest techniques for delivering Internet services into homes and businesses; it employs broadband signaling using radio frequencies over relatively short distances.

broadcast network   A communications network that transmits data in a broadcast fashion (when one workstation transmits data, all other workstations receive that data).

buffer preallocation   A possible solution to network congestion in which an application requests a certain level of transmission space and the network preallocates the necessary buffers.

burst rate   An agreed-upon rate between a customer and a frame relay provider; this agreement allows the customer to exceed the committed information rate by a fixed amount for brief moments of time.

bus/tree local area network   The first topology used when local area networks became commercially available in the late 1970s; it essentially consists of a single coaxial cable to which all workstations attach.

cable modem   A communications device that allows high-speed access to wide area networks, such as the Internet, via a cable television connection.

call filtering   A technology in which users can specify which telephone numbers are allowed to get through. All other calls will be routed to an attendant or voice mailbox.

capacity planning   A time-consuming operation in the process of computer network development that involves trying to determine the amount of network bandwidth necessary to support an application or a set of applications.

carrier sense multiple access with collision avoidance (CSMA/CA)   A contention-based medium access control protocol for wireless networks in which wireless workstations can only transmit at designated times, in an attempt to avoid collisions.

carrier sense multiple access with collision detection (CSMA/CD)   A contention-based medium access control protocol for bus and star-wired bus local area networks in which a workstation wanting to transmit can only do so if the medium is idle, otherwise it has to wait. Signal collisions are detected by transmitting workstations which then back off and retransmit.

cascading style sheets (CSS) Allow a Web page author to incorporate multiple features (fonts, styles, colors, and so on) in an individual HTML page.

Category 1 A designation for twisted pair wiring that can support analog telephone systems; rarely used today.

Category 2 A designation for twisted pair wiring that can support digital telephone systems such as T-1 and ISDN; rarely used today.

Category 3 A designation for twisted pair wiring that can support data transmissions up to 10 Mbps for 100 meters; rarely used for local area networks; now often used for analog telephone systems.

Category 4 A designation for twisted pair wiring that can support data transmissions up to 20 Mbps for 100 meters; rarely used today.

Category 5 A designation for twisted pair wiring that can support data transmissions up to 100 Mbps for 100 meters on local area networks.

Category 5e A designation for twisted pair wiring that can support data transmissions up to 125 Mbps for 100 meters on local area networks.

Category 6 A designation for twisted pair wiring that can support data transmissions up to 250 MHz for 100 meters.

Category 7 A designation for twisted pair wiring that can support data transmissions up to 600 MHz for 100 meters.

Category 1-7 (CAT 1-7) twisted pair A naming system used to categorize twisted pair wire by how far and how fast the wire may transmit data.

CDMA2000 1xRTT A "2.5 generation" cellular telephone signaling technique, based upon code division multiplexing.

central office (CO) Contains the equipment that generates a dial tone, interprets the telephone number dialed, checks for special services, and connects the incoming call to the next point.

centralized routing A technique for providing routing information that dictates that the routing information generated from the least-cost algorithm is stored at a central location within the network.

Centrex (central office exchange service) A service from local telephone companies through which up-to-date telephone facilities at the telephone company's central (local) office are offered to business users so that they do not need to purchase their own facilities.

certificate An electronic document, similar to a passport, that establishes your credentials when you are performing transactions on the World Wide Web.

certificate authority (CA) The specialized software on a network or a trusted third-party organization or business that issues and manages certificates.

certificate revocation list (CRL) A list of certificates that have been revoked either before or at their originally scheduled expiration date.

channel A path or connection typically supporting one user.

chip spreading codes The binary sequences assigned to devices so that they can perform code division multiple access.

ciphertext The data after the encryption algorithm has been applied.

circuit-switched network A communications network in which a dedicated circuit is established between sender and receiver and all data passes over this circuit.

class of service A definition of a type of traffic and the underlying technology that will support that type of traffic. Commonly found in Asynchronous Transfer Mode systems.

client/server system A distributed computing system consisting of a server and one or more clients that request information from the server.

cloud The network sub-structure of nodes (routers and switches) and high-speed links

coarse wavelength division multiplexing (CWDM) A less-expensive form of wavelength division multiplexing which involves the transfer of a small number of streams of data over a single optical fiber using multiple-colored laser transmitters.

coaxial cable A single wire wrapped in a foam insulation, surrounded by a braided metal shield, then covered in a plastic jacket.

codec A device that accepts analog data and converts it into digital signals. This process is also known as digitization.

Code Division Multiple Access (CDMA) A multiplexing technique used extensively by the military and cellular phone companies in which binary 1s and 0s are replaced with larger, unique binary sequences to allow multiple users to share a common set of frequencies. Based on spread spectrum technology.

code division multiplexing (CDM) (see code division multiple access)

collision The result of the signals from two or more devices colliding on a medium.

collision window The interval during which the signals in a CSMA/CD local area network propagate down the bus and back (the interval during which a collision can happen).

committed information rate (CIR) The data transfer rate that is agreed on by both customer and carrier in a frame relay network.

communications sub-network (or simply sub-network) The underlying physical system of nodes and communications links that support a network.

competitive local exchange carrier (CLEC) New providers of local telephone services (their creation was initiated by the Telecommunications Act of 1996).

compression The process of manipulating data such that it fits into a more compact space.

computer auditing A process in which a software program monitors every transaction within a system.

computer network An interconnection of computers and computing equipment that uses either wires or radio waves over small or large geographic areas.

computer simulation A software program used to simulate an often complex operation, such as simulating a nuclear explosion, or the addition of an additional runway at an airport.

computer-telephony integration (CTI) New telephone services and systems that combine more traditional voice networks with modern computer networks.

computer terminal A relatively non-intelligent device that allows a user to input data into a system or displays data from the system.

connection-oriented network application A type of network application that provides some guarantee that information traveling through the network will not be lost and that the information packets will be delivered to the intended receiver in the same order in which they were transmitted.

connection admission control The process of not letting a user or application create a connection before it can be guaranteed that the network can support the requested connection.

connectionless network application A type of network application that does not require a logical connection to be made before the transfer of data.

connectivity maps A series of figures used for modeling computer networks; types include wide area connectivity maps, metropolitan area connectivity maps, and local area overview and detailed connectivity maps.

constant bit rate (CBR) Used in Asynchronous Transfer Mode, a most expensive type of service that is similar to a current telephone system leased line.

consumer digital subscriber line (CDSL) A trademarked version of DSL that yields speeds that are a little slower than typical ADSL speeds.

contention-based protocol A first-come, first-served protocol—the first station to recognize that no one is transmitting data is the first station to transmit.

convergence The coming together of multiple communication concepts to form a single point. More precisely, technological convergence is the joining of two or more technologies into a single device; protocol convergence is the joining of two or more protocols into a single protocol; and industrial convergence is the merging of two or more companies into a single company.

cookie Data created by a Web server that is stored on the hard drive of a user's workstation.

corporate license An agreement that allows a software package to be installed anywhere within a corporation, even if the installation involves multiple sites.

crash protection software Software whose primary goal is to perform crash stalling (to try to keep the operating system running long enough to perform a graceful exit) on a workstation or network of workstations.

crosstalk An unwanted coupling between two different signal paths.

cryptography The study of creating and using encryption and decryption techniques.

CSU/DSU A hardware device about the size of an external modem that converts the digital data of a computer into the appropriate form for transfer over a 1.544-Mbps T-1 digital telephone line or over a 56-kbps to 64-kbps leased telephone line.

customized menuing systems A menu system that can be created and/or changed dynamically according to a user's profile or needs.

cut-through architecture A technology in a LAN bridge or switch that allows the data frame to exit the switch almost as soon as it begins to enter the switch.

cyclic redundancy checksum (CRC) An error detection technique that typically adds either 16 or 32 check bits to potentially large data packets and approaches 100 percent error detection.

data Entities that convey meaning within a computer or computer system.

data code The set of all textual characters or symbols and their corresponding binary patterns.

data communications The transfer of digital or analog data using digital or analog signals.

Data Encryption Standard (DES) A commonly employed encryption method used by businesses to send and receive secure transactions.

data link layer The layer of the OSI model that is responsible for taking the raw data and transforming it into a cohesive unit called a frame.

data network An interconnection of computers that is designed to transmit computer data.

data rate The speed at which data is transmitted between two devices; often referred to in bits per second (bps).

data transmission speed The number of bits per second that can be transmitted.

datagram The entity or packet of data transmitted in a datagram packet-switched network.

datagram packet-switched network A communications network in which packets are sent individually based upon some routing criteria, thus each data packet can follow its own, possibly unique, path through the communications network.

DCE (Data Communications Equipment or Data Circuit-Terminating Equipment) The device, such as a modem, that terminates a data transmission line.

decibel (db) A relative measure of signal loss or gain that is used to measure the strength of a signal.

dedicated segment network A local area network (or part of a local area network) in which a switch is used to interconnect two or more workstations. Since a hub is not involved, each workstation does not have to share the network capacity with other workstations.

de facto standard A standard that has not been approved by a standards-making organization but has become a standard through wide-spread use.

delay distortion The error that occurs when the velocity of propagation of a signal through a wire varies with the frequency of the signal.

delta modulation A method of converting analog data to digital signal in which the incoming analog signal is tracked and a binary 1 or 0, is transmitted, respectively, when the analog signal rises or falls.

demultiplexor A multiplexor that un-multiplexes the data stream and delivers the individual streams to the appropriate devices.

denial of service attacks A malicious hacking technique that bombards a computer site with so many messages that the site is incapable of performing its normal duties.

dense wavelength division multiplexing (DWDM) An expensive form of wavelength division multiplexing which involves the transfer of a high number of streams of data over a single optical fiber using multiple-colored laser transmitters.

deterministic protocol A protocol (such as a LAN medium access control protocol) in which a workstation can calculate (determine) when it will next get a turn to transmit.

differential Manchester code A digital encoding technique that transmits a binary 0 when there is a voltage change at the beginning of the bit frame and transmits a binary 1 when there is no voltage change at the beginning of the bit frame. This technique ensures that there is always a voltage transition in the middle of the bit frame.

Digital Advanced Mobile Phone Service (D-AMPS) The newer, digital equivalent of the original analog Advanced Mobile Phone Service.

digital data Entities that are represented by discrete waveforms, rather than continuous waveforms. Between a minimum value X and a maximum value Y, the discrete waveform takes on only a finite number of values.

digital signals The electric or electromagnetic encoding of data that is represented by discrete waveforms rather than continuous waveforms. Between a minimum value X and a maximum value Y, the discrete waveform takes on only a finite number of values.

digital signature A technology the uses public key cryptography to assign to a document a code for which only the creator of the document has the key.

digital subscriber line (DSL) A technology that allows existing twisted pair telephone lines to transmit multimedia materials and high-speed data.

digitization The process of converting an analog signal or data into digital data

Dijkstra's least-cost algorithm A procedure that determines the least-cost path from one node in a network to all other network nodes.

direct sequence spread spectrum An encoding technique that converts a binary 0 or 1 to a larger sequence of 0s and 1s.

discrete multitone (DMT) A modulation technique used in digital subscriber line.

disk mirroring A technique used in RAID systems in which data is duplicated onto two drives simultaneously in order to provide a backup for the data.

DisplayPort A digital audio/video interface between a computer and monitor or between a computer and a home-theatre system.

distributed routing A wide area network routing technique in which each node maintains its own routing table.

domain The primary unit of administration in all of the Windows network operating systems.

domain name The address that identifies a particular site on the Web.

Domain Name System (DNS) A large, distributed database of Internet addresses and domain names.

downlink The portion of a communications link between a satellite and a ground station in which data travels from the satellite down to the Earth.

DS-1 signaling The signaling technique used to transfer 1.544 Mbps over a T-1 system.

DSL Lite A form of consumer DSL that has lower transmission speeds and thus lower consumer costs.

DTE (Data Terminal Equipment) The device, such as a workstation or a terminal, that connects to a DCE.

Dynamic Host Configuration Protocol (DHCP) An Internet protocol that dynamically assigns Internet addresses to workstations as they request a connection to the Internet.

Dynamic HTML (D-HTML) A collection of new markup tags and techniques that can be used to create more flexible and more powerful Web pages.

e-commerce The term that has come to represent the commercial dealings of a business using the Internet.

e-mail bombing A malicious hacking technique in which a user sends an excessive amount of unwanted e-mail to someone.

EBCDIC   An 8-bit code allowing 256 possible combinations of textual symbols ($2^8 = 256$).

echo   The reflective feedback of a transmitted signal as the signal moves through a medium.

effective bandwidth   The bandwidth of a signal after noise and other factors such as environmental conditions have been applied.

EIA-232F standard   An interface standard for connecting a DTE to a voice-grade modem (DCE) for use on analog public telecommunications systems.

electrical component   A portion of an interface that deals with voltages, line capacitance, and other electrical characteristics.

electronic data interchange (EDI)   The processing of business orders, purchases, and payments using only electronic data transfers.

Electronic Industries Association (EIA)   An organization that creates standards and protocols for a wide variety of electronic devices.

electronic mail   The computerized version of writing a letter and mailing it at the local post office.

encapsulation   The process by which control information is added to a data packet as it moves through the layers of a communications module such as the OSI model or the TCP/IP protocol suite.

encryption algorithm   The computer program that converts plaintext into an enciphered form.

ENUM   The voice over IP standard that converts a telephone number into a fully qualified Internet address.

e-retailing   The selling of goods and services over the Internet.

error control   The process of detection an error and then taking some type of corrective action. There are three options: do nothing, return a message, and correct the error.

Ethernet   The first commercially available local area network system (and currently the most popular). Almost identical in operation to CSMA/CD.

even parity   A simple error detection method in which a bit is added to a character to produce an even number of binary 1s.

Evolution Data Only (EV-DO)   A variation of a third generation cellular telephone technology that is capable of supporting download rates of 300 kbps to 500 kbps.

explicit congestion control   A congestion control technique in which a transmitting station is told of congestion via a bit or bits within a message.

Extended Service Set (ESS)   In a wireless LAN topology, the collection of all the Basic Service Sets attached to the local area network through its access points.

eXtensible Hypertext Markup Language (XHTML)   A markup language that combines HTML, dynamic HTML, and XML into one standard.

eXtensible markup language (XML)   A subset of SGML, this specification describes how to create a Web-based document—the specification covers both the definition of the document (how the document should be displayed in the Web browser) and the contents of the document.

extranet   When an intranet is extended outside the corporate walls and made available to suppliers, customers, or other external agents.

failover   The process of a network reconfiguring itself if a network failure is detected.

failover time   The time necessary for a network to reconfigure or reroute should a given link or node fail.

Fast Ethernet   The group of 100-Mbps Ethernet standards designated by the IEEE 802.3u protocol.

fax processing   A computer-telephony integration (CTI) application in which a fax image that is stored on a LAN server's hard disk can be downloaded over a local area network, converted by a fax card, and sent out to a customer over a trunk line.

fax-back   A computer-telephony integration (CTI) application in which a user can dial in to a fax server, retrieve a fax by keying in a number, and send that fax anywhere.

Fiber Data Distributed Interface (FDDI) ring protocol   A medium access control protocol for a ring local area network topology that is based on token passing and uses fiber-optic cables.

fiber exhaust   What happens when a fiber-optic cable is transmitting at its maximum capacity.

fiber-optic cable   A thin glass cable, approximately a little thicker than a human hair, surrounded by a plastic coating. Fiber-optic cables transmit light pulses, can transfer data at billions of bits per second, experience very low noise, and are not susceptible to electromagnetic radiation.

Fibre Channel   A high-speed connection or series of buses used to interconnect processors and peripheral devices and transfer data at billions of bits per second

File Transfer Protocol (FTP)   One of the first services offered on the Internet, FTP's primary functions are to allow a user to download a file from a remote site to his or her computer and to upload a file from his or her computer to a remote site.

file server   A high-powered workstation on a local area network that acts as a repository for user and network files and is sometimes called a network server; a file server also holds all the files required by the network operating system.

filter   Examines the destination address of a frame and either forwards or does not forward the frame based on some address information stored within the bridge.

financial feasibility   The characteristic of a project that it can be completed as set forth within the budgetary constraints set by the company.

firewall   A system or combination of systems that supports an access control policy between two networks.

FireWire The name of a bus that connects peripheral devices, such as wireless modems and high-speed digital video cameras, to a microcomputer; it is currently IEEE standard 1394.

flooding A wide area network routing protocol in which each node takes the incoming packet and retransmits it onto every outgoing link.

forward error correction The process that enables a receiver, upon detecting an error in the arriving data, to correct the error without further information from the transmitter.

forward explicit congestion notification (FECN) A congestion control technique in which a network router informs a destination station (the receiver) about network congestion, and the receiver tells the originating station (the sender) to slow down the transfer of data.

frame A cohesive unit of raw data. The frame is the package of data created at the data link layer of the OSI model.

frame relay A commercially available packet-switched network that was designed for transmitting data over fixed lines (as opposed to dial-up lines).

free space optics A high-speed wireless transmission technique that uses line-of-sight lasers over short (less than 100 meters) distances.

frequency The number of times a signal makes a complete cycle within a given time frame.

frequency division multiplexing (FDM) The oldest and one of the simplest multiplexing techniques, FDM involves assigning non-overlapping frequency ranges to different signals or to each "user" of a medium.

frequency hopping spread spectrum A modulation technique in which data is transmitted over seemingly random frequencies in order to hide the transmissions from the enemy.

frequency shift keying A modulation technique for encoding digital data using various frequencies of an analog signal.

full-duplex connection A connection between a sender and receiver in which data can be transmitted from sender to receiver and from receiver to sender in both directions at the same time.

full-duplex switch A type of switch that allows for simultaneous transmission and reception of data to and from a workstation.

functional component The function of each pin or circuit that is used in a particular interface.

gateway A generic term for a device that interconnects two networks; can also be the device that connects a VoIP system to another network.

General Packet Radio Service (GPRS) A "2.5 generation" cellular telephone technology based on the second generation technology Global System for Mobile (GSM)

Communications; GPRS can transmit data at 30 kbps to 40 kbps.

generating polynomial An industry-approved bit string that is used to create the cyclic checksum remainder.

geosynchronous-Earth-orbit (GEO) satellite Satellites that are found 22,300 miles from the Earth and are always over the same point on Earth.

Gigabit Ethernet An Ethernet specification for transmitting data at 1 billion bits per second.

global positioning system (GPS) A system of satellites that can locate a user's position on Earth to within several meters.

Global System for Mobile (GSM) Communications A second generation mobile telephone technology that is based on a form of time division multiplexing.

guard band A set of unused frequencies between two channels on a frequency division multiplexed system.

H.323 A voice over IP standard that will probably be replaced with Session Initiation Protocol (SIP).

half-duplex connection A connection between a sender and receiver in which data can be transmitted from the sender to the receiver and from the receiver to the sender in only one direction at a time.

Hamming code A code that incorporates redundant bits such that, if an error occurs during transmission, the receiver may be able to correct the error.

Hamming distance The smallest number of bits by which character codes (in a character set such as ASCII) differ.

Hertz (Hz) Cycles per second, or frequency.

high-bit-rate DSL (HDSL) The earliest form of DSL; it provides a symmetric service with speeds usually equivalent to a T-1 service (1.544 Mbps).

highly elliptical orbit (HEO) satellite A satellite that follows an elliptical pattern and is designed to perform different operations based upon whether the satellite is far from or near the earth.

HiperLAN/2 A European-standard for wireless local area networks, is capable of transmitting data at 54 Mbps using the 5 GHz frequency range.

honeypot A trap that is set by network personnel in order to detect unauthorized use of a network resource.

hop count The counter associated with a packet as it moves from node to node within a wide area network. Every time the packet moves to the next node, the hop count is increased/decreased by 1.

hop limit Used with the flooding routing algorithm, this value is compared to the hop count of each packet as it arrives at a node. When a packet's hop count equals the hop limit, the packet is discarded.

hot swappable The capability of removing a device from a computer workstation without turning off the power to the workstation.

hub A device that interconnects two or more workstations in a star-wired bus local area network and broadcasts incoming data onto all outgoing connections.

hub polling A polling technique in which the primary polls the first terminal, which then passes the poll to the second terminal, and so on—thus, each successive terminal passes along the poll.

Hypertext Markup Language (HTML) A set of codes inserted into a document (Web page) that is used by a Web browser to determine how the document is displayed.

Hypertext Transfer Protocol (HTTP) An Internet protocol that allows Web browsers and servers to send and receive World Wide Web pages.

IEEE 802 suite of protocols A collection of protocols that define various types of local area networks, metropolitan area networks, and wireless networks. For example, the IEEE 802.3 protocol defines the protocol for CSMA/CD local area networks.

IEEE 802.11a A wireless local area network protocol that is capable of supporting a theoretical transmission speed of 54 Mbps in the 5 GHz frequency range.

IEEE 802.11b An older wireless local area network protocol that is capable of supporting a theoretical transmission speed of 11 Mbps in the 2.4 GHz frequency range.

IEEE 802.11g A wireless local area network protocol that is capable of supporting a theoretical transmission speed of 54 Mbps in the 2.4 GHz frequency range.

IEEE 802.11i A standard created by IEEE to support security in wireless local area networks.

IEEE 802.11n A standard created by IEEE to support wireless local area networks with transmission speeds as high, theoretically, as 600 Mbps.

implicit congestion control A congestion control technique in which the sender discovers network congestion by observing lost or delayed packets.

impulse noise A non-constant noise that is one of the most difficult errors to detect because it can occur randomly.

incumbent local exchange carrier (ILEC) A local telephone company that existed before the Telecommunications Act of 1996.

InfiniBand A high-speed serial connection or bus used to interconnect processors and peripheral devices and transfer data at billions of bits per second.

infrared transmission A special form of radio transmission that uses a focused ray of light in the infrared frequency range ($10^{12} - 10^{14}$ Hz).

Instant Messaging (IM) The sending and receiving of text messages between user workstations in real time.

Institute for Electrical and Electronics Engineers (IEEE) An organization that creates protocols and standards for computer systems, in particular, local area networks.

interframe space The time in which a workstation waits before transmitting on a wireless local area network. There are typically three different interframe spaces, depending upon the function to be performed.

Integrated Services Digital Network (ISDN) A service designed in the mid-1980s to provide an all digital worldwide public telecommunications network that would support telephone signals, data, and a multitude of other services for both residential and business users.

integrated voice recognition and response A system in which a user calling into a company telephone system provides some form of data by speaking in to the telephone and a database query is performed using this spoken information.

integrity checking The process by which a firewall observes transactions and their characteristics for irregularities.

interactive user license An agreement in which the number of concurrent active users of a particular software package is strictly controlled.

interactive voice response A system that enables a company to user a customer's telephone number to extract the customer's records from a corporate database when that customer calls the company.

interchange circuit The signal that is transmitted over a wire and is related to a particular pin in an interface connection.

interfacing The process of creating an interconnection between a peripheral and a computer.

interexchange carrier (IEC, IXC) The name given to long distance telephone companies after the divestiture of AT&T in 1984.

intermodulation distortion The noise that occurs when the frequencies of two or more signals mix together and create new frequencies.

International Organization for Standardization (ISO) An organization that creates protocols and standards for a wide variety of systems and functions.

International Telecommunication Union-Telecommunication Standardization Sector (ITU-T) An organization that creates protocols and standards for the support of telecommunications systems.

Internet Message Access Protocol (IMAP) An Internet protocol used to support the storage and retrieval of electronic mail.

Internet Control Message Protocol (ICMP) Used by routers and nodes, this protocol performs error reporting for the Internet Protocol.

Internet Protocol (IP) The software that prepares a packet of data so that it can move from one network to another on the Internet or within a set of networks in a corporation.

Internet software   The tool set of network software to support Internet-related services that include Web browsers, server software, and Web page publishing software, among other applications.

Internet2   Newer, very high-speed packet-switched wide area network that supplements the currently existing Internet and may eventually replace it.

internetworking   The interconnecting of multiple networks.

intranet   A TCP/IP network inside a company that allows employees to access the company's information resources through an Internet-like interface.

intrusion detection   The ability to electronically monitor data flow and system requests into and out of a system.

IP multicasting   The ability of a network server to transmit a data stream to more than one host at a time.

IPsec   A set of protocols, created by the Internet Engineering Task Force, that can provide for secure transmission using the Internet Protocol (IP).

IPv6   A more modern Internet Protocol that takes advantage of the current technology. Currently, most Internet systems are using IPv4.

iSCSI (Internet SCSI)   A protocol that supports the small computer systems interface (see SCSI) over the Internet. iSCSI allows for the interfaced devices to be in two widely different locations.

isochronous connection   A connection that provides guaranteed data transport at a pre-determined rate, which is essential for multimedia applications.

jitter   A kind of noise that can result from small timing irregularities during the transmission of digital signals and can become magnified as the signals are passed from one device to another.

JPEG (Joint Photographic Experts Group)   A technique commonly used to compress video images.

Kerberos   An authentication protocol that uses secret key cryptography and is designed to work on client/server networks.

key   The unique piece of information that is used to create ciphertext and then decrypt the ciphertext back into plaintext.

keylogger   A program, often malicious, that records each keystroke a user makes on a keyboard at a computer workstation.

lambda   In wavelength division multiplexing, the wavelength of each differently colored laser.

layer 2 protocol   A protocol that operates at the second layer, or data link layer, of the OSI seven layer model.

leaf object   An object in a hierarchical directory structure that is composed of no further objects and includes

entities such as users, peripherals, servers, printers, queues, and other network resources.

licensing agreement   A legal contract that describes a number of conditions that must be upheld for proper use of the software package.

linear projection   A capacity planning technique that involves predicting one or more network capacities based on the current network parameters and multiplying by some constant.

line-of-sight transmission   The characteristic of certain types of wireless transmission in which the transmitter and receiver are in visual sight of each other.

listserv   A popular software program used to create and manage Internet mailing lists.

local access transport area (LATA)   A geographic area, such as a large metropolitan area or part of a large state. Telephone calls that remain within a LATA are usually considered local telephone calls, while telephone calls that travel from one LATA to another are considered long distance telephone calls.

local area network (LAN)   A communication network that interconnects a variety of data communicating devices within a small geographic area and broadcasts data at high data transfer rates with very low error rates.

local exchange carrier (LEC)   The name given to local telephone companies after the divestiture of AT&T in 1984.

local loop   The telephone line that leaves your house or business; it consists of either four or eight wires.

logical connection   A non-physical connection between sender and receiver that allows an exchange of commands and responses.

logical design   A process of, or the final product, that maps how the data moves around a network from workstation to workstation.

logical link control (LLC) sublayer   A sublayer of the data link layer of the OSI model that is primarily responsible for logical addressing and for providing error control and flow control information.

Long Term Evolution (LTE)   A possible contender for the fourth generation of cellular telephone systems.

longitudinal parity   Sometimes called longitudinal redundancy check or horizontal parity, this type of parity check tries to solve the main weakness of simple parity, in which all even numbers of errors are not detected.

lossless compression   A compression technique in which data is compressed and then de-compressed such that the original data is returned—that is, no data is lost due to compression.

lossy compression   A compression technique in which data is compressed and then de-compressed, but this process does not return the original data—that is, some data is lost due to compression.

low-Earth-orbit (LEO) satellite   These satellites can be found as close as 100 miles from Earth and as far as 1000 miles from Earth.

managed hub   A hub in a local area network that possesses enough processing power that it can be managed from a remote location.

Management Information Base (MIB)   The database that holds the information about each managed device in a network that supports SNMP.

Manchester code   A digital encoding scheme that ensures that each bit has a signal change in the middle of the bit and thus solves the synchronization problem.

mean time between failures (MTBF)   The average time a device or system will operate before it fails.

mean time to repair (MTTR)   The average time necessary to repair a failure within the computer network.

mechanical component   One of the four parts of an interface; deals with items such as the connector or plug description.

media converters   Devices that convert cables and/or signals from one form to another.

media selection criteria   A checklist used when designing or updating a computer network that includes cost, speed, distance, right-of-way, expandability, environment, and security.

medium access control protocol   A protocol that allows a device (such as a workstation) to gain access to the medium (the transmission system) of a local area network.

medium access control (MAC) sublayer   A sublayer formed from the splitting of the data link layer of the OSI model, MAC works closely with the physical layer and contains a header, computer (physical) addresses, error detection codes, and control information.

Metro Ethernet   A data transfer service that can interconnect two businesses at any distance using standard Ethernet protocols.

metropolitan area network (MAN)   Networks that serve an area of 3 to 30 miles—approximately the area of a typical city.

micro-marketing   The marketing that is directed at consumers who use the Internet to purchase goods and services.

middle-Earth-orbit (MEO) satellite   Satellites that are in an orbit around Earth that is above LEO but below GEO, thus can be found 1000 miles to 22,300 miles from the Earth.

MILNET   The network for military use only that the Department of Defense broke apart from the Arpanet in 1983.

mobile malicious code   A virus or worm that is designed to get transported over the Internet.

mobile service area (MSA)   When mobile telephone service was first introduced in the U.S., the country was broken into mobile service areas, or markets.

modem   A device that modulates digital data on to an analog signal for transmission over a telephone line, then demodulates the analog signal back to digital data.

Modified Final Judgment   A court ruling in 1984 that required the divestiture, or breakup, of AT&T.

modulation   The process of converting digital data into an analog signal.

monoalphabetic substitution-based cipher   A fairly simple encryption technique which replaces a character or group of characters with a different character or group of characters.

MP3   A compression/encoding technique that allows a high-quality audio sample to be reduced to a much smaller-sized file.

MPEG (Motion Pictures Expert Group)   A technique used to compress motion picture images or moving video; MPEG is often an abbreviation for versions MPEG-1 and MPEG-2.

multimode transmission   A fiber-optic transmission technique that sends a broadly focused stream of light through thicker (62.5/125) fiber-optic cable.

multiple input multiple output (MIMO)   A technology used in wireless LANs in which sending and receiving devices have multiple antennas and transmit data over multiple streams in an effort to send data faster with fewer errors.

multiplexing   Transmitting multiple signals on one medium at essentially the same time.

multiplexor   The device that combines (multiplexes) multiple input signals for transmission over a single medium and then demultiplexes the composite signal back into multiple signals.

multipoint connection   A single wire with a mainframe connected on one end and multiple terminals connected on the other end.

Multipurpose Internet Mail Extension (MIME)   The protocol used to attach a document, such as a word processor file or spreadsheet, to an e-mail message.

multiprotocol label switching (MPLS)   A technique that enables a router to switch data from one path onto another path.

Multistation Access Unit (MAU)   A device in a token ring local area network that accepts data from a workstation and transmits this data to the next workstation downstream in the ring.

multitasking operating system   An operating system that schedules each task and allocates a small amount of time to the execution of each task.

NetWare Directory Service (NDS)   A database that maintains information on, and access to, every resource on the network, including users, groups of users, printers, data sets, and servers.

network access layer The lowest layer of the TCP/IP protocol suite; it defines both the physical medium that transmits the signal and the frame that incorporates flow and error control.

Network Address Translation (NAT) An Internet protocol that allows all workstations on a local area network to assume the identity of one Internet address.

network architecture A template that outlines the layers of hardware and software operations for a computer network and its applications.

network attached storage (NAS) A computer system attached to a network that provides both network storage and the file system that controls the storage. Not the same as storage area network.

network congestion A phenomenon that occurs in a network when too many data packets are moving through the network and that leads to a degradation of networks services.

network interface card (NIC) An electronic device, typically in the form of a computer circuit board, that performs the necessary signal conversions and protocol operations so that the workstation can send and receive data on the network.

network layer A layer in the OSI model and TCP/IP protocol suite that is responsible for creating, maintaining, and ending network connections.

network management The design, installation, and support of a network and its hardware and software.

network management protocol Facilitates the exchange of management information between network devices.

network-monitoring software Software designed to monitor a network and report usage statistics, outages, virus problems, and intrusions.

network operating system (NOS) A large, complex program that can manage all the resources that are commonly found on most local area networks, in addition to performing the standard functions of an operating system.

network server The computer that stores software resources such as computer applications, programs, data sets, and databases, and either allows or denies workstations connected to the network access to these resources.

network server license A license similar to the interactive user license, in which a software product is allowed to operate on a local area network server and to be accessed by one or more workstations.

network-network interface One of the types of connections in Asynchronous Transfer Mode (ATM); a network-network interface is created by a network and used to transfer management and routing signals.

node The computing devices that allow workstations to connect to the network and that decide which route a piece of data will follow next.

noise Unwanted electrical or electromagnetic energy that degrades the quality of signals and data.

nondeterministic protocol A local area network medium access control protocol in which you cannot calculate the time at which a workstation will transmit.

non-persistent algorithm A CSMA/CD persistence algorithm that, when sensing the medium busy, waits for a random amount of time, then tries to listen again.

nonreturn to zero inverted (NRZI) code A digital encoding technique that assigns a binary 1 from a binary 0 by the voltage change or lack of voltage change at the beginning of the bit.

nonreturn to zero-level (NRZ-L) code A digital encoding technique that assigns a binary 1 or binary 0 to a low or high voltage level, respectively.

Nyquist's theorem A theorem that states that the data transfer rate of a signal is a function of the frequency of the signal and the number of signal levels.

odd parity A simple error detection scheme in which a single bit is added to produce an odd number of binary 1s.

Open Shortest Path First (OSPF) protocol A routing algorithm used to transfer data across the Internet and is a form of a link state algorithm.

Open Systems Interconnection (OSI) model A template that consists of seven layers and defines a model for the operations performed on a computer network.

operating system The program that is initially loaded into computer memory when the computer is turned on; it manages all the other programs (applications) and resources (such as disk drives, memory, and peripheral devices) in a computer.

operational feasibility The characteristic of a project that it will operate as designed and implemented.

organizational unit (OU) An object in a hierarchical tree structure for a local area network operating system that is composed of further objects.

p-persistent algorithm A CSMA/CD persistence algorithm that governs the workstation to listen to the medium, and when it detects the medium is idle, transmit with probability p.

packet filter A router that has been programmed to filter out or allow to pass certain IP addresses or TCP port numbers.

packet-switched network A communications network that is designed to transfer all data between sender and receiver in fixed-sized packets.

parallel port A connection in which there are eight data lines transmitting an entire byte of data at one moment in time.

parity bit The bit added to a character of data to perform simple parity checking.

passive device A simple connection point between two runs of cable that does not regenerate the signal on the cable.

password The most common form of protection from unauthorized use of a computer system; often a string of letters, numbers, and symbols.

payback analysis A financial analysis technique that charts the initial costs and yearly recurring costs of a proposed system against the projected yearly income (benefits) derived from the proposed system.

PBX graphic user interface An interface in which different icons on a computer screen represent common PBX functions such as call hold, call transfer, and call conferencing, making the system easier for operators to use.

peer-to-peer networks Local area networks that may not have a server; most communications are from work-station to workstation.

perceptual encoding A compression technique applied to audio and video files in which aspects of the data with characteristics that are usually not noticed by the average person are removed from the data or compressed.

period The length, or time interval, of one cycle.

permanent virtual circuit (PVC) A fixed connection between two endpoints in a frame relay network. Unlike a telephone circuit, which is a physical circuit, a PVC is created with software routing tables, thus making it a virtual circuit.

personal area network (PAN) A network that involves wireless transmissions over a short distance, such as a few meters. Often used between devices such as personal digital assistants, laptop computers, portable music devices, and workstations.

Personal Communications System (PCS) The second generation cellular telephone technology; it is all digital and includes three competing (and incompatible) PCS technologies: TDMA, CDMA, and GSM.

pharming A Web-based attack in which a user seeking to visit a particular company's Web site is unknowingly redirected to a bogus Web site that looks exactly like that company's official Web site.

phase The position of the waveform relative to a given moment of time or relative to time zero.

phase shift keying A modulation technique for encoding digital data using various phases of an analog signal.

phishing A Web-based attack that involves sending the victim a e-mail that is designed to look like a legitimate request coming from a well-known company and thereby lure the victim into revealing private information.

photo diode A light source that is placed at the end of a fiber-optic cable to produce the pulses of light that travel through the cable.

photo receptor The device at the end of a fiber-optic cable that accepts pulses of light and converts them back to electrical signals.

photonic fiber A type of fiber optic cable with long continuous air tunnels through the glass through which a laser is fired from one end of the cable to the other.

physical connection The actual connection between sender and receiver at the physical layer where the digital content of a message (actual 1s and 0s) are transmitted.

physical design The pattern formed by the locations of the elements of a network, as it would appear if drawn on a large sheet of paper.

physical layer The lowest layer of the OSI model; it handles the transmission of bits over a communications channel.

piconet Another term for a personal area network. A collection of one or more devices interconnected in a small area via wireless communications.

piggybacking The concept of combining two or more fields of information into a single message, such as sending a message which both acknowledges data received and includes additional data.

ping storm A form of attack in which the Internet ping program is used to send a flood of packets to a server to make the server inoperable.

plain old telephone system (POTS) The basic telephone system.

plaintext Data before any encryption has been performed.

point-to-point connection A direct connection between a terminal and a mainframe computer.

Point-to-Point Protocol (PPP) A protocol used to connect two devices using a serial interconnection; often used to connect a user's microcomputer to an Internet service provider via a dial-up line.

polling The operation in which a mainframe prompts the terminals to see if they have data to submit to the mainframe.

polyalphabetic substitution-based cipher Similar to the monoalphabetic cipher, except that it uses multiple alphabetic strings to encode the plaintext rather than one alphabetic string.

Post Office Protocol (POP3) An Internet protocol used to store and retrieve electronic mail.

power over Ethernet (POE) A form of Ethernet LAN in which the electrical power to operate the device is transmitted over the data cabling such that a separate connection to an electrical outlet is not necessary.

presentation layer A layer of the OSI model that performs a series of miscellaneous functions that need to be carried out in order to present the data package properly to the sender or receiver.

Pretty Good Privacy (PGP) Encryption software that has become the de-facto standard for creating secure e-mail messages and encryption of other types of data files.

primary During polling, a mainframe computer is called the primary, and each terminal is called a secondary.

print server The local area network software that allows multiple workstations to send their print jobs to a shared printer.

Private Branch Exchange (PBX) A large computerized telephone switch that sits in a telephone room on the company property.

private line A leased telephone line that requires no dialing.

private VoIP A voice over IP system that is found within the confines of a company's system of networks and does not extend to the Internet.

procedural component One of the four components of an interface; it describes how the particular circuits are used to perform an operation.

propagation delay The time it takes for a signal to travel through a medium from transmitter to receiver.

propagation speed The speed at which a signal moves through a medium.

protocol A set of hardware and/or software procedures that allows communications to take place within a computer or through a computer network.

protocol analyzer A computer program that monitors a network 24 hours a day, seven days a week, and captures and records all transmitted packets.

proxy server A computer running proxy server software that acts as the "rare books librarian" into a corporate network.

public key cryptography One key encrypts the plaintext and another key decrypts the ciphertext.

public key infrastructure (PKI) The combination of encryption techniques, software, and services that involves all the necessary pieces to support digital certificates, certificate authorities, and public-key generation, storage, and management.

pulse amplitude modulation (PAM) Tracking an analog waveform and converting it to pulses that represent the wave's height above (or below) a threshold; part of pulse code modulation.

pulse code modulation (PCM) An encoding technique that converts analog data to a digital signal. Also known as digitization.

quadrature amplitude modulation (QAM) A modulation technique that incorporates multiple phase angles with multiple amplitude levels to produce numerous combinations, creating a bps that is greater than the baud rate.

quadrature phase shift keying A modulation technique that incorporates four different phase angles, each of which represents two bits: a 45-degree phase shift represents a data value of 11; a 135-degree phase shift represents 10; a 225-degree phase shift represents 01; and a 315-degree phase shift represents 00.

quality of service (QoS) The concept that data transmission rates, error rates, and other network traffic characteristics can be measured, improved, and (it is hoped) guaranteed in advance.

quantization error The error that is introduced during digitization. Also known as quantization noise.

quantization levels The divisions of the y-axis that are used in pulse code modulation.

quantizing noise The noise that occurs during digitization. When the reproduced analog waveform is not an accurate representation of the original waveform, it is said that quantizing noise has been introduced.

RAID (redundant array of independent disks) Describes how the data is stored on multiple disk drives.

rate-adaptive DSL (RADSL) A form of digital subscriber line in which the transfer rate can vary, depending on noise levels within the telephone line's local loop.

Real-Time Protocol (RTP) An application layer protocol that servers and the Internet use to deliver streaming audio and video data to a user's browser.

Real-Time Streaming Protocol (RTSP) An application layer protocol that servers and the Internet use to deliver streaming audio and video data to a user's browser.

redirection The technique of moving a data signal to an alternate path.

redundant array of independent disks See RAID.

reflection When a light wave bounces off a surface.

refraction The change in direction experienced by a light wave as it passes from one medium to another.

reliable service A network service that delivers packets to the receiver in the order in which they were transmitted by the sender, with no duplicate or lost packets.

reliability A calculation of the probability that a component or system will be operational for the duration of a transaction of time.

remote access software Allows a person to access all of the possible functions of a personal computer workstation from a mobile or remote location.

remote bridge The device that is capable of passing a data frame from one local area network to another when the two local area networks are separated by a long distance and thus connected by a wide area network.

remote login (Telnet) The Internet application that allows you to log in to a remote computer.

Remote Network Monitoring (RMON) A protocol that allows a network manager to monitor, analyze, and troubleshoot a group of remotely managed elements.

repeater   A device that regenerates a new signal by creating an exact replica of the original signal.

return on investment (ROI)   The business term for a "payback," which occurs when the revenue generated by a new project becomes greater than the expenses associate with that project. When developing a new project, companies often use financial analysis techniques to determine when the project will earn an ROI—in other words, pay for itself.

right-of-way   Permission to install a medium across public or private property.

roll-call polling   The polling method in which the mainframe computer (primary) polls each terminal (secondary), one at a time, in round-robin fashion.

root kit   A program, often malicious, that is stored deep within a user's operating system and is capable of redirecting user requests and performing errant operations.

round robin protocol   A protocol in which each workstation takes a turn at transmission, and the turns are uniformly distributed over all workstations.

router   The device that connects local area networks to wide area network and between transmission links within a wide area network.

Routing Information Protocol (RIP)   A protocol used to route data across the Internet.

RS-232   An older protocol designed for the interface between a terminal or computer (the DTE) and its modem (the DCE).

run-length encoding   A compression technique in which a commonly occurring symbol (or symbols) in a data set is replaced with a simpler character and a count of how many times that symbol occurs.

runts   Frames on a CSMA/CD local area network that are (probably due to a collision) shorter than 64 bytes.

sampling rate   The rate at which an analog input is sampled in order to convert it to a digital stream of 1s and 0s.

satellite microwave   A wireless transmission system that uses microwave signals to transmit data from a ground station to a satellite in space and back to another ground station.

scatternet   A collection of piconets.

SCSI   (see Small Computer System Interface)

secondary (1)   A channel in an RS-232 interface in which the data and control lines are equivalent in function to the primary data and control lines, except they are for use on a reverse or backward channel.

secondary (2)   The terminal or computer in a primary/secondary network connection. The mainframe computer is considered the primary.

Secure Sockets Layer (SSL)   An additional layer of software added between the application layer and the transport (TCP) layer that creates a secure connection between sender and receiver.

security assessment software   Software designed to assess the security weaknesses (and strengths) of a network.

selection   In a mainframe-terminal configuration, the process in which a mainframe (primary) transmits data to a terminal. The primary creates a packet of data with the address of the intended terminal and transmits the packet.

self-clocking   A characteristic of a signal in which the signal changes at a regular pattern, which allows the receiver to stay synchronized with the signal's incoming bit stream.

serial port   A connection on a computer that is used to connect devices such as modems and mice to personal computers.

server   A computer that stores the network software and shared or private user files.

server appliances   A specialized network server, such as a server that is specifically designed for database systems or for Web serving.

server blade   A network server that is contained on a printed circuit board that can be plugged into a rack with other server blades.

server virtualization   The process of making one computer (or server) act as if it were multiple computers (or servers) in order to isolate the operations of a server.

service level agreement (SLA)   A legally binding written document that can include service parameters offered in a service set up between a communications provider and its customer.

Session Initiation Protocol (SIP)   A standard created by the Internet Engineering Task Force for supporting voice over IP (the transfer of voice over the Internet).

session layer   A layer of the OSI model that is responsible for establishing sessions between users and for handling the service of token management.

Shannon's theorem   A theorem that demonstrates that the data rate of a signal is proportional to the frequency of the signal and its power level and inversely proportional to the signal's noise level.

shared network   A local area network in which all workstations immediately hear a transmission.

shared segment network   A local area network (or portion of a local area network) in which hubs interconnect multiple workstations. When one workstation transmits, all hear the signal and thus all are sharing the bandwidth of the network segment.

shielded twisted pair (STP)   Copper wire used for transmission of signals in which shielding is either wrapped around each wire individually, around pairs of wires, or around all the wires together in order to provide an extra layer of protection from unwanted electromagnetic interference.

shift keying   A technique in which digital data is converted to an analog signal for transmission over a telephone line.

signals   The electric or electromagnetic encoding of data. Signals are used to transmit data.

signature-based scanning   An anti-virus technique that works by recognizing the unique pattern of a virus.

Simple Mail Transfer Protocol (SMTP)   An Internet protocol for sending and receiving e-mail.

Simple Network Management Protocol (SNMP)   An industry standard created by the Internet Engineering Task Force; it was originally designed to manage Internet components but is now also used to manage wide area network and telecommunications systems.

simple parity   A simple error detection technique in which a single bit is added to a character in order to preserve an even number of 1s (even parity) or an odd number of 1s (odd parity).

simplex connection   A system that is capable of transmitting in one direction only, such as a television broadcast system.

single-mode transmission   A fiber optic transmission technique that sends a tightly focused stream of light through thinner (8.3/125) fiber-optic cable.

single-stranded coaxial cable   A type of coaxial cable in which there is a single wire surrounded by insulation.

single-user-multiple-station license   An agreement that allows a person to install a copy of a software program on multiple computers, for example, on his or her home computer as well as his or her work computer.

single-user-single-station license   An agreement that allows a person to install a single copy of a software program on only one computer.

site license   An agreement that allows a company to install copies of a software program on all the machines at a single site.

sliding window protocol   A protocol that allows a station to transmit a number of data packets at one time before receiving an acknowledgment.

slope overload noise   The noise that results during analog-to-digital conversion when the analog waveform rises or drops too quickly and the hardware tracking it is not able to keep up with the change.

small computer system interface (SCSI)   A specially designed interface that allows for a very high speed transfer of data between the disk drive and the computer.

Small Office / Home Office (SOHO) LAN   A local area network found in the home or small office.

smurfing   The name of an automated program that attacks a network by exploiting Internet Protocol (IP) broadcast addressing and other aspects of Internet operation.

sniffer   Software and hardware devices that can monitor a network to determine if there are invalid messages being transmitted, report network problems such as malfunctioning NICs, and detect traffic congestion problems; similar to network monitoring software.

SNMP manager   Controls the operations of a managed element and maintains a database of information about all the managed elements in a given network.

socket   A combination of IP address and TCP port number, used to recognize an application on a server.

spam   Unsolicited bulk e-mail (typically commercial in nature) that is becoming a major nuisance to corporate users as well as individuals.

spanning tree algorithm   An algorithm used by bridges and switches that looks at all possible paths within a network and creates a tree structure that includes *only* unique paths between any two points. Bridges and switches use this algorithm to avoid sending data across redundant paths (loops) within a network.

spectrum   The range of frequencies that a signal spans from minimum to maximum.

splitterless DSL   A form of digital subscriber line in which there is no POTS signal accompanying the DSL signal, thus there is no need for a splitter.

spoofing   A technique (commonly used by hackers) in which the data sender's identity is disguised, as in the case of an e-mail message that has a return address of someone other than the person sending the e-mail. A modem can also perform spoofing by mimicking older protocols that are rarely used today.

spread spectrum   A high security transmission technique that instead of transmitting the signal on one fixed frequency bounces the signal around on a seemingly random set of frequencies.

spyware   Malicious software that has been installed (often unknowingly) on a user's computer to monitor the user's actions.

star-wired bus local area network   The most popular configuration for a local area network; a hub (or similar device) is the connection point for multiple workstations and may be connected to other hubs.

star-wired ring local area network   A local area network configuration similar to the star-wired bus; a hub-like device interconnects the workstations and a token is passed from workstation to workstation indicating when a workstation can transmit.

start bit   Used in asynchronous transmission, a binary 0 that is added to the beginning of the character and informs the receiver that an incoming data character (frame) is arriving.

station   The device with which a user interacts in order to access a network; it contains the software application that allows someone to use the network for a particular purpose.

statistical time division multiplexing (stat TDM) A form of time division multiplexing in which the multiplexor creates a data packet of only those devices that have something to transmit.

steganography The technology of hiding data within another unrelated document.

stop bit Used in asynchronous transmission, a binary 1 that is added to the end of a character to signal the end of the frame.

Stop-and-wait error control An error control technique usually associated with a class of protocols, also called stop-and-wait, in which a single message is sent, then the sender waits for an acknowledgement before sending the next message.

storage area network (SAN) A storage system which allows users to store files on a network. The file system is not controlled by SAN but is left to the client. Not the same as network attached storage.

store-and-forward device A device that accepts a packet, temporarily stores it in a buffer, decodes the packet as required by the device, and forwards the packet onto the next device.

streaming audio and video The continuous download of a compressed audio or video file, which can then be heard or viewed on the user's workstation.

striping A concept used in RAID in which data is broken into pieces and each piece is stored on a different disk drive.

sub-network The underlying physical system of nodes and communications links that support a network.

subnet masking The process of dividing the host ID portion of an IP address (which consists of a network ID and host ID) further down into a subnet ID and a host ID.

surveillance A common security measure used to monitor key locations to deter vandalism and theft by using video cameras and intrusion detection.

switch A device that is a combination of a hub and a bridge: it can interconnect multiple workstations (like a hub), but can also filter out frames, thereby providing a segmentation of the network (like a bridge).

switched virtual circuit (SVC) A connection that enables frame relay users to dynamically expand their current PVC networks and establish logical network connections on an as-needed basis to end points on the same network or, through gateways, to end points on other networks.

symmetric connection A type of connection in which the transfer speeds in both directions are equivalent.

symmetric encryption A form of encryption in which the same key is used to encode and decode the data; often called private key encryption.

synchronization point Some form of backup points that are inserted into a long transmission to serve as

markers from which retransmission can be started, in case of errors or failures.

synchronous connection A technique for maintaining synchronization between a receiver and the incoming data stream.

Synchronous Digital Hierarchy (SDH) A high-speed synchronous time division multiplexing technology developed in Europe by ITU-T that uses fiber optic cables for high bandwidth transmission in the megabit to gigabit range for a wide variety of data types. Almost identical to SONET.

Synchronous Optical Network (SONET) A high-speed synchronous time division multiplexing technology developed in the United States by ANSI that uses fiber optic cables for high bandwidth transmission in the megabit to gigabit range for a wide variety of data types. Two common users of SONET are the telphone company and companies that provide an internet backbone service. Almost identical to SDH.

synchronous time division multiplexing (sync TDM) A multiplexing technique that gives each incoming source a turn to transmit, proceeding through the sources in round-robin fashion.

synchronous transport signals (STS) The signaling techniques used to support SONET transmissions when the data is transmitted in electrical form and not in optical form.

systems analyst A professional who is typically responsible for managing a project and following the SDLC phases, particularly the analysis and design phases.

systems development life cycle (SDLC) A methodology for a structured approach for the development of a business system; it includes the following phases: planning, analysis, design, implementation and maintenance.

T-1 multiplexing A type of synchronous time division multiplexing (involving T-1 multiplexors) where the data stream is divided into 24 separate digitized voice/data channels of 64 kbps each. Together, T-1 multiplexing and T-1 multiplexors provide a T-1 service.

T-1 multiplexor The device that creates a T-1 output stream that is divided into 24 separate digitized voice/data channels of 64 kbps each.

T-1 service An all digital telephone service that can transfer either voice or data at speeds up to 1.544 Mbps (1,544,000 bits per second).

tap A passive device that allows you to connect a coaxial cable to another continuous piece of coaxial cable.

TCP/IP protocol suite A model of communications architecture that incorporates the TCP/IP protocols, and has surpassed the OSI model in popularity and implementation.

technical feasibility The characteristic of a project that it can be created and implemented using currently existing technology.

Telecommunications Act of 1996 A major event in the history of the telecommunications industry that, among other things, opened the door for businesses other than local telephone companies to offer a local telephone service.

Telnet A terminal emulation program for TCP/IP networks, such as the Internet, that allows users to log in to a remote computer.

terminate-and-stay-resident monitoring Antivirus software that is activated and then runs in the background while users perform other computing tasks.

terrestrial microwave A transmission system that transmits tightly focused beams of radio signals from one ground-based microwave transmission antenna to another.

Text-to-speech and speech-to-text conversions Telephone systems that can digitize human speech and store it as a text file, and take a text file and convert it to human speech.

thick coaxial cable A coaxial cable that ranges in size from approximately 6 to 10 mm in diameter.

thin client A workstation computer that is connected to a network and has no floppy disk drive or hard disk storage.

thin coaxial cable A coaxial cable that is approximately 4 mm in diameter.

third-party call control A telephone feature which allows users to control a call (for example, set up a conference call) without being a part of the call.

tie lines Leased telephone lines that require no dialing.

time division multiple access (TDMA) A multiplexing technique used with PCS cell phones based upon time division multiplexing that divides the available user channels by time, giving each user a turn to transmit.

time division multiplexing (TDM) A multiplexing technique in which the sharing of a signal is accomplished by dividing the available transmission time on a medium among the medium's users.

time feasibility The characteristic of a project that it can be installed in a timely fashion that meets organizational needs.

time value of money A concept that states that one dollar today is worth more than one dollar promised a year from now, because today's dollar can be invested now and therefore accumulate interest.

timeout An action that occurs when a transmitting or receiving workstation has not received data or a response in a specified period of time.

token management A system that controls who talks when during the current session by passing a software token back and forth.

token ring A local area network that uses the ring topology for its hardware and a round robin protocol as its software.

topologies The physical layout or configuration of a local area network or a wide area network.

Transmission Control Protocol (TCP) The Internet protocol that turns an unreliable network into a reliable network, free from lost and duplicate packets.

transparent bridge An interconnection device designed for CSMA/CD LANs that observes network traffic flow and uses this information to make future decisions regarding frame forwarding.

transport layer The layer of software in the TCP/IP protocol suite and OSI model that provides a reliable end-to-end network connection.

transport layer security (TLS) A slightly updated version of Secure Sockets Layer (SSL).

transposition-based cipher An encryption technique in which the order of the plaintext is not preserved, as it is in substitution-based ciphers.

trees The more complex bus topologies, consisting of multiple cable segments that are all interconnected.

triple-DES A temporary solution for the shortcomings of DES security (which has now been replaced with AES) in which data is encrypted using DES three times; in many cases the first time by the first key, the second time by a second key, and the third time by the first key again.

Trojan horse A destructive piece of code that hides inside a harmless-looking piece of code, such as an e-mail message or an application macro.

trunk A telephone connection used by telephone companies that carries multiple telephone signals, is usually digital and high speed, and is not associated with a particular telephone number.

tunneling protocol The command set that allows an organization to create secure connections using public resources such as the Internet.

tweet A text-based message (or post) of no more than 140 characters sent over a twitter social-networking system.

twisted pair wire Two or more pairs of single-conductor copper wires that have been twisted around each other.

Twitter A free social-based networking system in which users exchange tweets (a message with a maximum of 140 characters).

ultra-wideband A transmission technique that sends data over a wide range of frequencies at low power so as to not interfere with other existing signals.

Unicode A character encoding technique that can represent all the languages on the planet.

unified messaging A telecommunication service that allows users to utilize a single desktop application to send and receive e-mail, voice mail, and fax.

Uniform Resource Locator (URL) An addressing technique that identifies files, Web pages, images, or any other type of electronic document that resides on the Internet.

uninstall software A program that works with the user to locate and remove applications that are no longer desired.

uninterruptible power supplies (UPS) Devices that can maintain power to a computer or device during a power failure for a period long enough to allow a safe shut-down to be performed.

Universal Mobile Telecommunications System (UMTS) A third-generation cellular telephone technology that is capable of supporting data transmission in hundreds of kilobits per second.

Universal Serial Bus (USB) A modern standard for interconnecting modems and other peripheral devices to microcomputers.

unmanaged hub A hub in a local area network that has little or no intelligence and cannot be controlled from a remote location.

unshielded twisted pair (UTP) The most common form of twisted pair, in which none of the wires are wrapped with a metal foil or braid.

unspecified bit rate (UBR) A class of service offered by ATM that is capable of transmitting traffic that may experience bursts of data, but does not make any promise about when the data may be sent. Plus, unlike available bit rate (ABR), UBR does not provide congestion feedback when there are congestion problems..

uplink The satellite connection from a ground station to the satellite.

Usenet A voluntary set of rules for passing messages and maintaining newsgroups.

User Datagram Protocol (UDP) A no-frills transport protocol that does not establish connections or watch for datagrams that have been in the network for too long (are beyond their hop limit).

user-network interface The connection between a user and the network in Asynchronous Transfer Mode (ATM).

utilities a type of network software that often operates in the background and supports one or more functions to keep the network or computer running at optimal performance.

V.90 standard A 56,000-bps dial-up modem standard approved by a standards-making organization rather than a single company; it is slightly incompatible with both x2 and K56flex.

V.92 standard An improvement of the V.90 standard that provides a higher upstream data transfer rate and also provides a call waiting service, in which a user's data connection is put on hold when someone calls the user's telephone number.

Variable Bit Rate (VBR) A class of service offered by ATM that is similar to frame relay service. VBR is used for real-time (or time-dependent) applications, such as sending compressed interactive video, and non-realtime (non-time dependent) applications, such as sending e-mail with large, multimedia attachments.

very high data rate DSL (VDSL) A form of digital subscriber line that is very fast format (between 51 and 55 Mbps) over very short distances (less than 300 meters).

Very Small Aperture Terminal (VSAT) A two-way data communications service performed by a satellite system in which the ground stations use non-shared satellite dishes.

Vigenére cipher Possibly the earliest example of a polyalphabetic cipher, created by Blaise de Vigenére in 1586.

virtual channel connection (VCC) Used in Asynchronous Transfer Mode, a logical connection that is created over a virtual path connection.

virtual circuit A packet-switched connection through a network that is not a dedicated physical connection, but is a logical connection created by using the routing tables located within each node/router along the connection.

virtual LAN (VLAN) A technique in which various workstations on a local area network can be configured via software and switches to act as a private segment local area network.

virtual path connection (VPC) Used in Asynchronous Transfer Mode to support a bundle of virtual channel connections (VCCs) that have the same endpoints.

virtual private network (VPN) A data network connection that makes use of the public telecommunication infrastructure, but maintains privacy through the use of a tunneling protocol and security procedures.

virus A small program that alters the way a computer operates without the knowledge of the computer's users, and often does various types of damage by deleting and corrupting data and program files, or altering operating system components, so that the computer operation is impaired or even halted.

voice network A type of network that is designed to support standard telephone calls.

voice over Frame Relay (VoFR) A technique for making telephone calls over the Internet that allows the internal telephone systems of companies to be connected using frame relay PVCs.

voice over Internet Protocol (VoIP) A technique for making telephone calls over the Internet.

voice over WLAN A system in which voice is digitized and then sent over the same signals used by a wireless local area network (WLAN).

VoIP gateway The device that converts an analog telephone call (voice and signals) into data packets (and vice versa) for traversal over an IP-based network.

war driver   A person who tries to pick up someone else's wireless LAN signals.

wavelength division multiplexing (WDM)   The multiplexing of multiple data streams onto a single fiber-optic cable through the use of lasers of varying wavelength.

Web server software   Software designed to store, maintain, and retrieve Web pages.

weighted network graph   A structure used for understanding routing that consists of nodes and edges in which the traversal of an edge has a particular cost associated with it.

white noise   A relatively constant type of noise, much like the static you hear when a radio is tuned between two stations.

wide area network (WAN)   An interconnection of computers and computer-related equipment that performs a given function or functions, typically uses local and long distance telecommunications systems, and can encompass parts of states, multiple states, countries, and even the world.

Wi-Fi Protected Access (WPA)   A set of security standards used to protect wireless LAN transmissions that is an improvement over Wired Equivalency Protocol (WEP) in that it provides dynamic key encryption (although it, too, uses a 40-bit key) and mutual authentication for wireless clients.

WiMAX   A broadband wireless transmission technology that is capable of transmitting signals for approximately 20–30 miles and at data rates in the range of millions of bits per second.

Wired Equivalency Protocol (WEP)   The first security protocol used to encrypt wireless LAN transmissions; it uses 40-bit long encryption keys that are static (as opposed to dynamic). Due to the existence of a number of weaknesses, WEP is being replaced with Wi-Fi Protected Access (WPA).

wireless   A short-hand term often used to denote the transmission of signals without the use of wires.

Wireless Application Protocol (WAP)   A set of protocols used to support the wireless transmission of web pages and other text-based Internet services to and from handheld and mobile devices.

Wireless Fidelity (Wi-Fi)   A protocol, also known as IEEE 802.11b, for wireless local area network technologies that transmit at speeds up to 11 Mbps.

wireless LAN   A network configuration that uses radio waves for intercommunication.

workstation   A personal computer or microcomputer where users perform computing work.

World Wide Web (WWW)   The collection of resources on the Internet that are accessed via the HTTP protocol.

worm   A special type of virus that copies itself from one system to another over a network, without the assistance of a human being.

xDSL   The generic name for the many forms of digital subscriber line (DSL).

ZigBee   A wireless transmission technology for the transfer of data between smaller, often embedded devices that require low data transfer rates and corresponding low power consumption.

Index

Applied Calculus

Fifth Edition

Stefan Waner
Hofstra University

Steven R. Costenoble
Hofstra University

BROOKS/COLE
CENGAGE Learning™

Australia • Brazil • Japan • Korea • Mexico • Singapore • Spain • United Kingdom • United States

BROOKS/COLE
CENGAGE Learning™

Applied Calculus, Fifth Edition
Stefan Waner, Steven R. Costenoble

Publisher: Richard Stratton

Senior Acquisitions Editor: Liz Covello

Associate Editor: Jeannine Lawless

Editorial Assistant: Lauren Hamel

Marketing Manager: Ashley Pickering

Marketing Coordinator: Erica O'Connell

Marketing Communications Manager:
Mary Anne Payumo

Content Project Manager: Susan Miscio

Senior Art Director: Jill Ort

Print Buyer: Diane Gibbons

Permissions Editor: Margaret
Chamberlain-Gaston

Production Service: MPS Content
Services

Text Designer: Henry Rachlin

Photo Manager: Don Schlotman

Photo Researcher: Pre-Press PMG

Cover Designer: Monica DeSalvo

Cover Image: © Getty Images/
Chip Forelli

Compositor: MPS Content Services

Library of Congress Control Number: 2009924247

Student Edition:
ISBN-13: 978-1-4390-4923-5
ISBN-10: 1-4390-4923-8

Brooks/Cole
20 Channel Center Street
Boston, MA 02210
USA

Cengage Learning products are represented in Canada by Nelson Education, Ltd.

For your course and learning solutions, visit
www.cengage.com.

Purchase any of our products at your local college store or at our preferred online store **www.ichapters.com.**

Printed in the United States of America
1 2 3 4 5 6 7 12 11 10 09